Functionalized Nanomaterials II

Functionalized Nanomaterials II

Applications

Edited by
Vineet Kumar
Praveen Guleria
Nandita Dasgupta
Shivendu Ranjan

CRC Press
Taylor & Francis Group
Boca Raton London New York

CRC Press is an imprint of the
Taylor & Francis Group, an **informa** business

First Edition published 2021
by CRC Press
6000 Broken Sound Parkway NW, Suite 300, Boca Raton, FL 33487-2742

and by CRC Press
2 Park Square, Milton Park, Abingdon, Oxon, OX14 4RN

© 2021 Taylor & Francis Group, LLC

CRC Press is an imprint of Taylor & Francis Group, LLC

ISBN: 978-0-8153-7049-9 (hbk)
ISBN: 978-0-3677-2366-8 (pbk)
ISBN: 978-1-351-02138-8 (ebk)

Typeset in Times
by Deanta Global Publising Services, Chennai, India

Contents

Preface

Nature is a self-sustainable system. It has been fulfilling the demands of several generations of living organisms. This has encouraged researchers to develop environmentally friendly and affordable novel materials. Efforts are being made to develop environmentally friendly functionalized nanomaterials with exceptional properties. Functionalized nanomaterials are useful for various applications.

This book provides comprehensive and detailed reviews on different applications of functionalized nanomaterials in different sectors. It also summarizes the safety concerns, risk assessment, risk-management, and regulations of nanoparticle toxicity. This volume contains 15 chapters focusing on applications of functionalized nanomaterials.

Chapters 1–2 discuss in detail the fabrication of functionalized amorphous and various other nanomaterials for drug delivery applications. **Chapter 3** discusses functionalized silica nanomaterials in context of targeted drug delivery for diagnosis and therapeutic purposes to cancer cells. **Chapter 4** focuses on solid lipid nanoparticles and nanostructured lipid carriers as a therapeutic tool. **Chapter 5** emphasizes the use of biopolymers as an alternative to non-renewable polymers for medical applications. **Chapters 6–8** explain the use of nanomaterials and nanocomposites in the fields of water remediation, drug delivery, tissue engineering, and catalysis. **Chapter 9** explains the application of quantum dots in solar cells, light-emitting diodes, sensors, and biomedical applications. **Chapter 10** discusses silica functionalization of metallic nanomaterials for various applications. **Chapter 11** focuses on the functionalization of silica nanoparticles for interdisciplinary applications. **Chapter 12** explains the application of various metal oxide nanomaterials in electronics. **Chapter 13** discusses the catalytic application of ferrite nano-composites. **Chapter 14** focuses on the applications of green fungus-synthesized nanoparticles. **Chapter 15** highlights the toxicity assessment of nanomaterials in a context for their safe human and environmental application.

This book will be beneficial for students, academics, researchers, industry personnel, and policymakers. It covers the recent trends and future applications of functionalized nanomaterials.

Vineet Kumar
Praveen Guleria
Shivendu Ranjan
Nandita Dasgupta

Editors

Vineet Kumar is currently working as Assistant Professor in the Department of Biotechnology, LPU, Jalandhar, Punjab, India. Previously, he was Assistant Professor at DAV University, Jalandhar, Punjab, India, and UGC-Dr DSK Postdoctoral Fellow (2013–2016) at the Department of Chemistry and Centre for Advanced Studies in Chemistry (CAS), Panjab University, Chandigarh, U.T., India. He has worked in different areas of biotechnology and nanotechnology in various institutes and universities, namely CSIR-Institute of Microbial Technology, Chandigarh, U.T., India; CSIR-Institute of Himalayan Bioresource Technology, Palampur, H.P., India; and Himachal Pradesh University, Shimla, H.P., India. His areas of interest include green synthesis of nanoparticles, nanotoxicity testing of nanoparticles and application of nanoparticles in drug delivery, food technology, sensing, dye degradation, and catalysis. He has published many articles in these areas featured in peer-reviewed journals. He is also serving as an editorial board member and reviewer for international peer-reviewed journals. He has received various awards like a senior research fellowship, best poster award, and postdoctoral fellowship, etc. He is currently in the final stage of editing two books each for CRC Press/Taylor & Francis Group, and Springer Nature.

Praveen Guleria is presently working as Assistant Professor in the Department of Biotechnology at DAV University, Jalandhar, Punjab, India. She has worked in the areas of plant biotechnology, plant metabolic engineering, and plant stress biology at the CSIR-Institute of Himalayan Bioresource Technology, Palampur, H.P., India. Her research interests include plant stress biology, plant small RNA Biology, plant epigenomics, and nanotoxicity. She has published several research articles in various peer-reviewed journals. She is also serving as an editorial board member and reviewer for certain international peer-reviewed journals. She has been awarded the SERB-Start Up Grant by DST, GOI. She was awarded the prestigious 'Bharat Gaurav Award' in 2016 by the India International Friendship Society, New Delhi. She has also received various awards like the CSIR/ICMR-Junior Research Fellowship and CSIR-Senior Research Fellowship, State-level merit scholarship awards. She is currently editing a book on nanotoxicity for Springer.

Nandita Dasgupta has vast working experience in micro-/nanoscience and is currently serving in VIT University, Vellore, T.N., India. She has been exposed to various research institutes and industries including CSIR-Central Food Technological Research Institute, Mysore, India, and Uttar Pradesh Drugs and Pharmaceutical Co. Ltd., Lucknow, India. Her areas of interest include micro-/nanomaterials fabrication and their applications in different fields, predominately medicine, food, the environment, agriculture, biomedical field, etc. She has published many books with Springer and has contracted a few with Springer, Elsevier, and CRC Press. She has also published many scientific articles in international peer-reviewed journals and is also serving as an editorial board member and referee for reputed international peer-reviewed journals. She has received the Elsevier Certificate for 'Outstanding Contribution' in Reviewing from Elsevier, the Netherlands. She has also been nominated for the Elsevier advisory panel for Elsevier, the Netherlands. She is the Associated Editor in *Environmental Chemistry Letters* – a Springer Nature journal of 3.59 Impact Factor. She has received several national and international awards and recognitions from different organizations.

Shivendu Ranjan is currently Director, Centre for Technological Innovations & Industrial Research (CTIIR), at SAIARD, Kolkata (an institute certified by Ministry of Micro, Small & Medium Entreprises, Govt of India). He is also working as Guest/Visiting Faculty at the National Institute of Pharmaceutical Education and Research-R (NIPER-R), Lucknow, Ministry of Chemicals and Fertilizers, Govt of India. Earlier, he worked as a scientist at the DST-Centre for Policy Research, Lucknow, supported by the Ministry of Science and Technology, Govt of India. He is also serving as a Senior Research Associate (Visiting) at the Faculty of Engineering & Built Environment, University of Johannesburg, Johannesburg, South Africa. He has major expertise in micro-/nanotechnology and is currently working at VIT University, Vellore, T.N., India. His area of research is multidisciplinary, which are, but not limited to, micro-/nanobiotechnology, nano-toxicology, environmental nanotechnology, nanomedicine, and nano-emulsions. He has published many scientific articles in international peer-reviewed journals. He has recently edited five books published by Springer and has contracted three books for Elsevier, and four books for CRC Press—all these books cover vast areas of applied micro/nanotechnology. He has vast editorial experience. In brief, he is serving as Associate Editor in *Environmental Chemistry Letters* (Springer Journal with 3.59 Impact Factor), and serving on the editorial panel in *Biotechnology and Biotechnological Equipment* (Taylor and Francis, 1.05 Impact Factor). He is also executive editor and expert board panel in several other journals. He has been recently nominated for the Elsevier Advisory Panel, the Netherlands. He has attained several awards and honors from different national and international organizations.

Contributors

Faisal Ahmad
Iris Worldwide
Gurugram, India

Shamim Ahmad
JCB University of Science & Technology
Haryana, India

Rayisa Beevi
Department of Pharmaceutics
RAK Medical and Health Sciences University
Ras Al Khaimah, UAE

Shalini Chaturvedi
Samarpan Science and Commerce College Gandhinagar
Gujarat, India

Subhasree Roy Choudhury
Institute of Nano Science and Technology
Punjab, India

Keat Theng Chow
Roquette Asia Pacific Pte. Ltd.
Singapore

Maria-Beatrice Coltelli
Department of Civil and Industrial Engineering
University of Pisa
Pisa, Italy
and
Inter University Consortium of Materials Science and
    Technology (INSTM)
Florence, Italy

Serena Danti
Department of Civil and Industrial Engineering
University of Pisa
Pisa, Italy

Pragnesh N. Dave
Department of Chemistry
Sardar Patel University
Gujarat, India

Atul Dev
Institute of Nano Science and Technology
Mohali, India

Varsha Dogra
Department of Chemistry
Panjab University
Chandigarh, India

Giovanna Donnarumma
Inter University Consortium of Materials Science and
    Technology (INSTM)
Florence, Italy
and
Department of Experimental Medicine
University of Campania "Luigi Vanvitelli"
Naples, Italy

Vinay Dwivedi
Department of Biotechnology
Naraina Vidyapeeth Engineering &
    Management Institute
Kanpur, India

Wen Chin Foo
Roquette Asia Pacific Pte. Ltd.
Singapore

Karyman Ahmed Fawzy Ghanem
Department of Pharmaceutics
RAK Medical and Health Sciences University
Ras Al Khaimah, UAE

Rajeev Gokhale
Roquette Singapore Pte. Ltd.
Helios, Singapore

Kohsuke Gonda
Graduate School of Medicine
Tohoku University
Sendai, Japan

Ankur H. Gor
Department of Chemistry
K S K V Kachchh University
Bhuj, India

Monika Gupta
Amity Institute of Biotechnology
Amity University Madhya Pradesh
Gwalior, India

Ayushi Jain
Department of Chemistry
Panjab University
Chandigarh, India

Rohini Kanwar
Department of Chemistry
Panjab University
Chandigarh, India

Surajit Karmakar
Institute of Nano Science and Technology
Punjab, India

Gurpreet Kaur
Department of Chemistry
Panjab University
Chandigarh, India

Yuet Mei Khong
AbbVie Inc.
Chicago, Illinois

Yoshio Kobayashi
Graduate School of Science and Engineering
Ibaraki University
Mito, Japan

Rajeev Kumar
Department of Environment Studies
Panjab University
Chandigarh, India

Sandeep Kumar
Department of Bio and Nano Technology
Guru Jambheshwar University of Science & Technology
Hisar, India

Vineet Kumar
Department of Biotechnology
Lovely Professional University
Punjab, India

Avinash Chandra Kushwaha
Institute of Nano Science and Technology
Habitat Centre
Mohali, India

Surinder Kumar Mehta
Department of Chemistry
Panjab University
Chandigarh, India

Raghvendra Kumar Mishra
Amity Institute of Biotechnology
Amity University Madhya Pradesh
Gwalior, India

Gianluca Morganti
ISCD Nanoscience Center
Rome, Italy

Pierfrancesco Morganti
ISCD Nanoscience Center
Rome, Italy
and
China Medical University
Shenyang, China

Mohammad Nadim Sardoiwala
Institute of Nano Science and Technology
Mohali, India

Shikha Sharma
Department of Botany
Post Graduate College for Girls
Chandigarh, India

Kulvinder Singh
Department of Chemistry
Maharaja Agrasen University
Baddi, India

Ankur Sood
Institute of Nano Science and Technology
Habitat centre
Mohali, India

Rajesh Singh Tomar
Amity Institute of Biotechnology
Amity University Madhya Pradesh
Gwalior, India

Shivani Uppal
Department of Chemistry
Panjab University
Chandigarh, India

Shahnaz Usman
Department of Pharmaceutics
RAK Medical and Health Sciences University
Ras Al Khaimah, UAE

Jalpa Vara
Department of Chemistry
K S K V Kachchh University
Bhuj, India

Shweta Wadhawan
Department of Chemistry
Panjab University
Chandigarh, India

Hemant K. S. Yadav
Department of Pharmaceutics
RAK Medical and Health Sciences
   University
Ras Al Khaimah, UAE

1

Nanoamorphous Drug Delivery Technology and an Exploration of Nanofabrication

Wen Chin Foo, Keat Theng Chow, Yuet Mei Khong, and Rajeev Gokhale

CONTENTS

1.1 Introduction to Nanotherapeutics

Nanofabrication is the design and process of constructing functional nanostructures with a focus on precise control of the nanometer scale and reproducible mass production (Gates et al. 2005, Betancourt and Brannon-Peppas 2006, Grove and Cortajarena 2016). Materials in the nanoscale range can exhibit radically different, often novel and improved physical, chemical, and biological properties compared to their macroscale counterparts. The concept of nanotechnology was first conceived in 1959 by the physicist Richard P. Feynman, who in his visionary lecture titled 'There's plenty of room at the bottom', discussed *'the problem of manipulating and controlling things on a small scale'* and envisioned that *'it would have an enormous number of technical applications'* (Feynman 1960). Indeed, the developments in nanotechnology have revolutionized technology as we know it, with a paradigm-shifting impact on the electronics/semiconductor, information technology, materials science, environment and energy industries, to name a few, and most importantly, modern medicine. The term 'nanomedicine' was coined to describe the medical applications of nanotechnology such as in vitro diagnostics, *in vivo* imaging, medical devices, biosensors, surgery, tissue engineering, therapeutics and drug delivery. The scope of this review is limited to the pharmaceutical application of nanomedicine i.e. nanotherapeutics. We present a brief overview on the current landscape of nanotherapeutics which includes large and small molecule drug delivery and therapy, before discussing in depth our focus which is nanoamorphous drug delivery technology for small molecule drugs.

1.1.1 Protein-Based Nanofabrication

Nanomaterials and nanostructures are commonly defined as having at least one dimension ≤ 100 nm (Gates et al. 2005). However, in nanotherapeutics, nanoscale materials up to 1000 nm are included since these size dimensions also afford optimized properties beyond that provided by bulk materials (Wagner et al. 2006). While nanotechnology has transformed the foundations of medicine in the 21st century, likewise biopharmaceuticals (biologics) have changed the landscape of the pharmaceutical industry in the same timeframe. In 2016, nine out of the top ten global bestselling pharmaceutical products were biologics (Lindsley 2017), as opposed to a global pharmaceutical market dominated by small molecule drugs in the past. Biologics cover a wide array of biological therapeutics, including proteins, and vaccines, in addition to cells and gene therapy (U.S. Food and Drug Administration n.d.). Thus, it follows that marrying the possibilities of nanotechnology with the therapeutic advantage and increasing knowledge of proteins should be the next frontier in nanotherapeutics.

Proteins exist in the nanoscale, hence naturally befit the field of nanotherapeutics (Erickson 2009). The first therapeutic protein, insulin was commercialized in 1922; however, it was the development of therapeutic monoclonal antibodies (mAb) which catalyzed the rapid rise of protein therapeutics and biologics in the 21st century (Watier and Reichert 2017). Human antibodies are one of the most robust and versatile protein structures capable of binding to a large range of biological targets with varying shapes, sizes, and properties. Their basic structure comprises two identical pairs of light and heavy chains linked together by disulfide bonds, which constitute two fragment antigen-binding (Fab) regions and one fragment crystallizing (Fc) region. The two Fab regions selectively bind targets (antigens), a property conferred by the complementarity-determining regions (CDR) of its variable domains, while the Fc region interacts with fragment crystallizable receptors (FcR) and complement component C1q to affect biological activity (Prabakaran and Dimitrov 2017). Although antibodies can be customized to bind diverse targets with high affinity, their large molecular size (~150 kDa) is associated with poor tumor penetration and renal clearance. These limitations, coupled with increasing knowledge of antibody structure-function relationships, availability of antibody sequence and structural data repositories, bioinformatics and *in silico* modeling have led to the advent of antibody mimetics. Antibody mimetics are alternative protein scaffolds engineered to afford antibody-like binding activity but with molecular weights smaller by an order of magnitude and typically better thermal, chemical, and proteolytic stability attributed to their homogeneous secondary structure. A successful example of antibody mimetics are monobodies which are derived from the 10th human fibronectin type III (FN3) domain. Amino acid diversification of the β-sheet and surface loop regions form a concave surface which binds to the convex surface of the target with maximal interface area (Koide et al. 2012). Monobodies are commercialized under the name Adnectins™ by Adnexus, now part of Bristol Myers Squibb, and its flagship drug pegdinetanib (Angiocept®) which functions as an antagonist of the vascular endothelial growth factor receptor-2 (VEGFR-2) has entered phase II clinical trials for glioblastoma treatment (Lipovšek 2010, Vazquez-Lombardi et al. 2015).

It is hypothesized that the potency of cytotoxic drugs can be enhanced by the selectivity of antibodies, thus antibody-drug conjugates (ADC) were developed to this end. ADCs selectively bind to overexpressed target antigens on tumor cells and promote internalization via the formation of a clathrin-coated pit. In the cytoplasm,

the ADC undergoes endosomal transport before finally arriving at the lysosome where acidic conditions and lysosomal enzymes result in cleavage of the antibody-drug linker and release of the cytotoxic drug (Howard 2017). Another successful application of engineered protein nanotherapeutics is in the Fc fusion proteins which comprise the Fc domain of an antibody fused to a non-antibody targeting domain via recombinant technology. The Fc domain confers an extended circulation half-life to the protein due to a recycling pathway mediated by the salvage neonatal Fc receptor (FcRn). Etanercept (Enbrel®), which fuses the tumor necrosis factor (TNF) receptor II to the Fc domain of human immunoglobulin G subclass 1 (IGG1), was the first Fc fusion protein to be approved in 1998 to treat autoimmune disorders (Strohl 2015).

The examples elucidated above illustrate the burgeoning field of protein therapeutics. Nevertheless, it is only in recent years that protein nanotechnology has become an emerging field. Protein nanotechnology refers to the utility of proteins as engineering materials for nanofabrication. This natural development comes as no surprise since proteins possess suitable characteristics making them ideal candidates for nanofabrication: (i) Proteins exist in the nanoscale; (ii) Proteins are the most versatile of biological building blocks with enormous possibilities for chemical functionalization and structural complexity; (iii) Proteins are nanotherapeutics per se, hence can effect therapeutic/synergistic functions; (iv) The incorporation of active proteins as those in molecular machines into nanostructures point to the exciting possibility of engineering nanomotors, although this topic is out of the scope of our discussion. The hierarchical structure of proteins engenders the complex architecture and specificity required for structure-activity relationships, however *in silico* techniques still fall short of the power needed for *de novo* protein design, thus current methods manipulate existing protein fragments to create novel structures by engineering specific interactions. Such constructs comprise oligomerizing domains which can be tethered via ligand binding or disulfide bridges, fused by DNA recombination, or designed to interact via novel interaction surfaces (Drobnak et al. 2016). These modular building blocks have been reported to self-assemble into cage-like nanoparticles with potential application for drug delivery (King et al. 2014). Another novel approach is designed protein origami which uses coiled-coil dimers as modular building blocks to form a cage-like polyhedron. The coiled-coil modules are joined by flexible peptide linkers which act as hinges at the vertices of the polyhedron (Ljubetič et al. 2017). Protein-polymer conjugates have also found useful application as bio-hybrid materials for nanofabrication by leveraging on the structural and functional complexity afforded by proteins, and the diversity for chemical functionalization of polymeric materials. Perhaps, the ideal situation is one where the protein is both the nanotherapeutic *and* the nanostructure. This is exemplified by the fabrication of nanostructures consisting of only immune signals (antigen and adjuvant) without any polymeric carrier, termed immune polyelectrolyte multilayers ('iPEM') as they were fabricated using a layer-by-layer electronic process (Chiu et al. 2016).

1.1.2 Nanocarriers and Nano-Drug Particles

Pharmaceutical nanoparticles are defined as particles in the submicron size range of 1–1000 nm (Wagner et al. 2006, Tinkle et al. 2014) that are purposefully engineered to produce desired properties such as enhanced bioavailability of poorly soluble drugs, targeted delivery, or controlled/sustained drug release, all which can lead to better therapeutic outcomes in patients, including reduction of toxicity (US Food Drug Administration 2014, Jog and Burgess 2017, Merisko-Liversidge and Liversidge 2008, Torchilin 2007, Xu and Burgess 2012). Nanocarrier systems such as liposomes (Janssen Products LP 2015, Talon Therapeutics Inc. 2012), micelles (Kataokaa et al. 2001), carbon nanotubes (Liu et al. 2008, Pastorin 2009), and polymeric nanoparticles (Brigger et al. 2012) (Figure 1.1) have found wide application in drug delivery due to their capacity for functionalization of carrier molecules (Torchilin 2012). The drug is incorporated into the nanocarrier which can be engineered to confer a specific or combination of properties which include but are not limited to enhanced drug solubilization, targeted delivery, increased intracellular penetration, prolonged circulation times, stimuli-sensitivity, and controlled drug release (Ke et al. 2014, Torchilin 2008). It follows that functionalized nanocarriers have much value in the fields of cancer therapeutics, contrast imaging, gene delivery, cosmetic, dermal and transdermal applications, vaccination, central nervous system drug delivery, and ocular drug delivery, amongst others (Torchilin 2007, Brigger et al. 2012, Thassu et al. 2007). Except for dermal/transdermal applications, nanocarrier systems are largely administered parenterally to preserve the totality of the drug-cum-carrier entity which is necessary to effect desired drug release properties in the systemic circulation as mentioned earlier.

Nano-drug particles (Figure 1.1) are distinguished from nanocarrier systems in that they constitute pure drug particles in the nano-size range. This simple but universal formulation strategy is increasingly viewed as the pharmaceutical industry's answer to solubility challenges presented by poorly-water soluble drugs belonging to the Biopharmaceutics Classification System (BCS) II (Chin et al. 2014). Modern techniques such as high throughput screening, combinatorial chemistry, *in silico* modeling and chemogenomics have advanced to identify more effective active pharmaceutical moieties but also resulted in an increased number of poorly soluble drugs (Bajorath 2002, Bleicher et al. 2003, Lipinski 2002) – up to 40% of marketed drugs and 90% of discovery pipeline drugs face solubility challenges (Kalepu and Nekkanti 2015). Traditional solubilizing techniques such as salt formation, cosolvent systems, and complexation with cyclodextrin are limited by difficulty in selecting an optimal salt form, solvent toxicity, and specific molecular properties, respectively. Thus, it comes as no surprise that pharmaceutical companies are also riding the wave of nanotechnology to solve bioavailability issues of poorly soluble drugs for oral delivery.

Nanonization of drug particles (both crystalline and amorphous) confers solubility advantages due to an increase in

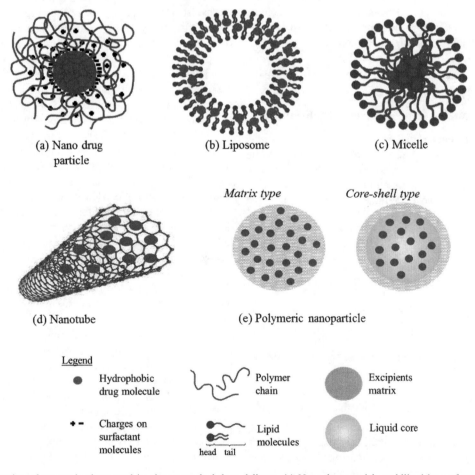

(a) Nano drug particle

(b) Liposome

(c) Micelle

(d) Nanotube

(e) Polymeric nanoparticle

Matrix type   Core-shell type

Legend

● Hydrophobic drug molecule

＋－ Charges on surfactant molecules

〜 Polymer chain

● Lipid molecules
head   tail

● Excipients matrix

● Liquid core

FIGURE 1.1 Illustration of nanotechnology used in pharmaceutical drug delivery. **(a)** Nano-drug particle stabilized by surface-adsorbed surfactant/polymer molecules. Examples of drug-containing nanocarriers are: **(b)** liposome, **(c)** micelle, **(d)** nanotube, and **(e)** polymeric nanoparticle: matrix-type or core-shell structure (Image not drawn to scale).

specific surface area and consequent gain in interfacial energies – the theoretical aspects of which are discussed below. However, these high interfacial energies also translate into a tendency for nano-drug particles to aggregate with dire consequences for bioavailability if the aggregation is irreversible. To inhibit aggregation, nano-drug particles are stabilized by surface-adsorbed surfactant and/or polymer molecules in suspension. Key differences between nano-drug particles and surfactant-/polymer-type nanocarrier systems can be explained by their particle formation methods. For the former, solid nano-drug particles are precipitated from solution and provide a surface for the adsorption of in-solution stabilizer molecules at the solid–liquid interface. This allows high drug loadings since only a small amount of stabilizing excipients relative to drug is needed to form an adsorbed layer on particle surfaces. In the case of the nanocarriers, formation of surfactant-/polymer-type nanocarrier structures precede or at the very least, occur simultaneously with entrapment of the drug *within* the nanocarrier (Tan et al. 2010, Laouini et al. 2012, Mittal 2010, Chan et al. 2010). Drug loadings are typically low to maintain kinetic stability (Mohammed et al. 2004, Kashi et al. 2012). Another notable difference between surface-stabilized nano-drug particles and polymeric nanocarriers relates to their different applications. Nano-drug particles are stabilized by water-soluble stabilizers to facilitate rapid redispersion of the dried product into its primary particles for immediate drug release and absorption. On the other hand, polymeric nanocarriers are commonly formulated with poorly soluble/gelling polymers which are frequently functionalized to afford specific targeting or modified drug release profiles (Kumari et al. 2010).

1.2 Bioavailability Advantages of Nanoamorphous Drug Delivery Technology

When particle size is reduced to the nano range, *dissolution rate* increases due to the increased surface area and a reduction in the thickness of the diffusion layer, as described by the Nernst–Brunner equation (Dokoumetzidis and Macheras 2006), which can be simplified into the Noyes–Whitney equation.

$$\frac{dC}{dt} = k \ (C_s - C) \tag{1.1}$$

where dC/dt is dissolution rate, $k = \dfrac{DS}{Vh}$, D is the diffusion coefficient, S is the surface area, V is the volume of the dissolution medium, h the thickness of the diffusion layer, C_S is the saturation solubility and C is the instantaneous concentration at time t.

The Ostwald–Freundlich equation relates particle curvature to an increase in *saturation solubility* of nanosized particles:

$$\ln \frac{C_r}{C_\infty} = \frac{2\gamma V_m}{rRT} \tag{1.2}$$

where C_r is the saturation solubility of the nanosized particle with radius r, C_∞ is its bulk solubility (solubility for a macroscopic phase corresponding to a particle with infinite radius), γ is the interfacial tension, V_m is the molar volume of the dispersed compound, R is the gas constant and T is the absolute temperature.

However, this effect is modest at 15%, as has been theoretically predicted by Kesisoglou et al. (Kesisoglou et al. 2007) and experimentally confirmed by Van Eerdenbrugh et al. (2010). Thus, nanocrystalline drugs are useful to improve bioavailability of dissolution rate-limited BCS II compounds but may be inadequate for formulation of BCS II drugs where enhancement of saturation solubility may play a bigger role.

Amorphous solid dispersions (ASD) which constitute molecular dispersions of amorphous drug in a polymer matrix, are widely used with good results to overcome bioavailability challenges of BCS II drugs via the enhancement of saturation solubility approach. The amorphous form of the drug is a thermodynamically metastable state with high chemical potential energy compared to its crystalline counterpart, and is analogous to a high-energy 'spring' which can be propelled into a supersaturated state with predicted apparent solubilities of 10- to 1600-fold of the equilibrium solubility, although experimental values are much lower due to recrystallization from solution (Brouwers et al. 2009, Hancock and Parks 2000). In theory, increasing the saturation solubility also has a positive impact on the dissolution rate in accordance with Equation 1.1. The high energy state of the amorphous form can be viewed as a double-edged sword – on one hand, it confers favorable dissolution/solubility properties; on the other hand, the thermodynamically metastable amorphous form has a propensity to convert to the more stable crystalline form. The ASD delivery strategy prevents recrystallization of the drug via two stabilization mechanisms: 1) Solubilization of the drug in the polymer below its solubility limit results in a thermodynamically stable system (Kyeremateng et al. 2014); 2) High glass transition temperature (T_g) polymers increase the T_g of the ASD formulation. The system is kinetically stabilized if the storage temperature is sufficiently lower than the T_g since molecular mobility is hindered at temperatures lower than 50°C below T_g (Hancock et al. 1995).

The ASD approach is currently favored by the industry to formulate poorly water-soluble drugs, as apparent by the high number of U.S. FDA-approved drug products formulated with this technique (Brough and Williams 2013, Kawabata et al. 2011, Jog and Burgess 2017). The availability of large-scale manufacturing equipment, technical experience, and low manufacturing costs underlie the industry preference for ASD formulations. Nevertheless, the ASD delivery strategy is not without its limitations – the rate of recrystallization of the amorphous drug form can be accelerated during storage or transport due to exposure to elevated temperature and humidity conditions (Rumondor et al. 2011, Sarode et al. 2013, Rumondor and Taylor 2010). To mitigate stability issues, drug loading can be lowered so as not to exceed the saturation solubility of the drug in the polymer to maintain a thermodynamically stable system. Although the pure amorphous drug has a high intrinsic dissolution rate due to its high saturation solubility, dissolution rate of macro-sized ASD powders is controlled by the polymeric carrier – usually slow-eroding excipients such as hydroxypropyl methyl cellulose (HPMC) or copovidone (Huang and Dai 2014, Tao et al. 2009). The polymeric carrier must first dissolve to release drug molecules embedded in the matrix. Consequently, this slow dissolution velocity results in recrystallization of the dissolved drug on the surface of the undissolved solid phase and a consequent dip in supersaturation levels.

Nanoamorphous drug particles marry the benefits of high saturation solubility attributable to higher Gibbs free energy of the amorphous state, with the dissolution advantage of nanosized particles. Compared to ASD formulations, nanoamorphous formulations dissolve rapidly owing to their high surface area, leaving no active sites available for solvent-mediated recrystallization to occur. Another advantage of nanoamorphous formulations is the potentially higher drug loading since lower amounts of stabilizing excipients are needed to adsorb at particle surfaces, compared to ASD formulations where the drug is dispersed in a polymer matrix (Cheow et al. 2014). Controlled-precipitated, amorphous drug nanoparticles of itraconazole (50% drug loading) with a size range of 200–600 nm achieved maximum supersaturation levels of 64-fold in 20 mins in contrast to the solid dispersion formulation (<1190 μm) of equal drug loading which achieved lower supersaturation of approximately 44-fold in 3 hours (Matteucci et al. 2007). In vitro dissolution studies by Kumar et al. (2014c) showed that nanoamorphous itraconazole had a saturation solubility approximately 10.5 times higher than the nanocrystalline formulation, and a superior saturation rate compared to the macroamorphous formulation. Nanoamorphous itraconazole reached saturation at 30 mins compared to macroamorphous itraconazole which did not plateau after 120 mins. The same authors performed in vivo rat studies which showed that nano- and macroamorphous itraconazole formulations, driven by their higher saturation solubilities, had lower T_{max} (4 hours) as opposed to nano- (8 hours) and macrocrystalline formulations (10 hours).

The challenges of fabricating nanoamorphous drug particles are its high tendency to agglomerate/aggregate due to high surface energies, and its susceptibility towards recrystallization. Thus, nanofabrication efforts to date are focused on stabilizing these nanoamorphous drug forms from the start of production all the way through to the shelf life of the product. Careful control of processing parameters and the application of physical surface functionalization are critical to ensure the quality and stability of nanoamorphous drug products. In the following sections, we first discuss the theoretical considerations underpinning a stable nanoamorphous drug form. This foundation is then applied to explain the main nanofabrication techniques and surface functionalization approaches used to generate nanoamorphous drug preparations. As nanoamorphous formulations to date are still largely investigational, we also review the biopharmaceutical advantage of these formulations as an impetus to drive continual formulation efforts. The potential for industrial manufacture is evaluated with a summary of proprietary technology and patents related to nanoamorphous drugs. Last but not least, we conclude by presenting our outlook on the future of nanoamorphous drug technology in the wider landscape of nanotherapeutics.

1.3 Fabrication of Nanoamorphous Drug Particles

The typical processing workflow of a nanoamorphous drug product involves the fabrication of nanoamorphous drug particles as a suspension followed by a drying step to transform the amorphous nanosuspension into a solid form (Figure 1.2).

The mobility of drug molecules and particles in amorphous nanosuspensions presents several physical and chemical stability issues. Sedimentation/creaming may occur, especially if large

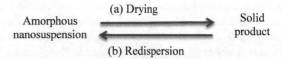

FIGURE 1.2 Diagrammatic representation of critical events of a nanoamorphous drug product. (a) Amorphous nanosuspensions are dried to give a solid product. (b) Redispersion of the dried product to give the original nanosuspension is necessary. This may be performed as an additional step prior to product administration, or it may happen **in vivo** after direct ingestion of the solid product.

particles beyond the colloidal range are present (Abdelwahed et al. 2006b). In addition, particle collision arising from Brownian motion can lead to aggregation. Amorphous nanosuspensions are also highly susceptible to particle growth via Ostwald ripening and recrystallization due to the presence of solvents and water. Furthermore, hydrolysis of water-labile drugs may also occur.

In view of the numerous stability issues associated with amorphous nanosuspensions and the advantage of convenience with pharmaceutical solids, preservation of nanoparticles in solid form is preferred. A key criterion of a successful nanoamorphous drug formulation is the ability of the dried product to redisperse into an amorphous nanosuspension with minimal changes to the original particle size and solid-state properties (Figure 1.2(**b**)). Care must be taken to preserve these two quality attributes throughout processing, storage, and redispersion since even small differences will greatly affect bioavailability.

1.3.1 Theoretical Considerations

Successful fabrication of stable nanoamorphous drug particles in suspension (amorphous nanosuspension) requires good understanding of the underlying thermodynamic principles. This information will aid the robust and stable production of nanosuspensions, ranging from particle formation, prevention of particle growth, and inhibition of recrystallization.

Fabrication of amorphous nanosuspensions is typically performed using the bottom-up approach, commonly known as the precipitation method. Most bottom-up methods involve a phase transformation of the drug due to the change in supersaturation. Researchers have tried to explain the precipitation process using the classical nucleation theory (CNT) and one of the most commonly used CNT is proposed by Becher and Döring (1935), whereby it is proposed that the nucleation rate per unit volume, J can be obtained by (Sugimoto 2007):

$$J = \frac{3D\varnothing^2}{2d^2 v_o \left(\pi k T \varphi^3\right)^{1/2}} \exp\left(\frac{-\Delta G^*}{kT}\right) \quad (1.3)$$

Where D is the diffusion coefficient of solute, d is the diameter of the monomer, v_o is the molecular volume of the solvent, $\varnothing$ is the supersaturation parameter ($\equiv kT \ln S$; $S \equiv$ supersaturation ratio, c/c_∞), φ is the surface energy of a monomer unit, ΔG^* is the energy barrier for nucleation, k is the Boltzmann constant, and T is the absolute temperature, given by:

$$\Delta G^* = \frac{4\varphi^3}{27\varnothing^2} \quad (1.4)$$

Equations (**1.3**) and (**1.4**) show that nucleation can be influenced by many factors, including degree of supersaturation, solid–liquid interfacial energy, diffusivity, polarity, and multiple elements' interaction (Beck et al. 2010, Sinha et al. 2013). Good control and balance of these factors can create conditions to facilitate rapid production of particles with minimal growth to achieve fine particles with narrow distribution (Viçosa et al. 2012). Polymers and surfactants can be introduced to increase the supersaturation, introduce interaction to reduce particle growth and reduce the interfacial energy to assist the rate of nuclei formation, leading to the formulation of ultrafine particles (Bi et al. 2015). The introduction of additional energy through mixing can also reduce particle growth and agglomeration. The addition of energy increases the diffusivity and increases the rate of nuclei formation, thereby reducing particle growth and agglomeration (Dalvi and Dave 2010). Basically, the faster supersaturation is consumed and converted into nuclei, the higher the success of fine nanoparticle formation and slower the growth rate.

Nucleation and particle growth is recognized as a concomitant process. The interaction between the surfactant, polymer, and drug is critical to prevent recrystallization and subsequent particle growth (Bi et al. 2015, Sinha et al. 2013). Ostwald ripening is a common challenge in nanosuspension systems whereby the difference between the bulk concentration and the local solubility results in a transfer of material from small to larger particles, and subsequently a growth in mean particle size over time (Ostwald 1900). The Ostwald ripening rate can be described according to the classical theory of Liftshitz–Slyozov–Wagner (LSW) as:

$$\frac{d\langle d \rangle^3}{dt} = \frac{64\gamma V_m^2 c_\infty D}{9RT} \tag{1.5}$$

where $\langle d \rangle^3$ is the particle diameter, γ is the interfacial tension, V_m is the molar volume of the dispersed compound, c_∞ is its bulk solubility (i.e. the molecular concentration that is in thermal equilibrium with a macroscopic bulk phase), D is the diffusion coefficient in the solvent, R is the gas constant, and T is the absolute temperature (Lindfors et al. 2006). The high interfacial energy of smaller particles leads to increased solubility compared to the larger particles. Therefore, the bulk concentration is typically supersaturated with respect to the larger particles and below the solubility for smaller particles. Consequently, larger particles grow and smaller particles dissolve, resulting in an overall increase in particle size.

The interplay between saturation solubility and particle radius can be described using the Ostwald-Freundlich equation (Equation 1.2). It has been shown in previous work (Lindfors et al. 2006) that the addition of another compound that is insoluble in the solvent to the nanoparticle can inhibit Ostwald ripening. This effect is due to the radius dependence on the molar chemical potential that is governed by:

$$\mu_1(r) = \mu_1^0 + \frac{2\gamma V_m}{r} + RT \ln\{1 - x_2\} \tag{1.6}$$

where '1' and '2' refer to the drug and additive respectively, μ refers to the chemical potential, x_2 refers to the fraction of the second component in the particle. It was proposed that, as the proportion of the second component decreases, the radius of the particle will continue to increase until a critical point and terminate growth, thereby inhibiting Ostwald ripening. For inhibition to occur, the drug and the additive must be miscible and incorporated into the particle.

In addition to Ostwald ripening, particles can also grow via coagulation, agglomeration, or aggregation. It was proposed that colloidal stability can be described using the Derjaguin–Landau–Verwey–Overbeek (DLVO) theory, which describes the interaction between surfaces of particles suspended in a liquid, encompassing various surface interactive forces including van der Waals attractive force, electrostatic repulsive force, and steric effects (Horn and Rieger 2001, Bi et al. 2015). The selection of stabilizers is critical to present an acceptable surface environment that prevents particle growth via this route. It has been reported that selection of hydrophilic surfactants can help reduce Gibbs free energy and introduce electrostatic repulsion, thereby offering more stable nanoparticles. However, the presence of polymeric stabilizers is shown to incur a more significant inhibition to aggregation via steric hindrance (Bi et al. 2015).

Another thermodynamic challenge in amorphous nanosuspension systems is recrystallization. An amorphous system typically has higher free energy, and therefore higher solubility and/or dissolution rate, than its crystalline counterparts. In an ASD system, nucleation needs to take place before recrystallization. This does not occur until a certain degree of supersaturation is achieved to overcome the activation energy (Baghel et al. 2016). By selecting appropriate polymers acting as precipitation inhibitors, recrystallization can be prevented by expanding the metastable phase where nucleation does not occur. This prolonged metastable phase is described as a 'spring with parachute', in comparison to a short metastable phase in the 'spring', where nucleation can occur spontaneously in a supersaturated solution (Brouwers et al. 2009). Similarly, in an amorphous nanosuspension system, whereby the amorphous nanoparticles have even higher solubility than ASD powders, suitable precipitation inhibitors must be introduced to maintain a longer metastable phase. The mechanism of crystal nuclei formation is similar to the precipitation process as described previously, except that at this point, it is essential to drive down the rate of nuclei formation.

Once nuclei are formed, crystal growth will happen rapidly in an amorphous nanosuspension system (Lindfors et al. 2007). De Yoreo and Vekilov explained that the driving force for recrystallization is governed by the change in chemical potential of a crystallizing particle (De Yoreo and Vekilov 2003). In bomineralogy, they explained that the ease of a molecule leaving or attaching to a surface for recrystallization is dependent on the surface morphology, whereby a rougher surface can inhibit recrystallization rate. The concept of recrystallization as an effect of surface integration is also described by Lindfors and colleagues (Lindfors et al. 2007). In their studies, a characteristic length, λ, given by:

$$\lambda = \frac{D}{k_+} \tag{1.7}$$

where D = monomer diffusion coefficient in solvent, k_+ = attachment rate constant, was introduced to determine if a process is diffusion-limited or driven by surface kinetics. They found that for crystal growth, the value of λ increases above the particle radii and it is therefore a result of slow surface integration kinetics. In their seeding experiments, it was shown that the crystalline particles are growing by a constant number of crystalline particles being surface integrated.

1.3.2 Nanofabrication Techniques

Amorphous nanosuspensions are generated via the bottom-up approach which involves precipitation of nanoparticles from a drug solution (Jog and Burgess 2017). Compared to the top-down approach for generating nanocrystalline drugs, bottom-up methods are energy-efficient and less time-consuming. Broadly, nanoamorphous particles can be fabricated via precipitation from a solution by three fundamental approaches: changing the solubility of the drug (controlled precipitation); formation of a nanoemulsion (emulsion templating); by solvent evaporation (atomization techniques). The basic principles of nanofabrication by each of these approaches are elucidated below, while a summary of technical details, advantages, and limitations of common methods used can be found in Table 1.1. For a more comprehensive and in-depth discussion on the various manufacturing techniques, case studies, and literature overview of nanoamorphous particles, readers are referred to the review paper by Jog and Burgess (Jog and Burgess 2017).

1.3.2.1 Controlled Precipitation

Controlled precipitation of the drug in the form of amorphous nanoparticles is achieved by the dual effort of reducing drug solubility and application of micromixing. Precipitation of nanoparticles (nanoprecipitation) can best be described by the LaMer mechanism of nucleation and growth (Thanh et al. 2014). Firstly, the drug in solution is transformed rapidly to a condition of supersaturation with the addition of an antisolvent or changes in pH and temperature. This abrupt change causes drug molecules in solution to undergo 'burst nucleation', followed by a growth phase.

Two critical quality attributes of an amorphous nanosuspension are particle size distribution and its amorphous nature. To this end, rapid micromixing is required to attain high supersaturation levels rapidly throughout the system to increase nucleation rate at the expense of particle growth and crystallization (Chan and Kwok 2011). This 'burst-nucleation' effect greatly depletes the concentration of free drug in solution, thus suppressing particle growth to give smaller particle sizes. Short mixing times (time required for mixing to achieve a homogenous solution) are also important to achieve uniform concentrations throughout the system to ensure uniform nucleation and particle growth rates which directly impact on particle size distribution (Nogi et al. 2012, Groß and Koöhler 2010).

Controlled precipitation processes usually result in amorphous nanosuspensions, since the arrangement of freshly precipitated drug molecules into an orderly crystalline lattice is inhibited during this rapid quench-precipitation process. Stabilizers are utilized to adsorb at the surface of formed nanoparticles – this occupies surface active sites to prevent further deposition of drug molecules, thus resulting in inhibition of particle growth and recrystallization.

1.3.2.2 Emulsion Templating

The formation of a nanosuspension may be based on a nanoemulsion template. Initially, an o/w nanoemulsion is formed which contains the hydrophobic drug in a water-immiscible organic solvent (oil phase) stabilized by emulsifiers in the aqueous phase. The organic solvent may be evaporated to produce a nanosuspension or the nanoemulsion may be directly spray dried to give a solid product (Margulis-Goshen et al. 2010). For the former, solvent removal methods which have so far been explored in the literature are vacuum distillation (Kumar et al. 2014c) and heat evaporation by introduction of the nanoemulsion into a heated aqueous medium (Bosselmann et al. 2012). As the solvent is evaporated, stabilizers in the aqueous phase adsorb onto surfaces of newly formed drug nanoparticles. It is hypothesized that each emulsion droplet gives rise to a discrete nanoparticle – this is substantiated by experimental results which show resulting nanoparticles in the same size range as emulsion droplets (Bosselmann et al. 2012, Sjöström et al. 1993).

In general, it is easier to control nanoparticle size distribution by this approach since it can be tailored according to the size of the emulsion droplets. As long as droplet coalescence does not occur, particle growth is well-controlled since the drug is solubilized within emulsion droplets. Surfactants/amphiphilic stabilizers which adsorb at the oil–water interphase form an outer shell to encapsulate each nanoparticle as it precipitates during solvent evaporation, thus ensuring intimate drug-stabilizer interaction. This is in contrast to the controlled precipitation method where careful optimization is required to achieve optimal supersaturation levels to ensure a fine balance between nucleation and particle growth.

1.3.2.3 Atomization Techniques

Atomization techniques for nanofabrication involve two steps: 1) Atomization of the drug solution to form fine droplets, and 2) Droplet-to-particle conversion by solvent evaporation. The key determinant of particle size is atomization. Droplet size directly impacts on particle size and is controlled by atomizer parameters (e.g. centrifugal force, frequency, pressure, etc.) which vary with equipment type (Nandiyanto and Okuyama 2011). In addition, feed rate and properties of the drug solution such as surface tension, density, and viscosity also affect droplet size. Traditional atomizing techniques are limited by the inability to generate droplets small enough to give dried particles in the submicron range. Recently, the nano spray dryer B-90 introduced by Büchi is able to produce nanoparticles with narrow size distribution by utilizing an innovative vibrating-mesh technology (Schmid et al. 2011, Li et al. 2010, Heng et al. 2011). Nevertheless, this technology is relatively new with limited experimental results reported in literature.

The second step in the process involves evaporation of the solvent to form solid particles. Careful optimization of process parameters is required to ensure sufficient drying time. An optimal evaporation rate has to be established – too low an evaporation rate will result in inefficient drying leading to aggregation of sticky particles and wall deposition. On the other hand, if the evaporation rate is too high, particles with hollow internal structures are generated which have low mechanical strength

TABLE 1.1

Methods Used for Production of Amorphous Nanoparticles

Technique	Description	Process Parameters	Advantages	Limitations	Examples of API
Controlled precipitation					
Antisolvent precipitation (including sonoprecipitation)	The API is solubilized in a water-miscible organic solvent, and then added to an antisolvent (usually an aqueous solution containing stabilizers) under stirring. When ultrasonication is applied to provide rapid micromixing (bath sonication or probe sonication), the process is classified as sonoprecipitation. The organic solvent is then removed by vacuum distillation, heat evaporation or membrane dialysis.	• Solvent-to-antisolvent ratio • Rate of mixing • Process temperature	• Universal method	• Solvent toxicity issues • Solvent removal step necessary to enhance nanosuspension stability	ABT-102 (BCS II) (Jog et al. 2016) Felodipine (Lindfors et al. 2006) AZ68 (Sigfridsson et al. 2007) Hydroxycamptothecin (Pu et al. 2009) Bicalutamide (Lindfors et al. 2006) Cefuroxime axetil (Dhumal et al. 2008, Zhang et al. 2006) Celecoxib (Liu et al. 2010) Ezetimibe (Thadkala et al. 2014) Indomethacin (He et al. 2013) Israpidine (Tran et al. 2014) Curcumin (Aditya et al. 2015) Itraconazole (Chen et al. 2008, Mugheirbi et al. 2014, Matteucci et al. 2007) Nifedipine (Lindfors et al. 2006)
Acid-base neutralization	Drugs which are weak acids or bases have pH-dependent solubility. The drug is dissolved in an aqueous solution of pH in which it has high solubility. This solution is then neutralized by addition to an aqueous solution containing stabilizers which is of a pH of low drug solubility. Rapid mixing is also required to ensure uniform nanoprecipitation.	• Acid-to-base ratio • Rate of mixing • Process temperature	• No organic solvents required	• Only applicable to drugs with pH-dependent solubility	Itraconazole (Cheow et al. 2014, Mou et al. 2011, Chen et al. 2008) Azithromycin (Hou et al. 2012)
Emulsion templating					
o/w nanoemulsions	The API is solubilized in a water-immiscible organic solvent and added to an aqueous solution containing surfactants. O/w nanoemulsions are formed by application of high-pressure homogenization or sonication. These nanoemulsions may be subjected to solvent removal to produce nanosuspensions, or directly spray dried to obtain solid nanoparticles.	• Solvent-to-antisolvent ratio • Rate of mixing • Process temperature	• Easier optimization – particle size may be tailored according to emulsion droplet size	• Solvent toxicity issues • Solvent removal step necessary to enhance nanosuspension stability	Celecoxib (Margulis-Goshen et al. 2010) Itraconazole (Kumar et al. 2014a, Bosselmann et al. 2012) Cholesteryl acetate (Sjöström et al. 1993)
Atomization methods					
Evaporative precipitation	The drug in organic solution is atomized into a heated aqueous solution containing stabilizers. Nanoprecipitation is effected by solvent evaporation and contact with the aqueous solution which functions as an antisolvent.	• Atomization parameters (pressure, nozzle size) • Process temperature	• No additional solvent removal step necessary	• Problems with residual solvent • Difficulty in obtaining very fine and uniform atomization droplet sizes which directly impacts on particle size distribution	Danazol (Vaughn et al. 2005) Cyclosporin A (Chen et al. 2002)

(Continued)

TABLE 1.1 (CONTINUED)

Methods Used for Production of Amorphous Nanoparticles

Technique	Description	Process Parameters	Advantages	Limitations	Examples of API
Nano spray-drying	The Büchi Nano Spray Dryer B-90 can generate particles in the submicron range due to its efficient atomization and particle separation technology. Spray drying can be performed in two configurations: i) The drug in an organic solution containing stabilizers is directly spray dried; ii) A nanoemulsion or nanosuspension containing the drug and stabilizers is spray dried.	• Operational parameters (Inlet/ outlet temperatures, feed rate, gas flow rate) • Solids concentration	• A solid product is obtained without an additional drying step • Fine aerosol droplets of uniform distribution can be obtained • Small sample quantities in the milligram range are possible	• Limited work on pharmaceutical drugs with this new technology	Cyclosporin A (Schafroth et al. 2012) Dexamethasone (Schafroth et al. 2012) Bovine serum albumin (Lee et al. 2011)
Supercritical fluid (SCF) technology	In the rapid expansion of supercritical solutions (RESS) technique, the drug is dissolved in supercritical carbon dioxide (SCO_2). Atomization of the drug solution into a depressurized vessel results in nanoprecipitation due to the rapid expansion of SCO_2. If the drug is insoluble in SCO_2, the gas can be used as an antisolvent to induce nanoprecipitation from a drug solution in a process called the supercritical antisolvent (SAS) method. Rapid micromixing is achieved due to high diffusivity of SCF.	• Operational parameters (pressure, temperature, feed rate, gas flow rate)	• Low polarity of SCO_2 allows dissolution of hydrophobic drugs • Rapid removal of SCF and solvent without additional drying step	• Complex and costly equipment required • Scale-up is tricky as nanoparticle properties are extremely sensitive to changes in operational parameters	Itraconazole (Lee et al. 2005) Atorvastatin calcium (Kim et al. 2008) Cefuroxime axetil (Varshosaz et al. 2009)

(Vehring 2008). The high evaporation rate inhibits solute condensation in the receding droplet and leads to formation of an outer crust on the particle. In general, evaporative processes with high drying efficiencies will produce amorphous nanoparticles since there is insufficient time for the drug to precipitate as an ordered crystalline lattice.

1.3.3 Drying

Drying of amorphous nanosuspensions into powders is crucial for maintaining product stability against moisture which can promote particle aggregation, Ostwald ripening, and recrystallization. The primary goals of drying nanoamorphous formulations are to preserve particle size comparable to that of the nanosuspension and to preserve the solid-state stability (amorphous state) of the product.

While the aim of drying is to preserve the stability of nanoamorphous drug particles, the irony is that the drying process itself may induce detrimental physical changes to the amorphous nanosuspension. As water is removed from the nanosuspension, the concentration of solubilized drug in the aqueous phase increases to high supersaturation levels that are difficult to sustain and may lead to drug precipitation. Exposure to high temperatures during drying accelerates the kinetics of drug recrystallization directly, and also in an indirect manner by increasing the supersaturation ratio, as explained in Equation 1.3. Furthermore, nanoparticle aggregation can be induced by their close proximity due to the reduced volume of the continuous phase (Wang et al. 2005, Van Eerdenbrugh 2008). Evaporation of water from the surface of the nanosuspension creates a flux, bringing constituent nanoparticles with it and in close contact with each other. Stabilizing excipients in the nanosuspension may also contribute to *destabilization* during the drying process: increased concentration of surfactants leads to higher drug solubility and consequent Ostwald ripening (Brouwers et al. 2009); pH changes caused by concentration of buffer salts can lead to solubility changes in ionizable drugs and affect the charge of electrostatic stabilizers (Chin et al. 2014); chain entanglement of polymeric stabilizers can cause irreversible agglomeration (Kim and Lee 2010).

These numerous stability pitfalls may be overcome by incorporation of dispersants (sugars) in the formulation to prevent nanoparticle aggregation (described in detail in Section 4), and also selection of a suitable drying method.

1.3.3.1 Spray-Drying and Freeze-Drying

Current literature indicates that spray-drying and freeze-drying methods are most commonly used to dry amorphous nanosuspensions (Mugheirbi et al. 2014, Cheow et al. 2014, 2015, Matteucci et al. 2007, Kumar et al. 2014c, Jog et al. 2016, Bosselmann et al. 2012, Liu et al. 2010, Wang et al. 2016). The availability of scale-up technology is an added advantage as both methods are widely used in the industry.

Several studies have been conducted to investigate the effect of spray-drying process parameters on the particle size of redispersed spray-dried nanoparticles. Kumar et al. found that inlet temperature should be maintained lower than drug melting point (or glass transition temperature for nanoamorphous particles) to prevent melting/glass-to-liquid transitions on the nanoparticle surface and resulting particle aggregation (Kumar et al. 2014b). Zhang et al. found that further reduction in inlet temperatures below the drug melting point gave spray-dried nanoparticles with improved redispersibility (Zhang et al. 2014). This behavior can be explained by the Péclet number which describes the interplay of solvent evaporation and diffusion of nanoparticles in the droplet. When Péclet numbers are large, which is the case with higher inlet temperatures (hence increased evaporation rates), the droplet surface recedes faster than the diffusion of nanoparticles towards the core, giving rise to surface enrichment and a greater chance of particle aggregation (Vehring 2008). Beyond a critical temperature, attractive capillary forces generated during solvent evaporation become sufficiently large to overcome interparticle repulsions in the nanoparticle-dense droplet surface, forcing nanoparticle aggregation and causing the outer shell to buckle (Biswas et al. 2016, Sugiyama et al. 2006). Nevertheless, although it is established that inlet temperature should be low to prevent particle aggregation, it must still be sufficiently high to maintain drying efficiency; failing to do so, paradoxically, may result in aggregation. This is consistent with the findings from Jog et al. who observed larger particle sizes at lower inlet temperatures and high feed flow rates (Jog et al. 2016). An increase in feed flow rates results in less efficient solvent evaporation and a drop in outlet temperatures, both which induce particle aggregation. With regard to formulation parameters, it was observed that lower drug-to-stabilizer ratios and incorporation of low molecular weight dispersants prevent particle aggregation during drying (Kumar et al. 2014b, Jog et al. 2016, Kumar et al. 2014c).

Freezing is the first step in freeze drying and temperatures must be maintained below the glass transition temperature of the cryoconcentrated region (Tg') to ensure complete solidification of the frozen sample. Rapid (flash) freezing of the amorphous nanosuspension with liquid nitrogen immobilizes constituent nanoparticles and in combination with formulation dispersants (sugars) which form a vitrified matrix, prevents particle aggregation (Beirowski et al. 2011). Such supercooling is theorized to result in the formation of smaller ice crystals compared to slow freezing, thus minimizing mechanical stress on the nanoparticles (Lee et al. 2009, Abdelwahed et al. 2006b). However, the reality is that the mechanism of irreversible particle aggregation during drying is not so straightforward. Different freezing rates also result in variation of internal structures of the frozen cake, indirectly affecting particle aggregation during drying (Chung et al. 2012). Studies in this area have given conflicting results – Lee et al. found that fast freezing rates favored redispersibility of freeze-dried nanosuspensions, as consistent with conventional understanding of the mechanistics of freeze drying (Lee et al. 2009). However, the same authors also found that under certain conditions, slow freezing rates resulted in less particle aggregation. This was in agreement with Chung et al. who showed that slow freezing rates gave better nanoparticle redispersibility with polyethylene glycol (PEG) as the cryoprotectant (Chung et al. 2012). This unique effect of freezing rate was attributed to the kinetics of freezing – freezing rates which are too high cause nanoparticles or cryoprotectant molecules with low diffusivity to be trapped in ice crystals instead of the cryoconcentrate region. If either one component is trapped while the other is not, such an inhomogeneous distribution will cause drug nanoparticles to come in direct contact with each other, thus facilitating irreversible particle aggregation.

Primary drying which involves the removal of ice via sublimation should be performed at temperatures below the collapse temperature of the product but high enough to maintain an efficient sublimation rate. When drying temperature approaches Tg' of the frozen sample, viscous flow is promoted to a point where the reduced viscosity is unable to support the macrostructure of the sample. The point where structural collapse occurs is the collapse temperature, which is typically about 2 °C above Tg' (Abdelwahed et al. 2006b). During secondary drying, unfrozen or bound water molecules are removed from the product. This residual water content would otherwise lead to nanoparticle aggregation via increased molecular mobility and crystallization of stabilizing dispersants (Abdelwahed et al. 2006a).

1.3.3.2 Dewatering Methods

Simple dewatering methods such as centrifugation and filtration of the nanosuspension, followed by drying of the pellet/residue are also described in literature for the preparation of dried amorphous nanoparticles. As these methods do not allow the inclusion of a dispersant (sugar) component in the formulation (dispersants [sugar] would be solubilized in the discarded aqueous phase), irreversible particle aggregation/ agglomeration will inevitably occur due to fusion of drug nanoparticles or surface chain entanglement of polymeric stabilizers. This premise is further substantiated by the lack of a redispersion step of the dried product, common in all such publications (Dhumal et al. 2008, Mugheirbi et al. 2014, Zhang et al. 2006, Chen et al. 2006). Particle sizing after drying, if performed, was carried out using a 'dry' method such as electron microscopy, which is unable to distinguish if irreversible particle agglomeration had indeed occurred.

Matteucci et al. have proposed a simple but elegant solution to this problem (Matteucci et al. 2008). They flocculated nanosuspensions stabilized with HPMC and poloxamer 407 with the addition of sodium sulfate. The loose flocs were then easily filtered and dried under vacuum to obtain a powder which redispersed in water to give its constituent nanoparticles. The authors claimed high drug loadings up to 78% with the salt flocculation method: much higher than spray- or freeze-drying methods due to the unnecessity of a matrix former. Nevertheless, the washing/filtration steps involved result in product loss and complicate determination of the final formulation constitution.

1.4 Functionalization

Functionalization refers to the surface modification of nanoparticles to impart new and improved physicochemical and functional properties. Functional moieties can be attached to the nanoparticle via chemical grafting or physical adsorption. The latter mechanism is employed by nano-drug particles for stabilization, as will be discussed in the section below. On the other hand, nanocarriers are typically functionalized by the addition of surface groups via covalent bonding. The most common example of nanocarrier functionalization is the surface addition of PEG chains to nanocarriers to improve circulation half-life by affecting steric protection against protein opsonization and clearance by phagocytic cells (Klibanov et al. 1990). PEGylated nanocarriers may be further functionalized by the engineering of a stimuli-sensitive dePEGylation mechanism, in which the polymer linkage cleaves when exposed to acidic pH in the tumor tissue or lysosome to facilitate drug release (Lee et al. 2008). Targeting can be achieved via attachment of specific targeting ligands to the nanoparticle surface, such as mAbs which bind specific antigens, transferrin, or transferrin receptor ligands which bind transferrin receptors commonly overexpressed on tumor cells, and folate which attaches to overexpressed folate receptors on cancerous cells to promote tumor penetration (Torchilin 2012). Protein-polymer conjugation methods including both covalent and non-covalent conjugation, the latter of which is exemplified by the biotin/avidin complex, are applied to the design of the protein/nanocarrier targeting system. A comprehensive review is provided by Carter et al. (2016). Nanocarriers have also found utility as contrast agents for medical imaging by means of attachment of contrast-generating elements, usually heavy metal atoms, anchored to the nanoparticle surface via a metal-chelating moiety (Torchilin 2012).

In contrast to nanocarriers, surface modification of nanoamorphous drug particles involve physical functionalization by the adsorption of additives, also known as formulation excipients. As nanoamorphous drug particles are designed to dissolve rapidly in the gastrointestinal fluid to facilitate oral absorption of the solubilized drug, there is limited need for sophisticated functional properties such as targeting, stimuli-sensitive release or nanoparticle longevity, although these potential developments in the future cannot be discounted. Current functionalization strategies are focused on stabilizing the drug particles and facilitating rapid dissolution via the use of water-soluble stabilizers.

As described in detail in Section 1.2, the physical stability of nanoamorphous drug formulations can be reflected in the successful inhibition/retardation of:

- Particle aggregation
- Particle growth
- Drug recrystallization

Any of the above may occur in the stages of nanosuspension preparation, drying, and storage. Formulation efforts to date have focused on the incorporation of additives (Figure 1.3) to preserve the extremely delicate physical stability inherent in

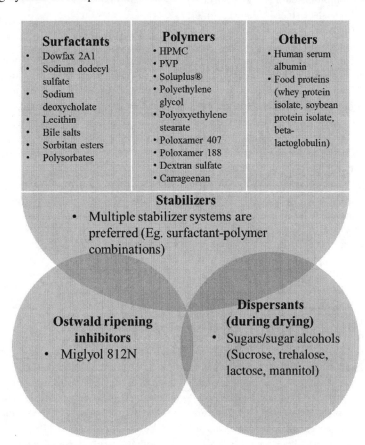

FIGURE 1.3 Overview of additives used to stabilize nanoamorphous drug formulations.

nanoamorphous formulations. Stabilizers can be used together with dispersant(s) (sugars) and/or Ostwald ripening inhibitor(s). Ostwald ripening inhibitors are drug-specific, as a homogeneous composite mixture in the amorphous particles needs to be formed between the drug and the inhibitor in order to achieve effective inhibition (Malkani et al. 2014, Lindfors et al. 2006).

1.4.1 Stabilizers

Stabilizers adsorb onto the surface of nanoamorphous drug particles to prevent particle aggregation and Ostwald ripening in the nanosuspension. This can be affected in two ways: electrostatic stabilization and steric stabilization.

In electrostatic stabilization, surfactants adsorb onto the particle surface to impart a charge (Thorat and Dalvi 2012). This inhibits agglomeration due to charge repulsion among particles of similar charges. Nevertheless, excessive use of surfactants should be avoided. Although the incorporation of surfactants help facilitate redispersion and dissolution of the dried product by increasing wettability (Hörter and Dressman 2001, Van Eerdenbrugh et al. 2008b), in a similar mechanism, they can also result in solubilization of the hydrophobic drug, especially during drying when reduced volumes give rise to elevated concentrations, leading to prominent Ostwald ripening or recrystallization (Brouwers et al. 2009). Surfactant concentration should be maintained below the critical micelle concentration (CMC) as the formation of micelles will lead to two subpopulations: micelles (containing solubilized drug) and drug nanoparticles. The physical stability of the drug nanoparticle population will be compromised since less surfactant molecules will be available for surface stabilization (Lin and Alexandridis 2002, Deng et al. 2010).

Non-ionic, amphiphilic polymers are utilized for steric stabilization while polyelectrolytes provide electrosteric stabilization of drug nanoparticles (Lourenco et al. 1996). As a general rule, polymer molecules adsorb onto the particle surface via an anchor segment, while the hydrophilic tail portion extends into the dispersing medium (Wu et al. 2011, Thorat and Dalvi 2012). The long, dense polymer chains create a steric barrier to reduce van der Waals attraction forces between particles. In addition, solvation of the hydrophilic polymer tail – which is the stabilizing segment – creates an excluded volume around each particle due to the rise in osmotic pressure when steric layers of two approaching particles encroach upon each other (Studart et al. 2007). It follows that stabilizers with larger size and higher molecular weights will give thicker adlayers thus better stabilization. Nevertheless, this effect is complicated by polymer chain entanglement during drying as well as slow dissolution associated with long-chain polymers (Matteucci et al. 2007, Kim and Lee 2010). Increasing polymer concentrations increase particle surface coverage and adlayer thickness until the critical flocculation concentration is reached, beyond that which further increase in osmotic pressure leads to depletion or bridging flocculation (Thorat and Dalvi 2012). On another note, addition of polymers also increases nanosuspension viscosities, thus suppressing Ostwald ripening and particle aggregation by reducing diffusion of drug molecules and particle mobility (Chin et al. 2014). From the supersaturation perspective, incorporation of polymers, most notably HPMC, HPMCAS, and PVP have been found to act as precipitation inhibitors (Gao and Shi 2012).

Combinations of surfactants, polymers or polyelectrolytes are commonly used to provide synergistic electrostatic, steric, and electrosteric stabilization. However, this stabilization synergy depends on the molecular packing of stabilizer molecules in the adlayer. In general, combinations of neutral and ionic stabilizers allow closer packing to give better surface stabilization by reducing charge repulsion between the ionic molecules (Wu et al. 2011, Thorat and Dalvi 2012).

Selection of an appropriate stabilizer has to be tailored to the drug of interest. A comprehensive review by Thorat et al. of the various drug, polymer, and solvent interactions involved in stabilization can be utilized to aid this selection (Thorat and Dalvi 2012). Scanning and colloidal probe microscopy have been identified as feasible stabilizer screening methods based on drug-stabilizer interactions, surface adsorption properties, and particle–particle interactions (Wu and da Rocha 2007, Verma et al. 2009, Lin et al. 2003).

Figure 1.3 shows an overview of electrostatic and steric stabilizers used for the preparation of pharmaceutical amorphous nanosuspensions from literature.

1.4.2 Ostwald Ripening Inhibitor

Lindfors et al. showed that incorporation of a component which is miscible with the active pharmaceutical ingredients (API) of interest but immiscible with the bulk medium successfully inhibited Ostwald ripening in amorphous nanosuspensions (Lindfors et al. 2006). An example of a compound used as such an inhibitor is shown in Figure 1.3.

1.4.3 Dispersants

It is essential to convert the nanosuspension to a solid form to preserve the stability of dosage form. Many methods of converting the nanosuspension into solid form have been adopted; the commonly used methods include spray coating, granulation, spray drying, and freeze drying. The drying process typically subjects the nanosuspension to additional stress, e.g. thermal stress in spray drying and cryo stress in freeze drying, which affects the performance of the nanoparticles (Kumar et al. 2014a). Another concern during drying is aggregation, which may lead to subsequent poor dissolution performance (Van Eerdenbrugh et al. 2008a). In the case of amorphous nanosuspensions, aggregation may trigger recrystallization (Kumar et al. 2014c). Van Eerdenbrugh et al. reported that the tendency for agglomeration of crystalline nanosuspensions after drying was not dependent on the drying method but significantly influenced by the surface hydrophobicity of the API (Van Eerdenbrugh et al. 2008b). Less hydrophobic surface API which adsorbed less surfactant (TPGS) was found to be more stable. One of the strategies employed to prevent aggregation is to introduce a matrix former or a dispersant, which are most commonly sugars, due to their dispersible properties and established usage as cryoprotectants. The effectiveness of the matrix former to prevent aggregation is dependent on the type of drying process, type of matrix former, the API, and the hydrophobicity of the nanoparticle. (Kumar et al. 2014a, Van Eerdenbrugh et al. 2008a)

Kumar et al. investigated the effects of different sugars on the aggregation of crystalline nanosuspensions during spray drying and freeze drying. It was found that small sugar alcohols were more effective in preventing aggregation than high molecular weight sugars due to the improved interaction between the small molecular weight sugars and the surfactant, leading to more uniform protection of the nanoparticles (Kumar et al. 2014a). Similar findings were observed with amorphous nanosuspensions whereby low molecular weight sugars inhibited aggregation and prevented recrystallization (Kumar et al. 2014c). Larger disaccharides and mannitol were found to have minimal interaction with the API and were therefore ineffective aggregation inhibitors. Although sugars have been commonly used as a dispersant, it was found that increasing concentrations of sucrose led to higher agglomeration of crystalline itraconazole nanosuspension dried by freeze drying. On the contrary, an alternative matrix former, Avicel®PH101, was found to improve dissolution with increasing concentration (Van Eerdenbrugh et al. 2008c). It was proposed that an insoluble matrix is able to better inhibit agglomeration due to less complex physical changes during the freeze-drying process compared to sucrose. The same group also investigated other alternative matrix formers including Avicel®PH101, Fujicalin®, Aerosil®200, and Inutec®SP1(Van Eerdenbrugh et al. 2008a). They reported that agglomeration inhibition is more effective with small particles such as Aerosil®200 and surface active Inutec®SP1 as a result of improved coverage of the matrix former on the nanoparticles, improved distribution of the matrix former in the powder, and improved wetting of the particles. Other alternative dispersants such as microcrystalline cellulose-carboxymethyl cellulose sodium (Dan et al. 2016) and clay (Dong et al. 2014) were also investigated and drug products were found to have excellent redispersibility properties.

1.5 Biopharmaceutical Considerations

The benefit of nanocrystalline formulations has been highlighted in a plethora of literature and the various nanocrystalline formulations currently on the market have attested to their clinical usefulness. However, nanoamorphous formulations are still largely investigational primarily due to the added complexity in processing and challenges faced in stabilizing these formulations in nano size and in amorphous form. A discussion on the biopharmaceutical performance of nanoamorphous formulations amidst concerns about solid state purity and limitations of current characterization techniques has been presented by Chow et al. (2016). This section will focus on reviewing the biopharmaceutical aspect of nanoamorphous formulations and evaluate how well the in vitro enhancement is being translated into in vivo efficacy.

The in vitro performances of spray-dried itraconazole formulations in various states were reported by Kumar et al. (2014c) as follows: nanoamorphous (200 nm) > macro-amorphous ($\leq$0.4mm) > nanocrystalline (280 nm) > macro-crystalline (10 µm). All formulations were stabilized with Dowfax2A1 and trehalose, except for the nanocrystalline formulation which was stabilized with PVP40 and SDS. The nanoamorphous formulation demonstrated a higher rate of supersaturation, that is 30 min to attain 42 µg/ml, equivalent to 10.5× itraconazole C_{eq} (equilibrium solubility of crystalline drug = 4 µg/ml) in pH 1.2 media. In contrast, the macro-amorphous formulation took 180 min to

attain its supersaturated concentration (exact concentration value was not reported). When dosed orally to Sprague–Dawley rats as a single dose, the nanoamorphous formulation demonstrated 2.5× higher in vivo exposure over 24 hours against the macroamorphous formulation due to the smaller particle size of the former. The nanoamorphous and nanocrystalline formulations demonstrated similar AUC and this was attributed to the bioadhesive property imparted by the PVP polymer in the nanocrystals resulting in longer retention at the gastrointestinal tract for drug absorption. However, the nanoamorphous formulation exhibited higher C_{max} and shorter T_{max} than the nanocrystalline formulation driven by its higher rate of supersaturation.

Yang et al. (2010) compared the in vitro and in vivo performances of nanocrystalline and nanoamorphous itraconazole in the context of pulmonary delivery. The nanoamorphous formulation was prepared using ultra rapid freezing followed by freeze drying, and stabilized using lecithin and mannitol at 58.8% drug load whereas the nanocrystalline formulation with the same composition was wet-milled. The $d_v(0.1)$, $d_v(0.5)$, $d_v(0.9)$ of the redispersed nanoamorphous powder was 100 nm, 250 nm, 590 nm, and nanocrystalline powder was 160 nm, 572 nm, 1220 nm, respectively. In simulated lung fluid at pH 7.4, the redispersed amorphous nanosuspension loaded at 100x non-sink, achieved 26× supersaturation against itraconazole C_{eq} at 15 min versus the nanocrystalline suspension which achieved only 3× supersaturation at 30 min. The cumulative extent of supersaturation over 24 hours for the nanoamorphous formulation was 4.7× higher than the nanocrystalline formulation. This was directly translated into enhanced pulmonary absorption of the nanoamorphous formulation demonstrating the same order of magnitude of bioavailability enhancement (3.8× AUC_{0-24h}) in Sprague–Dawley rats. The plasma concentration of the nanoamorphous formulation was significantly higher at all time points. Both formulations demonstrated similar aerosol performance as characterized by Andersen Cascade Impactor and similar lung deposition immediately post dosing as shown by the wet lung tissue analysis. Increased in vivo absorption was attributed to higher supersaturation in lung lining fluids resulting in a high concentration gradient for absorption. The solid solution state of the nanoamorphous formulation presented higher maximal achievable surface area for supersaturation. The slower absorption of nanocrystalline particles was deemed to have triggered increased lung clearance by alveolar macrophages (active against particles of 1–3 um), as shown by similar residual drug levels after 24 hours.

Sigfridsson et al. (Sigfridsson et al. 2007) compared the pharmacokinetics of an investigational BCS II compound, AZ68 formulated as a solution in a PEG400/DMA cosolvent system, amorphous nanosuspension stabilized by PVP K30, SDS, Miglyol 812N and mannitol, and crystalline nanosuspension stabilized by PVP K30 and Aerosol OT. The amorphous and crystalline nanosuspensions which were produced via antisolvent precipitation in ultrasonic bath and wet milling, gave particle sizes of 125 nm and 201 nm, respectively. Intravenous and oral dosing in Sprague-Dawley rats afforded similar exposures in all the three formulations. The absolute oral bioavailability, was calculated based on the i.v. data. T_{max} was significantly different in the following order: solution (0.19h) < nanoamorphous (1.3h) < nanocrystalline (3.5h). Drug absorption from a solution was most efficient since a dissolution step was not required but potential instability issues compromised its practical application. The slower absorption of nanocrystalline suspension versus the nanoamorphous suspension was attributed to the greater particle size and the need to overcome its crystal lattice energy. This study presented amorphous and crystalline nanosuspensions as attractive formulation approaches for AZ68. However, the pharmacokinetics data presented showed that amorphous nanosuspension had an edge over its crystalline counterpart, although it may not be statistically significant.

Carvedilol nanosuspensions (mean size 212 nm) produced using antisolvent precipitation cum ultrasonication with alpha-tocopherol succinate and SDS as stabilizers demonstrated superior in vivo performance to the dispersed commercial tablet in orally dosed Wistar rats whereby the respective C_{max} and AUC_{0-36h} was 3.3× and 2.9× higher, and T_{max} 3× lower (Liu et al. 2012). This observation was in parallel with the 2.4× and 3.8× dissolution enhancement of the nanosuspension at pH 1 and pH 6.8, respectively. In addition, direct nanoparticle uptake by M-cells in the Peyer's patches was noted as another possibility behind the enhanced absorption of the nanosuspensions. The enhancement of this interesting but yet poorly characterized pathway was proposed as a means to avoid first-pass metabolism of carvedilol.

TPGS (vitamin E polyethylene glycol succinate) -stabilized nanosuspension of cyclosporin A (mean size 357 nm) which was applied dermally in a HPMC hydrogel matrix exhibited enhanced penetration of 6.3–6.8× in pig's ear skin as compared to its micronized counterpart (Romero et al. 2016). A nano size of about 350 nm was reported to be sufficient to increase adhesion, dissolution, and saturation solubility in the skin, hence resulting in enhancement of bioavailability.

Nanoamorphous cefuroxime axetil prepared by sonoprecipitation (mean size 80 nm) demonstrated 1.5× and 2× higher oral bioavailability in Wistar rats than amorphous particles prepared by spray drying (mean size 10 μm) and unprocessed cefuroxime axetil (20–50 μm), respectively (Dhumal et al. 2008). This was in agreement with in vitro dissolution enhancement of 1.4× in the spray-dried, and 2.2× in the unprocessed particles in acidic medium over 150 min.

In view of the added complexity, pursuance of nanoamorphous formulations is only worthwhile when there is clear biopharmaceutical advantage over nanocrystalline formulations. The in vitro and in vivo successes demonstrated by the nanoamorphous formulations in studies described above warrant continued efforts to make nanoamorphous formulations a clinical success. The success of nanoamorphous formulations hinges on the solubility advantage gained from its nano size and high energy amorphous state. It is to be noted that the solubility advantage of amorphous formulations will only materialize into clinically meaningful biopharmaceutical consequences if the formulation achieves a solubility ratio of 2× or higher relative to the reference formulations (Hancock and Parks 2000). The good alignment between in vitro and in vivo data in the studies reported above reiterate the importance of in vitro studies as the first screening step in biopharmaceutical evaluations.

While the enhanced in vivo activity of nanoparticulate formulations will improve therapeutic efficacy, a safety concern has been raised regarding the high potency of nanosized particles once they enter the body, either via a drug delivery system or environmental exposure. Such a concern is especially pronounced for pulmonary delivery as particles in the nanosize ranges are known to be able evade the normal clearance mechanism to be translocated directly into the systemic circulation and target other organs (Yang et al. 2008). It should be noted that many studies evaluating nanoparticles toxicity were conducted on inorganic or insoluble materials. The actual biological impact of soluble and biodegradable carriers commonly used on therapeutic agents is yet to be clearly elucidated and it could invoke a remarkably different physiological response from inorganic/insoluble materials (Bosselmann and Williams 2012). However, nanoamorphous drug particles are to be distinguished from conventional nanocarrier systems. Nanocarrier systems, administered via direct entry into the blood stream, are meant to be absorbed as a whole (i.e. drug cum excipient molecules), often for additional functions such as organ/cell targeting. In contrast, nanoamorphous drug particles which are administered orally are designed to improve oral bioavailability by increased dissolution/ solubility in gastrointestinal fluids and hence, the main entity targeted for absorption into the systemic circulation is the drug molecule itself. Safety evaluation on nanomaterials has concluded that nanocrystalline drugs can be regarded generally as safe because their inherent fast dissolution rates and/or solubility will make them comparable with conventional soluble drugs post-absorption (Chin et al. 2014).

1.6 Proprietary Technologies, Patents, and Marketed Drug Products

1.6.1 Proprietary Technologies

* **Hydrosol technology** [Patent No. US 5389382 A] (List and Sucker 1995). Preparation of hydrosols via antisolvent precipitation was patented by List and Sucker in the 1980s and subsequently acquired by Sandoz / Novartis. It is based on the classical precipitation process known as *via humida paratum*, used by early chemists to prepare ointments. The invention provides for nanosuspensions of cyclosporin A (stabilized by gelatin) and dihydropyridines (stabilized by ethyl cellulose) suitable for intravenous use. These nanosuspensions may be lyophilized to give a solid product which is re-suspendable in an aqueous medium. Particle sizes given in the examples ranged from 100–300 nm; however, the crystalline/amorphous nature of the solid nanoparticles was not elucidated.

* **Nanomorph®** [Patent No. US 5968251 A] (Auweter et al. 1999). A process to produce amorphous nanosuspensions of carotenoids by temperature-modulated antisolvent precipitation was initially developed by Auweter et al. (priority date 1996); the intellectual property of which is owned by BASF. For example, beta-carotene nanosuspensions of particle size 200nm, stabilized by a surfactant and protective colloid (gelatin) were spray dried to give cold-water dispersible powders with 70–100% amorphous content. Marketed products produced by this technique are Lucantin® and Lucarotin® by BASF which are carotenoid food dyes used in the nutrition industry. This technology

is also being advertised as a drug delivery platform under the brand name Nanomorph® by Soliqs/ Abbott. Nevertheless, there has yet to be any drug product in the market manufactured by this technique.

- **Solumer™** [Patent No. US 6878693 B2] (Goldshtein 2005). A patent was filed in 2001 to describe inclusion complexes (nanoparticles) of amorphous drug entrapped within water-soluble, amphiphilic polymers. These inclusion complexes, termed 'solumers', were prepared by antisolvent precipitation of a drug solution in organic solvent into a heated polymer solution, thus effecting immediate removal of the organic solvent via evaporation. The resulting nanosuspension may be dried by lyophilization. This method gave itraconazole nanosuspensions of approximately 200 nm in particle size which was shown to be amorphous by PXRD and DSC. The Solumer™ technology incorporating spray drying is marketed as a drug delivery platform by SoluBest Ltd. for BCS Class II and IV drugs. Several products have been formulated with this platform – resveratrol in the pre-marketing stage; fenofibrate in the advanced clinical stage; and albendazole in animal studies.
- **Nanocrystal®** [Patent No. US 5145684 A] (Liversidge et al. 1992). The Nanocrystal® technology owned by Elan Pharma (now Alkermes) is the most successful nanotechnology so far with several licensed products in the market. It is based on wet milling of the macro-sized drug in the presence of surface stabilizers to produce crystalline nanosuspensions. Surprisingly, Bosch et al. [Patent No. US 6656504 B1] (2003) described surface-stabilized, amorphous nanoparticles of Cyclosporin A with particle size < 300 nm obtained using high energy pearl milling techniques. This 'top-down' generation method was claimed to confer the advantage of lower stabilizer-to-drug ratios and higher solid content – thus better drug loading compared to conventional controlled precipitation techniques used to generate amorphous nanosuspensions.

1.6.2 Marketed Products

- **SangCya®** by SangStat Medical Corporation. Cyclosporin 100mg/mL Oral Solution, USP (modified) – *withdrawn from market* [Patent No. US 5827822 A] (Floc'H and Merle 1998). Introduced in 1998, SangCya® oral solution was formulated as a stable dispersion of cyclosporin A in ethanol (solvent) and propylene glycol (co-solvent) containing polysorbate 80 (surfactant) (RxMed n.d.). The oral solution may be mixed with an aqueous medium prior to administration or directly ingested – both administration methods were claimed to result in the formation of an amorphous nanosuspension of cyclosporine A. Reconstitution with water produced nanoparticles in the range of 200–400nm with the presence of aggregates, while direct administration was expected to give an exclusive nanoparticle population of 600 nm. The product claimed to be bioequivalent to the innovator Neoral® oral solution (microemulsion formulation by Novartis) when mixed with apple juice, as recommended in its labeling (U.S. Food and Drug Administration 2013). Novartis filed a patent infringement lawsuit against SangStat in February 1999 claiming that SangCya® employed hydrosol technology owned by Novartis (PR Newswire 27 Jul 2000). Nevertheless, the product was subsequently withdrawn from the market in July 2000 after clinical studies showed non-bioequivalence to Neoral® oral solution contrary to what was claimed (U.S. Food and Drug Administration 2013).

1.6.3 Other Patents

The number of patents filed related to nanoamorphous drugs have burgeoned since the landmark patent on hydrosols was filed by List and Sucker in 1988. A summary of the more notable patents concerning nanoamorphous formulations is presented in Table 1.2.

Despite the robust patent landscape, progress from patent invention to product innovation has been excruciatingly slow – hampered by stability issues and lack of scale-up technology. On the contrary, nanocrystalline drugs have met with tremendous success in a relatively short time. It took only nine years since Elan's Nanocrystal® patent (1991) for the first commercial nanocrystal product, Rapamune® to be marketed. There are currently 17 licensed drug products utilizing nanocrystalline drug technology in the market (Jog and Burgess 2017). .

1.7 Summary and Future Perspectives

The advent of nanotherapeutics opens up possibilities for improved disease management and treatment. Nanofabrication of protein therapeutics involves protein engineering to modulate specificity, affinity, stability and biological activity. Protein therapeutics possess high specificity and potency but suffer poor biodistribution due to their large molecular size and inherent instability. Challenges in protein therapeutics delivery revolve around two fundamental aspects: 1) Physiological stability issues, i.e. stability of therapeutic proteins in the gastrointestinal environment, uptake, and clearance by immune macrophages in circulation, as well as redox stability in the local tissue and intracellular environments; 2) Size and polarity preventing efficient transport across pharmacologically important biological membranes, such as the intestinal epithelium, blood-brain barrier, and the nuclear membrane (Wu et al. 2017). Overcoming these challenges will enable the development of oral biologics, and parenteral drug delivery systems to access to the central nervous system and intracellular targets. On the other hand, the significance of nanofabrication for small molecule drugs lies in the design of drug delivery systems, including nanocarrier and nano-drug systems which confer additional properties such as improved bioavailability, reduced toxicity, modified release. Protein nanotechnology has emerged as a promising new field for drug delivery. Protein nanotechnology refers to the fabrication of nanostructures using proteins as engineering materials, thus imparting useful protein-related functionalities such as

TABLE 1.2

Patents Related to Production of Nanoamorphous Drug Particles

Priority Year	Patent No.	Assignee/ Inventor	Production Method	Process Technology	Product
1998	US 6177103 B1 (Pace et al. 2001)	Rtp Pharma	Atomization	Rapid expansion of supercritical solutions (RESS)	Nano-drug particles (solid state not specified)
1999	WO 2001015664 A2 (York et al. 2001)	Bradford Particle Design Limited; Bristol-Myers Squibb Company	Antisolvent precipitation	Solution Enhanced Dispersion by Supercritical fluids (SEDS)	Amorphous nano-composite particles
2001	US 6,756,062 B2 (Johnston et al. 2004)	University of Texas	Evaporative precipitation into aqueous solution (EPAS)	Atomization	Nanoamorphous drug particles
2002	US 20060013869 A1 (Ignatious and Sun 2006)	Smithkline Beecham Corporation	Electrospinning of drug/ polymer melt or solution	Electrospinning	Polymeric nanofibers containing amorphous drug
2007	US 2008/0299210 A1 (Wei et al. 2008)	Johnson & Johnson	Particle size reduction of amorphous bulk material	Wet milling, high speed homogenization, hydrodynamic cavitation, ultrasonication etc.	Nanoamorphous drug particles
2009	US 2012/0251595 A1 (Williams and Chow 2012)	University of Texas	Emulsion templating + ultra-rapid freezing	Homogenization + thin film freezing	Nanoamorphous drug particles
2011	US 2013/0095198 A1 (Keck 2013)	PharmaSol GmbH	Particle size reduction	High pressure homogenization	Nanoamorphous drug particles
2014	WO 2016016665 A1 (Temtem et al. 2016)	Hovione International Ltd	Antisolvent precipitation	Microfluidization	Amorphous nano-composite particles

self-assembly, targeting, or even therapeutic/ synergistic activity (Gerrard 2013).

In this chapter, we have highlighted the efficacy of nano-drug formulations in addressing the issue of low bioavailability of poorly soluble drugs. The clinical success exhibited by nano-crystalline formulations has culminated in several marketed products. Nanocrystalline has an advantage of better stability as compared to nanoamorphous formulations. Scientific data demonstrating superior in vivo efficacy of nanoamorphous versus nanocrystalline formulations is evident in the literature. Therefore, the potential of further therapeutic outcome improvement by nanoamorphous products is worth pursuing. The future of nanoamorphous formulations is dependent on the ability to maintain physical stability and address scale-up challenges associated with the bottom-up manufacturing approach.

Intense effort gearing towards formulation and process optimization has generated a large knowledge base for producing stable nanoamorphous formulations. To bring us a step closer to realizing the potential of nanoamorphous formulations, increased effort directed towards larger scale end-to-end nanoamorphous products manufacturing (from particle formation to final drying) is imperative.

Acknowledgments and Disclosures

All authors contributed to the development of the content. The authors, Roquette, and AbbVie reviewed and approved the publication.

REFERENCES

Abdelwahed, W., G. Degobert, and H. Fessi. (2006a). "Investigation of nanocapsules stabilization by amorphous excipients during freeze-drying and storage." *European Journal of Pharmaceutics and Biopharmaceutics*, 63(2), 87–94.

Abdelwahed, W., G. Degobert, S. Stainmesse, and H. Fessi. (2006b). "Freeze-drying of nanoparticles: Formulation, process and storage considerations." *Advanced Drug Delivery Reviews*, 58(15), 1688–713. doi: 10.1016/j.addr.2006.09.017.

Aditya, N.P., H. Yang, S. Kim, and S. Ko. (2015). "Fabrication of amorphous curcumin nanosuspensions using beta-lactoglobulin to enhance solubility, stability, and bioavailability." *Colloids and Surfaces B: Biointerfaces*, 127, 114–21. doi: 10.1016/j.colsurfb.2015.01.027.

Auweter, Helmut, Heribert Bohn, Herbert Haberkorn, Dieter Horn, Erik Luddecke, and Volker Rauschenberger. (1999). *Production of carotenoid preparations in the form of coldwater-dispersible powders, and the use of the novel carotenoid preparations*. Google Patents.

Baghel, Shrawan, Helen Cathcart, and Niall J. O'Reilly. (2016). "Polymeric amorphous solid dispersions: A review of amorphization, crystallization, stabilization, solid-state characterization, and aqueous solubilization of biopharmaceutical classification system class II drugs." *Journal of Pharmaceutical Sciences*, 105(9), 2527–44.

Bajorath, Jürgen. (2002). "Integration of virtual and high-throughput screening." *Nature Reviews Drug Discovery*, 1(11), 882–94.

Beck, Christian, Sameer V. Dalvi, and Rajesh N. Dave. (2010). "Controlled liquid antisolvent precipitation using a rapid mixing device." *Chemical Engineering Science*, 65(21), 5669–75.

Becker, R., and W. Döring. (1935). "Kinetische behandlung der keimbildung in übersättigten dämpfen." *Annalen der Physik*, *416*(8), 719–52.

Beirowski, J., S. Inghelbrecht, A. Arien, and H. Gieseler. (2011). "Freeze-drying of nanosuspensions, 1: Freezing rate versus formulation design as critical factors to preserve the original particle size distribution." *Journal of Pharmaceutical Sciences*, *100*(5), 1958–68. doi: 10.1002/jps.22425.

Betancourt, Tania, and Lisa Brannon-Peppas. (2006). "Micro- and nanofabrication methods in nanotechnological medical and pharmaceutical devices." *International Journal of Nanomedicine*, *1*(4), 483.

Bi, Yanping, Jingjing Liu, Jianzhu Wang, Jifu Hao, Fei Li, Teng Wang, Hong Wei Sun, and Fengguang Guo. (2015). "Particle size control and the interactions between drug and stabilizers in an amorphous nanosuspension system." *Journal of Drug Delivery Science and Technology*, *29*, 167–72.

Biswas, Priyanka, D. Sen, S. Mazumder, C.B. Basak, and P. Doshi. (2016). "Temperature mediated morphological transition during drying of spray colloidal droplets." *Langmuir*, *32*(10), 2464–73.

Bleicher, Konrad H., Hans-Joachim Böhm, Klaus Müller, and Alexander I. Alanine. (2003). "Hit and lead generation: Beyond high-throughput screening." *Nature Reviews Drug Discovery*, *2*(5), 369–78.

Bosch, H. William, Kevin D. Ostrander, and Douglas C. Hovey. (2003). *Nanoparticulate compositions comprising amorphous cyclosporine and methods of making and using such compositions*. Google Patents.

Bosselmann, S., M. Nagao, K.T. Chow, and R.O. Williams, 3rd. (2012). "Influence of formulation and processing variables on properties of itraconazole nanoparticles made by advanced evaporative precipitation into aqueous solution." *AAPS PharmSciTech*, *13*(3), 949–60. doi: 10.1208/s12249-012-9817-0.

Bosselmann, S., and R.O. Williams. (2012). "Has nanotechnology led to improved therapeutic outcomes?" *Drug Development and Industrial Pharmacy*, *38*(2), 158–70. doi: 10.3109/03639045.2011.597764.

Brigger, Irène, Catherine Dubernet, and Patrick Couvreur. (2012). "Nanoparticles in cancer therapy and diagnosis." *Advanced Drug Delivery Reviews*, *64*, 24–36. doi: 10.1016/j.addr.2012.09.006.

Brough, C., and R.O. Williams, 3rd. (2013). "Amorphous solid dispersions and nano-crystal technologies for poorly water-soluble drug delivery." *International Journal of Pharmaceutics*, *453*(1), 157–66. doi: 10.1016/j.ijpharm.2013.05.061.

Brouwers, Joachim, Marcus E. Brewster, and Patrick Augustijns. (2009). "Supersaturating drug delivery systems: The answer to solubility: Limited oral bioavailability?" *Journal of Pharmaceutical Sciences*, *98*(8), 2549–72.

Carter, Nathan A., Xi Geng, and Tijana Z. Grove. (2016). "Design of self-assembling protein-polymer conjugates." In *Protein-Based Engineered Nanostructures* (pp. 179–214). Springer.

Chan, H.K., and P.C. Kwok. (2011). "Production methods for nanodrug particles using the bottom-up approach." *Advanced Drug Delivery Reviews* *63*(6), 406–16. doi: 10.1016/j.addr.2011.03.011.

Chan, Juliana M., Pedro M. Valencia, Liangfang Zhang, Robert Langer, and Omid C. Farokhzad. (2010). "Polymeric nanoparticles for drug delivery." *Cancer Nanotechnology: Methods and Protocols*, *624*, 163–75.

Chen, Jian-Feng, Ji-Yao Zhang, Zhi-Gang Shen, Jie Zhong, and Jimmy Yun. (2006). "Preparation and characterization of amorphous cefuroxime axetil drug nanoparticles with novel technology: High-gravity antisolvent precipitation." *Industrial & Engineering Chemistry Research*, *45*(25), 8723–7.

Chen, W., B. Gu, H. Wang, J. Pan, W. Lu, and H. Hou. (2008). "Development and evaluation of novel itraconazole-loaded intravenous nanoparticles." *International Journal of Pharmaceutics*, *362*(1–2), 133–40. doi: 10.1016/j.ijpharm.2008.05.039.

Chen, Xiaoxia, Timothy J. Young, Marazban Sarkari, Robert O. Williams, and Keith P. Johnston. (2002). "Preparation of cyclosporine A nanoparticles by evaporative precipitation into aqueous solution." *International Journal of Pharmaceutics*, *242*(1), 3–14.

Cheow, W.S., T.Y. Kiew, and K. Hadinoto. (2015). "Amorphous nanodrugs prepared by complexation with polysaccharides: carrageenan versus dextran sulfate." *Carbohydrate Polymer*, *117*, 549–58. doi: 10.1016/j.carbpol.2014.10.015.

Cheow, W.S., T.Y. Kiew, Y. Yang, and K. Hadinoto. (2014). "Amorphization strategy affects the stability and supersaturation profile of amorphous drug nanoparticles." *Molecular Pharmaceutics*, *11*(5), 1611–20. doi: 10.1021/mp400788p.

Chin, W.W., J. Parmentier, M. Widzinski, E.H. Tan, and R. Gokhale. (2014). "A brief literature and patent review of nanosuspensions to a final drug product." *Journal of Pharmaceutical Sciences*, *103*(10), 2980–99. doi: 10.1002/jps.24098.

Chiu, Yu-Chieh, Joshua M Gammon, James I Andorko, Lisa H Tostanoski, and Christopher M Jewell. (2016). "Assembly and immunological processing of polyelectrolyte multilayers composed of antigens and adjuvants." *ACS Applied Materials & Interfaces*, *8*(29), 18722–31.

Chow, Keat Theng, David Cheng Thiam Tan, and Rajeev Gokhale. (2016). "Small is big: Is nanoamorphous better than amorphous solid dispersion and nanocrystalline in pharma?" *Research & Reviews: Journal of Pharmaceutics and Nanotechnology*, *4*(3), 11–4.

Chung, N.O., M.K. Lee, and J. Lee. (2012). "Mechanism of freeze-drying drug nanosuspensions." *International Journal of Pharmaceutics*, *437*(1–2), 42–50. doi: 10.1016/j.ijpharm.2012.07.068.

Dalvi, Sameer V., and Rajesh N. Dave. (2010). "Analysis of nucleation kinetics of poorly water-soluble drugs in presence of ultrasound and hydroxypropyl methyl cellulose during antisolvent precipitation." *International Journal of Pharmaceutics*, *387*(1), 172–9.

Dan, J., Y. Ma, P. Yue, Y. Xie, Q. Zheng, P. Hu, W. Zhu, and M. Yang. (2016). "Microcrystalline cellulose-carboxymethyl cellulose sodium as an effective dispersant for drug nanocrystals: A case study." *Carbohydrate Polymer*, *136*, 499–506. doi: 10.1016/j.carbpol.2015.09.048.

De Yoreo, James J., and Peter G. Vekilov. (2003). "Principles of crystal nucleation and growth." *Reviews in Mineralogy and Geochemistry*, *54*(1), 57–93.

Deng, J., L. Huang, and F. Liu. (2010). "Understanding the structure and stability of paclitaxel nanocrystals." *International Journal of Pharmaceutics*, *390*(2), 242–9. doi: 10.1016/j.ijpharm.2010.02.013.

Dhumal, R.S., S.V. Biradar, S. Yamamura, A.R. Paradkar, and P. York. (2008). "Preparation of amorphous cefuroxime axetil nanoparticles by sonoprecipitation for enhancement of bioavailability." *European Journal of Pharmaceutics and Biopharmaceutics*, *70*(1), 109–15. doi: 10.1016/j.ejpb.2008.04.001.

Dokoumetzidis, A., and P. Macheras. (2006). "A century of dissolution research: From Noyes and Whitney to the biopharmaceutics classification system." *International Journal of Pharmaceutics*, *321*(1–2), 1–11. doi: 10.1016/j.ijpharm.2006.07.011.

Dong, Y., W.K. Ng, J. Hu, S. Shen, and R.B. Tan. (2014). "Clay as a matrix former for spray drying of drug nanosuspensions." *International Journal of Pharmaceutics*, *465*(1–2), 83–9. doi: 10.1016/j.ijpharm.2014.02.025.

Drobnak, Igor, Ajasja Ljubetič, Helena Gradišar, Tomaž Pisanski, and Roman Jerala. (2016). "Designed protein origami." In *Protein-based Engineered Nanostructures* (pp. 7–27). Springer.

Erickson, Harold P. (2009). "Size and shape of protein molecules at the nanometer level determined by sedimentation, gel filtration, and electron microscopy." *Biological Procedures Online*, *11*(1), 32.

Feynman, Richard P. (1960). "There's plenty of room at the bottom." *Engineering and Science*, *23*(5), 22–36.

Floc'H, R., and C. Merle. 1998. *Cyclosporin a formulations as nanoparticles*. Google Patents.

Gao, P., and Y. Shi. (2012). "Characterization of supersaturatable formulations for improved absorption of poorly soluble drugs." *AAPS Journal*, *14*(4), 703–13. doi: 10.1208/s12248-012-9389-7.

Gates, Byron D., Qiaobing Xu, Michael Stewart, Declan Ryan, C Grant Willson, and George M. Whitesides. (2005). "New approaches to nanofabrication: Molding, printing, and other techniques." *Chemical Reviews*, *105*(4), 1171–96.

Gerrard, Juliet A. (2013). "Protein nanotechnology: What is it?" In *Protein Nanotechnology: Protocols, Instrumentation, and Applications*, Humana Press, New York (2nd ed., pp. 1–15).

Goldshtein, Rina. (2005). *Hydrophilic complexes of lipophilic materials and an apparatus and method for their production*. Google Patents.

Groβ, Gregor Alexander, and Johann Michael Koöhler. (2010). "Residence time distribution and nanoparticle formation in microreactors." In *Microfluidic Devices in Nanotechnology: Fundamental Concepts*, John Wiley & Sons, (pp. 317–40).

Grove, Tijana Z., and Aitziber L. Cortajarena. (2016). "Protein design for nanostructural engineering: General aspects." In *Protein-based Engineered Nanostructures* (pp. 1–5). Springer.

Hancock, Bruno C., and Michael Parks. (2000). "What is the true solubility advantage for amorphous pharmaceuticals?" *Pharmaceutical Research*, *17*(4), 397–404.

Hancock, Bruno C., Sheri L. Shamblin, and George Zografi. (1995). "Molecular mobility of amorphous pharmaceutical solids below their glass transition temperatures." *Pharmaceutical Research*, *12*(6), 799–806.

He, W., Y. Lu, J. Qi, L. Chen, F. Hu, and W. Wu. (2013). "Food proteins as novel nanosuspension stabilizers for poorly water-soluble drugs." *International Journal of Pharmaceutics*, *441*(1–2), 269–78. doi: 10.1016/j.ijpharm.2012.11.033.

Heng, D., S.H. Lee, W.K. Ng, and R.B. Tan. (2011). "The nano spray dryer B-90." *Expert Opinion on Drug Delivery*, *8*(7), 965–72. doi: 10.1517/17425247.2011.588206.

Horn, Dieter, and Jens Rieger. (2001). "Organic nanoparticles in the aqueous phase: Theory, experiment, and use." *Angewandte Chemie International Edition*, *40*(23), 4330–61.

Hörter, D., and J.B. Dressman. (2001). "Influence of physicochemical properties on dissolution of drugs in the gastrointestinal tract." *Advanced Drug Delivery Reviews*, *46*(1), 75–87.

Hou, C.D., J.X. Wang, Y. Le, H.K. Zou, and H. Zhao. (2012). "Preparation of azithromycin nanosuspensions by reactive precipitation method." *Drug Development and Industrial Pharmacy*, *38*(7), 848–54. doi: 10.3109/03639045.2011.630394.

Howard, Philip W. (2017). "Antibody–drug conjugates (ADCs)." *Protein Therapeutics*, *1*, 271–309.

Huang, Y., and W.G. Dai. (2014). "Fundamental aspects of solid dispersion technology for poorly soluble drugs." *Acta Pharmaceutica Sinica B*, *4*(1), 18–25. doi: 10.1016/j.apsb.2013.11.001.

Ignatious, F., and L. Sun. (2006). *Electrospun amorphous pharmaceutical compositions*. Google Patents.

Janssen Products LP. (2015). "Doxil(R) [Full prescribing information]." https://www.doxil.com/shared/product/doxil/doxil-prescribing-information.pdf.

Jog, R., and D.J. Burgess. (2017). "Pharmaceutical amorphous nanoparticles." *Journal of Pharmaceutical Sciences*, *106*(1), 39–65. doi: 10.1016/j.xphs.2016.09.014.

Jog, R., S. Kumar, J. Shen, N. Jugade, D.C. Tan, R. Gokhale, and D.J. Burgess. (2016). "Formulation design and evaluation of amorphous ABT-102 nanoparticles." *International Journal of Pharmaceutics*, *498*(1–2), 153–69. doi: 10.1016/j.ijpharm.2015.12.033.

Johnston, K.P., R.O. Williams, T.J. Young, and X. Chen. (2004). *Preparation of drug particles using evaporation precipitation into aqueous solutions*. Google Patents.

Kalepu, Sandeep, and Vijaykumar Nekkanti. (2015). "Insoluble drug delivery strategies: Review of recent advances and business prospects." *Acta Pharmaceutica Sinica B*, *5*(5), 442–53.

Kashi, T.S., S. Eskandarion, M. Esfandyari-Manesh, S.M. Marashi, N. Samadi, S.M. Fatemi, F. Atyabi, S. Eshraghi, and R. Dinarvand. (2012). "Improved drug loading and antibacterial activity of minocycline-loaded PLGA nanoparticles prepared by solid/oil/water ion pairing method." *International Journal of Nanomedicine*, *7*, 221–34. doi: 10.2147/IJN.S27709.

Kataokaa, Kazunori, Atsushi Haradaa, and Yukio Nagasakib. (2001). "Block copolymer micelles for drug delivery design, characterization and biological significance." *Advanced Drug Delivery Reviews*, *47*, 113–31.

Kawabata, Y., K. Wada, M. Nakatani, S. Yamada, and S. Onoue. (2011). "Formulation design for poorly water-soluble drugs based on biopharmaceutics classification system: Basic approaches and practical applications." *International Journal of Pharmaceutics*, *420*(1), 1–10. doi: 10.1016/j.ijpharm.2011.08.032.

Ke, X., V.W. Ng, R.J. Ono, J.M. Chan, S. Krishnamurthy, Y. Wang, J.L. Hedrick, and Y.Y. Yang. (2014). "Role of non-covalent and covalent interactions in cargo loading capacity and stability of polymeric micelles." *Journal of Controlled Release*, *193*, 9–26. doi: 10.1016/j.jconrel.2014.06.061.

Keck, C. (2013). *Nanocrystals and amorphous nanoparticles and method for production of the same by a low energy process*. Google Patents.

Kesisoglou, F., S. Panmai, and Y. Wu. (2007). "Nanosizing--oral formulation development and biopharmaceutical evaluation." *Advanced Drug Delivery Reviews*, *59*(7), 631–44. doi: 10.1016/j.addr.2007.05.003.

Kim, M.S., S.J. Jin, J.S. Kim, H.J. Park, H.S. Song, R.H. Neubert, and S.J. Hwang. (2008). "Preparation, characterization and in vivo evaluation of amorphous atorvastatin calcium nanoparticles using supercritical antisolvent (SAS) process." *European Journal of Pharmaceutics and Biopharmaceutics*, *69*(2), 454–65. doi: 10.1016/j.ejpb.2008.01.007.

Kim, S., and J. Lee. (2010). "Effective polymeric dispersants for vacuum, convection and freeze drying of drug nanosuspensions." *International Journal of Pharmaceutics*, *397*(1–2), 218–24. doi: 10.1016/j.ijpharm.2010.07.010.

King, Neil P., Jacob B. Bale, William Sheffler, Dan E. McNamara, Shane Gonen, Tamir Gonen, Todd O. Yeates, and David Baker. (2014). "Accurate design of co-assembling multi-component protein nanomaterials." *Nature*, *510*(7503), 103–8.

Klibanov, Alexander L., Kazuo Maruyama, Vladimir P. Torchilin, and Leaf Huang. (1990). "Amphipathic polyethyleneglycols effectively prolong the circulation time of liposomes." *FEBS Letters*, *268*(1), 235–7.

Koide, Akiko, John Wojcik, Ryan N. Gilbreth, Robert J. Hoey, and Shohei Koide. (2012). "Teaching an old scaffold new tricks: Monobodies constructed using alternative surfaces of the FN3 scaffold." *Journal of Molecular Biology*, *415*(2), 393–405.

Kumar, S., R. Gokhale, and D.J. Burgess. (2014a). "Sugars as bulking agents to prevent nano-crystal aggregation during spray or freeze-drying." *International Journal of Pharmaceutics*, *471*(1–2), 303–11. doi: 10.1016/j.ijpharm.2014.05.060.

Kumar, S., J. Shen, and D.J. Burgess. (2014c). 'Nano-amorphous spray dried powder to improve oral bioavailability of itraconazole." *Journal of Controlled Release*, *192*, 95–102. doi: 10.1016/j.jconrel.2014.06.059.

Kumar, Sumit, Rajeev Gokhale, and Diane J. Burgess. (2014b). "Quality by design approach to spray drying processing of crystalline nanosuspensions." *International Journal of Pharmaceutics*, *464*(1), 234–42.

Kumar, Sumit, Xiaoming Xu, Rajeev Gokhale, and Diane J. Burgess. (2014d). "Formulation parameters of crystalline nanosuspensions on spray drying processing: A DoE approach." *International Journal of Pharmaceutics*, *464*(1), 34–45.

Kumari, A., S.K. Yadav, and S.C. Yadav. (2010). "Biodegradable polymeric nanoparticles based drug delivery systems." *Colloids and Surfaces B: Biointerfaces*, *75*(1), 1–18. doi: 10.1016/j.colsurfb.2009.09.001.

Kyeremateng, S.O., M. Pudlas, and G.H. Woehrle. (2014). "A fast and reliable empirical approach for estimating solubility of crystalline drugs in polymers for hot melt extrusion formulations." *Journal of Pharmaceutical Sciences*, *103*(9), 2847–58. doi: 10.1002/jps.23941.

Laouini, Abdallah, Chiraz Jaafar-Maalej, Imen Limayem-Blouza, S. Sfar, Catherine Charcosset, and Hatem Fessi. (2012). "Preparation, characterization and applications of liposomes: State of the art." *Journal of Colloid Science and Biotechnology*, *1*(2), 147–68.

Lee, Eun Seong, Zhonggao Gao, and You Han Bae. (2008). "Recent progress in tumor pH targeting nanotechnology." *Journal of Controlled Release*, *132*(3), 164–70.

Lee, M.K., M.Y. Kim, S. Kim, and J. Lee. (2009). "Cryoprotectants for freeze drying of drug nano-suspensions: Effect of freezing rate." *Journal of Pharmaceutical Sciences*, *98*(12), 4808–17. doi: 10.1002/jps.21786.

Lee, S.H., D. Heng, W.K. Ng, H.K. Chan, and R.B. Tan. (2011). "Nano spray drying: A novel method for preparing protein nanoparticles for protein therapy." *International Journal of Pharmaceutics*, *403*(1–2), 192–200. doi: 10.1016/j.ijpharm.2010.10.012.

Lee, Sibeum, Kyungwan Nam, Min Soo Kim, Seoung Wook Jun, Jeong-Sook Park, Jong Soo Woo, and Sung-Joo Hwang. (2005). "Preparation and characterization of solid dispersions of itraconazole by using aerosol solvent extraction system for improvement in drug solubility and bioavailability." *Archives of Pharmacal Research*, *28*(7), 866–74.

Li, X., N. Anton, C. Arpagaus, F. Belleteix, and T.F. Vandamme. (2010). "Nanoparticles by spray drying using innovative new technology: The Buchi nano spray dryer B-90." *Journal of Controlled Release*, *147*(2), 304–10. doi: 10.1016/j.jconrel.2010.07.113.

Lin, Yi, Ge-Bo Pan, Gui-Jin Su, Xiao-Hong Fang, Li-Jun Wan, and Chun-Li Bai. (2003). "Study of citrate adsorbed on the Au (111) surface by scanning probe microscopy." *Langmuir*, *19*(24), 10000–3.

Lin, Yining, and Paschalis Alexandridis. (2002). "Temperature-dependent adsorption of Pluronic F127 block copolymers onto carbon black particles dispersed in aqueous media." *The Journal of Physical Chemistry B*, *106*(42), 10834–44.

Lindfors, Lennart, Pia Skantze, Urban Skantze, Mikael Rasmusson, Anna Zackrisson, and Ulf Olsson. (2006). "Amorphous drug nanosuspensions. 1. Inhibition of Ostwald ripening." *Langmuir*, *22*(3), 906–10.

Lindfors, Lennart, Pia Skantze, Urban Skantze, Jan Westergren, and Ulf Olsson. (2007). "Amorphous drug nanosuspensions. 3. Particle dissolution and crystal growth." *Langmuir*, *23*(19), 9866–74.

Lindsley, Craig W. (2017). *New 2016 Data and Statistics for Global Pharmaceutical Products and Projections through 2017*. ACS Publications.

Lipinski, Calf. (2002). "Poor aqueous solubility-an industry wide problem in drug discovery." *American Pharmaceutical Review*, *5*(3), 82–5.

Lipovšek, D. (2010). "Adnectins: Engineered target-binding protein therapeutics." *Protein Engineering Design & Selection*, *24*(1–2), 3–9.

List, M., and Sucker, H.. (1995). Hydrosols of pharmacologically active agents and their pharmaceutical compositions comprising them.

Liu, D., H. Xu, B. Tian, K. Yuan, H. Pan, S. Ma, X. Yang, and W. Pan. (2012). "Fabrication of carvedilol nanosuspensions through the anti-solvent precipitation-ultrasonication method for the improvement of dissolution rate and oral bioavailability." *AAPS PharmSciTech*, *13*(1), 295–304. doi: 10.1208/s12249-011-9750-7.

Liu, Yinghui, Changshan Sun, Yanru Hao, Tongying Jiang, Li Zheng, and Siling Wang. (2010). "Mechanism of dissolution enhancement and bioavailability of poorly water soluble celecoxib by preparing stable amorphous nanoparticles." *Journal of Pharmacy & Pharmaceutical Sciences*, *13*(4), 589–606.

Liu, Z., K. Chen, C. Davis, S. Sherlock, Q. Cao, X. Chen, and H. Dai. (2008). "Drug delivery with carbon nanotubes for in vivo cancer treatment." *Cancer Research*, *68*(16), 6652–60. doi: 10.1158/0008-5472.CAN-08-1468.

Liversidge, G.G., K.C. Cundy, J.F. Bishop, and D.A. Czekai. (1992). *Surface modified drug nanoparticles*. Google Patents.

Ljubetič, Ajasja, Fabio Lapenta, Helena Gradišar, Igor Drobnak, Jana Aupič, Žiga Strmšek, Duško Lainšček, Iva Hafner-Bratkovič, Andreja Majerle, and Nuša Krivec. (2017). "Design of coiled-coil protein-origami cages that self-assemble in vitro and in vivo." *Nature Biotechnology*, *35*(11), 1094.

Lourenco, Cristina, Maribel Teixeira, Sérgio Simões, and Rogério Gaspar. (1996). "Steric stabilization of nanoparticles: Size and surface properties." *International Journal of Pharmaceutics*, *138*(1), 1–12.

Malkani, A., A.A. Date, and D. Hegde. (2014). "Celecoxib nanosuspension: Single-step fabrication using a modified nanoprecipitation method and in vivo evaluation." *Drug Delivery and Translational Research*, *4*(4), 365–76. doi: 10.1007/s13346-014-0201-3.

Margulis-Goshen, K., E. Kesselman, D. Danino, and S. Magdassi. (2010). "Formation of celecoxib nanoparticles from volatile microemulsions." *International Journal of Pharmaceutics*, *393*(1–2), 230–7. doi: 10.1016/j.ijpharm.2010.04.012.

Matteucci, M.E., J.C. Paguio, M.A. Miller, R.O. Williams III, and K.P. Johnston. (2008). "Flocculated amorphous nanoparticles for highly supersaturated solutions." *Pharmaceutical Research*, *25*(11), 2477–87. doi: 10.1007/s11095-008-9659-3.

Matteucci, Michal E., Blair K. Brettmann, True L. Rogers, Edmund J. Elder, Robert O. Williams III, and Keith P. Johnston. (2007). "Design of potent amorphous drug nanoparticles for rapid generation of highly supersaturated media." *Molecular Pharmaceutics*, *4*(5), 782–93.

Merisko-Liversidge, Elaine M., and Gary G. Liversidge. (2008). "Drug nanoparticles: Formulating poorly water-soluble compounds." *Toxicologic Pathology*, *36*(1), 43–8.

Mittal, Vikas. (2010). *Advanced Polymer Nanoparticles: Synthesis and Surface Modifications*: CRC Press.

Mohammed, A.R., N. Weston, A.G. Coombes, M. Fitzgerald, and Y. Perrie. (2004). "Liposome formulation of poorly water soluble drugs: Optimisation of drug loading and ESEM analysis of stability." *International Journal of Pharmaceutics*, *285*(1–2), 23–34. doi: 10.1016/j.ijpharm.2004.07.010.

Mou, D., H. Chen, J. Wan, H. Xu, and X. Yang. (2011). "Potent dried drug nanosuspensions for oral bioavailability enhancement of poorly soluble drugs with pH-dependent solubility." *International Journal of Pharmaceutics*, *413*(1–2), 237–44. doi: 10.1016/j.ijpharm.2011.04.034.

Mugheirbi, N.A., K.J. Paluch, and L. Tajber. (2014). "Heat induced evaporative antisolvent nanoprecipitation (HIEAN) of itraconazole." *International Journal of Pharmaceutics*, *471*(1–2), 400–11. doi: 10.1016/j.ijpharm.2014.05.045.

Nandiyanto, Asep Bayu Dani, and Kikuo Okuyama. (2011). "Progress in developing spray-drying methods for the production of controlled morphology particles: From the nanometer to submicrometer size ranges." *Advanced Powder Technology*, *22*(1), 1–19. doi: 10.1016/j.apt.2010.09.011.

Nogi, Kiyoshi, Makio Naito, and Toyokazu Yokoyama. (2012). *Nanoparticle Technology Handbook*. Elsevier.

Ostwald, Wilhelm. (1900). "Über die vermeintliche Isomerie des roten und gelben Quecksilberoxyds und die Oberflächenspannung fester Körper." *Zeitschrift für Physikalische Chemie*, *34*, 495–503.

Pace, G.W., M.G. Vachon, A.K. Mishra, I.B. Henrikson, and V. Krukonis. (2001). *Processes to generate submicron particles of water-insoluble compounds*. Google Patents.

Pastorin, G. (2009). "Crucial functionalizations of carbon nanotubes for improved drug delivery: A valuable option?" *Pharmaceutical Research*, *26*(4), 746–69. doi: 10.1007/s11095-008-9811-0.

PR Newswire. (27 Jul 2000). "Novartis announces settlement with sangstat medical corp." Accessed 20 May 2016. http://www.prnewswire.com/news-releases/novartis-announces-settlement-with-sangstat-medical-corp-72568087.html.

Prabakaran, Ponraj, and Dimiter S Dimitrov. (2017). "Human antibody structure and function." *Protein Therapeutics, 1*, 51–84.

Pu, X., J. Sun, Y. Wang, Y. Wang, X. Liu, P. Zhang, X. Tang, W. Pan, J. Han, and Z. He. (2009). "Development of a chemically stable 10-hydroxycamptothecin nanosuspensions." *International Journal of Pharmaceutics*, *379*(1), 167–73. doi: 10.1016/j.ijpharm.2009.05.062.

Romero, G.B., A. Arntjen, C.M. Keck, and R.H. Muller. (2016). "Amorphous cyclosporin A nanoparticles for enhanced dermal bioavailability." *International Journal of Pharmaceutics*, *498*(1–2), 217–24. doi: 10.1016/j.ijpharm.2015.12.019.

Rumondor, Alfred C.F., and Lynne S. Taylor. (2010). "Effect of polymer hygroscopicity on the phase behavior of amorphous solid dispersions in the presence of moisture." *Molecular Pharmaceutics*, *7*(2), 477–90.

Rumondor, Alfred C.F., Håkan Wikström, Bernard Van Eerdenbrugh, and Lynne S Taylor. (2011). "Understanding the tendency of amorphous solid dispersions to undergo amorphous–amorphous phase separation in the presence of absorbed moisture." *AAPS PharmSciTech*, *12*(4), 1209–19.

RxMed. (n.d.). "Pharmaceutical information: Sangcya oral solution." Accessed 20 May 2016. http://www.rxmed.com/b.main/b2.pharmaceutical/b2.1.monographs/CPS-%20Monographs/CPS-%20(General%20Monographs-%20S)/SangCya%20ORAL%20SOLUTION.html.

Sarode, Ashish L., Harpreet Sandhu, Navnit Shah, Waseem Malick, and Hossein Zia. (2013). "Hot melt extrusion for amorphous solid dispersions: Temperature and moisture activated drug–polymer interactions for enhanced stability." *Molecular Pharmaceutics*, *10*(10), 3665–75.

Schafroth, N., C. Arpagaus, U.Y. Jadhav, S. Makne, and D. Douroumis. (2012). "Nano and microparticle engineering of water insoluble drugs using a novel spray-drying process." *Colloids and Surfaces B Biointerfaces*, *90*, 8–15. doi: 10.1016/j.colsurfb.2011.09.038.

Schmid, K., C. Arpagaus, and W. Friess. (2011). "Evaluation of the nano spray dryer B-90 for pharmaceutical applications." *Pharmaceutical Development and Technology*, *16*(4), 287–94. doi: 10.3109/10837450.2010.485320.

Sigfridsson, K., S. Forssen, P. Hollander, U. Skantze, and J. de Verdier. (2007). "A formulation comparison, using a solution and different nanosuspensions of a poorly soluble compound." *European Journal of Pharmaceutics and Biopharmaceutics*, *67*(2), 540–7. doi: 10.1016/j.ejpb.2007.02.008.

Sinha, Biswadip, Rainer H Müller, and Jan P Möschwitzer. (2013). "Bottom-up approaches for preparing drug nanocrystals: formulations and factors affecting particle size." *International Journal of Pharmaceutics*, *453*(1), 126–41.

Sjöström, Brita, Björn Bergenståhl, and Bengt Kronberg. (1993). "A method for the preparation of submicron particles of sparingly water-soluble drugs by precipitation in oil-in-water emulsions. II: Influence of the emulsifier, the solvent, and the drug substance." *Journal of Pharmaceutical Sciences*, *82*(6), 584–9.

Strohl, William R. (2015). "Fusion proteins for half-life extension of biologics as a strategy to make biobetters." *BioDrugs*, *29*(4), 215–39.

Studart, Andre R., Esther Amstad, and Ludwig J. Gauckler. (2007). "Colloidal stabilization of nanoparticles in concentrated suspensions." *Langmuir*, 23(3), 1081–90.

Sugimoto, Tadao. (2007). "Underlying mechanisms in size control of uniform nanoparticles." *Journal of Colloid and Interface Science*, 309(1), 106–18.

Sugiyama, Yoichi, Ryan J Larsen, Jin-Woong Kim, and David A. Weitz. (2006). "Buckling and crumpling of drying droplets of colloid–polymer suspensions." *Langmuir*, 22(14), 6024–30.

Talon Therapeutics Inc. (2012). "Marqibo(R) [full prescribing information]." http://www.marqibo.com/pi/.

Tan, J.P., S.H. Kim, F. Nederberg, K. Fukushima, D.J. Coady, A. Nelson, Y.Y. Yang, and J.L. Hedrick. (2010). "Delivery of anticancer drugs using polymeric micelles stabilized by hydrogen-bonding urea groups." *Macromolecular Rapid Communications*, 31(13), 1187–92. doi: 10.1002/marc.201000105.

Tao, J., Y. Sun, G.G. Zhang, and L. Yu. (2009). "Solubility of small-molecule crystals in polymers: D-mannitol in PVP, indomethacin in PVP/VA, and nifedipine in PVP/VA." *Pharmaceutical Research*, 26(4), 855–64. doi: 10.1007/s11095-008-9784-z.

Temtem, M., R. Pereira, J. Vicente, F. Gasper, and I. Duarte. (2016). *A method of preparing amorphous solid dispersion in submicron range by co-precipitation.* Google Patents.

Thadkala, K., P.K. Nanam, B. Rambabu, C. Sailu, and J. Aukunuru. (2014). "Preparation and characterization of amorphous ezetimibe nanosuspensions intended for enhancement of oral bioavailability." *International Journal of Pharmaceutical Investigation*, 4(3), 131–7. doi: 10.4103/2230-973X.138344.

Thanh, N.T., N. Maclean, and S. Mahiddine. (2014). "Mechanisms of nucleation and growth of nanoparticles in solution." *Chemical Reviews* 114(15), 7610–30. doi: 10.1021/cr400544s.

Thassu, Deepak, Michel Deleers, and Yashwant Vishnupant Pathak. (2007). *Nanoparticulate Drug Delivery Systems* (Vol. 166). CRC Press.

Thorat, Alpana A., and Sameer V. Dalvi. (2012). "Liquid antisolvent precipitation and stabilization of nanoparticles of poorly water soluble drugs in aqueous suspensions: Recent developments and future perspective." *Chemical Engineering Journal*, 181–182, 1–34. doi: 10.1016/j.cej.2011.12.044.

Tinkle, S., S.E. McNeil, S. Muhlebach, R. Bawa, G. Borchard, Y.C. Barenholz, L. Tamarkin, and N. Desai. (2014). "Nanomedicines: Addressing the scientific and regulatory gap." *Annals of the New York Academy of Sciences*, 1313, 35–56. doi: 10.1111/nyas.12403.

Torchilin, Vladimir. (2008). *Multifunctional Pharmaceutical Nanocarriers* (Vol. 4). Springer.

Torchilin, Vladimir P. (2007). "Targeted pharmaceutical nanocarriers for cancer therapy and imaging." *AAPS Journal*, 9(2), E128–47.

Torchilin, Vladimir P. (2012). "Multifunctional nanocarriers." *Advanced Drug Delivery Reviews*, 64, 302–15.

Tran, T.T., P.H. Tran, M.N. Nguyen, K.T. Tran, M.N. Pham, P.C. Tran, and T.V. Vo. (2014). "Amorphous isradipine nanosuspension by the sonoprecipitation method." *International Journal of Pharmaceutics*, 474(1–2), 146–50. doi: 10.1016/j.ijpharm.2014.08.017.

U.S. Food and Drug Administration. (2013). "Safety alert for human medical products. SangCya (Cyclosporine Oral Solution, USP (MODIFIED))." Last Mo dified 14 Aug 2013. http://www.fda. gov/Safety/MedWatch/SafetyInformation/SafetyAlertsforHumanMedicalProducts/ucm173083.htm.

U.S. Food and Drug Administration. (2014). *Guidance for Industry. Considering Whether an FDA-Regulated Product Involves the Application of Nanotechnology.* U.S. Department of Health and Human Services.

U.S. Food and Drug Administration. (2015). "What are 'biologics' questions and answers." Last Modified 8 May 2015 accessed 12 Jan 2018. https://www.fda.gov/AboutFDA/CentersOffices/OfficeofMedicalProductsandTobacco/CBER/ucm133077.htm.

Van Eerdenbrugh, B. (2008). "Top-down production of nanocrystals." Katholieke Universiteit Leuven.

Van Eerdenbrugh, B., L. Froyen, J. Van Humbeeck, J.A. Martens, P. Augustijns, and G. Van Den Mooter. (2008a). "Alternative matrix formers for nanosuspension solidification: Dissolution performance and X-ray microanalysis as an evaluation tool for powder dispersion." *European Journal of Pharmaceutical Sciences*, 35(4), 344–53. doi: 10.1016/j.ejps.2008.08.003.

Van Eerdenbrugh, B., L. Froyen, J. Van Humbeeck, J.A. Martens, P. Augustijns, and G. Van den Mooter. (2008b). "Drying of crystalline drug nanosuspensions-the importance of surface hydrophobicity on dissolution behavior upon redispersion." *European Journal of Pharmaceutical Sciences*, 35(1–2), 127–35. doi: 10.1016/j.ejps.2008.06.009.

Van Eerdenbrugh, B., S. Vercruysse, J.A. Martens, J. Vermant, L. Froyen, J. Van Humbeeck, G. Van den Mooter, and P. Augustijns. (2008c). "Microcrystalline cellulose, a useful alternative for sucrose as a matrix former during freeze-drying of drug nanosuspensions: A case study with itraconazole." *European Journal of Pharmaceutics and Biopharmaceutics*, 70(2), 590–6. doi: 10.1016/j.ejpb.2008.06.007.

Van Eerdenbrugh, Bernard, Jan Vermant, Johan A. Martens, Ludo Froyen, Jan Van Humbeeck, Guy Van den Mooter, and Patrick Augustijns. (2010). "Solubility increases associated with crystalline drug nanoparticles: Methodologies and significance." *Molecular Pharmaceutics*, 7(5), 1858–70.

Varshosaz, J., F. Hassanzadeh, M. Mahmoudzadeh, and A. Sadeghi. (2009). "Preparation of cefuroxime axetil nanoparticles by rapid expansion of supercritical fluid technology." *Powder Technology*, 189(1), 97–102. doi: 10.1016/j.powtec.2008.06.009.

Vaughn, J.M., X. Gao, M.J. Yacaman, K.P. Johnston, and R.O. Williams, 3rd. (2005). "Comparison of powder produced by evaporative precipitation into aqueous solution (EPAS) and spray freezing into liquid (SFL) technologies using novel Z-contrast STEM and complimentary techniques." *European Journal of Pharmaceutics and Biopharmaceutics*, 60(1), 81–9. doi: 10.1016/j.ejpb.2005.01.002.

Vazquez-Lombardi, Rodrigo, Tri Giang Phan, Carsten Zimmermann, David Lowe, Lutz Jermutus, and Daniel Christ. (2015). "Challenges and opportunities for non-antibody scaffold drugs." *Drug Discovery Today*, 20(10), 1271–83.

Vehring, R. (2008). "Pharmaceutical particle engineering via spray drying." *Pharmaceutical Research* 25(5), 999–1022. doi: 10.1007/s11095-007-9475-1.

Verma, S., B.D. Huey, and D.J. Burgess. (2009). "Scanning probe microscopy method for nanosuspension stabilizer selection." *Langmuir*, 25(21), 12481–7. doi: 10.1021/la9016432.

Viçosa, Alessandra, Jean-Jacques Letourneau, Fabienne Espitalier, and Maria Inês Re. (2012). "An innovative antisolvent precipitation process as a promising technique to prepare ultrafine rifampicin particles." *Journal of Crystal Growth*, *342*(1), 80–7.

Wagner, Volker, Bärbel Hüsing, Sibylle Gaisser, and Anne-Katrin Bock. (2006). Nanomedicine: Drivers for development and possible impacts. European Commission (EC) Joint Research Centre (JRC) Institute for Prospective Technological Studies (IPTS).

Wang, Baohe, Wenbo Zhang, Wei Zhang, Arun S. Mujumdar, and Lixin Huang. (2005). "Progress in drying technology for nanomaterials." *Drying Technology*, *23*(1–2), 7–32. doi: 10.1081/drt-200047900.

Wang, Y., X. Han, J. Wang, and Y. Wang. (2016). "Preparation, characterization and in vivo evaluation of amorphous tacrolimus nanosuspensions produced using CO2-assisted in situ nanoamorphization method." *International Journal of Pharmaceutics*, *505*(1–2), 35–41. doi: 10.1016/j.ijpharm.2016.03.056.

Watier, Hervé, and Janice M. Reichert. (2017). "Evolution of antibody therapeutics." *Protein Therapeutics*, 25–49.

Wei, M., S. Xu, and A.C. Lam. (2008). *Stable nanosized amorphous drug*. Google Patents.

Williams, R.O., and K.T. Chow. (2012). *Emulsion template method to form small particles of hydrophobic agents with surface enriched hydrophilicity by ultra rapid freezing*. Google Patents.

Wu, Herren, Carl Webster, Judy Paterson, Sandrine Guillard, Ron Jackson, and Ralph Minter. (2017). "Future horizons and new target class opportunities." *Protein Therapeutics*, 661–700.

Wu, L., J. Zhang, and W. Watanabe. (2011). "Physical and chemical stability of drug nanoparticles." *Advanced Drug Delivery Reviews*, *63*(6), 456–69. doi: 10.1016/j.addr.2011.02.001.

Wu, Libo, and Sandro RP da Rocha. (2007). "Biocompatible and biodegradable copolymer stabilizers for hydrofluoroalkane dispersions: A colloidal probe microscopy investigation." *Langmuir*, *23*(24), 12104–10.

Xu, Xiaoming, and Diane J. Burgess. (2012). "Liposomes as carriers for controlled drug delivery." In *Long Acting Injections and Implants* (pp. 195–220). Springer.

Yang, W., K.P. Johnston, and R.O. Williams, 3rd. (2010). "Comparison of bioavailability of amorphous versus crystalline itraconazole nanoparticles via pulmonary administration in rats." *European Journal of Pharmaceutics and Biopharmaceutics*, *75*(1), 33–41. doi: 10.1016/j.ejpb.2010.01.011.

Yang, W., J. I. Peters, and R.O. Williams, 3rd. (2008). "Inhaled nanoparticles: A current review." *International Journal of Pharmaceutical*, *356*(1–2), 239–47. doi: 10.1016/j.ijpharm.2008.02.011.

York, P., S.A. Wilkins, R.A. Storey, S.E. Walker, and R.S. Harland. (2001). *Coformulation methods and their products*. Google Patents.

Zhang, J.Y., Z.G. Shen, J. Zhong, T.T. Hu, J.F. Chen, Z.Q. Ma, and J. Yun. (2006). "Preparation of amorphous cefuroxime axetil nanoparticles by controlled nanoprecipitation method without surfactants." *International Journal of Pharmaceutics* *323*(1–2), 153–60. doi: 10.1016/j.ijpharm.2006.05.048.

Zhang, Xin, Jian Guan, Rui Ni, Luk Chiu Li, and Shirui Mao. (2014). "Preparation and solidification of redispersible nanosuspensions." *Journal of Pharmaceutical Sciences*, *103*(7), 2166–76.

2

Synthesis of Nanomaterials for Drug Delivery

Hemant K. S. Yadav, Shahnaz Usman, Karyman Ahmed Fawzy Ghanem, and Rayisa Beevi

CONTENTS

2.1 Introduction

It is a great challenge in the field of science and medicine to revitalize or restore the existing or new drug moieties to treat tissues or organs for normal functions. Researchers are centering their focus on different techniques and materials for augmenting medical diagnostics and surgical issues especially in the fields of cancer, tumors, and HIV, etc. In this respect, nanomaterial approaches are much more efficient than conventional molecules. In the development of new drug delivery systems, nanotechnologies can prove to be powerful tools to solve most of the problems of regenerative medicine. The term 'nano' indicates one billionth or 10^{-9} units.[1,2] Nanomaterials are significantly useful for bio diagnosis of pathogens due to their size and surface functionalization. Moreover, nanomaterials provide the advantage of in vivo detection as well as enhanced targeted drug delivery. This is achieved through their ability to circulate for a longer time, multiple binding capacity, and designed clearance pathways. In addition to this, properly protected nanomaterials can be safely used in the fabrication of different functional systems, providing reduced toxicity, high targeting capacity, and proper clearance from the body.

Presently, several works have been done to overcome the challenges that occur in a drug delivery system/targeted drug delivery system by the advances and fabrication of nanomaterials. Added to this, nanomaterials have the ability to protect drugs from getting degraded in the gastrointestinal tract and preventing first-pass metabolism. The technology enables the transportation of drugs with poor solubility, along with enhancing oral bioavailability, due to their specific uptake mechanisms such as absorptive endocytosis. As a result, the drugs have the advantage of controlled release as well as a longer circulation time, which in turn leads to less fluctuation in the plasma levels and reduces side effects. The size of nanostructures provides efficient drug delivery to the desired sites of action as the nanoparticles can easily penetrate the tissues. Nanotechnology provides dosage forms with improved acceptability and performance, thus improving safety, effectiveness, patient compliance, as well as reducing the healthcare cost.

2.2 Nanomaterials

Conveying of the therapeutic agents to the site of action is a major problem in the treatment of diseases. This problem can be overcome by controlling drug delivery. With the help of a controlled drug delivery system, the drug can be shifted to the site of action without any undesirable side effects. Along with this, a drug delivery system provides protection to the drug from fast degradation or clearance, ultimately enhancing the concentration of the drug in the target tissues. Thus, a small concentration of the drug is sufficient for the treatment.[3]

Nanomaterials are substances which have a particle size smaller than 1 micrometer in at least one dimension. However, the size of the atomic and the building blocks should be approximately 0.2nm of the substance to be considered as a nanomaterial. Bulk crystals with lattice spacing of nanometers are considered as an example of nanomaterial. Nanomaterial research takes materials' scientific approach (design and discovery of new materials) to nanotechnology, augmenting improvements in both the synthesis and metrology of the material which was developed to support research in micro-fabrication.

Nano-scaled materials often have unique electronic, optical, thermal, mechanical, and catalytic properties.[4]

Nanomaterials provide size reduction of atoms and molecules up to nanoscale. These nanoscale materials reveal a large specific surface area and size-dependent quantum incarceration effects. Nanomaterials are available in a crystalline or amorphous form. Also, the surface of nanomaterials can be used as carriers for gases or liquid droplets. In addition to the solid, liquid, gaseous, and plasma states, nano particulate matter is considered as a separate state of matter due to its distinct properties; for example, quantum size effects and a large surface area. For instance, carbon nanotubes, fullerenes are considered as crystalline nanoparticles, whereas diamond and graphite are considered as traditional crystalline solid forms. Nanoscale of molecules significantly modifies the chemical and physical properties of materials. The emerging use and careful measurement of nanotechnologies expand the use of nanomaterials as therapeutic and diagnostic tools. Moreover, nanoparticles can be used as drug carriers, in order to deliver chemotherapeutic agents specifically to cancer cells without causing any damage to a drug's substance.[5] Innovations in the science of materials have resulted in the development of biocompatible, biodegradable, stimuli-responsive, and targeted delivery systems.

The structural features of nanomaterials lie in between those of bulk material and atoms, where the majority of micro-structured materials have properties similar to their corresponding bulk materials. The size of nanomaterials makes them characteristically different from atoms and bulk materials. The small size of nanomaterials results in an extremely large surface area to volume ratio which helps in the development of more 'surface' dependent material properties. If the size of the nanomaterial is comparable to the length, then the material is more affected by the above-mentioned surface properties, which results in the modification of the properties of the bulk materials.

2.3 Classification of Nanomaterials

A nanomaterial consists of single, fused, aggregated, or agglomerated forms with spherical, tubular, and irregular shapes.

According to Richard W. Siegel (1994), nanostructured materials are classified as:[6]

1. Zero-dimensional (spheres and clusters).
2. One-dimensional (nanofibers, nanowires, and nanorods).
3. Two-dimensional (nanofilms, nanoplates, and networks).
4. Three-dimensional nanomaterials.

Nanomaterials can be broadly classified into four groups according to shape (dimensionality) of the crystallites and chemical composition.

2.3.1 Clusters or Powders (MD=0)

Clusters are inorganic particles made up of a few hundred or a few dozen atoms. Clusters are characteristically different from the corresponding macro-crystalline material. Oswald (1915) was the first to realize that nanoscale particles should possess interesting as well as novel properties that depend on shape and size to a great extent.[7]

There are different types of forces which help in keeping the elemental clusters held together, depending on the nature of the constituting atoms.

For example:

1) Van-der-waal forces weakly hold the inert gas clusters together, e.g. helium.
2) Strong directional covalent bonds hold the semiconductor clusters together, e.g. silicon.
3) Fairly strong delocalized non-directional bonds hold the metallic clusters together, e.g. sodium.

Cluster nanomaterials are considered as an original class of nanostructured solids with specific properties and different structure, and hence their classification lies between crystalline and amorphous materials. The random stacking of nano grains results in cluster nanomaterials which are characterized by a short-range grain size without a long-range order. The physiochemical properties of cluster nanomaterials are controlled by both the interactions between adjacent grains as well as the intrinsic properties of the nano grains. Cluster nanomaterials are prepared through the deposition of clusters onto a solid substrate which is highly porous in nature. This in turn reduces the density of the cluster nanomaterial to about one-half of the density of the corresponding bulk material.[8]

2.3.2 Nano Fibers, Nanowires, and Nanorods (Multilayers, MD=1)

Nanowires, nanorods, fibers of fibrils, whiskers: all these come under the category of one-dimensional structures. All these are characterized as a nanometer size in one of the dimensions, this direction depends on the growth of crystal structure which produces quantum confinement in the material and changes their surface properties. Nanomaterials with a 1D coherence are more suitable for the construction of active nano devices and interconnect rather than zero-dimensional (0D) amorphous nanoparticles. Inorganic 1D nanomaterial has been widely used as a building block in many kinds of optoelectronic integrations along with their organic counterparts. However, the organic 1D nanomaterial is constructed from small functional molecules having unique potential applications in nanoscale devices. Their high structural tunability, reaction activity, and processability provide great opportunities to miniaturized optoelectronic chips based on organic 1D nanostructure, since they are usually assembled from molecular units with weak intermolecular interactions, such as hydrogen bonds, π–π stacking, and van der Waals force. These weak interactions allow more facile and mild conditions in the fabrication of high quality organic 1D nanostructures rather than those in the construction of their inorganic counterparts[9–11]

2.3.3 Ultrafine Grained over Layers or Buried Layers (<50nm; MD=2)

These materials are actually obtained by severe plastic deformation of conventional coarse-grained metals and converted into grain sizes within the sub-micrometer or even into the nanometer range. Nano films, nanoplates, and nano disks are a good example of 2D quantum and can be easily converted from micro to nano electronics. The ultrafine nanomaterial has a different chemical composition including different interfaces as compared to the forming matter. The ultrafine nanomaterials are obtained if there is a variation in the composition which occurs between interface regions and the crystallites. In this case, one type of molecules gets partially segregated into the interfacial regions. Subsequently, the structural modulation (crystals/interfaces) is coupled to the local chemical modulation; for example, ceramic of alumina with gallium in its interface.[12]

Recently a material was produced by co-milling of aluminum oxide with gallium. This procedure helped to produce nanometer-sized aluminum oxide crystals which are separated by a network of non-crystalline layers of gallium.[12] The thickness of the gallium boundaries between the aluminum oxide crystals vary depending on the gallium content, i.e. a monolayer of up to seven layers.

2.3.4 Nanomaterials Composed of Equiaxed Nanometer-Sized Grains (MD=3)

The members of this group of nanomaterials are made up of nanometer-sized grains which are dispersed in a matrix consisting of different chemicals. It is in the form of layers, rods, or equiaxed crystallites. For instance, nickel aluminide is made up of nanometer-sized nickel aluminide particles which are dispersed in a nickel matrix. This is generated by annealing the supersaturated nickel aluminide solid solution.[13,14]

2.4 Nanotechnology

The science of nanotechnology deals with processes occurring at a nano-size range. It involves preparation and development of material structure by having a control over the size and shape at nano scale. The conversion of material to nanoscale leads to many changes; for example, the reformation of gold into a nanoscale shifts the melting point from 200°C to 1068°C and color changes from yellow to blue to violet, together with the change in its catalytic activity.[15] Functional proteins may be categorized as nanoparticles. Nanoparticles also form the basis of some biological systems as in a different locomotory function. The butterfly's wings which look colorful are due to light being bounced off nanoscale layers. The colors observed during sunset are also due to nanoparticles.[16] The imaging of organs is usually done by using super paramagnetic iron oxide having less than a 50μm size. It can be used for treating complicated brain disorders.[17]

2.4.1 Synthesis and Processing of Nanomaterial

The two general methods of synthesis of nanomaterial are the top-down approach and bottom-up approach. The top-down approach is a process in which the bulk material is broken down into finer pieces forming nanoparticles. On the other hand, bottom-up approach is a process in which atoms are assembled or combined together to form clusters from which nanoparticles are prepared. The main principle behind the top-down approach is the dissociation of the bulk material into nanoparticles, whereas the principle of the bottom-up approach is based on the building up of nanoparticles from atomic level through the formation of clusters.[18,19]

For instance, mechanical grinding is a common example of top-down approach, whereas colloidal dispersion is a typical example of bottom-up approach. Lithography is considered a hybrid of both the top-down and bottom-up approaches, since the growth of the thin film is a bottom-up process and etching is a top-down process.[20] Solvated metal dispersion, laser ablation, electric arc reduction, and high vacuum evaporation are examples of the various methods that can be employed in the synthesis of nanomaterial using a top-down approach. The bottom-up approach includes methods such as hydrothermal, solvothermal, inert gas condensation, and sol-gel process.[18,19]

The bottom-up approach and top-down approach play an important role in the field of nanotechnology. The bottom-up approach is the most commonly used method in the synthesis of nanomaterial; it is favored over the use of the top-bottom approach, since nanoparticles synthesized by the bottom-up approach have a more precise and accurate monodispersity and particle size. Another reason for preferring the bottom-up approach over the top-down approach is that the top-down approach results in surface imperfection (Figure 2.1).[18,19]

2.4.2 Physical Methods Used in the Synthesis of Nanomaterial

A wide range of physical methods are used in the fabrication of nanomaterial; for example, condensation-evaporation, microwave, laser ablation, mechanical milling, and sputter disposition. One of the disadvantages of using physical methods in the synthesis of nanomaterials is the difficulty in producing ultrafine particles. Moreover, physical methods involve the use of expensive equipment and costly vacuum systems.[21] Furthermore, the main limitation of using physical methods in the synthesis of nanomaterials is the production of less quantity and poor-quality nanoparticles as compared to chemical methods. However, the use of mechanical grinding is advantageous over other physical methods as it is a cheap, simple process that enables the

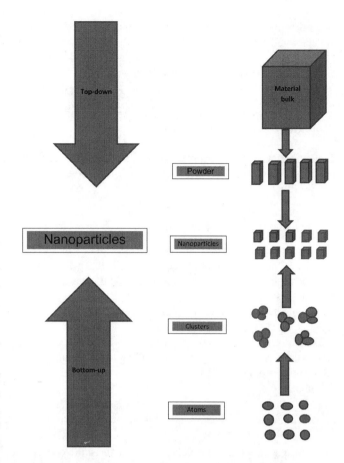

FIGURE 2.1 Schematic illustration of bottom-up and top-down approaches.

large-scale production of nanomaterials. The main advantages of using physical methods include: the production of uniformly distributed nanoparticles; lack of solvent contamination; rapid process; and non-usage of toxic chemicals.[21]

2.4.3 Chemical Methods Used in the Synthesis of Nanomaterial

There are various types of chemical methods that can be employed in the synthesis of nanomaterials. Chemical methods of synthesis are employed in the fabrication of polymeric nanomaterials, where different polymers and their derivatives are used. For example, poly (lactic-co-glycolic acid) copolymer (PLGA) and polyethylene glycol (PEG) are used in the synthesis of polymeric nanomaterial.[21] Functional polymeric nanoparticles are prepared by methods such as nano-precipitation, which is also known as solvent deposition. The polymers employed for this purpose are PLGA-COOH and PLGA-PEG-COOH. Dendrimers are three-dimensional nanostructures that are highly branched and have a wide range of applications in drug delivery. Dendrimers are synthesized by two main approaches; converged and divergent methods of synthesis.[22] The synthesis of dendrimers is a controlled step-by-step method that includes both polymer chemistry and molecular methods. The fabrication process of dendrimers involves the outward growth of dendrimers from a multifunctional core material. The first-generation dendrimer is formed through the reaction of monomers that contain one reactive and two dormant molecules with the core molecule. The choice of method of synthesis depends upon the type of monomer to be used and the structure of the target polymer.[21,22] The advantages of using the chemical method in the synthesis of nanomaterials include the production of large quantities of nanomaterials in a relatively shorter period of time, with good control over the particle size and size distribution. On the other hand, chemical methods are expensive, energy demanding, and involve the use of toxic chemicals and organic solvents which produce hazardous waste products that are harmful to the biological systems, as well as to the environment.[21]

2.4.4 Biological Methods Used in the Synthesis of Nanomaterial

Recently, there has been a growing interest in implementing green chemistry in the synthesis of nanomaterials, in order to reduce waste products and obtain processes that are sustainable. The use of biological methods in the synthesis of nanomaterials plays a crucial role in medicine as well as technology; for example, the use of magnetotactic bacteria in the preparation of magnetic nanoparticles.[21] In previous decades, only prokaryotic members were used for their ability to reduce toxic, insoluble metal ions to nontoxic, soluble metal salts. However, recently, eukaryotic members such as plants, diatoms, and algae have been used for the same purpose as prokaryotes, as their ability to reduce metal ions to metal nanoparticles was recognized.[21]

The advantages of using biological approaches in the synthesis of nanomaterials include improved biocompatibility, fast synthesis of nanomaterials in neutral pH and under ambient temperatures. In addition to this, biological methods are less toxic and produce nanoparticles of a controlled size and morphology.[21] Biomaterials possessing reducing potential are used in the biodirect synthesis of metal nanoparticles. Biomaterials such as phytochemicals, enzymes, or moieties derived from natural sources such as plants, yeast, fungi, bacteria, and actinomycetes are used for both their reducing potential and stabilizing capability.[21–23]

Microorganisms produce nanoparticles by taking in the desired metal ion from the surrounding, reducing metal ions to elemental metals through the action of various metabolites and cellular enzymes. The nanoparticles produced by microbes are classified as intracellular or extracellular according to the site of production in the microorganism. Intracellular as well as extracellular inorganic nanoparticles can be produced by both unicellular and multicellular organisms (Figure 2.2).[21,22]

2.4.4.1 Top-Down Approach

The top-down approach is classified into three main categories, which are: grinding system, mechanochemical, and mechanical alloying method. The grinding system is further divided into dry grinding and wet grinding.[22]

2.4.4.2 Mechanical Grinding

Mechanical grinding is a good example of a top-down approach that is used in the synthesis of nanomaterials. Mechanical attrition is based on the principle of breaking down coarse material into finer particles as a result of severe plastic deformation. Mechanical grinding is the most popular method used in the synthesis of nanomaterials as it is simple, requires relatively inexpensive equipment, and can be applied for almost all classes of materials.[22]

The main advantage of using mechanical grinding is that it is easy to scale up. However, the main disadvantage of mechanical

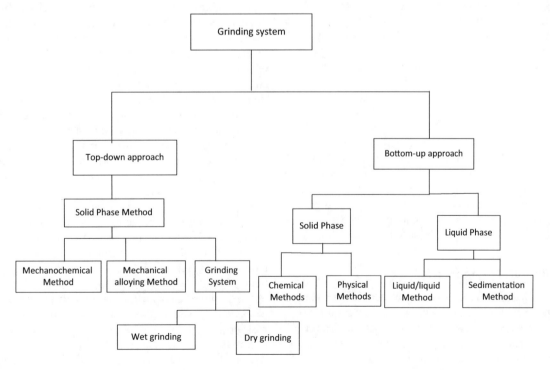

FIGURE 2.2 Classification of methods of synthesis of nanomaterials.

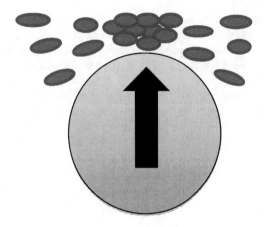

FIGURE 2.3 Schematic representation of mechanical grinding.

grinding is contamination of the prepared nanomaterial from the atmosphere or milling media. The problem of contamination can cause this method to be dismissed and overlooked, at least for some materials.[23]

The mechanism of mechanical milling is based on the principle that energy is transferred from the steel ball or refractory to the powder, which in turn, results in shear stress which is responsible for the formation of nanomaterials. This energy is generated by high-energy planetary ball, shaker, or tumblers. The transferred energy depends on the vibrational or rotational speed of the balls, the number and size of the balls, as well as the ball size-to-powder mass ratio. In addition to this, it also depends on the milling atmosphere and duration of milling (Figure 2.3).[23]

2.4.4.3 Dry Grinding Process

In dry grinding, the bulk material is broken down into finer particles using compression force, shock, or by friction using methods such as hammer milling, jet milling, roller milling, shear milling, shock shear milling, tumbling milling, and using a ball mill. In the dry grinding process, pulverization is accompanied by particle condensation, which in turn makes it difficult to produce particles sizes of less than 3 μm.[24]

2.4.4.4 Wet Grinding Process

In wet grinding process, equipment such as tumbling ball mill, planetary ball mill, centrifugal fluid mill, vibratory ball mill, agitation bead mills, annular gap beads mill, fluid conduit bead mill, and a wet jet mill are used to produce the nanomaterials. The wet grinding process is favored over the dry grinding process as it prevents condensation of the formed nanoparticles and hence, helps in producing highly dispersed nanoparticles.[24]

2.4.4.5 Bottom-Up Approach

The bottom-up approach can be classified into two main categories, which are solid phase and liquid phase. The solid phase is divided into chemical methods and physical methods. The chemical methods include chemical vapor deposition (CVD) and thermal decomposition methods, and the physical method includes physical vapor deposition (PVD).[24]

The liquid phase method is sub-divided into two types, namely sedimentation method and liquid-liquid methods. As a rough classification, the bottom-up approach can be divided into gas phase methods and liquid phase methods. The gas phase method can be broadly divided into two, namely physical vapor deposition (PVD) and chemical vapor deposition (CVD). The gas phase method is preferred over liquid phase methods as it reduces the formation of organic impurities to a minimum. However, the gas phase methods require the use of complicated vacuum equipment which is expensive and of low production capacity.[20]

The chemical vapor deposition process can produce ultra-fine particles of a particle size less than 1μm through a series of chemical reactions which take place in a gaseous state. Under controlled reaction conditions, the chemical vapor deposition method can produce nanoparticles with a particle size that range between 10 to 100nm. Heat sources such as chemical flame, plasma process, electric furnace, and laser are used in high temperature reactions of the chemical vapor deposition process. The physical vapor deposition process involves rapid cooling of the vapors formed after evaporating a solid or liquid material, which results in the formation of nanomaterials. The materials are evaporated either by the simple thermal decomposition method or by the arc method. The thermal decomposition method is preferred over the arc method since it is more effective in metal oxide production, as well as other particle types.[20]

Liquid phase methods have been extensively used in the preparation of nanomaterials. The liquid phase method is divided into two main categories, which are sedimentation methods and liquid-liquid methods. A common example of the liquid-liquid method is the chemical reduction of metal ions. The chemical reduction method is advantageous over other methods as it involves the facile formation of particles of various shapes; for example, nanowires, nanoplates, nanorods, nanoprisms, and hollow nanoparticles. The chemical reduction method allows fine tuning of both the size and shape of nanoparticles by changing the dispersing agent, reducing agent, temperature, and reaction time. In the chemical reduction method, the metal ions are reduced, i.e. oxidation number is reduced to zero. One of the main advantages of using the chemical reduction method is that it is cost-effective and produces large quantity of nanoparticles in a short period of time.[25] Moreover, high quality nanoparticles can be produced using the microwave radiation in a shorter period of time which is an advantage over the chemical reduction method.

The chemical reduction method is a direct reducing method since it involves the direct use of a reducing agent. However, other reducing methods (indirect methods) such as gamma radiation, liquid plasma, ultrasonic waves, and photoreduction can also be used in the preparation of nanomaterial. These methods do not involve the addition of reducing agents, and hence it reduces the occurrence of impurities in the prepared nanomaterials. Liquid-liquid methods also include methods such as solvothermal synthesis, spray pyrolysis, and spray drying as well as supercritical fluid.[25]

The sedimentation method is used to prepare metal oxide nanoparticles using methods such as the sol-gel technique. The basic principle behind the sol-gel technique involves the use of a hydrolysis process to convert metal alkoxide into a sol. The sol is converted into a gel by polycondensation. Wet methods are preferred over dry methods as wet methods result in a high dispersion of nanoparticles as compared to the dry methods. However, if the prepared nanoparticles dry, aggregation takes place soon after drying. Subsequently, the process used in the solid phase method is employed to re-disperse the nanoparticles.[25]

There are numerous methods for the synthesis of nanomaterial; however, there are certain features that are common for all the methods and should apply to all the processes and devices used in the synthesis of nanomaterial.[25] All processes and devices must fulfill the conditions listed below:

1. The particle shape, particle size, composition distribution, size distribution, as well as crystal structure, should be controlled.
2. The nanoparticles prepared should be of high purity, i.e. low levels of impurities.
3. It should control the aggregation of nanoparticles.
4. It should stabilize the structure, physical properties, and reactants.
5. It should be of high reproducibility.
6. It should result in larger mass production at a low cost.

2.5 Synthesis of Nanostructured Materials

Nanomaterials are classified into four categories on the basis of dimension, which are: zero-dimensional, one-dimensional, two-dimensional, and three-dimensional nanomaterials. The dimension of nanomaterials is responsible for their unique chemical and physical properties, which make them different from bulk material. There are various methods developed for the synthesis and fabrication of nanomaterials of controlled shape, size, structure, and dimensionality. These methods can be either physical or chemical. The description of both physical and chemical methods of synthesis of 0D, 1D, 2D, and 3D nanomaterials are discussed in this section.

2.5.1 Physical Methods

2.5.1.1 Evaporation Techniques

Evaporation techniques are commonly employed in thin film deposition. For instance, evaporation techniques like thermal evaporation, sputtering, and plasma techniques are the most commonly used techniques in the synthesis and preparation of 0D, 1D, 2D, and 3D nanomaterials.[26]

2.5.1.2 Thermal Evaporation

Thermal evaporation technique was the first technique used to evaporate the metallic material in inert-gas condensation technique. Thermal evaporation is a common method of the physical vapor deposition (PVD) technique and is used in thin film deposition. It is mostly used in the synthesis of nanowires and nanobelts.[26] Thermal evaporation technique involves the use of pure materials to coat various objects. The object to be coated is referred to as substrate. The applied coating is referred to as film; the thickness of this film can range from angstroms to microns. Moreover, the coated layer can be of a single material or multiple

materials in a layered structure. The applied coat can be of pure metals, non-metals, oxides, or nitrides.[26,27]

Thermal evaporation involves the use of a high vacuum chamber in which the source material is vaporized by heating at an elevated temperature sufficient to produce vapor pressure. The produced vapor pressure raises the vapor cloud of the source material inside the chamber; the vapor cloud condenses over the substrate, forming a thin film.[26,27]

A horizontal furnace is commonly used to carry out the thermal evaporation technique. The main components of the furnace are gas supply, alumina tube, and a rotary pump system, along with a control system. The right side of the alumina tube is connected to the pump system, whereas a view window is placed at the left side of the alumina tube – it is used to observe the growth process. The gas is pumped in from the left side of the tube and pumped out from the right side; both sides of the tube are sealed using O-rings made of rubber. The source material is placed on the boat in the center of the alumina tube, where the temperature is very high. The substrate is inverted and placed in the top of the chamber facing the heating source, so that its surface gets coated with the coating material.[28]

There are two methods used in thermal evaporation technique to vaporize the source material. The first method is known as filament evaporation. Filament evaporation involves the use of a filament or an electrical resistive heating element. There are various types of resistive evaporation filaments, but the most commonly used is the boat type. It is made up of a metal piece that can resist high temperatures, such as tungsten.[27,28] The boats have troughs in which the material to be evaporated is kept. The second method of heating the source material is the E-beam evaporation or the electron beam technique. This method involves the use of hot filament that boils off electrons in order to generate an electron beam of high energy, which heats up the material kept in the hearth. The use of a resistive evaporation filament requires the use of low voltage, whereas the electron beam method requires the use of high voltage.[28]

2.5.1.3 Sputtering Technique

The sputtering technique is used for depositing thin films of a material onto the substrate. It belongs to the conventional thin film deposition technique. Here, in the first process, gaseous plasma is generated and the ions from this are accelerated into the target material. The ions erode the target material by the process of energy transfer, which is then ejected as neutral particles. These neutral particles may be individual atoms or a cluster of atoms or molecules. The ejected particles will travel only in a straight line unless there is any interference such as any surface or particles. The substrate to be coated with the film is placed in the chamber, and they will be coated onto the substrate. The sputtering technique is mostly performed on a cold substrate and the pressure is kept low.[29,30]

There are mainly three steps involved in the sputtering technique:[29]

a. Energetic particles are produced by glow discharge.
b. Momentum transfer from an indirect energetic projectile into the target.
c. Condensation of sputtered molecules.

Compared to other chemical methods, nanomaterials formed by the sputtering technique have fewer impurities. This is because targets of pure material are used to generate the atom vapor. Similarly, sputtering allows the formation of alloy nanoparticles with easier control over their composition, as this method allows simultaneous sputtering of different target materials.[31]

2.5.1.4 Plasma Method

Plasma is referred to as either hot or cold plasma. If the plasma is ionized completely, it is known as 'hot plasma' while the term 'cold plasma' is used if only a small fraction is ionized. The cold plasma method is used mainly for large-scale production of nanowires. The equipment used for the cold plasma method consists of an inductively coupled coil which is driven by a 13.56 MHz radio frequency (RF) power supply, along with a horizontal quartz tube furnace. The method is also known as the RF plasma method since the radio frequency is used here. In the cold plasma method, the starting material is kept in a pestle and placed in an evacuated chamber. The evacuated chamber is wrapped around with high voltage RF coils which will help to heat the metal above its melting point during the reaction. In the subsequent step that follows, helium gas is pumped into the system, forming a high temperature plasma region within the coil. The metal vapor nucleates on the helium gas atoms. They diffuse up to a collector rod (colder) where the formation of nanoparticles takes place.[32]

2.5.1.5 Inert-Gas Condensation

Inert-gas condensation technique is an example of the bottom-up approach. It is one of the oldest and easiest techniques used in the synthesis of nanomaterials. The basic principle behind gas phase synthesis involves homogenous nucleation, which is followed by condensation and coagulation. The inert-gas condensation technique is used to prepare a wide range of metallic nanoparticles; for example, Fe, Mn, Zn, Co, and Mo.[33] Moreover, this technique can also be used in the synthesis of metallic alloys such as Fe–Cu and Fe–Ni. Inert-gas condensation is a two-step process which involves the following:

1. Evaporation of the metallic or inorganic material at low pressure in an inert gas medium, i.e. Helium, Xenon, or Argon using an evaporation source.
2. Rapid and controlled condensation of the vaporized material, which allows the formation of particles of the required size.

The material can be evaporated using various methods which include resistive heating, laser vaporization, thermal evaporation, sputtering, plasma heating, or electric arc discharge. The vapors of the material are rapidly condensed by collecting them on a cold surface. This technique can be carried out using solid or gaseous material and it is done in an inert atmosphere of Helium, Xenon, or Argon.[33]

The inert-gas condensation technique is carried out in an ultra-high vacuum chamber which is previously evacuated using a turbo-molecular pump to a high vacuum level at pressure below 10^{-5}Pa and the chamber is backfilled at low pressure with inert gas.[33,34] The vaporized particles collide with the inert gas particles, resulting in loss of kinetic energy. This collision in turn limits the mean free path of gas particles, which leads to supersaturation over the source of vapors. The high level of supersaturation leads to nucleation, thus the vapors rapidly form clusters which grow by coalescence and agglomeration. The formed clusters condense over the liquid N_2 cooled surface, forming nanoparticles.[34]

The ultra-high vacuum chamber is fitted with a scraper and a compaction unit. The scraper is used to remove the collected powder particles, whereas the collected nanophase particles are consolidated using the compaction unit. The type of inert gas used, as well as the pressure of the chamber, influences the nucleation and growth of the formed nanoparticles. In reactive condensation, oxygen gas is introduced into the inert gas medium which allows the formation of nanosized ceramic particles.[34,35]

One of the main advantages of using inert-gas condensation method is that it can be used in the synthesis of wide range of nanomaterials; for example, ceramics, metal, and metal oxide nanoparticles, intermetallic compounds, alloys, composites, and semiconductors. Added to this, particle size can be controlled via this technique by controlling factors such as pressure, temperature, and type of inert gas. On the other hand, one of the main challenges faced in inert-gas condensation technique is the difficulty in stopping agglomeration, which can be prevented by optimizing the synthesis conditions of nanomaterial.[36]

2.5.1.6 Pulsed Laser Ablation

Pulsed laser ablation technique is a thin film deposition technique used in the synthesis of a wide range of nanomaterials. The basic principle of this technique involves evaporating the sample material using photons, which are emitted by a short-pulsed high energy laser beam in a chamber filled with reagent gas.[37] The evaporated particles are condensed over a support, forming nanoparticles. As the vapor of the sample material travel from the evaporation source to the substrate, it interacts with the gas particles producing the required compound. For instance, if oxygen gas is used, then oxides are produced.[37,38] However, nitrides are produced if ammonia or nitrogen gas are used, whereas carbides are formed in case of using methane. The size distribution of the formed nanomaterial as well as its elemental composition depends on the power of the laser pulse, temperature, and the gas composition used in the chamber.[38]

2.5.1.7 Lithography

Precise yet complicated two- or three-dimensional structures can be obtained using the technique of lithography.[39] There are various ways to perform lithography such as optical or photolithography which utilizes light, electron beam lithography using electrons, ion beam lithography using ions, and X-ray lithography using X-ray.[40] Resolution, registration, and throughput are the three main components to be considered in any of the lithographic techniques. Resolution may be defined as the smallest imprint that can be printed. The process of obtaining an integrated structure by aligning one layer to another is known as 'registration'. Balance between both cost-effectiveness and rate of production is referred to as 'throughput'.[41,42]

All the lithographic techniques share similar technical approaches and fundamentals though they employ different radiation sources. In lithography, the target material is applied onto a silicon substrate, after which a polymeric resist layer is applied to this substrate using spin coating. Through a mask containing a predetermined pattern, an energy beam is shined onto the resist. The resist is further subjected to a development step. Those regions which are exposed to radiations are either protected (negative resist) or sensitized (positive resist). Through etching process, the resist, either negative or positive is removed. Once the etching process is completed, the resist is removed, hence transferring the pattern which was inscribed by mask to the target.[41]

2.5.1.8 Optical or Photolithography

In photolithography, light is used to transfer patterns onto any substrate. It is basically a top-down approach. Mostly visible and ultraviolet radiations are employed in photolithography. The substrate used to produce photo-resistant coating should be free of contaminants. The contaminants may include dust from atmosphere, solvent strains, microorganisms, and aerosol particles, etc. The process must be carried out in clean rooms which are free from airborne particles, and should have strict control over temperature, air, pressure, vibrations, humidity, and lightning. Silicon wafer is mostly used as substrate for biomedical applications due to ideal characteristics of being flat, rigid, smooth, and low-cost. For patterning biomaterials such as proteins, infrared light is preferred over ultraviolet light. IR is less photodamaging than UV and also has deeper penetration.[39]

There are generally two approaches used in photolithography to expose wafers, which are shadow printing and projection printing. Projecting printing is the technique which is most widely used. Shadow printing is further subdivided into two types namely, contact or contact mode printing, and proximity printing. In contact printing, the mask is placed above resist and there is no wafer mask gap. It does not require any magnification, and the resolution is limited to approximately 500nm. In this type of configuration, the mask degrades, which leads to a loss in the planarity. While using proximity printing, the mask and resist lie in close proximity and there is no magnification. The resolution used here is lower. The accuracy of the pattern transfer process is limited due to diffraction effects. In projection printing, the image is projected through a mask and is reduced by a factor of four to ten times on the resist. Here, resolution is about 70nm and diffraction limits accuracy.[41]

2.5.1.9 Electron Beam Lithography

The process of using an electron beam in order to generate patterns on the surface is known as electron beam lithography. This method allows making features in the sub-micrometer regime and is also a way to beat the diffraction limit of light.[42] Both wave-like and particle-like property is possessed by the electron and their wavelength is in the order of a few tenths of an angstrom. Hence diffraction considerations do not limit their resolution. The factor that will limit the resolution here is the back scattering from the underlying substrate and the forward scattering of electrons in the resist layer.[18]

As an electron beam enters a solid material or polymeric film, it undergoes elastic scattering, leading to loss in energy. The electron scattering occurs either by elastic collisions or inelastic collisions. Energy is lost only during inelastic collisions, whereas elastic collisions only result in a direction change. Due to the scattering, the electrons spread out and penetrate into the solid. This will result in transversal or lateral flux which is normal for the incident beam direction, causing an exposure of resist at points remote from the point of electron incidence, leading to development of resist images wider than expected. The factors determining the magnitude with which electron scattering will take place are the atomic number, density of the substrate and resist, and also the velocity of electrons.[18]

2.5.1.10 X-Ray Lithography (XRL)

In X-ray lithography, X-ray is used as a source of radiation in order to produce a high-resolution pattern replication onto the resist. The wavelength used in this method is suited for nanoscale work, minimizes the scattering, and also maximizes the resist absorption. But X-ray lithography is an expensive technique and is time-consuming.[41]

2.5.1.11 Spray Pyrolysis

Spray pyrolysis method belongs to the solution process. It is used mostly in producing metal and metal oxide powders. In this process, heat is applied to convert micro-sized precursor liquid droplets into solids. Precursor aerosol molecules are prepared by introducing the precursor solution into the heating zone with the help of a carrier gas. A nebulizer is used to produce the droplets by utilizing a transducer. The size of the particle is controlled by controlling parameters such as nebulizer energy, relative vapor pressure of gases, and the furnace temperature.[43]

The precursor solution is sprayed into the thermal zone where the evaporation of the solvent takes place and each particle undergoes a reaction to form the product. This method can be used effectively in producing spherical, dense particles which fall into the size range of about 100nm to 1,000nm.[44] Spray pyrolysis is also known as aerosol decomposition synthesis.

The advantages of the spray pyrolysis method are listed below:

1. This method is useful in producing multicomponent nanoparticles because solutions of different metal salts can be mixed and sprayed into the reaction zone.

2. A wide variety of available solution chemistries are used here which helps in compartmentalizing the solution into unique droplets. This helps in retaining excellent stoichiometry on the surface of the particle formed, which is of particular use when single and mixed metal oxides are being synthesized.

3. It enables us to obtain a wide range of particle morphologies.[45]

This technique is only suitable for those materials which can withstand heat without undergoing any chemical degradation. The main steps involved in spray pyrolysis are as follows:

- From the precursor solution micro-sized droplets are produced.
- The solvent is evaporated.
- Solute undergoes condensation.
- Decomposition and reaction of the solute followed by sintering of solid particles.[43]

In spray pyrolysis, the atomization of the precursor solution to the diffusion layer takes place which then undergoes thermolysis to form a nonporous structure and finally it passes through a sintering furnace where they coalesce into a solid mass. Atomization is the process of the formation of droplets and their dispersion into a gaseous environment. The size of the droplet formed is the key determinant of the size of the final product, i.e. solid particle.[46]

The atomization rate determines the process scalability and the residence time of droplets (in the hot zone) is influenced by droplet velocity. Spray pyrolysis allows the formation of the solid particles in the size range ranging from a few micrometers to hundreds of micrometers depending on two factors, which are droplet size and solute concentration.[46]

The next step in the process is the evaporation of the solvent from the solid particle. During the process of evaporation, solvent molecules diffuse away from the surface; there is a change in the droplet temperature, solute diffuses towards the center of the droplet and subsequent change in the size of droplets takes place. Precipitation or drying is the next step where the solute precipitates out and the solvent is evaporated.[46]

2.5.1.12 Flame Spray Pyrolysis (FSP)

Flame spray pyrolysis is a variant of the spray pyrolysis technique. In this method, heat energy required for the decomposition of the precursor molecule is produced in situ by the combustion process.[43]Here, the liquid precursor is sprayed directly into the flame, and synthesis of the particles occurs within the flame by means of a combustion reaction. This enables one to employ a low vapor pressure precursor.[46]

In the process of flame spray pyrolysis, the metal precursor which is dissolved in a suitable solvent (liquid feed) along with an oxidizing gas is sprayed onto the flame. The spray undergoes combustion, converting the precursor into nanosized particles.[47,48]

Factors which affect the product formation are:

- a. Flame configuration.
- b. Temperature.
- c. Oxidant composition.
- d. Added additives.

The main components of a flame spray pyrolysis are:

- a. An atomizer – This assists in creating droplets from the liquid feed administered. There are mainly three types of atomizers which are ultrasonic nebulizer, two-fluid nozzle, and electro sprayer. An ultrasonic nebulizer uses high frequency vibrations to produce sprays and to convert liquid into mist. which then enters the flame. The two-fluid nozzle works by using an atomizing gas to disperse liquid into the spray. An electro sprayer forms an aerosol from liquid using a high voltage.
- b. Burner – This generates the flame required for the process.
- c. Collector – This is used to collect the final end product.[49]

Silica and titania powders can be synthesized by the process of flame spray pyrolysis when metal-organic precursors are employed. Similarly, carbon black is synthesized using heavy residual oil as the precursor.[46] Flame spray pyrolysis is a flexible process[47] and allows strict control of the size and morphology of nanomaterial formed and this is determined by the precursor concentration and dispersion gas flow rate. This method enables the use of low-cost precursors to obtain a product with narrow size distribution and high purity.[50]

2.5.1.13 Sonochemical Reduction

In the process of sonochemical reduction, an ultrasound radiation frequency ranging between 20 KHz and 10 MHz is used to nucleate a chemical reaction. Transient localized hot zones with a high temperature and pressure are generated by an acoustic cavitation process. The sonochemical precursor undergoes distraction and forms nanoparticles due to the abrupt change in temperature and pressure.[51]

The ultrasound having a wavelength of 1 to 10,000nm is generated in a liquid-filled reaction vessel using a magnetostrictive or piezoelectric transducer. The chemical effects of the ultrasound come because of cavitation and not due to a direct interaction with precursor. As the liquid is sonicated, they are subjected to alternating expansive and compressive acoustic waves which create bubbles – and the bubbles so formed oscillate. These oscillating bubbles are able to accumulate ultrasound energy effectively and they grow in size, typically up to a few microns. Mostly spherical nanoparticles are produced by the method of sonochemical reduction.[38]

2.5.2 Chemical Methods

2.5.2.1 Lyotropic Liquid Crystal (LLC) Templates

The technique using lyotropic liquid crystal templates has a wide range of applications in drug delivery, tissue engineering, catalysis, and optical elements such as lasers and lenses. Lyotropic liquid crystal templates produce nanomaterials with anisotropic morphologies and with dimensions that make them good candidates for the previously mentioned applications. Lyotropic liquid crystal templates are used in the direct synthesis of nanowires, nanorods, and nanoplates.[52] Lyotropic liquid crystal templates include three phases which are bicontinuous cubic phase, hexagonal, and lamellar phases. This technique is used to prepare all dimensions of nanomaterials, i.e. 0D, 1D, 2D, and 3D.[53] If the morphology of the template material is adjusted, the production of nanomaterials of controlled morphology on both a micro- and nanoscale is possible. The limitations of using this technique include the complexity of the synthesis process and risk of contamination. Moreover, calcinations are used to remove the templates in most cases, which in turn makes the materials expensive.[53]

2.5.2.2 Chemical Vapor Deposition (CVD)

Chemical vapor deposition (CVD) is a chemical process which involves the deposition of a thin film of solid onto a substrate or a heated surface due to a chemical reaction that is carried out in the vapor phase. The deposited species can be either atoms, molecules, or a mixture of both, thus CVD belongs to vapor-transfer processes which are atomistic in nature. The deposited solid film can be in the form of 0D, 1D, 2D, and 3D nanomaterial. This technique is commonly used in the semiconductor industry as it produces high performance and high purity thin films.[54]

The apparatus used in CVD is made up of a reaction chamber, gas supply system, and exhaust system. The reactants can be in the solid state, liquid state, or gaseous state at room temperature, but they are delivered to the reaction chamber in the vapor or gaseous form. The precursor reactants are pumped into the reactor or the reaction chamber and get adsorbed onto the substrate surface. The reactants undergo a chemical reaction at the surface of the substrate, thus forming a thin film of solid. The by-products formed at the end of the reaction are removed by the flowing gas out of the reaction chamber (desorption of by-product vapors). This chemical reaction is carried out at elevated temperature, i.e. 900°C. The quality of the produced nanomaterial is influenced by the rate of reaction, concentration of the reactants, and temperature of the reaction.[54,55]

In chemical vapor deposition, the product can be formed by either homogenous nucleation or heterogeneous nucleation depending on its application. Homogenous nucleation refers to the formation of the nuclei in the vapor phase before getting deposited onto the substrate surface and it is not incorporated in the crystal structure of the film, whereas heterogeneous nucleation refers to the formation of the nuclei selectively onto the substrate surface and it gets incorporated into the film structure easily. Chemical vapor deposition is used in the synthesis of nano-composite powders such as silicon carbide and silicon nitride composite powder at 1400°C using methane, silane, tungsten hexafluoride, and hydrogen as a gas supply source.[54,55]

2.5.2.3 Types of Chemical Vapor Deposition Reactions

There are three types of CVD reactions, which are: type I reactions, type II reactions, and type III reactions. Type I reactions include pyrolysis and reduction, whereas type II includes oxidation and compound formation. Type III reactions include disproportionation and reversible transfer reactions.[55]

2.5.2.3.1 Pyrolysis Reaction

In pyrolysis reactions, a hot substrate is used to thermally decompose a gaseous species. Pyrolysis can be used to deposit titanium, aluminum, molybdenum, lead, zirconium, iron, silicon, boron, carbon, germanium, aluminum oxide, and silicon dioxide.

2.5.2.3.2 Reduction

Reduction is the reaction in which hydrogen gas is used to reduce halides, oxyhalides, and carbonyl halides. This chemical reaction can also be used to deposit compounds such as aluminum oxide, tantalum pentoxide, titanium dioxide, zinc oxide, and tin (IV) oxide.

2.5.2.3.3 Oxidation

Oxidation refers to the reactions in which oxygen is used to form an oxide. It can be used to deposit zinc oxide, titanium dioxide, tin (IV) oxide, tantalum pentoxide, SnO_2, and Ta_2O_5.

$$SiH_4(g) + O_2(g) \rightarrow SiO_2(s) + 2H_2(g) \text{ at } 450°C$$

2.5.2.3.4 Compound Formation

Compound formation reaction refers to the formation of a variety of nitride, boride, and carbide films.

$$SiCl_4(g) + CH_4(g) \rightarrow SiC(s) + 4HCl(g)$$

$$BF_3(g) + NH_3(g) \rightarrow BN(s) + 4HF(g)$$

2.5.2.3.5 Disproportionation

Disproportionation reactions refer to the simultaneous increase and decrease in the oxidation number of an element forming two new species.

$$2GeI_2(g) \rightarrow Ge(s) + GeI_4(g)$$

$$TiCl_2(g) \rightarrow Ti(s) + TiCl_4(g)$$

2.5.2.3.6 Reversible Transfer

Reversible transfer reaction is also known as chemical transport reaction; it is a temperature dependent reaction, i.e. deposition or etching is dependent on the temperature.

The main advantages of using chemical vapor deposition technique include the formation of uniform nanoparticle coating, high reproducibility, the ability to deposit substances that are hard to evaporate, and the possibility of high growth rates. However, the disadvantages of using CVD include the complexity of the process, use of high temperature, corrosive, and toxic gases. Furthermore, scale-up using CVD is challenging and difficult.[54,55]

2.5.2.4 Laser Chemical Vapor Deposition (LCVD)

Laser chemical vapor deposition is a derivative of chemical vapor deposition. This technique is mostly used in the synthesis of carbon nanotubes. The difference between CVD and LCVD is that the heat source in CVD is replaced by a laser beam. Laser chemical vapor deposition is of two types: pyrolytic and photolytic.[56] In pyrolytic LCVD, a laser beam is used to induce the chemical reaction that results in the chemical vapor deposition. Pyrolytic laser chemical vapor deposition is influenced by the wavelength of the laser beam. This technique can be used if small and localized deposits are desired.[56,57] In the photolytic laser chemical vapor deposition technique, photodecomposition of the sample material takes place, and the sample material deposits onto the substrate. Added to this, the laser beam is set in a parallel position to the substrate. Furthermore, both the techniques, i.e. pyrolytic and photolytic LCVD can be combined together. Combining both techniques is referred to as photo physical LCVD. In such a case, twin beam or a single beam can be used to activate the combined processes.[57]

2.5.2.5 Electrochemical Deposition (ED)

Electrochemical deposition is a type of electrolysis in which solid material is deposited on an electrode. This method is also known as the electrodeposition technique. The main process involved here is the diffusion of charged species through a solution by the application of an external electrical field and the reduction of charged growth species at the deposition surface or electrode. This method is applicable only to electrically conductive materials.[58]

There are two different processes by which electrodeposition of metals can be done. In the first method, the electrons are provided by an external power supply. Another technique used is an electroless method in which the deposition process takes place by the virtue of the presence of a reducing agent in solution.[59]

Nanocomposites are formed when the deposition is confined within the pores of template membrane, whereas nanorods or nanowires are formed when the template is removed.[48] The electrochemical deposition technique employs an electric current as the driving force to deposit the thin film onto the substrate. The main components of this method are:

1. A substrate.
2. Two or three electrodes.
3. The electrolyte solution, which is an anionic solution which consists of solutes of metal salts.
4. A source of current.[58]

Deposition potential, deposition time, electric current, temperature, pH of the solution, nature of the substrate, as well as the nature of the solution are the factors that affect the growth of nanostructures. The quality, size, and distribution of nanomaterial are determined by the factors such as nature, type, and concentration of electrolyte solution and the amount of oxygen dissolved in the solution.[58]

The factors determining nucleation of nanostructures on electrode surface are specific free surface energy, crystal structure, adhesion energy, crystallographic lattice mismatch at the nucleus substrate interface, and the electrode surface's lattice orientation. The major determinants of the final size distribution of electrodeposited materials are the kinetics of nucleation and growth. The nucleation in process of electrodeposition can be either instantaneous nucleation or progressive nucleation. In the case of progressive nucleation, the number of nuclei formed depends directly on the time of electrode deposition, while in instantaneous nucleation, nuclei are formed instantaneously on substrate and grow subsequently with the increasing time of electrode deposition.[60]

Among the electrodes, the cathode is made up of metal or metal oxides and anode is made up of inert materials. The current used in deposition can be either a direct current or an alternating current. Deposition processes are influenced by solvent polarity, current density, electrode separation, and the temperature. The certain advantages of using electrochemical deposition over other chemical methods are:[61]

a. The deposition process can be controlled kinetically and thermodynamically by controlling the electric current. Hence it does not require the use of elevated temperatures for the reaction.
b. The applied potential controls the oxidation reduction potential.
c. Electrode deposition is a low-cost process.
d. The method produces nano-composites which possess improved thermo-mechanical properties.
e. The method is eco-friendly.
f. At low expense energy, the method offers high yield
g. By changing the deposition potential, deposition time, and substrate, the growth of nanomaterial can be optimized.
h. It enables the production of nanomaterials with more stimulating properties and with control over size growth, morphology, and crystal quality by merging this method with self-assisted templates.[60,61]

The mechanism by which electro deposition process proceeds is as follows. The application of an applied potential reduces the cation in solution at the surface of an electrode, followed by injection of an electron from the electrode into the cation. On reduction, the cation forms an adatom which migrates to a favorable site and binds onto it. The reduction process continues further and the adatoms produced subsequently will start aggregating together to form a nucleus of electrodeposit. The substrate will be completely deposited with the electrodeposit if there is a required concentration of metal ions in the solution and if sufficient potential is applied.[62,63]

2.5.2.6 Electroless Deposition

Electroless deposition belongs to the chemical bottom-up process. In the electroless deposition process, a material from the surrounding phase is plated onto the surface of a template using a chemical agent. Electroless deposition does not require an electromotive force and is a non-galvanic process. This process requires a metal salt and a reducing agent and is an autocatalytic method.[64]

The major difference between electrochemical and electroless deposition is that, in the case of electrochemical deposition, the deposited material should be electrically conductive while the electroless deposition process does not require an electrically conductive material. Also, the deposition begins at the bottom electrode in the case of electrochemical deposition, whereas in electroless deposition the deposition starts from the pore wall. Solid nanorods or nanowires are produced by employing the electrochemical deposition process and the length of nanowires or nanorods produced can be controlled by the deposition time.

Electroless deposition results in nanotubules or hollow fibrils, the length of which depends on the length of the deposition channel or pores.[64]

An example of formation of an Ag film on a substrate by the electroless deposition process is given below.

$$AgNO_3 + KOH \rightarrow AgKOH + KNO_3$$

$$AgOH + 2\ NH_3 \left[Ag(NH_3)_2 \right]^{+1} + OH^-$$

$$\left[Ag(NH_3)_2 \right]^{+1} + OH^- + H_2COAg_{(s)} + 2NH_3 + H_2O$$

2.5.3 Hydrothermal Method

In recent years, intensive studies were conducted on the hydrothermal method and it has been applied in the field of nanotechnology to obtain nanomaterials. In the last decade, the hydrothermal method has become one of the most powerful techniques used in transforming various inorganic compounds and in treating raw materials for various technological applications.[65]

Hydrothermal synthesis refers to synthesis of nanoparticles under conditions of elevated temperature and high vapor pressure with the reaction taking place in an aqueous solution. The method of hydrothermal synthesis is based on wet chemical synthesis and is carried out in a sealed reactor or autoclave.[66] In the process of hydrothermal synthesis, crystal growth is observed from substances (which are not soluble at normal pressure and temperature) when high temperature and pressure are employed.[67]

Yet another broader term which is used to describe hydrothermal synthesis is solvothermal synthesis. Solvothermal synthesis may be defined as a process where a chemical reaction takes place in a supercritical or near supercritical condition in the presence of a suitable solvent. Depending on the type of solvent used, the reaction may be termed 'allothermal', 'glycothermal', or even 'ammonothermal'. In all the above-mentioned processes, a particular solvent is used to dissolve a material and is allowed to crystallize in a closed system. Use of different solvents in the process will help to lower the temperature or pressure required for the reaction.[68]

The thermal hydrolysis reaction is regarded as a combination of dehydration and hydrolysis reactions because the equilibrium is seen to shift towards the hydroxide or oxide side when metallic salts are heated up.[69]

$$M^{x+} + xOH^- = M(OH)_x$$

$$M(OH)_x = MOH_{x/2} + (x/2)H_2O$$

Hydrothermal synthesis is the method by which hydroxides or oxides are synthesized by making use of this equilibrium shift. In supercritical hydrothermal synthesis, the reaction zone is supercritical water.[70]

Hydrothermal synthesis is usually carried out in an autoclave and the reaction proceeds as the temperature rises. The final product obtained by this method consists of the desired product along with other compounds generated during the temperature rise, even when the process reaches a critical point. To avoid this problem, an alternate flow-type rapid temperature rise reactor has been developed. Here the feed liquid is heated rapidly to a supercritical state. A high-pressure liquid pump is used to supply a metal salt solution, and the water (which has been heated to supercritical state) is fed through a high temperature furnace in the other separated line, i.e. the reaction occurs in the supercritical water directly. At the exit of the reactor, a cooling water jacket is used which feeds cooling water directly into the line to quench the reaction.[69]

This method has the capability to synthesize fine metal oxides of both single component and also composite oxides. Compared to nanoparticles synthesized under subcritical conditions, the particles had a smaller size and were seen to be single crystalline in most cases when supercritical hydrothermal synthesis was employed.[70]

Under supercritical conditions, solvent properties of the compounds are seen to vary; for example, solubility and dielectric constant. At the critical point, a drastic change in the properties of water is observed and this is seen to change the equilibrium and speed of reaction that takes place in the aqueous medium. The critical temperature of water is 374°C and the critical pressure is 22.1MPa.[70]

At room temperature, the dielectric constant of water is 78, which will decrease with the increasing temperature and decreasing pressure. Under supercritical conditions, the dielectric constant is below 10, similar to a polar organic solvent. That is, supercritical water provides a reaction condition which is favorable for the formation of particles as it enhances the rate of reaction and large supersaturation according to the nucleation theory (due to a decrease in solubility). Supercritical water is suitable for the synthesis of hybrid nanoparticles due to its low dielectric constant which provides specific properties which induce low solubility of inorganics, whereas there is high solubility for organics.[67] For a supercritical fluid, small change in pressure or temperature changes density to a great extent, as the kinetic energy and the molecular forces are antagonized competitively near the balancing point. Change in the density of water also alters the properties of fluid such as molecular distance, as it is the molecular interactions that determine such properties.

Supercritical water is in a state of high-density steam and hence the injection of air into water does not lead to formation of bubbles, instead water mixes completely with air forming a homogenous oxidizing or reducing atmosphere.[69] It is observed that at an elevated temperature and pressure, water and oil produce a uniform phase. The reason for the above is the reduction of the dielectric constant, as a result of which it loses water-like properties.[70] Rapid nucleation and fast growth of particles is observed at a high reaction temperature. The factors which determine the geometry of the formed product is the reaction time, temperature, solvents, and metal precursors. At a given reaction condition, the rate of nucleation is seen to be faster at a high reaction temperature and the particles formed are small. Conversely, larger particles are formed when the reaction time is longer.[66]

A suitable reaction zone for nanoparticles is provided by supercritical hydrothermal synthesis. The process is seen to proceed easily in a medium with a lower dielectric constant value. Keeping the pressure constant, change in temperature will decrease dielectric constant. This is seen to accelerate the hydrothermal synthesis above supercritical region. On the other hand, a high degree of supersaturation is observed when the solution is heated up to the supercritical point in the mixing zone, as it reacts quickly and solubility of the product formed is low in a supercritical state.[70]

2.5.4 Solvothermal Method

Bottom-up colloidal synthetic methods such as the solvothermal method and wet chemical reaction are applied in nanotechnology to synthesize nanoparticles which are of uniform size. The solvothermal method is in a way similar to the hydrothermal method, except in the type of solvent being used. The solvent used in the solvothermal method is non-aqueous, unlike the hydrothermal method where the solvent used was water.[71] In this method (or hydrothermal method), the reagents which are taken in appropriate stoichiometric ratios are dissolved in a suitable solvent (in the case of the solvothermal method) or water (in the case of the hydrothermal method) which is then stirred constantly to mix them. A suitable alkali is used to adjust pH if necessary. An inert atmosphere and stabilizing agents are also used. The contents of the reaction vessel (autoclave) are sealed, which is then subjected to the required temperature and for a specific time. The product can be obtained after cooling the contents by centrifugation process or it can be separated magnetically. The product obtained is washed thoroughly with water and alcohol, ether or acetone, in order to remove any impurities present in it. The final product is then dried at 60–100°C in an oven or by using air drying or vacuum drying.[72]

This method uses moderate to high temperature of about 100–1000°C and a pressure of about 1–10000atm. Using non-aqueous solvents provides an additional advantage of being able to use much higher temperatures for the reaction, as a variety of organic solvents with a high boiling point can be used. In comparison to the hydrothermal method, the solvothermal method was seen to give better control in the crystallinity, size, and morphology of TiO$_2$ nanoparticles when used. Solvothermal methods help to synthesize nanoparticles having a narrow size distribution and dispersity.[71]

Synthesis of crystal or particle by solvothermal method depends mainly on the solubility of the material in the hot solvent under high pressure. Generally chemical reactions such as oxidation, hydrolysis, reduction, metathesis, and thermolysis are used in the synthesis of nanoparticles depending upon the starting material used, precursor complex, and final product. Temperature and pressure are the two main parameters which differentiate solvothermal process from other processes.[72]

Yuhui Cao et al. used titanic acid nanobelts (TAN), and as a precursor, prepared anatase TiO$_2$ nanocrystals exposed by (001) facets in HF-C$_4$H$_9$OH mixed solution using the method of solvothermal synthesis. Varying the amount of HF helps to adjust the percentage of exposure of (001) facets.[73]

Wen et al. prepared an ultra-long single crystalline TiO$_2$ nanowires by using commercial Degussa P25, NaOH, and absolute ethanol as precursors at temperature range of 170–200°C. Wen et al. also reported the synthesis of bamboo-shaped Ag-doped TiO$_2$ nanowire heterojunctions using solvothermal thermal synthesis by mixing 0.2M titanium butoxide in ethanol, 10M NaOH solution, and silver nitrate at a temperature of 200°C/24 h.[71,74]

The use of high boiling point solvents in the synthesis of nanomaterials is limited due as they are generally toxic, expensive, and it is sometimes difficult to dissolve simple salts in such solvents. The abovementioned problems can be overcome using the method of solvothermal synthesis. Single crystal particles and grains can be developed without any defect and also improved crystallization using the technique of solvothermal synthesis. This method offers several advantages, such as providing:

- Increase in solubility and crystallization.
- Control over morphology, size, and phase change.

By using suitable reagents and solvents, good quality nanoparticles can be synthesized using this method. Low boiling point solvents are used in the solvothermal process. The properties of synthesized nanomaterials largely depend on the size and shape of nanoparticles. In order to achieve desired control of size, crystallinity, and shape for nanoparticles, the experimental conditions are required to be controlled. In order to increase the rate of nucleation (to control growth) parameters such as reaction time, pressure, temperature of the reaction, type of starting material, solvent type, and type of surfactant need to be adjusted. This method is relatively safe and economical. Solvothermal methods make it convenient to prepare nanostructures at a lower reaction temperature than the traditional solid/vapor phase reaction.[72]

If the solvent used in preparing the nanoparticle is water, the method is termed the hydrothermal method. Some advantages of using hydrothermal method are:

- Nanoparticles which are ultrafine, pure, and relatively strain-free with narrow size distribution, sinter ability, and high reactivity can be synthesized.
- Crystalline products are obtained without any post-treatment.
- Since the reactions are conducted in a closed system, the method is eco-friendly. The products can be further isolated on cooling to room temperature.[75]

Using microwave heating in solvothermal method has an additional advantage compared to a conventional method as it is faster, cleaner, and energy efficient. It enhances the rate of reaction up to twofold, which helps in saving time and energy. The microwave-assisted solvothermal method has been used in the synthesis of zeolites such as LTA, faujasite, SOD, and ZSM-5.[76]

2.5.5 Sol-Gel Process

The sol-gel process belongs to the class of wet chemical synthesis. The methodology of producing small particles is called the sol-gel process and it is an important method to synthesize nanomaterials. This method is mostly employed in synthesizing metal oxides.[76]

The sol-gel process can be defined as a chemical reaction where an ion or molecular compound forms a three-dimensional network via the formation of oxygen bond between the ions. The reaction is also accompanied by the loss of water or other smaller molecules. In general, the sol-gel process is a polycondensation reaction which gives a three-dimensional network.[77]

In the first step of the sol-gel process, a monomer or the starting material is converted into a sol (colloidal solution) which acts as the precursor for further gel formation. The gel so formed consists of polymers or discrete particles. The sol-gel method being a low temperature process gives control over the product composition and is economically feasible, making it a preferred method for nanomaterial synthesis. Chlorides and metal alkoxides are used commonly as the precursor, which is then hydrolyzed and polycondensed to produce colloids. Materials synthesized by this method find application in numerous fields such as in energy, optics, sensors, space, and separation technology like chromatography and in medicine.[76]

Gel can be formed in several ways in the sol-gel process. In certain situations, a small change in the reaction condition results in products with different structures, though the precursor used is similar in both cases.[78]

There are basically three steps namely, hydrolysis, condensation, and polymerization in the preparation of nanoparticles through the sol-gel process. Metal organic compounds are generally used as the precursor for this method.[79] The presence of organic ligands (such as $-OCH_3$ or $OC2H_5$) which bind onto the metal or metalloid atom makes metal alkoxides a preferred precursor for the sol-gel process. These groups present on them can react with water easily.[80]

In the first step, a hydrolysis reaction occurs in which a species (M–OH) which is unstable is formed and it undergoes further reaction with other species.[77] The factors which affect the hydrolysis reaction are nature of solvent, nature of alkyl group, water-to-alkoxide molar ratio temperature, and the presence of a catalyst (base or acid).[81] In the step that follows, the unstable M–OH group reacts with another molecule of M–OH or M–OR group to form a condensation product (M–O–M) along with the elimination of water or alcohol. This results in the formation of a three-dimensional network. The solid particles hence formed will be suspended in the liquid forming the sol. They further cross-link with each other to form the gel.[77]

The preparation of mono-sized nanoparticles of silica is an important application of the sol-gel process. Particles with varying size can be determined by controlling the process of hydrolysis and polymerization.[72]

In a sol-gel process, at a precipitation preventing pH colloidal particles are mixed with water to form a colloidal solution. The metal alkoxide precursor $M(OR)_4$ undergoes hydrolysis and polycondensation reaction on coming in contact with water. This results in the formation of colloidal dispersion of small particles which is then converted into a 3-D network of corresponding inorganic oxide. With further progression, 3-D networks start to form from the colloidal particles. The property of the so-formed gel is determined by the size of the particles and condensed silica.[82] The viscosity will increase as the gelation process proceeds. During the gelation process, fibers are spun together with a controlled change in viscosity. Gelation will result in agglomeration with colloidal particles. The agglomeration is due to electrical interactions among the components.[83]

Excess solvent is removed from the complex network through the process of drying. Cracks may be produced on the gels if capillary stress develops (if pores are small) during drying. To avoid this problem, small pores are eliminated by hypercritical evaporation or by addition of surfactants. Larger pore size and a stronger network will reduce the stress, which will help in reducing cracking. The gel formed is then heated at an elevated temperature in order to cause densification of gel.[76]

Once the formation of gel is complete, the solvent is removed. Depending on the method used in removing the solvent, different types of gels, namely cryogels, xerogels, or aerogels can be obtained. Methods of drying employed include ambient drying, freeze drying, or supercritical drying. Xerogels are formed when the process of drying causes shrinking in the gel and if no shrinking is observed, then an aerogel is formed. Xerogels are mostly obtained by employing ambient drying, whereas supercritical drying results in aerogels.[81,84]

Certain advantages of using the sol-gel process are listed below:

1. The method is versatile.
2. This method allows preparation of highly purified materials.
3. It provides the provision of an easy way for the introduction of trace elements.
4. Allows for the synthesis of special materials.
5. It helps in saving energy as the process employs a low processing temperature[68]

The method of the sol-gel process offers molecular level mixing. It is also capable of improving the chemical homogeneity of the composite formed here. Since a low temperature is used, materials that are liable to degradation at high temperature such as growth factors, biomolecules, and proteins can be included.[85]

2.5.6 Laser Pyrolysis

Laser pyrolysis belongs to a category of gas phase synthesis.[86] The basis of laser pyrolysis process is the interaction between laser photon, gaseous reactant, and a sensitizer. The role of the sensitizer is to function as an energy transfer agent which transfers energy onto the reactants by collision. The precursor used for the reaction can be either in gaseous or liquid form. If a liquid precursor is being used, then it must be atomized in the synthesis region so as to improve its interaction with a laser beam.[87]

In laser pyrolysis, the compound will undergo excitation by absorbing the energy of laser radiation. This energy is transmitted to the reaction medium thus increasing the temperature of the reaction medium rapidly. The reactants undergo decomposition on being exposed to high temperature. These precursors on dissociation form nanoparticles which then undergo a sudden cooling effect. Growth of the particle is stopped by the sudden drop in temperature yielding nanosized particles. The physical and chemical uniformity of the particles is obtained by adjusting parameters such as flow rate, pressure, laser power, and types of precursor.[88]

2.6 Applications of Nanomaterials in Drug Delivery

The innovation of nanotechnology has led to numerous advances in a number of fields including robotics, physics, engineering, biology, chemistry, and most significantly, in medicine. In the field of medicine, nanomaterial plays a role in the drug delivery system and for the development of cures for various ailments. These distinct physiochemical characteristics of nanomaterials have attracted researchers to use them as drug delivery devices.[89]

In the present era, nanoscale size drug delivery systems have gained vast significance over conservative therapy to get optimum benefits without involving the other parts of the body. This technique leads to an increment in the surface area that assists in binding, adsorbing, and carrying with other compounds such as drug, probes, and proteins. The active drug moities themselves can be converted to a nanoscale size range.[90] Generally, unapproachable areas such as cancer cells, inflamed tissues, etc. can be cured by nanosized device systems in a better way, due to their enhanced permeability and retention effect (EPR); thus this technique can be effectively used for the administration of various molecules like genes, proteins, etc.[91]

In 2007 Sahoo et al., suggested that these materials can also be used to target the reticuloendothelial cells and enable a natural system for treating intracellular infections by facilitating passive targeting of a drug to the macrophages of the liver and spleen,[92] But the nanomaterials used for this purpose should be soluble, nontoxic, and biocompatible as well as physiologically available for extend. They should not obstruct blood vessels and should be less invasive and less toxic.[93] However, they are sensitive to enzymatic and hydrolytic degradation in the gastrointestinal tract, thereby, they need a protecting drug that helps in bypassing the 'first-pass' metabolism in the liver. Those nanomaterials which are coated with hydrophilic polymers generally stay in the circulation for a longer time. Hence they are better candidates for increasing the efficacy of short half-life drugs and can be also be used to deliver a drug as a sustained release formulation.[94] The premature loss of drug through rapid clearance and metabolism can also be prevented.[92] They also have the capability to increase retention time due to bio-adhesion properties. The study done by Couvreur et al (2013) shows that nanomaterials can be used for the proper designing and development of new drug delivery systems and reformulating existing drugs to enhance their therapeutic effect, patent protection, patient-compliance, safety of drugs, as well as lowering costs within the health care system.[17,92]

Any drug delivery nanoparticle possesses three main regions: the core material, therapeutic payload, and surface modifiers, which help the carrier to accumulate in a specific location for targeted delivery of drugs. For use in drug delivery for various conditions, a suitable carrier should be selected. Some particles have the ability to load only hydrophilic or hydrophobic particles, whereas some of them possess the ability to load both hydrophilic and hydrophobic groups. To be employed as a drug delivery vehicle, they must meet the criteria of being nontoxic and non-immunogenic. They should be eliminated from the body easily so as to avoid complications of toxicity on accumulation and subsequent side effects. Once the carrier reaches the destination, there should be a proper mechanism to release the active drug component.[95]

Currently, various types of nanomaterial-based systems are being extensively studied and used in medicine as a drug delivery vehicle. Some of the nanomaterial-based drug delivery systems such as polymeric nanomaterials, dendrites, liposomes, carbon nanomaterials are used for drug delivery.

2.6.1 Liposomes

The morphology of liposome is similar to that of our cellular membranes and they are able to incorporate a large variety of substances within them, making liposomes an ideal candidate for drug delivery. Liposomes can incorporate both hydrophilic and lipophilic drugs. They are found to be biologically inert, nontoxic, and are immunogenic.[96]

The relative stability of liposomes is both advantageous as well as disadvantageous. Any drug carrier must remain stable during the transport but must release the active drug particle on reaching the target site. Being more stable may hinder the release of drug particles on to the target. The encapsulation process is employed in incorporating the drug molecules into the liposomes. The factors which control the drug release from liposomes are composition of liposome, pH, osmotic gradient, and surrounding environment. There is a variety of mechanisms developed for triggering the release of the drug molecules from liposomes.[97]

One mechanism to release drugs is to design liposomes which degrade in a pH-dependent environment. Tumors or inflamed tissues will be more acidic than normal tissues and this can be used as a trigger factor to release drug molecules. Liposomes are designed in such a way that they will be stable under physiological conditions and once they are exposed to an acidic environment, they will get destabilized releasing the drug. Similar to pH, temperature-sensitive liposomes can also be developed which will release the drug moieties on exposure to heat either due to the disease itself (hyperthermia) or on application of heat externally. Various other mechanisms are also available for drug release. Liposomal formulations have been developed for anticancer drugs, neurotransmitters, antibiotics, and many more.[97]

2.6.2 Silica Materials

Biocompatibility, ease in terms of functionalization, and a highly porous framework are some of the advantages offered by these materials for use in drug delivery. Xerogels or mesoporous silica particles are two classifications of silica materials which are used in drug delivery.[98]

Silica xerogels offer high porosity and an increased surface area. Silica materials are favored over many other inorganic nanoparticles for biological purposes. It is the synthesis parameters which determine the porous structure, i.e. the shape and pore dimensions. Silica xerogels are synthesized mainly using the sol-gel process. In order to modify the xerogel for use in drug release, certain parameters are controlled. Ratio of reagents used, temperature at which reaction is carried out, catalyst concentration, and pressure of drying are those synthesis conditions needed to be controlled. There are numerous drugs which have been incorporated into silica xerogels, some of which are phenytoin, diclofenac, nifedipine, and heparin.[98]

The two types of mesoporous silica nanomaterials (MSN) used are MCM-41 and SBA-15. Chemical or physical adsorption is the mechanism used to load drug molecule into MSN and the drug is released into the target site by means of diffusion. When compared to xerogels, superior properties are exhibited by mesoporous silica nanomaterials. They have a more homogenous structure and lower polydispersity than xerogels. Also, they have a much larger surface area for the adsorption of the therapeutic agent. A wide range of drug classes such as antibiotics, anti-cancer, heart disease drugs, etc. can be loaded into MSN. In addition, both larger and smaller molecules can also be loaded into MSN.[98]

2.6.3 Carbon Nanomaterials

Nanotubes and nanohorns are the two categories of carbon nanomaterials used in drug delivery systems. Drug molecules can be incorporated into carbon nanotubes via three different mechanisms. Encapsulation of a drug molecule within the interior of a carbon nanotube is one method adopted to incorporate the active drug component. Another method used involves the attachment of the therapeutic agent to the surface of carbon nanotubes which are functionalized (f-CNT). Drugs can also be incorporated via chemical adsorption onto the surface or in the space present between carbon nanotubes. The adsorption can be through formation of bonds via electrostatic, hydrophobic, π–π interaction, and H-bonds. Different kinds of nanotubes such as single-walled or multi-walled carbon nanotubes are used for drug delivery.[98]

The most preferred mechanism among the three techniques is encapsulation. The advantage of using encapsulation is that it prevents the risk of drug degradation during transport and will release the drug only under specific conditions. An example of a drug encapsulated into nanotubes is cisplatin and the CNT used is multiwall carbon nanotubes (MW CNT). Carbon nanohorns are easy to prepare, are low-cost, and of high quality. The mechanism of drug immobilization involves either adsorption of a drug on to nanohorns or nanoprecipitation of the therapeutic agent with nanohorns.[98]

2.6.4 Solid Lipid Nanoparticles

Solid lipid nanoparticles made up of biocompatible and biodegradable materials provide an alternative drug delivery system for polymeric particulate carriers. Their size range is in the submicron range, between 50 and 1,000nm. These carriers use physiological lipids or lipid molecules enhancing compatibility with body tissues. They can be used to load both hydrophilic and lipophilic drugs.[99]

The mechanism of drug release from solid lipid nanoparticles occurs due to the degradation of lipid matrix in vivo or by diffusion of drug molecules through the matrix. The release rate of drug molecules from the matrix is influenced by the solubility of drug in lipid, drug lipid interactions, particle size, and the temperature employed during the synthesis of solid lipid nanoparticle. During preparation, the high temperature employed will solubilize the drug in the aqueous phase, hence promoting drug localization at the surface region.[100]

2.6.5 Magnetic Nanoparticles

Magnetic nanoparticles have been studied extensively for their unique physical property and for their ability to function at both a molecular as well as cellular level. As a drug delivery agent, magnetic nanoparticles function either by magnetic drug targeting or through active targeting via attachment to high affinity ligands. Magnetic nanoparticles help to direct therapeutic agents to the site of action using magnetic attraction or by specific targeting of the disease biomarker. In addition, they help to reduce the dosage and side effects associated with the drug therapy by blocking the uptake of drug by tissues other than the target site.[101]

Most of the time, these systems are recognized and cleared by the reticuloendothelial system before reaching the desired site leading to a reduction in their efficacy. They also exhibit an inability to cross the biological barriers such as a blood brain barrier or vascular endothelium, posing an additional challenge for their use in drug delivery. Once given IV, their fate in our body is determined by factors such as their size, surface chemistry, charge, and morphology, because these parameters affect the pharmacokinetics and distribution of these systems directly. Size reduction and grafting non-fouling polymers are some of the methods adopted to enhance the effectiveness and increase the distribution rates in the body.[101]

2.6.6 Dendrimers

Dendrimers are macromolecules with a three-dimensional structure which are highly branched and star-shaped with dimensions in the nanometer scale. A central core, an interior dendritic structure, and an exterior surface with functional groups are the three main components of a dendrimer structure. Properties of dendrimers like their small size, versatility, monodispersity, and stability make them appropriate candidates for drug delivery. Dendrimers are used widely in targeted and controlled drug delivery, gene delivery, and for certain industrial applications.[102]

The three-dimensional structures and presence of a functional group on the surface of dendrimers help to load drug molecules for delivery into various bodily tissues. Two mechanisms are used in loading dendrimers with drug molecules; they are either incorporated within the interior by encapsulation process or attached to the functional groups on dendrimer surface via electrostatic or covalent bonds.[103]

Drug delivery at the target site occurs in two ways. In the first method, drug molecules are released by degradation of the covalent bond between drug molecules and dendrimer in vivo. This bond breaking takes place in the presence of any suitable enzyme or environment condition favoring bond breaking. In the second mechanism, drug release is triggered by changes in temperature or pH and this occurs in the outer shell of receptors or in the cavities of the core. There are a broad number of drugs which have been delivered using dendrimers. Examples of some drugs which have been incorporated into dendrimers are 5-flurouracil, pyridine, paclitaxel, and docetaxel.[103]

2.6.7 Polymeric Nanomaterials

While employing polymeric nanomaterials in drug delivery, the chosen polymer should be bio-compatible and biodegradable. Polymeric nanoparticles are colloidal particles that fall into the size range of submicron size. The therapeutic agent to be delivered can be encapsulated or embedded within the matrix of the polymer. Any drug molecule which is entrapped within the polymeric matrix is protected from degradation by either enzyme or through hydrolysis. This structure contains a polymer matrix within the core shell which allows encapsulation of drug molecules. Varying the composition of hydrophilic and hydrophobic groups enables to modify the core shell structure. Examples of some of the polymers employed in drug delivery are polycaprolactone, polylactic acid, alginic acid, and chitosan. They are potential candidates for drug delivery due to their biocompatibility and biodegradability.[97]

Several advantages are provided by polymeric nanoparticles in comparison to free drugs such as:

- Increased bioavailability.
- Enhanced carrier capacity.
- They release drug molecules in a controlled fashion.
- Different routes of administration can be used.
- Enhanced permeability.
- Increased stability.

Drugs delivered via polymeric nanoparticles have an enhanced ability to target specific tissues. By choosing appropriate polymers, polymer lengths, surfactants, and solvents, various properties of nanoparticles such as size, zeta potential, and drug release patterns can be controlled.[97]

2.7 Conclusion

Applying nanomaterials in drug delivery has been an extensively discussed topic. A wide range of nanomaterials are being developed for use in drug delivery nowadays. Several methods have been developed to synthesize nanomaterials. Some of these methods such as laser ablation, CVD, inert gas condensation, hydrothermal synthesis, evaporation techniques, sputtering technique, and electrochemical deposition have been explained in the chapter. In comparison to conventional drug carrier systems, nanomaterials are seen to offer improved therapeutic and pharmacological effects. Also, these systems are seen to be more selective and effective than conventional drug delivery systems. Nanomaterials, when used as a drug delivery vehicle, will protect the drug molecules from degradation in addition to providing targeted and controlled release of drugs. Several clinical studies are taking place in this field in order to expand the use of nanomaterial in drug delivery, due to their potential benefits as a drug delivery vehicle.

REFERENCES

1. Sergeev, G.B. (2003). Cryochemistry of metal nanoparticles. *Journal of Nanoparticle Research*, 5(5–6), 529–537.
2. Sergeev, Gleb B. and Tatyana I. Shabatina. (2008). Cryochemistry of nanometals. *Colloids and Surfaces A: Physicochemical and Engineering Aspects*, 18, 313–314.
3. Nevozhay, D., U. Kanska, R. Budzynska, J. Boratynski. (2007). Current status of research on conjugates and related drug delivery systems in the treatment of cancer and other diseases. *Postępy Higieny i Medycyny Doświadczalnej*, 61, 350–360.
4. Hübler, A.W., and O. Onyeama. (2010). Digital quantum batteries: Energy and information storage in nanovacuum tube arrays. *Complexity*, 15(5), 48–55.
5. Ochekpe, N.A., P.O. Olorunfemi, and N.C. Ngwuluka. (2009). Nanotechnology and drug delivery—part 1: Background and applications. *Tropical Journal of Pharmaceutical Research*, 8(3), 265–274.
6. Siegel, R.W. (1994). Nanophase materials. In Trigg, G.L. (Eds.), *Encyclopedia of Applied Physics* (Vol. 11, pp. 173–200). Weinheim, Germany: Wiley-VCH.
7. Oswald, W. (1915). *The World of Neglected Dimensions*. Dresden, Germany.
8. Paillard, V., P. Melinon, J.P. Perez, V. Dupuis, A. Perez, and B. Champagnon. (1993). Diamondlike carbon films obtained by low energy cluster beam deposition: Evidence of a memory effect of the properties of free carbon clusters. *Physical Review Letters*, 71(25), 4170–4173.
9. Lieber, C.M. (2003). Nanoscale science and technology building a big future from small things. *MRS Bulletin*, 28, 486–488.
10. Cushing, B.L., V.L. Kolesnichenko, and C.J. O'Connor. (2004). Recent advances in the liquid-phase syntheses of inorganic nanoparticles. *Chemical Reviews*, 105, 3893.
11. Rao, C.N.R., and A.K. Cheetham. (2001). Science and technology of nanomaterials: Current status and future prospects. *Journal of Materials Chemistry*, 11, 2887–2894.
12. Konrad, H., J. Weissmüller, J. Hempelmann, R. Birringer, C. Karmonik, and H. Gleiter. (1998). Kinetics of Gallium Films Confined at Grain-Boundaries. *Physical Reviews B: Condensed Matter*, 58(4), 2142–2149.
13. Gleiter, H. (2000). Nanostructured materials: Basic concepts and microstructure. *Acta Materialia*, 48(1), 1–29.
14. Gleiter, H. (1995). Nanostructured materials: State of the art and perspectives. *Nanostructured Materials*, 6(1–4), 3–14.
15. European Commission. (2012). Communication from the Commission to the European Parliament, the Council and the European Economic and Social Committee. Second Regulatory Review on Nanomaterials. SWD (2012) 288 final.
16. Svenson, S. (2007). Multifunctional nanoparticles for drug delivery applications. In *The Nanotech Revolution in Drug Delivery*. Cientifica Ltd.
17. Couvreur, P. (2013). Nanoparticles in drug delivery: Past, present and future. *Advanced Drug Delivery Reviews*, 65(1), 21–23.
18. Cao, Guozhong. (2011). *Nanostructures and Nanomaterials: Synthesis, Properties, and Applications*. Singapore: World Scientific.
19. Sanyal, Amtiva, and Tyler B. Norsten. (2004). Nanoparticle polymer Ensembles. In Vincent M. Rotello (Ed.), *Nanoparticles: Building Blocks for Nanotechnology* (pp. 201–205). Berlin: Springer.
20. Alagarasi, A. (2009). *Nanomaterials*. Narosa Publishing House.
21. Kumar, Chityal Ganesh, Yedla Poornachandra, and Sujitha Pombala. (2017). Therapeutic nanomaterials: From a drug delivery perspective. In Ecaterina Andronescu, and Alexandru Mihai Grumezescu (Eds.), *Nanostructures for Drug Delivery* (pp. 2–17). The Netherlands: Elsevier.
22. Abbasi, Elham, Sedigheh Fekri Aval, and Abolfazl Akbarzadeh. (2014). Dendrimers: synthesis, applications, and properties. *Nanoscale Research Letters*, 9(1), 247.
23. Vithiya, K., and S. Sen. (2011). Biosynthesis of nanoparticles. *International Journal of Pharmaceutical Sciences and Research*, 2(11), 2781–2785.

24. Yadav, Thakur Prasad, Ram ManoharYadav, and Dinesh Pratap Singh. (2012). Mechanical milling: A top down approach for the synthesis of nanomaterials and nanocomposites. *Nanoscience and Nanotechnology, 22*(3), 22–48.

25. Horikoshi, Satoshi. (2013). *Microwaves in Nanoparticle Synthesis: Fundamentals and Applications.* Germany: Wiley.

26. Nalwa, H.S. (2000). *Nanostructured Materials and Nanotechnology.* Cambridge, MA: Academic.

27. Siegel, R.W., Ramasamy, S., Hahn, H., Zongquan, L., Ting, L., and Gronsky, R.J. (1988). *Materials Research, 3,* 1367.

28. Yu, Shu-Hong, and Wei-Tang Yoa. (2013). Inorganic nanobelt material. In Charles M. Lukehart, and Robert A. Scott (Eds.), *Nanomaterials: Inorganic and Bioinorganic Perspectives* (pp. 145–150). Hoboken, NJ: Wiley.

29. Kumar, Narendra. (2016). Nanomaterials: General synthetic approaches. In Sunita Kumbhat (Ed.), *Essentials in Nanoscience and Nanotechnology* (pp. 29–74). Hoboken, NJ: Wiley.

30. Roy, Arup. (2015). Synthesis techniques of nanomaterials. In Jayanta Bhattacharya (Ed.), *Nanotechnology in Industrial Wastewater Treatment* (pp. 35–42) London: IWA.

31. Nie, M., Sun, K., and Meng, D. (2009). Formation of metal nanoparticles by short-distance sputter deposition in a reactive ion etching chamber. *Journal of Applied Physics, 106*(5), 054314.

32. Rafiei, Saeedeh, and A.K. Haghi. (2015). *Nanomaterials and Nanotechnology for Composites.* Canada: CRC Press.

33. Suryanarayana, C., and Prabhu, B. (2007). *Nanostructured Materials: Processing, Properties and Applications.* New York: William Andrew.

34. Koch, Carl, Ilya Ovid'ko, and Sudipta Seal. (2007). Processing of structural nanocrystalline material structural. In Stan Veprek (Ed.), *Nanocrystalline Materials: Fundamentals and Applications* (pp. 26–29). Cambridge, UK: Cambridge University Press.

35. Gleiter, H. (1989). *Progress in Material Science, 33,* 223.

36. Siegel, R.W. (1991). *Annual Review of Material Science, 21,* 559.

37. Schneider, Christof W. and Thomas Lippert. (2010). Laser ablation and thin film deposition. In Peter Schaaf (Ed.), *Laser Processing of Materials: Fundamentals, Applications and Developments* (pp. 89–91). Germany: Springer.

38. Tiwari, Jitendra N., Rajanish N. Tiwari, and Kwang S. Kim. (2012). Zero-dimensional, one-dimensional, two-dimensional and three-dimensional nanostructured materials for advanced electrochemical energy devices. *Progress in Materials Science, 57,* 724–803.

39. Tran, K., and Nguyen, T. (2017). Lithography-based methods to manufacture biomaterials at small scales. *Journal of Science: Advanced Materials and Devices, 2*(1), 1–14.

40. Daraio, C., and Jin, S. (2011). Synthesis and Patterning Methods for Nanostructures Useful for Biological Applications. *Nanotechnology for Biology and Medicine* (pp. 27–44).

41. Hornyak, Gabor L. (2008). *Introduction to Nanoscience.* Boca Raton, FL: CRC Press.

42. Chattopadhyay, K.K. (2009). *Introduction to Nanoscience and Nanotechnology.* New Delhi: PHI Learning.

43. Rao, C., Thomas, P., and Kulkarni, G. (2007). *Nanocrystals: Synthesis, Properties and Applications.* Berlin: Springer.

44. Skandan, Ganesh, and Amit Singhal. (2006). *Nanomaterials Handbook.* Boca Raton, FL: CRC Press.

45. Gouma, Pelagia-Irene. (2010). Resistive gas sensing using nanomaterials. In Pelagia-Irene Gouma (Ed.), *Nanomaterials for Chemical Sensors and Biotechnology* (pp. 13–20). Singapore: Pan Stanford Publishing.

46. Madou, M. (2011). *Manufacturing Techniques for Microfabrication and Nanotechnology.* Boca Raton, FL: CRC Press.

47. Thiébaut, B. (2011). Flame spray pyrolysis: A unique facility for the production of nanopowders. *Platinum Metals Review, 55*(2), 149–151.

48. Mäkelä, J., J. Haapanen, J. Harra, P. Juuti, and S. Kujanpää (2017). Liquid flame spray: A hydrogen-oxygen flame based method for nanoparticle synthesis and functional nanocoatings. *KONA Powder and Particle Journal, 34,* 141–154.

49. Gagan, J., and G. Pelagia-Irene. (2017). Flame spray pyrolysis processing to produce metastable phases of metal oxides. *JOJ Material Science, 1*(2), 1–5.

50. Solero, G. (2017). Synthesis of nanoparticles through flame spray pyrolysis: Experimental apparatus and preliminary results. *Nanoscience and Nanotechnology, 7*(1), 21–25.

51. Ashby, M., P. Ferreira, and D. Schodek. (2009). *Nanomaterials, Nanotechnologies and Design.* Burlington, MA: Butterworth-Heinemann.

52. Cuiqing. Wang, Dairong Chen, and Xiuling Jiao. (2009). Lyotropic liquid crystal directed synthesis of nanostructured materials. *Science and Technology of Advanced Materials, 10*(2).

53. Karanikolos, G.N., P. Alexandridis, R. Mallory, A. Petrou, and T.J. Mountziaris. (2005). *Nanotechnology, 16,* 2372.

54. Kern, W. (2012). *Microelectronic Materials and Processes.* Germany: Springer.

55. Creighton, J.R. and P. Ho. (2001). *Chemical Vapor Deposition.* US: ASM International.

56. Bondi, S.N., W.J. Lackey, R.W. Johnson, X. Wang, and Z.L. Wang. (2006). Laser assisted chemical vapor deposition synthesis of carbon nanotubes and their characterization. *Carbon, 44,* 1393–1403.

57. Wang, H., and Lu, Y.F. (2008). Core-shell photonic band gap structures fabricated using laser-assisted chemical vapor deposition. *Journal of Applied Physics, 103,* 013113.

58. Majid, Abdul. (2018). *Cadmium based II-VI Semiconducting Nanomaterials: Synthesis Routes and Strategies.* Switzerland: Springer.

59. Catherine, M. (2018). *Electrochemical synthesis of silver nanoparticles for applications in nitrate detection, catalysis and antibacterial activity* (PhD dissertation, National University of Ireland).

60. Bera, D., S. Kuiry, and S. Seal. (2004). Synthesis of nanostructured materials using template-assisted electrodeposition. *Journal of Operations Management, 56*(1), 49–53.

61. Baoxiyang, Wang, Dongfeng Xue, and Yong Shi. (2008). Titania ID nanostructured material: Synthesis, properties and appliations. In Wesley V. Prescott, Arnold I. Schwartz (Eds.), *Nanorods, Nanotubes, and Nanomaterials Research Progress* (pp. 171–180). New York: Nova Science.

62. Sheridan, Eoin. (2019). *Electrodeposited nanoparticles: Properties and photocatalytic applications* (PhD thesis, Dublin City University).

63. Gurrappa, I., and L. Binder. (2008). Electrodeposition of nanostructured coatings and their characterization: A review. *Science and Technology of Advanced Materials, 9*(4), 043001.

64. Dadras, Siamak. (2012). Ultrasound assisted synthesis of nano-materials. In Dong Chen (Ed.), *Handbook on Applications of Ultrasound: Sonochemistry for Sustainability* (pp. 75–90). Boca Raton, FL: CRC Press.

65. Joshi, Upendra A., and Jae Sung Lee. (2008). Surfactant free, solution phase synthesis of nano structured metal oxide. In Xiaohua J. Huang (Ed.), *Nanotechnology Research: New Nanostructures, Nanotubes and Nanofibers* (pp. 92–94). New York: Nova Science.

66. Konecny, Andrew, Jose Covarrubias, and Hongwang Wang. (2017). Magnetic nanoparticles design and application in magnetic hyperthermia. In Stefan H. Bossmann, and Hongwang Wang (Eds.), *Magnetic Nanomaterials: Applications in Catalysis and Life Sciences* (pp. 32–33). London, UK: Royal society of chemistry.

67. Hayashi, Hiromichi, and Yukiya Hakuta. (2010). Hydrothermal synthesis of metal oxide nanoparticles in supercritical water. *Materials, 3*(12), 3794–3817.

68. Azimi, S. (2013). Sol-gel synthesis and structural characterization of nano-thiamine hydrochloride structure. *Nanotechnology*, 1–4.

69. Arai, Y. (2002). Material processing using supercritical fluid. In T. Sako, and Y. Takebayashi (Eds.), *Supercritical Fluids: Molecular Interactions, Physical Properties, and New Application* (pp. 331–340). Berlin: Springer.

70. Fukumori, Yoshinobu. (2012). Structural control of nanoparticles. In Kiyoshi Nogi, Masuo Hosokawa, Makio Naito, and Toyokazu Yokoyama (Eds.), *Nanoparticle Technology Handbook* (pp. 61–64). Amsterdam: Elsevier.

71. Sajith, N.V., Soumya B. Narendranath, Sheetu Jose, and Pradeepan Perriyat. (2018). TiO2 nanomaterials a future prospect. *Photocatalytic Nanomaterials for Environmental Applications, 27*, 1–47.

72. Niaz, Shahida B. (2016). Solvothermal/hydrothermal synthetic methods for nanomaterials. In Sher Bahadar Khan (Ed.) *Development and Prospective Applications of Nanoscience and Nanotechnology* (Vol 1, pp. 180–184). Sharjah: Bentham Science.

73. Cao, Y., L. Zong, Q. Li, C. Li, J. Li. and J. Yang. (2017). Solvothermal synthesis of TiO 2 nanocrystals with {001} facets using titanic acid nano belts for superior photocatalytic activity. *Applied Surface Science, 391*, 311–317.

74. Wen, B., Liu, C., Liu, Y. (2005). Solvothermal synthesis of ultralong single-crystalline TiO2 nanowires. *New Journal of Chemistry, 29*(7), 969.

75. Kumar, Amit, Nishtha Yadav, Monica Bhatt, Neeraj K. Mishra, Pratibha Chaudhary, and Rajeev Singh. (2015). Sol-gel derived nanomaterials and its applications: A review. *Research Journal of Chemical Sciences, 5*(12), 98–105.

76. Gnanakumar, G. (2016). Zeolites and Composites. In Visakh P.M (Ed.), *Nanomaterials and Nanocomposites* (pp. 197–200). Germany: Wiley-VCH.

77. Kickelbick, Guido. (2014). Introduction to sol gel nanocomposites. In Massimo Guglielmi, Guido Kickelbick, and Alessandro Martucci (Eds.), *Sol-Gel Nanocomposites* (pp. 1–10). New York: Springer.

78. Danks, A., S. Hall, and Z. Schnepp. (2016). The evolution of 'sol–gel' chemistry as a technique for materials synthesis. *Materials Horizons, 3*(2), 91–112.

79. Sajjadi, Seyed Pooyan. (2005). Sol-gel process and its application in Nanotechnology. *Journal of Polymer Engineering and Technology, 13*, 38–41.

80. Muhammed, Mamoun. (2003). Engineering of nanostructured materials. In Thomas Sakalakos, Ilya A. Ovid'ko, and Asuri K. Vasudevan (Eds.), *Nanostructures: Synthesis, Functional Properties and Applications* (pp. 44–58). Dordrecht, The Netherlands: Springer.

81. Salaheldeen, Elnashaie S., F. Danafar, and Hashemipour Rafsanjani H. (2015). *Nanotechnology for Chemical Engineers*. Singapore: Springer.

82. Schmidt, H., E. Geiter, M. Mennig, H. Krug, C. Becker, and R. Winkler. (1998). The sol-gel process for nano-technologies: new nanocomposites with interesting optical and mechanical properties. *Journal of Sol-Gel Science and Technology, 13*, 397–404.

83. Ian, Esther H., Bruce Dunn, and Jeffrey I. Zink. (2006). Nanostructured systems for biological materials. In Tuan Vo-Dinh (ed.), *Protein Nanotechnology: Protocols, Instrumentation, and Applications* (pp. 53–56). Totowa, NJ: Humana Press.

84. Błaszczyński, T., A. Ślosarczyk, and M. Morawski (2013). Synthesis of silica aerogel by supercritical drying method. *Procedia Engineering, 57*, 200–206.

85. Owens, G., R. Singh, F. Foroutan, M. Alqaysi, C. Han, C. Mahapatra, H. Kim, and J. Knowles. (2016). Sol–gel based materials for biomedical applications. *Progress in Materials Science, 77*, 1–79.

86. Chiruvolu, Shiv, W. Li, M. Ng, K. Du, N.K. Ting, W.E. Mcgovern, N. Kambe, R. Mosso, and K. Drain. (2006). Laser pyrolysis: A platform technology to produce nanoscale materials for a range of product applications. NSTI Nanotechnology Conference and Trade Show – NSTI Nanotech 2006 Technical Proceedings. 1.

87. Rashdan, S. (2018). Nanoparticles for biomedical application. In Paulo Davim, J. (Ed.), *Biomaterials and Medical Tribology: Research and Development* (pp. 50–60. Cambridge: Wood Head Publishing.

88. Commissariat a l'Energie Atomique. (2012). *Synthesis of Nanoparticles by Laser Pyrolysis* (US; US 8,097,233 B2).

89. Namdeo, M., S. Saxena, R. Tankhiwale, M. Bajpai, Y. Mohan, and S. Bajpai. (2008). Magnetic nanoparticles for drug delivery applications. *Journal of Nanoscience and Nanotechnology, 8*(7), 3247–3271.

90. Hadjipanayis, C.G., R. Machaidze, M. Kaluzova, L. Wang, A.J. Schuette, H. Chen, X. Wu, and H. Mao. (2010). EGFRvIII antibody-conjugated iron oxide nanoparticles for magnetic resonance imaging-guided convection-enhanced delivery and targeted therapy of glioblastoma. *Cancer Research, 70*(15), 6303–6312.

91. Jong, W.H.D., and P.J.A. Borm. (2008). Drug delivery and nanoparticle applications and hazards. *International Journal of Nanomedicine, 3*(2), 133–149.

92. Sahoo, S.K., S. Parveen, J.J. Panda. (2007). The present and future of nanotechnology in human healthcare. *Nanomedicine: Nanotechnology, Biology and Medicine, 3*(1), 20–31.

93. Webster, D.M., P. Sundaram, and M.E. Byrne. (2013). Injectable nanomaterials for drug delivery: Carriers, targeting moieties, and therapeutics. *European Journal of Pharmaceutics and Biopharmaceutics, 84*(1), 1–20.

94. Chakroborty, G., N. Seth, and V. Sharma. (2013). Nanoparticles and nanotechnology: clinical, toxicological, social, regulatory and other aspects of nanotechnology. *Journal of Drug Delivery & Therapeutics, 3*(4), 138–141.

95. Dalbhanjan, Rachana R., and Snehal D. Bomble. (2013). Biomedical approach of nanomaterials for drug delivery. *International Journal of Chemistry and Chemical Engineering*, *3*(2), 95–100.

96. Bozzuto, G., and A. Molinari. (2015). Liposomes as nanomedical devices. *International Journal of Nanomedicine*, *10*(1), 975–999.

97. Andronescu, E., and A. Grumezescu. (2017). *Nanostructures for Drug Delivery*. Saint Louis: Elsevier Science.

98. Wilczewska, Agnieszka Z., Katarzyna Niemirowicz, Karolina H. Markiewicz, and Halina Car. (2012). Nanoparticles as drug delivery systems. *Pharmacological Reports*, *64*(5), 1020–1037.

99. Emeje, Martins Ochubiojo, Ifeoma Chinwude Obidike, Ekaete Ibanga Akpabio, and Sabinus Ifianyi Ofoefule. (2012). Nanotechnology in drug delivery. In *Recent Advances in Novel Drug Carrier Systems*. New York: CRC Press.

100. Manjunath, K., J. Reddy, and V. Venkateswarlu. (2005). Solid lipid nanoparticles as drug delivery systems. *Methods and Findings in Experimental and Clinical Pharmacology*, *27*(2), 127.

101. Sun, C., J. Lee, and M. Zhang. (2008). Magnetic nanoparticles in MR imaging and drug delivery. *Advanced Drug Delivery Reviews*, *60*(11), 1252–1265.

102. Rai, A., R. Tiwari, P. Maurya, and P. Yadav. (2016). Dendrimers: A potential carrier for targeted drug delivery system. *Pharmaceutical and Biological Evaluations*, *3*(3), 275–287.

103. Tripathy, Surendra, and Malay K. Das. (2013). Dendrimers and their applications as novel drug delivery carriers. *Journal of Applied Pharmaceutical Science*, *3*(9), 142–149.

3

Fabrication of Multifunctional Silica Nanocarriers for Cancer Theranostics

Avinash Chandra Kushwaha, Ankur Sood, and Subhasree Roy Choudhury

CONTENTS

3.1 Introduction

Over the last decade nanotechnology has gained immense importance, especially in the field of health care wherein the novel strategies have been utilized to fabricate drug delivery systems (Ragelle et al., 2017; Blanco et al., 2015). Diverse types of nanoparticle-based drug delivery systems have been designed for diverse biomedical applications (Lu et al., 2010). An efficient drug delivery system aids in the localized release of therapeutic agents, thereby facilitating the targeted delivery of drugs at specified regions. These systems could facilitate a decrease in the frequency of dosing which, in turn, reduces the side effects of the therapeutic agents used. Most of the common and potent anticancer drugs are hydrophobic in nature and possess a poor aqueous solubility which accounts for major limitations, including a low bioavailability, poor solubility, and slow absorption of these drugs. With advances in the field of nanotechnology, targeted delivery of such therapeutic agents is significantly revolutionized. The high surface area-to-volume ratio, along with the potential to overcome physiological barriers have helped considerably to overcome the undesirable properties of drugs (Watermann and Brieger, 2017). Also, the fields of nanotechnology have increased the fabrication of new drug delivery systems for a better and more effective drug efficacy. Both inorganic as well as organic nanoparticles have been synthesized for their utility in delivering therapeutics to targeted areas. Among them, silica-based nanoparticles have gained considerable attention owing to their facile synthetic routes

and durability (Kato et al., 2010). Silica-based quantum dots are currently being evaluated in clinical trials in cancer patients for advanced imaging. However, the inadequate loading efficiency of drugs associated with silica-based quantum dots restricts its usage as an effective drug delivery system (Phillips et al., 2014). Due to their porous structure, the mesoporous silica nanoparticles overcome the limitation of low encapsulation efficiency, thereby making it a favorable candidate as efficient drug carriers. Mesoporous silica nanoparticles have displayed promising features of their usefulness in biomedical applications, especially in constructing efficient drug delivery systems (Dave and Chopda, 2014). In 2001, Professor Vallet-Regi introduced mesoporous silica nanoparticles for the first time in drug delivery applications, due to their remarkable adsorption potential, along with their potential in controlling host-guest interactions. Since then, mesoporous silica nanoparticles have been extensively used for loading diverse therapeutics for drug delivery and imaging applications (Manzano and Vallet-Regí, 2019). Materials are classified as mesoporous in nature, owing to the presence of pores with a diameter ranging from 2–50nm. Mobil Composition of Matter No. 41 (MCM 41) and Santa Barbara Amorphous–15 (SBA–15) were the initial silica materials produced approximately three decades ago, with hexagonal cylindrical pores in the case of MCM 41, and a large pore size associated with SBA–15 (Mehmood, 2017). These nanomaterials were effectively used as drug delivery carriers in various pathological conditions. The monodispersed mesoporous silica nanoparticles with a size distribution ranging from 80–500nm, large surface area of 700–1000 m^2/g, pore volume of 0.5–2.5 cm^3/g and tuneable pore diameter of 2–6nm, along with high stability, biocompatibility, low toxicity, and rigid matrix are some of the advantages which make it a potential drug delivery system (Bowen Yang et al., 2019). Additionally, tailoring the response of mesoporous silica nanoparticles through the introduction of stimuli-responsive behavior gives mesoporous silica nanoparticles an extra edge over other nanoparticulate-based drug delivery systems (Kresge et al., 1992).The mesoporous silica nanoparticles can easily be modified on their inner as well as outer matrix, based on the postulated applications and also provide opportunities to release loaded cargo under stimuli conditions such as internal pH, temperature, redox potential, and active biomolecules. Mesoporous silica nanoparticles also provide the opportunity to be functionalized through multiple moieties to counter the premature drug release, thereby rendering multiple stimuli responsive behavior (Bagheri et al., 2018). In this report, diverse strategies to functionalize mesoporous silica nanoparticles, along with the mechanism of action are discussed. Insight is given into recent advances in stimuli-responsive approaches of mesoporous silica nanoparticle-based drug delivery system. The multifunctional nature of mesoporous silica nanoparticles associated with the functionalization of multiple moieties rendering response to multiple stimuli concurrently, is also discussed to mark the success of mesoporous silica-based nanomaterials in the field of drug delivery and imaging.

3.2 Multifunctional Nanomaterials for Cancer Therapeutics

Cancer is a heterogeneous disease which affects the major population around the globe. The current treatments of cancer include chemotherapy, which is associated with side effects which in turn hinders the efficacy of the drugs in use. With the help of targeted drug delivery, the efficiency of cancer-curing drugs could be improved, along with minimizing their side effects. Delivering the drugs to specific areas could be achieved through stimuli-responsive drug delivery systems. The fabrication of stimuli-responsive carriers are based on the exceptional properties of cancer cells/tissues/microenvironments when compared to the normal cell (Watermann and Brieger, 2017). The ideal characteristics of stimuli-responsive nanoplatforms include: (i) extreme selectivity for tumor microenvironment, and (ii) precession in stimuli responsiveness (Baek et al., 2015). Researchers have developed drug delivery systems in response to stimuli such as pH, temperature, enzymatic activity, etc. In order to attain effectiveness to the fabricated delivery system, these systems are designed with new modifications and functionalization to achieve a multi-stimulus response for cancer applications (Bagheri et al., 2018). Herein, a detailed explanation is discussed for fabrication procedures, along with the diversity in functionalization of mesoporous silica-based nanoplatforms for cancer therapeutics.

3.2.1 pH-Responsive Mesoporous Silica Nanoparticles

Mesoporous silica nanoparticle-based drug delivery systems can be controlled and manipulated through specific stimuli to generate precise responses. The stimulus can be broadly classified into two types, namely endogenous stimulus and exogenous stimulus (Mura et al., 2013). Endogenous stimuli are based on the differences in the microenvironment of normal tissues, when compared to tissues with a pathological condition, such as cancer tissues. These differences include: increased/decreased activity of certain enzymes, change in intercellular/intracellular pH, temperature, higher redox potential (Lin Zhu and Torchilin, 2012). Conversely, exogenous stimuli are generated as a response to physical alterations, like temperature changes, change in electric fields, ultrasound, and magnetic fields. Considering a tumor condition, stimuli in response to change in the pH condition is immensely exploited by researchers, due to differences in the pH of the microenvironment of tumours (~6), when compared to the pH of a normal tissue (~7.4). The difference in the pH across a tumor tissue further broadens when intracellular organelles, such as endosomes (pH=5.5) and lysosomes (pH<5.5) are compared to the extracellular microenvironment of a tumor. This abnormal change in pH values around a tumor region, along with capabilities of mesoporous silica nanoparticles, as a potent drug carrier provides opportunities for fabricated pH-responsive mesoporous silica nanoparticles as drug delivery systems for theranostic applications in the field of oncology (Xing et al., 2012). Wen et al. (2016) studied the conversion of cerium oxide nanoparticles into cerium ions in the reduction environment, like tumor for the development of multifunctional cerium oxide coated mesoporous silica nanoparticles. Cerium oxide nanoparticles could provide a fluorescent off-on platform because of the efficient fluorescence-quenching property of loaded entities, and the effective fluorescence storage capacity during release. The design of triple responsive cerium oxide-coated mesoporous silica nanoparticles were created for the stimuli of an intracellular glutathione and tumor acidic environment that degraded cerium oxide coating and the loading of photosensitizer, hematoporphyrin for light responsiveness of the system, along with the anticancer drug doxorubicin, for therapeutic effect (Wen et al., 2016).

In another approach, mesoporous silica nanoparticles were synthesized using tetraethyl orthosilicate (TEOS) as a silica source, while using a cetyltrimethylammonium bromide (CTAB) template, and were conjugated with hematoporphyrin-3-Aminopropyl triethoxysilane to fabricate hematoporphyrin-conjugated mesoporous silica nanoparticles. Here, N, N′-dicyclohexylcarbodiimide was used as a condensing agent to mediate the specific reaction between the carboxylic group of hematoporphyrin and the amino group of 3–aminopropyl triethoxysilane to synthesize hematoporphyrin–3–aminopropyl triethoxysilane. Further, doxorubicin was loaded to hematoporphyrin mesoporous silica nanoparticles before coating with cerium oxide nanoparticles using cerium nitrate as a precursor. This nano-system presented a comparatively high release of doxorubicin in the acidic environment, along with the presence of glutathione and irradiation by 650nm laser. A general understanding of the multi-functionalization of mesoporous silica nanoparticles with a pH-sensitivity is represented in Figure 3.1.

3.2.1.1 Polyelectrolyte Gatekeepers as pH-Responsive Agents for Multi-Functional Mesoporous Silica Nanoparticles

Polyelectrolytes are defined as polymers with repeating units of electrolyte groups. These polyelectrolytes are either covalently bonded or adsorbed to the surface of mesoporous silica

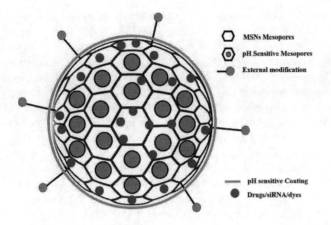

FIGURE 3.1 Functionalized pH-responsive mesoporous silica nanoparticles for drug delivery in cancer cells.

nanoparticles in order to achieve a stimulus-based response under different pH conditions. These polyelectrolytes are very tightly bonded to mesoporous silica nanoparticles' surface at neutral and weak basic conditions, thereby blocking the pores filled with therapeutic agents. When encountered with an acidic pH environment (in the case of cancer cells), these polyelectrolytes are removed through the process of swelling or coiling, resulting in the delivery of loaded therapeutics (Fu et al., 2003). A unique type of pH-responsive mesoporous silica nanoparticle-based drug delivery system was designed by Fenget et al., with polyelectrolyte multilayers composed of polyallylamine hydrochloride and sodium polystyrene sulfonate with the help of a layer-by-layer technique. The overall structure of this unique system consisted of eight polymeric layering, imparting an enhanced encapsulation efficiency. The polyelectrolyte multilayer mesoporous silica nanoparticles provide maximum encapsulation efficiency, together with an improved and sustained release of the drug (doxorubicin) when encountering an acidic environment. Uniform distribution of doxorubicin-loaded polyelectrolyte multilayer-mesoporous silica nanoparticles was seen in cytoplasm within 6 hours in HeLa cells, with the release of the drug in nucleus observed for up to 12 hours post-treatment. It was evident from in vitro studies and analysis that doxorubicin-loaded polyelectrolyte multilayer-mesoporous silica nanoparticles internalized into the lysosomes, where low pH level triggers a doxorubicin release from the cytoplasm to the nucleus (Feng et al., 2013). In another approach, poly[2-(diethylamino) ethyl methacrylate] was selected to functionalize the mesoporous

silica nanoparticles surface through atom transfer of a radical polymerization of [2-(diethylamino)ethyl methacrylate] (Sun et al., 2010). Under neutral and alkaline conditions, poly [2-(diethylamino) ethyl methacrylate] chains are in an aggregation state, owing to the polymer chain interaction blocking the pores of mesoporous silica nanoparticles. As the system encounters acidic conditions, the tertiary amines present in poly[2-(diethylamino) ethyl methacrylate] is protonated, resulting in polymer chain stretching, due to electrostatic repulsions and robust chain-solvent interaction facilitating the drug release. Another triple responsive (pH, reduction and light) nanoplatform was fabricated with modification of hallow mesoporous silica nanoparticles by pH-sensitive polymer, poly [2-(diethylamino)ethyl methacrylate], cerium oxide and a photosensitizer hematoporphyrin. The conjugation of hollow-mesoporous silica nanoparticles and poly [2-(diethylamino)ethyl methacrylate] was decomposed due to the internalization of hollow mesoporous silica nanoparticles into tumor cells due to the presence of a reduction-cleavable disulphide bond and light-cleavable o-nitrobenzyl ester bond to release the drug (Wen et al., 2016). The polyelectrolyte-functionalized mesoporous silica nanoparticles release the drug under the acidic environment of tumor, as represented in Figure 3.2.

3.2.1.2 Supramolecular Nanovalve-Based pH-Responsive Mesoporous Silica Nanoparticles

With advances in the field of supramolecular chemistry, supramolecular assemblies made in the form of 'nanovalves', machines could be utilized in response to stimulus with diverse origin such as chemical, light, and electrical. This machine includes a stalk molecule covalently bound to the silica surface that is immobile in nature, along with a mobile cyclic molecule that encircles the stalk molecules through non-covalent interactions (Yang et al., 2011). Under optimum conditions, the binding constant between the mobile cap and immobile stalks fade, which results in unblocking of the nanotunnels present in the mesoporous silica nanoparticles. This method provides opportunities in fabricating a mesoporous silica nanoparticle-based drug delivery system with acidic pH responsiveness. In an approach to this concept, β-cyclodextrin nanovalves were functionalized as mobile caps with the mesoporous silica nanoparticles. β-cyclodextrin is believed to be responsible for the acidic environment of endosomes in cancerous cells. In this approach, N-methyl benzimidazole was selected as stalks with the optimized pK_a of 5.67 to

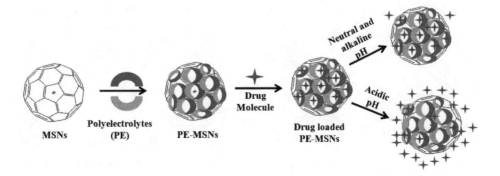

FIGURE 3.2 Acidic pH of cancer microenvironment leading to drug release from 'Polyelectrolyte gatekeepers'.

strongly bind with the cap made of β-cyclodextrin rings, at pH 7.4. At an acidic pH of around 6.0, β-cyclodextrin is removed from its position, resulting in delivery of the cargo loaded in the nanochannels of mesoporous silica nanoparticles (Meng et al., 2010). This approach helps in releasing the cargo molecules at specific pH stimuli present in a tumor micro environment, demonstrating typical pH-responsive behavior of the system. Additionally, Liu et al. (2019) fabricated a multifunctional nanoplatform for the targeted therapy and diagnosis which could respond to collective external and internal stimuli of the magnetic field, near infrared irradiation, and acidic environment in cancer, and release the anti-cancer drug doxorubicin. Here, the mesoporous silica nanoflowers were synthesized with a large pore size of ~25nm to enhance the drug encapsulation efficiency, along with the increased space for metal binding. These large mesopores enabled the in situ synthesis of Au nanoparticles. The interiors of mesoporous silica nanoparticles were functionalized with amino groups and provided an improved binding efficiency to doxorubicin. The number of amino groups was found to be correlated to the drug loading efficacy and also aids in binding between amino groups and the doxorubicin molecule. The exterior of silica was modified with β-cyclodextrin, which has been used extensively for pH-responsive applications. In turn, β-cyclodextrin can exhibit competitive adsorption of doxorubicin to the external surface and can nullify the surface binding of the drug. Interestingly, these systems were created separately before loading of the drug, thus providing more credibility to the system compared to others. At low pH, due to the protonation of the amino groups and the removal of β-cyclodextrin, release of the drug was evident in the release assays. Au@FiNF, FiNF, and Au@FiMNF were reacted with doxorubicin in a glass reactor under vacuum with mild stirring followed by incubation at 25°C for 48 hours, allowing the amino group to form new coordination bonds with doxorubicin. The doxorubicin release behavior for DOX@F_{10}NF, DOX@Au@F_{10}NF, DOX@Au@F_{10}MNF, and DOX@F_{10}MNF in neutral (pH 7.4) and acidic (pH 5.0) environment was assayed and found that acidic pH release was 60% in 36 hours, compared to no significant release in a neutral pH condition. The release was monitored in the presence or absence of an alternating magnetic field and near infrared light irradiation (Liu et al., 2019). These systems have immense application in cancer diagnosis and therapeutics, as the tumor surroundings tend to possess an acidic nature and this triple responsive (pH, of

the alternating magnetic field, and the near infrared light) nanosystem could prove to be a real asset in the drug delivery field. Another study performed by Du et al. (2012) explained the synthesis of a biocompatible pH-responsive nanovalve-based mesoporous silica nanoparticle-based drug delivery system utilizing α-cyclodextrin as caps, along with *p*-anisidino linkers behaving as stalks bounded on the surface of mesoporous silica nanoparticles. As acidic pH is encountered in the lysosomes/endosomes of cancer cells, the nitrogen of *p*-anisidino linkers is protonated, resulting in dislocation of α-cyclodextrin caps, thereby releasing the cargo (Du et al., 2012; Ke-Ni Yang et al., 2014). The drug/cargo release under the acidic environment of tumor requires the opening of mesoporous silica nanoparticles supramolecular nanovalves, as represented in Figure 3.3.

3.2.1.3 Acid-Sensitive Linkers for pH-Responsive Mesoporous Silica Nanoparticles

The acid-sensitive (or pH-sensitive) linkers are prone to cleavage by the alteration in pH values, thus providing opportunities to fabricate pH-responsive mesoporous silica nanoparticles. Diverse kinds of bonds including acetal, hydrazine, and ester bonds could be broken with a decrease in pH value, which in turn provides opening for designing advanced drug delivery systems. These pH-sensitive linkers associated with big molecules could be attached to the pore entrance of mesoporous silica nanoparticles, thereby hindering any drug being released from the nanotunnels. In another approach, a drug could directly be bounded to pH-sensitive linkers which, upon cleavage under optimum pH conditions, release the drug from the nanotunnels of mesoporous silica nanoparticles. Gao et al. (2012) utilized polypseudorotaxanes as pH-sensitive linkers to block the inner pores of mesoporous silica nanoparticles for drug encapsulation purposes. The polypseudorotaxanes consisted of polyethylene glycol chains, along with α-Cyclodextrin with polyethylene glycol playing the role of guest polymer for α-CD, which is the host in this case. Polyethylene glycol imparted longevity for in vivo circulation, thus preventing non-specific adsorption during the circulation time of a designed drug delivery system. Polypseudorotaxanes come with benzoic-imine bonds in their structure that is hydrolysed in acidic conditions and receives a positive charge due to the amine group present, which is helpful for cellular internalization. Polypseudorotaxane-mesoporous silica nanoparticle-based

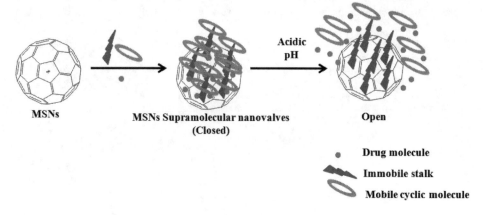

FIGURE 3.3 Supramolecular nanovalves opening in acidic pH to release therapeutic agent at malignant location.

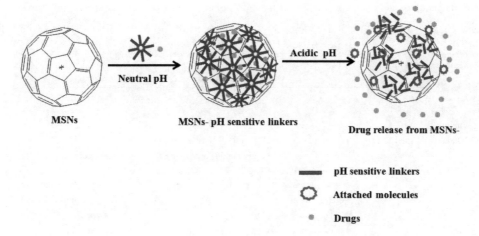

MSNs

MSNs- pH sensitive linkers

Drug release from MSNs-

pH sensitive linkers

Attached molecules

Drugs

FIGURE 3.4   pH-sensitive linkers mediate the drug release from mesoporous silica nanoparticles at tumor location.

delivery systems remain stable under a neutral environment. During the encounter of polypseudorotaxane-mesoporous silica nanoparticles with the acidic conditions of endosomes/lysosomes (pH~5.0) of cancer cells, hydrolysis of benzoic-imine bonds took place that resulted in removal of the polypseudorotaxanes caps, thereby enhancing the drug release from the system (Gao et al., 2012). The drug release under the acidic environment of a tumor requires the functionalization of mesoporous silica nanoparticles with pH-sensitive linker molecules, as represented in Figure 3.4.

3.2.1.4 pH-Sensitive SERS Positive Mesoporous Silica Nanoparticles for Cancer Application

The construction of Janus nanoparticles consisting of a silver nanoparticle head and mesoporous silica nanoparticles as a body provide extreme opportunities for biomedical applications. In this system, individual properties were retained even after combining the two components. To further explore this strategy, Shao et al. (2016) constructed a dual applicative silica-based material for theranostic purpose, taking advantage of SERS for imaging and pH-responsiveness for drug delivery. The silver nanoparticles were synthesized by reducing silver nitrate with glucose in the presence of polyvinyl pyrrolidone, which was followed by the synthesis of Janus Ag-mesoporous silica nanoparticles

by using tetraethyl orthosilicate as a silica source and cetyltrimethylammonium bromide as a template in an alternative sol-gel process. The proposed mechanism for the formation of the Janus Ag-mesoporous silica nanoparticles was marked for the change in the total surface energy ($\Delta\sigma$). This nanoplatform was transformed into the pH-responsive delivery of the anti-tumor drug doxorubicin, through the functionalization of the pores' surface of Janus Ag-mesoporous silica nanoparticles by the carboxylate functional group. The drug release was enhanced and over 40% of release in 24 hours was witnessed. A SERS signal of doxorubicin appeared strongly in cytoplasm, and went undetected in the nucleus due to the presence of doxorubicin loaded Ag-mesoporous silica nanoparticles in the cytosol (Shao et al., 2016). Together, these results lead to a SERS-traceable and pH-sensitive drug delivery system for biomedical application for cancer. Figure 3.5 shows a general aspect of SERS-responsive mesoporous silica nanoparticles and Table 3.1 shows the pH-sensitive polymers used for the functionalization.

3.2.2 Biomolecular-Responsive Drug Release from Mesoporous Silica Nanoparticles

Biomolecules like enzymes, glucose, aptamers, and antigens are most commonly used to prompt drug release (Schlossbauer et al.,

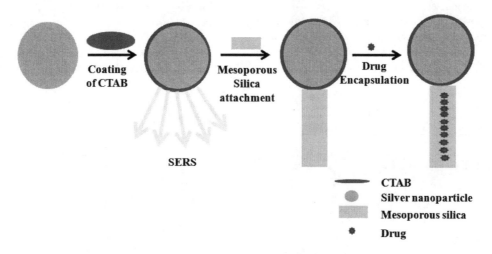

Coating of CTAB

Mesoporous Silica attachment

Drug Encapsulation

SERS

CTAB

Silver nanoparticle

Mesoporous silica

Drug

FIGURE 3.5   SERS-positive mesoporous silica nanomaterial for cancer theranostics.

TABLE 3.1

List of Diverse Strategies Employed for pH-Responsive Mesoporous Silica Nanoparticle-Based Delivery Systems with Different Functionalities

pH-Sensitive Moiety on Mesoporous Silica Nanoparticles	Other Modifications	Therapeutic Agents	Refs
Polyelectrolyte Multilayers Poly Allylamine hydrochloride and Sodium Poly (Styrene Sulfonata)	----	Doxorubicin	Feng et al. (2013)
Cerium Oxide Nanoparticles and PDEAEMA	Hematoporphyrin (HP) for NIR light responsive	Doxorubicin	Wen et al. (2016)
β-cyclodextrin	AuNPs for NIR Irradiation and Magnetite (Fe_3O_4) NPs for AMF responsive	Doxorubicin	Liu et al. (2019)
α-cyclodextrin	*p*-anisidino linkers	Propidium Iodide	Du et al. (2012)
Polyacrylic Acid	Fe_2O_3 NPs core	Doxorubicin	Wu et al. (2012)
Polypseudorotaxanes	PEG along with α-CD	Doxorubicin	Gao et al. (2012)
Carboxylate functional group	Janus Ag-mesoporous silica nanoparticles for SERS activity	Doxorubicin	Shao et al. (2016)

2009). Biomolecules give a quick response to internal changes and body stimuli. Atypical enzymatic activities are generally observed in tissues involved in pathological conditions. These conditions and responses could be targeted using enzyme-stimulated nano-gates that help in blocking the mesopores of mesoporous silica nanoparticles, thereby aiding in a sustained drug release. In a typical strategic design, mesoporous silica nanoparticles loaded with $[Ru(bipy)3]^{2+}$ dye and capped with antibodies have shown the potential to regulate the opening and closing of nanotunnels present in mesoporous silica nanoparticles when exposed to specific antigens like sulfathiazole (Climent et al., 2009). In order to achieve this regulation, the surface of mesoporous silica nanoparticles was functionalized with the help of (4-(4-aminobenzenesulfonylamino) benzoic acid, a class of haptens, which could be easily recognized by a specific antibody. The haptens bind to the antigen-binding sites of the antibody, which in turn is capped to the surface of mesoporous silica nanoparticles. On exposure of antibody-capped mesoporous silica nanoparticles to specific antigen sulfathiazole, the antibody incorporated on the surface of mesoporous silica nanoparticles, due to its antigen specificity binds to the antigen and thereby gets dislocated from mesoporous silica nanoparticles. This results in the release of cargo loaded inside the pores of mesoporous silica nanoparticles.

In another strategy, nucleic acid aptamers were used to cap the pores of mesoporous silica nanoparticles and the advantage of the target specificity of aptamers were explored to achieve an effective stimuli-responsive drug release system (Chun-Ling Zhu et al., 2011). Aptamers are single-stranded short oligonucleotide sequences with high specificity to particular targets. Stability and biodegradability, being less prone to denaturation, easy to obtain, and simplicity for modification are some advantages associated with aptamers when compared to an antibody, making them a potent choice for aiding in drug delivery strategies. An aptamer-mediated mesoporous silica nanoparticles delivery system was designed with the pores of mesoporous silica nanoparticles capped with adenosine triphosphateaptamer tagged with gold nanoparticles. When exposed to a pathological environment associated with cancer cells, the increased adenosine triphosphate production in cancer cells displaces the aptamer-tagged gold nanoparticles, thereby allowing release of the cargo molecule loaded in the mesopores of mesoporous silica nanoparticles. A list of recently used biomolecules to functionalize the mesoporous silica nanoparticles for the therapeutic delivery in cancer has been provided in Table 3.2.

3.2.2.1 Hyaluronidase-Responsive Mesoporous Silica Nanoparticles

Zhang et al.'s work (2019) led to the construction of hybrid mesoporous silica nanoparticles with responsive ability towards intracellular stimuli of glutathione and hyaluronidase. Thesemesoporous silica nanoparticles were functionalized with hyaluronic acid and polyethyleneimine for co-delivery of bcl-2 siRNA and doxorubicin. Apart from conventional techniques for mesoporous silica nanoparticle fabrication, a biodegradable hybrid mesoporous silica nanoparticle was also synthesized from tetraethyl orthosilicate and bis[3-(triethoxysilyl)propyl]

TABLE 3.2

List of Biomolecule-Functionalized Mesoporous Silica Nanoparticle-Based Delivery Systems

Biomolecules	Other Modifications	Therapeutic Agents	Refs
(4-(4-aminobenzenesulfonylamino) benzoic acid,	Sulfathiazole	$[Ru(bipy)3]^{2+}$ dye	Climent et al. (2009)
Nucleic Acid Aptamers	Adenosine triphosphate and gold nanoparticles	Fluorescein dye	Chun-Ling Zhu et al. (2011)
Hyaluronic acid	Polyethyleneimine	bcl-2 siRNA and Doxorubicin	Zhang et al. (2019)
Hyaluronic acid	Desthiobiotin–streptavidin complex	Doxorubicin	Zhang et al. (2016)
Thymidine	Poly A (Adenine) tail	Doxorubicin	Li et al. (2019)

tetrasulfide. Cetyltrimethylammonium chloride and triethanolamine were reacted at 95°C, followed by addition of tetraethyl orthosilicate to form a core layer of mesoporous silica nanoparticles, then a mixed solution of tetraethyl orthosilicate and bis[3-(triethoxysilyl)propyl] tetrasulfide was used to create the outer hybrid layer of hollow-mesoporous silica nanoparticles. The collected hollow-mesoporous silica nanoparticles were transformed into hollow-mesoporous silica nanoparticles/hyaluronic acid/polyethyleneimine for the designated application, and conjugation of hyaluronic acid and polyethyleneimine was achieved by the ionic adsorption method before loading the doxorubicin and bcl-2 siRNA. The nanocarriers were sensitive to glutathione which could cleave the disulphide bridges and lead to degradation of the hybrid mesoporous silica nanoparticles as confirmed by the irregular shapes of the nanoparticles when kept in the presence of glutathione for 14 days. Moreover, these systems were responsive to hyaluronidase due to the presence of its substrate, hyaluronic acid in the nanoparticles (Zhang et al., 2019). In addition, the doxorubicin release was very sensitive to the presence of glutathione and hyaluronidase at pH of (7.4 and 5). The presence of the upregulated conditions of CD44 and biotin in cancer could be utilized for the targeted therapeutics. The CD44 receptor helps the internalization of hyaluronic acid through receptor-mediated endocytosis, and hyaluronidase degrades hyaluronic acid used as an enzyme substrate. Biotin binds specifically to streptavidin and a modified form of biotin, desthiobiotin, less tightly binds to streptavidin but with an efficient binding specificity. A study conducted by Zhang et al. established a mesoporous dual responsive drug delivery of doxorubicin for colon cancer using intracellular hyaluronidase and a biotin–streptavidin complex as gatekeepers. The fabrication of this nano-system initiated with the modification of mesoporous silica nanoparticles-NH$_2$ on the external surface of mesoporous silica nanoparticles with N-Hydroxysuccinimide-desthiobiotin to make desthiobiotin-mesoporous silica nanoparticles, followed by the blocking of pores with streptavidin using the desthiobiotin-streptavidin interaction. Also, the pores were blocked by the biotin modified hyaluronic acid to achieve the tumor targeting (Zhang et al., 2016).

3.2.3 Heat and NIR-Responsive Mesoporous Silica Nanoparticles for Cancer Therapy

In nucleic acid chemistry, there are four bases, namely Adenine (A), Guanine (G), Cytosine (C), and Thymine (T), where A forms two hydrogen bonds with T, and G forms three hydrogen bonds with C. This bonding is very specific and stable under normal conditions but break under heating conditions. Following these mechanisms, Li et al. (2019) recently published an article exploiting the base pairing rule of nucleic acid to fabricate a heat and near infrared light-responsive mesoporous silica nanoparticle system (Li et al., 2019). The mesoporous silica nanoparticles were amino functionalized through (3-aminopropyl) triethoxysilane, then EDC-NHS chemistry was used to conjugate the thymine. Indocyanine green was utilized as a near infrared light-responsive dye, and doxorubicin was used as a model anticancer drug which were co-encapsulated post-synthesis of mesoporous silica nanoparticles-T. Further, the thymidine-functionalized surface of mesoporous silica nanoparticles was trailed to attach the poly

A through the base complementarity. Upon irradiation of the 808 NIR laser, the rise in temperature (50°C) destabilized the base pairing of T and poly A, and released the doxorubicin to perform its function. However, no significant doxorubicin was released in the absence of light, suggesting that the nano-system be near infrared light and is heat-responsive.

3.2.4 Ultrasound-Responsive Mesoporous Silica Nanoparticles

In order to design an effective drug delivery system, better adsorption property of molecules is very important. As a drug delivery system, mesoporous silica nanoparticles provide the significant property of enhanced adsorption, making mesoporous silica nanoparticles adequate for biomedical applications. Mesoporous silica nanoparticles could be easily functionalized due to the presence of silanol groups. Functionalization allows covalent attachment of molecules that could help the introduction of stimuli responsiveness to mesoporous silica nanoparticles, along with improvement in drug loading and a better release profile (Hoffmann et al., 2006). Ultrasound is mainly produced with the help of piezoelectric transducers that are capable of transforming electric signals to a mechanic wave. Ultrasound is currently employed in many biomedical applications, including tumor ablations (Kennedy, 2005; Uchida et al., 2012). Ultrasound mainly produces two types of effects – thermal and mechanical – when applied to living cells/tissues. Thermal effects are produced as a result of energy adsorbed by tissues during the movement of ultrasound that results in an increased temperature around the micro-environment around the tissues. When applied across a tumor tissue for ablation, this phenomenon is known as hyperthermia. Another effect of ultrasound interaction with tissues is that of mechanical effects, mainly cavitation, which is based on the development of gas bubbles in the tissue fluids (van den Bijgaart et al., 2017; Schroeder et al., 2009).

Recently, ultrasound-stimulated mesoporous silica nanoparticles have been developed by introducing the acetal group to the pores of mesoporous silica nanoparticles. The strategy was to transform the drug delivery system from hydrophobicity to hydrophilicity, based on the polymer compositions. The acetal groups are cleavable with the help of ultrasound which exposes the hydrophilic polymer (polyethylene glycol) used in combination. The strategy not only allowed enhanced drug loading but also presented the opportunity to release therapeutics at a specific time. An interesting feature associated with this type of drug delivery system was the introduction of thermo-responsive behavior through the polymer used in combination with acetals that were used as copolymer. The pores of a mesoporous silica nanoparticle-based delivery system remained capped at 37°C and opened at 40°C, thus making it possible to design polymer-coated mesoporous silica nanoparticles with grafted copolymers on their surface. The mechanism for opening the pores for drug release was linked to the mechanical effect of ultrasounds (cavitation) (Paris et al., 2015). It was also demonstrated that increasing the temperature alone was not capable of releasing the drug from a mesoporous silica nanoparticle system ensuring security of the system where an increase in temperature around the tissues was not a requirement for drug release. In turn, the thermo-responsive linker helped in detaching the polyethylene glycol chains, helpful in restricting the aggregations of mesoporous silica nanoparticles when exposed to optimum temperatures. This leads to a cellular uptake of positively charged mesoporous silica nanoparticles, which is a very important step to ensure the proper delivery of therapeutics to designated regions (Paris et al., 2018). A separate approach was taken to develop pH-sensitive dual-responsive polydopamine-coated mesoporous silica nanoparticles for drug delivery. A self-polymerization technique was used to graft polydopamine shell onto the surface of the mesoporous silica nanoparticles under alkaline conditions. Doxorubicin was loaded inside the system to study its release profile and efficacy as a potent drug carrier system (Wang et al., 2018). Similar studies were done using paclitaxel and folic acid-functionalised beta-cyclodextrin has also been used to close the pores of a mesoporous silica nanoparticle-mediated drug delivery system (Lv et al., 2017).

3.2.5 Magnetic-Field Responsive Drug Release from Mesoporous Silica Nanoparticles

Apart from stimuli from thermal, redox potential, and biomolecular, the magnetic field could also offer great potential in the field of drug release from mesoporous silica nanoparticles. Iron oxide nanoparticles, due to their superparamagnetic properties, offer opportunities to magnetically direct the delivery system to target-specific areas (Yanes and Tamanoi, 2012). Iron oxide nanoparticles are present mainly in two forms, namely magnetite (Fe_3O_4) and haematite (Fe_2O_3), with magnetite being more explored in the field of magnetic field-mediated drug delivery. Apart from a magnetically directed drug delivery system, iron oxide nanoparticles could also offer the advantage of being a potent contrast agent for magnetic resonance imaging (MRI), thereby aiding in this field. Another polymer ubiquitously used to develop a pH-sensitive nanoplatform is polyacrylic acid; Wu et al. (2012) fabricated a nano-based multifunctional drug delivery system using polyacrylic acid-modified magnetic mesoporous silica nanoparticles. The magnetic hematite nanoparticles were fabricated with the hydrothermal method and coated with silica materials to obtain hematite-coated SiO2 microspheres, followed by another coating using tetraethyl orthosilicate and n-octadecyltrimethoxysilane polymerizations. Further, magnetic

mesoporous silica nanoparticles were recovered by the removal of organic groups of n-octadecyltrimethoxysilane and reduction of hematite to magnetite. The pH-sensitivity was introduced by surface modification of the magnetic-mesoporous silica nanoparticles by -NH$_2$ group with the support of ammonium persulfatein toluene, followed by reaction with 2-bromoisobutyryl bromide for the conjugation of an atom transfer radical polymerization initiator. This magnetic mesoporous silica nanoparticles-initiator dissolved in 2-butonol and isopropanol followed by charging with copper bromide, N,N,N′,N″,N″-Pentamethyl diethylenetriamine, and tertbutyl acrylate to synthesize nanocomposites of poly tetrabutyl acrylate-anchored magnetic-mesoporous silica nanoparticles. Polytetraputyl acrylate chains were then hydrolyzed to finalize the product, and magnetic mesoporous silica nanoparticles–polyacrylic acid and collected by a permanent magnet (Wu et al., 2012).

In a recent study, the overexpression of glycan sialyl Lewis on the cancer cell surface, compared to the non-cancerous cell, was exploited as one of the innovative ways of targeting the cancerous cells. Das et al. (2018) successfully targeted the cancerous cells, HepG2, through sialyl Lewis binding, while exhibiting a simultaneous delivery of 5-fluorouracil and imaging via fluorescence and magnetic resonance. This was achieved through the construction of a multifunctional nanoplatform with the core of magnetic gadolinium oxide and iron oxide having a shell structure of mesoporous silica-functionalized with carbon quantum dots modified by boronic acid. These boronic acid-capped carbon dots exhibit the 'turn on' fluorescence imaging of sialyl Lewis-positive cancerous cells, HepG2 when these carbon dots are unattached from the silica backbone, due to low pH, induce the ionization of boronic and carboxylic acid. This phenomenon unprotects the silica from carbon dots and subsequently releases the anticancer drug 5-fluorouracil. In addition, this nanomaterial proved to be comparable with commercial MRI contrast agents in terms of r1 and r2 relaxivities, which was 10 mM^{-1}s^{-1} and 165 mM^{-1}s^{-1}. The construction of the nano-system started with the synthesis of gadolinium oxide and iron oxide nanoparticles by a solvothermal method using iron-(III)acetylacetonate, gadolinium (III) 2,4-pentanedionate hydrate, oleyl amine, oleic acid, 1,2-dodecanedioland benzyl ether at 190°C. This oleic acid-stabilized gadolinium oxide and iron oxide nanoparticles were

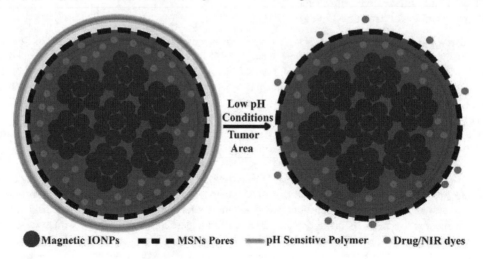

FIGURE 3.6   Functionalized magnetic-mesoporous silica nanoparticles for cancer theranostics applications.

TABLE 3.3

List of Functionalized Magnetic Mesoporous Silica Nanoparticle-Based Delivery Systems

Magnetic Core	Other Modifications	Therapeutic Agents	Refs
Gadolinium oxide and iron oxide (GdIO) NPs	Carbon quantum dots modified by boronic acid (B, N, S doped) (BNSCQD)	5-fluorouracil (5-FU)	Das et al. (2018)
Hematite nanoparticles (Fe$_2$O$_3$ NPs)	PAA modified	Doxorubicin	Wu et al. (2012)
Magnetite (Fe$_3$O$_4$) NPs for AMF responsive	AuNPs for NIR Irradiation and β-cyclodextrin (β-CD) for pH-sensitivity	Doxorubicin	Liu et al. (2019)

shelled with silica material using cetyltrimethylammonium bromide as a template, and tetraethyl ortho silicate as a silica precursor, followed by external surface amine functionalization using (3-aminopropyl) tetraethyl o-silicate. Further, this nanoformulation prepared by a standard method, was reacted with amine-modified magnetic gadolinium oxide-iron oxide at mesoporous silica nanoparticles and purified before loading 5-fluorouracil (Das et al., 2018). Strategies have been devised in which iron oxide nanoparticles are capped on the surface of mesoporous silica nanoparticles, the pores of which are loaded with therapeutics (Kim et al., 2012; Sood et al., 2017).

In the presence of a magnetic field, targeting of specific tissues is possible. Imaging of the iron oxide nanoparticle-capped mesoporous silica nanoparticle delivery system could also be possible post-treatment to assess the exact status and location using magnetic resonance imaging. A general understanding of the multi-functionalization of magnetic-mesoporous silica nanoparticles towards the application in cancer detection and therapeutics is shown in Figure 3.6. Table 3.3 lists the magnetic-responsive mesoporous silica nanoparticles with different modifications.

3.3 Mesoporous Silica Nanoparticles in Clinical Trials for Cancer

The Food and Drug Administration (FDA) has categorized silica as 'generally recognized as safe' (GRAS), and completed the clinical trial as well (NCT03667430). According to the website ClinicalTrials.gov, several human trials for silica materials are completed for ridge deficiency (NCT03897010), acute diarrhea (NCT03633344), intrabony periodontal defect (NCT03651908). However, there is a lack of significant human clinical trials for mesoporous silica nanoparticles in cancer theranostics. Nonetheless, the targeted and functionalized silica nanoparticles, non-radioactive cRGDY-PEG-Cy5.5-Cornell dots for sentinel lymph node mapping, has entered phase 1 and phase 2 of clinical trials for breast and colorectal cancer; and head and neck cancer, respectively (NCT01266096, NCT02106598).

3.4 Conclusion

Mesoporous silica nanoparticles have been used extensively for many biomedical applications. High loading efficiency, biocompatibility, and biodegradability have marked the success of mesoporous silica nanoparticles so far. Capabilities to be selectively functionalized from inner and outer regions have increased the suitability of mesoporous silica nanoparticles as

an efficient drug delivery system for cancer, along with performing other functions of diagnostics, imaging, and targeted therapy. While many nanoparticulate-based drug delivery systems have been fabricated and various strategies have been devised, mesoporous silica nanoparticle-based drug delivery systems are among the few to enter clinical trials. The stimuli-responsive behavior of mesoporous silica nanoparticles towards various endogenous (pH, temperature, enzyme activity) and exogenous (ultrasound, light, magnetic) stimuli have been achieved through the successful attachment of different functional groups. Multiple moieties could be conjugated to mesoporous silica nanoparticles in order to achieve multiple forms of stimuli-responsive behavior. These nano-systems provide a new era of drug delivery systems regarding the treatment of wide ranging types of cancer.

REFERENCES

Baek, S., Singh, R.K., Khanal, D., Patel, K.D., Lee, E.J., Leong, K.W., et al. (2015). Smart multifunctional drug delivery towards anticancer therapy harmonized in mesoporous nanoparticles. *Nanoscale*, 7(34), 14191–14216. doi:10.1039/c5nr02730f.

Bagheri, E., Ansari, L., Abnous, K., Taghdisi, S.M., Charbgoo, F., Ramezani, M., et al. (2018). Silica based hybrid materials for drug delivery and bioimaging. *Journal of Controlled Release*, 277, 57–76. doi:10.1016/j.jconrel.2018.03.014.

Blanco, E., Shen, H., & Ferrari, M. (2015). Principles of nanoparticle design for overcoming biological barriers to drug delivery. *Nature Biotechnology*, 33, 941. doi:10.1038/nbt.3330.

Climent, E., Bernardos, A., Martínez-Máñez, R., Maquieira, A., Marcos, M.D., Pastor-Navarro, N., et al. (2009). Controlled delivery systems using antibody-capped mesoporous nanocontainers. *Journal of the American Chemical Society*, 131(39), 14075–14080. doi:10.1021/ja904456d.

Das, R.K., Pramanik, A., Majhi, M., & Mohapatra, S. (2018). Magnetic mesoporous silica gated with doped carbon dot for site-specific drug delivery, fluorescence, and MR imaging. *Langmuir*, 34(18), 5253–5262. doi:10.1021/acs.langmuir.7b04268.

Dave, P.N., & Chopda, L.V. (2014). A review on application of multifunctional mesoporous nanoparticles in controlled release of drug delivery. *Materials Science Forum*, 781, 17–24. doi:10.4028/www.scientific.net/MSF.781.17.

Du, L., Song, H., & Liao, S. (2012). A biocompatible drug delivery nanovalve system on the surface of mesoporous nanoparticles. *Microporous and Mesoporous Materials*, 147(1), 200–204. doi:https://doi.org/10.1016/j.micromeso.2011.06.020.

Feng, W., Zhou, X., He, C., Qiu, K., Nie, W., Chen, L., et al. (2013). Polyelectrolyte multilayer functionalized mesoporous silica nanoparticles for pH-responsive drug delivery:

Layer thickness-dependent release profiles and biocompatibility. *Journal of Materials Chemistry B, 1*(43), 5886–5898. doi:10.1039/c3tb21193b.

Fu, Q., Rao, G.V.R., Ista, L.K., Wu, Y., Andrzejewski, B.P., Sklar, L.A., et al. (2003). Control of molecular transport through stimuli-responsive ordered mesoporous materials. *Advanced Materials, 15*(15), 1262–1266. doi:10.1002/adma.200305165.

Gao, Y., Yang, C., Liu, X., Ma, R., Kong, D., & Shi, L. (2012). A multifunctional nanocarrier based on nanogated mesoporous silica for enhanced tumor-specific uptake and intracellular delivery. *Macromolecular Bioscience, 12*(2), 251–259. doi:10.1002/mabi.201100208.

Hoffmann, F., Cornelius, M., Morell, J., & Fröba, M. (2006). Silica-based mesoporous organic–inorganic hybrid materials. *Angewandte Chemie International Edition, 45*(20), 3216–3251. doi:10.1002/anie.200503075.

Kato, K., Suzuki, M., Tanemura, M., & Saito, T. (2010). Preparation and catalytic evaluation of cytochrome c immobilized on mesoporous silica materials. *Journal of the Ceramic Society of Japan, 118*(1378), 410–416. doi:10.2109/jcersj2.118.410.

Kennedy, J.E. (2005). High-intensity focused ultrasound in the treatment of solid tumours. *Nature Reviews Cancer, 5*(4), 321–327. doi:10.1038/nrc1591.

Kim, J.-E., Shin, J.-Y., & Cho, M.-H. (2012). Magnetic nanoparticles: An update of application for drug delivery and possible toxic effects. *Archives of Toxicology, 86*(5), 685–700. doi:10.1007/s00204-011-0773-3.

Kresge, C.T., Leonowicz, M.E., Roth, W.J., Vartuli, J.C., & Beck, J.S. (1992). Ordered mesoporous molecular sieves synthesized by a liquid-crystal template mechanism. *Nature, 359*(6397), 710–712. doi:10.1038/359710a0.

Li, X., Wang, X., Hua, M., Yu, H., Wei, S., Wang, A., et al. (2019). Photothermal-triggered controlled drug release from mesoporous silica nanoparticles based on base-pairing rules. *ACS Biomaterials Science & Engineering, 5*(5), 2399–2408. doi:10.1021/acsbiomaterials.9b00478.

Liu, F., Huang, P., Huang, D., Liu, S., Cao, Q., Dong, X., et al. (2019). Smart "on-off" responsive drug delivery nanosystems for potential imaging diagnosis and targeted tumor therapy. *Chemical Engineering Journal, 365*, 358–368. doi:10.1016/j.cej.2019.02.037.

Lu, J., Liong, M., Li, Z., Zink, J.I., & Tamanoi, F. (2010). Biocompatibility, biodistribution, and drug-delivery efficiency of mesoporous silica nanoparticles for cancer therapy in animals. *Small, 6*(16), 1794–1805. doi:10.1002/smll.201000538.

Lv, Y., Cao, Y., Li, P., Liu, J., Chen, H., Hu, W., et al. (2017). Ultrasound-triggered destruction of folate-functionalized mesoporous silica nanoparticle-loaded microbubble for targeted tumor therapy. *Advanced Healthcare Materials, 6*(18), 1700354. doi:10.1002/adhm.201700354.

Manzano, M., & Vallet-Regí, M. (2019). Ultrasound responsive mesoporous silica nanoparticles for biomedical applications. *Chemical Communications, 55*(19), 2731–2740. doi:10.1039/c8cc09389j.

Mehmood A.G.H., Yaqoob S., Gohar U.F., & Ahmad, B. (2017). Mesoporous silica nanoparticles: A review. *Journal of Developing Drugs, 6*, 174.

Meng, H., Xue, M., Xia, T., Zhao, Y.-L., Tamanoi, F., Stoddart, J.F., et al. (2010). Autonomous in vitro anticancer drug release from mesoporous silica nanoparticles by pH-sensitive nanovalves. *Journal of The American Chemical Society, 132*(36), 12690–12697. doi:10.1021/ja104501a.

Mura, S., Nicolas, J., & Couvreur, P. (2013). Stimuli-responsive nanocarriers for drug delivery. *Nature Materials, 12*, 991. doi:10.1038/nmat3776.

Paris, J.L., Cabañas, M.V., Manzano, M., & Vallet-Regí, M. (2015). Polymer-grafted mesoporous silica nanoparticles as ultrasound-responsive drug carriers. *ACS Nano, 9*(11), 11023–11033. doi:10.1021/acsnano.5b04378.

Paris, J.L., Villaverde, G., Cabañas, M.V., Manzano, M., & Vallet-Regí, M. (2018). From proof-of-concept material to PEGylated and modularly targeted ultrasound-responsive mesoporous silica nanoparticles. *Journal of Materials Chemistry B, 6*(18), 2785–2794. doi:10.1039/c8tb00444g.

Phillips, E., Penate-Medina, O., Zanzonico, P.B., Carvajal, R.D., Mohan, P., Ye, Y., et al. (2014). Clinical translation of an ultrasmall inorganic optical-PET imaging nanoparticle probe. *Science Translational Medicine, 6*(260), 260ra149. doi:10.1126/scitranslmed.3009524.

Ragelle, H., Danhier, F., Préat, V., Langer, R., & Anderson, D.G. (2017). Nanoparticle-based drug delivery systems: A commercial and regulatory outlook as the field matures. *Expert Opinion on Drug Delivery, 14*(7), 851–864. doi:10.1080/1742 5247.2016.1244187.

Schlossbauer, A., Kecht, J., & Bein, T. (2009). Biotin–avidin as a protease-responsive cap system for controlled guest release from colloidal mesoporous silica. *Angewandte Chemie International Edition, 48*(17), 3092–3095. doi:10.1002/anie.200805818.

Schroeder, A., Kost, J., & Barenholz, Y. (2009). Ultrasound, liposomes, and drug delivery: Principles for using ultrasound to control the release of drugs from liposomes. *Chemistry and Physics of Lipids, 162*(1), 1–16. doi: 10.1016/j.chemphyslip.2009.08.003.

Shao, D., Zhang, X., Liu, W., Zhang, F., Zheng, X., Qiao, P., et al. (2016). Janus silver-mesoporous silica nanocarriers for SERS traceable and pH-sensitive drug delivery in cancer therapy. *ACS Applied Materials & Interfaces, 8*(7), 4303–4308. doi:10.1021/acsami.5b11310.

Sood, A., Arora, V., Shah, J., Kotnala, R.K., & Jain, T.K. (2017). Multifunctional gold coated iron oxide core-shell nanoparticles stabilized using thiolated sodium alginate for biomedical applications. *Material Science and Engineering: C, 80*, 274–281. doi:10.1016/j.msec.2017.05.079.

Sun, J.-T., Hong, C.-Y., & Pan, C.-Y. (2010). Fabrication of PDEAEMA-coated mesoporous silica nanoparticles and pH-responsive controlled release. *The Journal of Physical Chemistry C, 114*(29), 12481–12486. doi:10.1021/jp103982a.

Uchida, T., Nakano, M., Hongo, S., Shoji, S., Nagata, Y., Satoh, T., et al. (2012). High-intensity focused ultrasound therapy for prostate cancer. *International Journal of Urology, 19*(3), 187–201. doi:10.1111/j.1442-2042.2011.02936.x.

van den Bijgaart, R.J.E., Eikelenboom, D.C., Hoogenboom, M., Fütterer, J.J., den Brok, M.H., & Adema, G.J. (2017). Thermal and mechanical high-intensity focused ultrasound: Perspectives on tumor ablation, immune effects and combination strategies. *Cancer Immunology, Immunotherapy, 66*(2), 247–258. doi:10.1007/s00262-016-1891-9.

Wang, J., Jiao, Y., & Shao, Y. (2018). Mesoporous silica nanoparticles for dual-mode chemo-sonodynamic therapy by low-energy ultrasound. *Materials, 11*(10), 2041. doi:10.3390/ma11102041.

Watermann, A., & Brieger, J. (2017). Mesoporous silica nanoparticles as drug delivery vehicles in cancer. *Nanomaterials, 7*(7). doi:10.3390/nano7070189.

Wen, J., Yang, K., Xu, Y., Li, H., Liu, F., & Sun, S. (2016). Construction of a triple-stimuli-responsive system based on cerium oxide coated mesoporous silica nanoparticles. *Scientific Reports, 6*, 38931. doi:10.1038/srep38931.

Wu, H., Tang, L., An, L., Wang, X., Zhang, H., Shi, J., et al. (2012). pH-responsive magnetic mesoporous silica nanospheres for magnetic resonance imaging and drug delivery. *Reactive and Functional Polymers, 72*(5), 329–336. doi:10.1016/j.reactfunctpolym.2012.03.007.

Xing, L., Zheng, H., Cao, Y., & Che, S. (2012). Coordination polymer coated mesoporous silica nanoparticles for pH-responsive drug release. *Advanced Materials, 24*(48), 6433–6437. doi:10.1002/adma.201201742.

Yanes, R.E., & Tamanoi, F. (2012). Development of mesoporous silica nanomaterials as a vehicle for anticancer drug delivery. *Therapeutic Delivery, 3*(3), 389–404. doi:10.4155/tde.12.9.

Yang, B., Chen, Y., & Shi, J. (2019). Mesoporous silica/organosilica nanoparticles: Synthesis, biological effect and biomedical application. *Materials Science and Engineering: R: Reports, 137*, 66–105. doi:10.1016/j.mser.2019.01.001.

Yang, H., Lou, C., Xu, M., Wu, C., Miyoshi, H., & Liu, Y. (2011). Investigation of folate-conjugated fluorescent silica nanoparticles for targeting delivery to folate receptor-positive tumors and their internalization mechanism. *International Journal of Nanomedicine, 6*, 2023–2032. doi:10.2147/IJN.S24792.

Yang, K.-N., Zhang, C.-Q., Wang, W., Wang, P.C., Zhou, J.-P., & Liang, X.-J. (2014). pH-responsive mesoporous silica nanoparticles employed in controlled drug delivery systems for cancer treatment. *Cancer Biology & Medicine, 11*(1), 34–43. doi:10.7497/j.issn.2095-3941.2014.01.003.

Zhang, B., Liu, Q., Liu, M., Shi, P., Zhu, L., Zhang, L., et al. (2019). Biodegradable hybrid mesoporous silica nanoparticles for gene/chemo-synergetic therapy of breast cancer. *Journal of Biomaterials Applications, 33*(10), 1382–1393. doi:10.1177/0885328219835490.

Zhang, M., Xu, C., Wen, L., Han, M.K., Xiao, B., Zhou, J., et al. (2016). A hyaluronidase-responsive nanoparticle-based drug delivery system for targeting colon cancer cells. *Cancer Research, 76*(24), 7208–7218. doi:10.1158/0008-5472.CAN-16-1681.

Zhu, C.-L., Lu, C.-H., Song, X.-Y., Yang, H.-H., & Wang, X.-R. (2011). Bioresponsive controlled release using mesoporous silica nanoparticles capped with aptamer-based molecular gate. *Journal of the American Chemical Society, 133*(5), 1278–1281. doi:10.1021/ja110094g.

Zhu, L., & Torchilin, V.P. (2012). Stimulus-responsive nanopreparations for tumor targeting. *Integrative Biology, 5*(1), 96–107. doi:10.1039/c2ib20135f.

4

Solid Lipid Nanoparticles (SLNs) and Nanostructured Lipid Carriers (NLCs): Fabrication and Functionalization for Impending Therapeutic Applications

Rohini Kanwar, Shivani Uppal, and Surinder Kumar Mehta

CONTENTS

4.1 Introduction

Richard Feynman was the man who coined the term 'nano' in 1959 in his famous talk 'There's Plenty of Room at the Bottom'. Since then, a lot of research has been focused on unleashing the untamed potential of the nanoregime. It proved to be a revolutionary approach which resolved the current glitches in therapeutics (Sanchez and Sobolev, 2010). Nanotechnology diversification into the biomedical domain is quite impressive in comparison to conventional drug delivery systems. Their potential as nanowagons to carry diverse array of drugs and bioactives has been exploited enormously. To date, a number of nanocarriers have been developed e.g. micelles, vesicles, dendrimers, niosomes, liposomes, microspheres, polymersomes, polymeric nanoparticles, carbon nanotubes, silica nanoparticles, and many more (Mishra et al., 2010). Figure 4.1 shows different types of nano delivery carriers explored to date.

However, in 1990 researchers started exploring the field of lipid nanoparticles (LNs), beginning with solid lipid nanoparticles (SLNs) to nanostructured lipid carriers (NLCs). LNs are basically spheres or platelets in the size range of 10–1000nm, and comprise of a physiological lipid core (biodegradable and biocompatible) dispersed in an aqueous emulsifier solution (Pardeike et al., 2009; Montenegro et al., 2016). Various snags like poor aqueous solubility, low intestinal permeability, presystemic metabolism, P-gp efflux, gastro-intestinal degradation, and issues associated with permeability, etc. linked with most of the drugs can be efficiently tackled by the use of LNs and surface functionalized LNs (Chakraborty et al., 2009).

4.1.1 Solid Lipid Nanoparticles (SLNs)

SLNs, termed the first generation of LNs composed of lipids, are in solid form at room and body temperature (instead of liquid

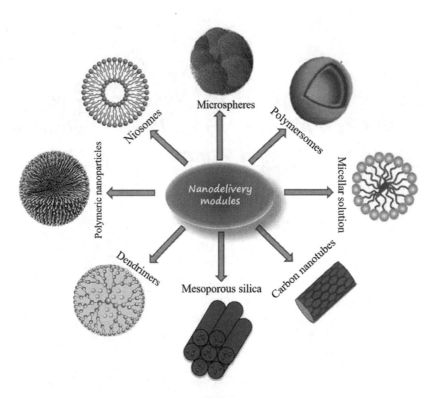

FIGURE 4.1   Structure of different assemblies of colloidal systems for drug delivery.

lipid as present in nano-emulsions) dispersed in aqueous emulsifier solution (Patel et al., 2013). SLNs are fully crystallized and possess an organized crystalline structure where drugs are accommodated within the lipid matrix. SLNs can have a single solid lipid or mixtures of solid lipids such as triglycerides, partial glycerides, fatty acids, steroids, or waxes, being stabilized by surfactant solution (Bunjes, 2011). The solid state of the lipid (similar to polymeric nanoparticles) assists in shielding the accommodated drugs against chemical degradation under harsh conditions by simply decreasing the mobility of drugs in the solid state which thereby leads to a controlled release. SLNs maintain various advantages such as high drug payload, easy scale-up and sterilization, increased physical stability, and excellent tolerance (Weber et al., 2014).

4.1.2 Nanostructured Lipid Carriers (NLCs)

NLCs are termed the second generation of SLNs in which the liquid lipid is added in the inner phase in association with the solid lipid to resolve the snags associated with SLNs. NLCs have achieved better drug solubilization in the presence of liquid lipid, higher drug loading capacity, slow release, and no need for the organic solvents in production, slower polymorphic transition, and a low crystallinity index in the mixture of lipids (Muller et al., 2002). A pictorial representation elucidating the difference between a SLN and NLC particle matrix structure is shown in Figure 4.2.

The present chapter showcases various preparation methods offered by SLNs and NLCs, in detail amplifying their merits and demerits. Surface modification as a panacea to overcome the hiccups of LNs as drug delivery nano wagons is explained and exemplified using hydrophilic modifiers.

4.2 Synthesis of Different Types of Lipid Nanoparticles

The synthesis of lipid nanoparticles using a green methodology depends on the energy requirements and the energy constraints of the process involved. Extensive optimizations are required to correlate and balance these two factors. The methods involved can be distinguished on the basis of their unique characteristics and benefits. Depending upon the distinctive energy requirements, percentage yield, ease of applicability, and feasibility, etc., various methods have been exploited to develop SLNs and NLCs (Ramteke et al., 2012).

A top-down approach is applied for the production of both SLNs and NLCs based on encapsulation of the active pharmaceutical ingredients (API) in the melted lipid and dispersing in the amphiphilic surfactant solution. A few different approaches employed for the preparation of SLNs and NLCs, along with their advantages and disadvantages, have been tabulated in Table 4.1 (Ramteke et al., 2012; Ekambaram et al., 2011).

4.2.1 High Energy Approaches

4.2.1.1 *High Pressure Homogenization (HPH) Technique*

The method involves the use of high pressure (100–2000 bar) to push the liquid using high-pressure homogenizers. This pressure results in high shear stress in the surface of the liquid, and the cavitation forces induced break down the size of accelerated particles into a sub-micrometer or nanometer range. It is considered to be the most reliable and powerful method for the large-scale production of LNs. This process can be carried out both at an

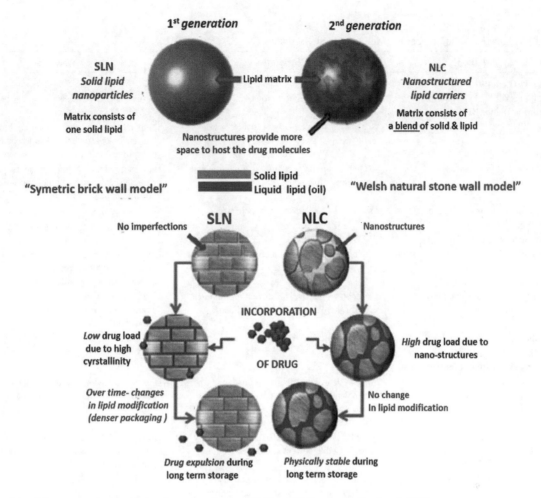

FIGURE 4.2 Pictorial representation elucidating the contrast between SLN and NLC structure (Patel et al., 2013).

elevated temperature (hot HPH) as well as below room temperature (cold HPH). Figure 4.3 shows the schematic representation of the hot and cold HPH method. Initially, the lipid matrix containing either a mixture of solid lipid and liquid lipid (in the case of NLCs) or only solid lipid is melted above the melting temperature of the lipid (Bevilacqua et al., 2007).

4.2.1.1.1 Hot HPH

A coarse pre-emulsion is formed using simple stirring to mix the melted lipid matrix with aqueous surfactant solution. HPH is then applied at an elevated temperature above the characteristic melting point of each lipid to obtain a hot o/w nano-emulsion. Further, the obtained nano-emulsion is cooled down to room temperature or refrigeration conditions to solidify the lipid droplets and formulate LNs.

4.2.1.1.2 Cold HPH

The major disadvantage linked with the hot HPH method is drug degradation at high temperature. To overcome this limitation, the cold HPH was exploited. This involves the solidification of the premixed API in the melted lipid matrix using liquid nitrogen or dry ice and milling up to micron level. The lipid microparticles are later mixed in an aqueous surfactant solution and subjected to HPH at room temperature to form LNs.

4.2.1.2 High Speed Homogenization (HSH) or Ultrasonication (US)

Homogenization and ultrasonication are well explored dispersing techniques which can also be employed together in synergy with each other for the fabrication of LNs. The melted lipid matrix (above the melting point of lipid) is dispersed into the aqueous surfactant solution by HSH followed by US.

Further, warm emulsion is cooled down below the crystallization temperature of the lipid and LNs' dispersion is made (Aditya et al., 2013). The schematic representation of the HSH method followed by US is given in Figure 4.4.

4.2.2 Low Energy Approaches

4.2.2.1 Micro-Emulsification

The micro-emulsification method is the simplest and one of the most explored methods for the formation of LNs of requisite size. It simply involves the mixing of the aqueous surfactant solution (surfactant, co-surfactant, and water) with the lipid matrix under constant stirring to develop a hot microemulsion. Later, this hot mixture is quenched into a high volume of cold water (2–3°C) to solidify the lipid droplets (Kanwar et al., 2016). A schematic illustration of the micro-emulsification technique is represented in Figure 4.5.

TABLE 4.1

Strengths and Weaknesses of the Preparation Methods for SLNs and NLCs

S.No.	Method	Strength	Weakness
I.	**High energy approaches**		
a)	High-pressure homogenization (HPH)	Promising dispersing technique	Energy intensive process
	Hot HPH	Ascendible, commercially available	Temperature actuated degradation of the drug and the carrier, co-existence of supercooled melts and crystalline modifications
	Cold HPH	No temperature induced drug degradation and crystalline-refinement	High energy input, high polydispersity, large particle size, abrasive homogenization conditions, unsubstantiated scalability
b)	Ultrasonication or High speed homogenization	Low shear stress and small particle size	Metal contamination, energy intensive process, poor entrapment efficiency, physical instability
II.	**Low energy approaches**		
c)	Microemulsion	Simple, reproducible, theoretical stability, low energy input, narrow size distribution	Low nanoparticle concentration, sensitivity, labor intensive, high concentration of surfactants/ co-surfactants, high dilution ratio
d)	Phase inversion temperature (PIT) technique	Less energy intensive, no solvent requirement, heat sensitive molecules can be aimed	Unstable emulsion, additional molecules can be assimilated to easily affect the inversion phenomena
e)	Membrane contractor method	Scalable, controllable size	Membrane clogging
III.	**Approaches with organic solvents**		
f)	Solvent emulsification -evaporation technique	Small particle size, effective in case of thermolabile drugs, high encapsulation efficiency, monodispersity, low energy input, low viscous system, avoid heat	Unstable emulsion, low dispersing ability, insolubility of lipids in organic solvents, needs additional solvent removal step
g)	Solvent emulsification-diffusion technique	Evade heat during synthesis, utilizes partially water-miscible solvent	Unstable emulsion, low dispersing ability, insolubility of lipids in organic solvents, demands additional solvent removal step
h)	Solvent injection technique	Easy to handle, demands no special instrument like HPH, small sized particles, increased lipid concentration	Particle size variability
i)	Supercritical fluid method	Dry powder of particles is obtained, evade the use of solvents, require mild temperature and pressure, CO_2 solution is regarded as a good choice of solvent	Extremely expensive

4.2.2.2 Phase Inversion Temperature (PIT)

The reversal of phase from o/w to w/o emulsion with change in temperature is called the PIT. To obtain the required LNs, the aqueous and non-aqueous phases are heated together above the PIT. Generally, three heating and cooling cycles are repeated, starting from room temperature to PIT, followed by the final step of dilution with cold water (0°C) (Heurtault et al., 2002). Figure 4.6 showcases a flow diagram of the PIT method.

4.2.2.3 Double Emulsion Technique

This technique is preferred for the loading of hydrophilic API and peptides in SLNs and NLCs. Firstly, primary w/o emulsion stabilized with suitable excipients is formulated by emulsifying the aqueous drug solution with the melted lipid phase (Figure 4.9). Then, the formed primary w/o emulsion is added to the hydrophilic emulsifier solution, leading to the formation of a double w/o/w emulsion. The latter is then isolated by filtration after the formation of SLNs/NLCs by continuous stirring (Yang et al., 2011). The steps involved in the double emulsion technique are shown in Figure 4.7.

4.2.2.4 Membrane Contractor Method

This method involves the use of a cylindrical membrane module to synthesize the SLNs/NLCs. The aqueous phase containing an emulsifier is circulated in the internal channel of the membrane, and melted lipid is pressed at the temperature above the melting point of lipid through membrane pores into internal water flow, letting the formation of small droplets which are swept away by the aqueous phase. Later the SLNs/NLCs are formed by cooling down the produced droplets of melted lipid to room temperature (Ahmed El-Harati et al., 2006). Figure 4.8 shows a schematic representation of the membrane contractor method.

4.2.3 Approaches with Organic Solvents

4.2.3.1 Solvent Emulsification-Evaporation Technique

Lipid is dissolved in a water-immiscible organic solvent with low boiling point (like chloroform, cyclohexane) and emulsified in an aqueous surfactant solution using HSH and the resulting coarse pre-emulsion is passed through HPH to obtain nano-emulsion.

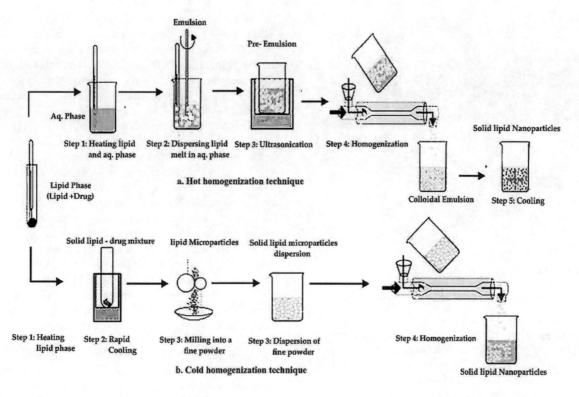

FIGURE 4.3 High-pressure homogenization (HPH) technique a) Hot HPH and b) Cold HPH. (Reproduced from Ganesan and Narayanasamy, 2017. Copyright 2020 Elsevier.)

After the solvent is evaporated using a rotary evaporator at 50–60°C, LNs are formed (Negi et al., 2014).

4.2.3.2 Solvent Emulsification-Diffusion Technique

Partially water-miscible solvents such as tetrahydrofuran, benzyl alcohol are employed to dissolve the lipid. Organic solvents are saturated with water for establishing the initial thermodynamic equilibrium between the solvent and water. Next, the organic solvent diffuses from an organic phase to aqueous phase, resulting in the solidification of the dispersed phase and crystallization of the lipid (Trotta et al., 2003).

4.2.3.3 Solvent Injection Technique

The lipid, dissolved in a water-miscible solvent (like methanol, ethanol, acetone, dimethyl sulfoxide) is injected into an aqueous solution of surfactant using an injection needle. Similar to the solvent emulsification-diffusion method, here, too, lipid crystallization is effected by the diffusion of the organic solvent from the organic phase to the aqueous phase (Schubert and Müller-Goymann, 2003).

A combined pictorial representation of solvent emulsification-evaporation, solvent emulsification-diffusion, and solvent injection methods is given in Figure 4.9.

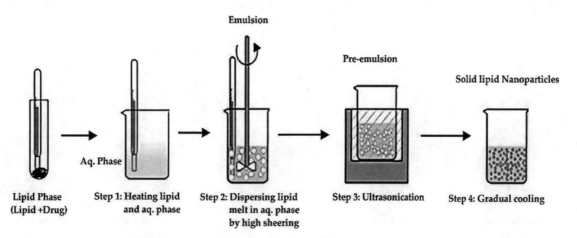

FIGURE 4.4 High shear homogenization/ultrasonication technique. (Reproduced from Ganesan and Narayanasamy, 2017. Copyright 2020 Elsevier.)

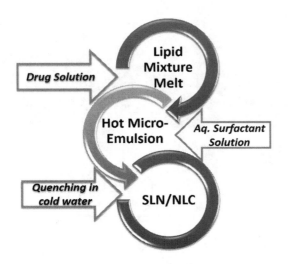

FIGURE 4.5 Micro-emulsification technique.

4.2.3.4 *Supercritical Fluid (SCF) Method*

In this method, the organic phase is prepared by solubilizing the drug and lipid in an organic solvent (such as chloroform) in the presence of an appropriate emulsifier. After dispersing the organic phase into an aqueous emulsifier solution, the formed mixture is successively passed through a HPH to form an o/w emulsion. Later, the o/w emulsion is introduced from one end of the extraction column and the supercritical fluid (maintained in constant physiological conditions) counter-currently (at a constant flow rate) from another end.

SLNs/NLCs dispersions are prepared by continuous extraction of the solvent from the o/w emulsions (Chattopadhyay et al., 2007). The schematic representation of supercritical fluid (SCF) is shown in Figure 4.10.

4.3 Functionalization of Different Types of Lipid Nanoparticles

The prime objective of any successful nanocarrier is to target the API at the desired location. Most drugs do not possess requisite physicochemical properties to reach or be taken up by target cells. In order to target the API to the specific tissue sites, LNs have gained considerable attention in the constantly growing field of nanobiomedicine, owing to their unique attributes like

biocompatibility, ease of preparation, ability to protect the drug, and accessibility to different routes of administration (Gaspar et al., 2017). However, this modern formulation approach suffers from the hindrances of drug expulsion and initial burst release. Therefore, their surface needs to be modified.

Hence, site-specific drug delivery systems with surface modifications have been of great interest, focusing on overcoming various biological barriers such as blood brain barrier (BBB), intestinal mucous barrier, gastrointestinal barrier, etc., thus improving the drug performance (Lahkar and Das, 2013). Figure 4.11 illustrates the need of surface modification.

Ideally, providing a perfect environment to the API is an arduous task, since the distribution of drugs is governed and limited by the anatomical barriers. Altering the surface characteristics by coating the LNs improves plasma stability and bio-distribution of the entrapped drugs, besides increasing their storage stability. Modification is also done to augment the pharmacokinetics and bioavailability of drugs so as to target the site of action via LNs. It is the surface characteristics of LNs that determine the ability of modification and subsequently, the nature of the modifier and the type of interaction between the nanocarrier and modifier (Redhead et al., 2001). The mechanism of particle-cell interaction inherently depends on the constitution of the opsonic component. Surface functionalization is also responsible for the increased permeation and resistance to chemical degradation, possibility of co-delivery of various therapeutic agents, or stimuli responsiveness. To evade the phagocytic uptake and regulate bio-distribution parameters of drugs for their prolonged blood circulation, the surface properties of colloidal drug carriers are altered by using several techniques. Figure 4.12 demonstrates the merits of surface modification (Mahapatro and Singh, 2011).

Predominantly, the LNs undergo a swift intestinal transit, owing to their weak mucous anchorage. As a consequence, there is an insufficient uptake at the desired sorption site. Hence, the primary requisiteness of surface modification resides in its ability to reduce the uptake of LNs by the reticuloendothelial system (RES). Numerous strategies have been devised to overcome the above mentioned impediments and enhance the residence time of LNs in the gastrointestinal tract (Li et al., 2009).

Over the past two decades, several passive or active functionalization approaches have been effectually applied to LNs to overcome the paucity of the API. A myriad of ligands can be functionally attached, including various internalizable ligands, specific targeted peptides, saccharide ligands, or therapeutic

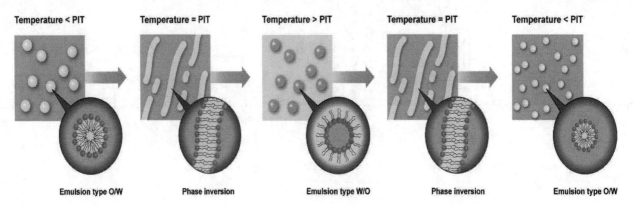

FIGURE 4.6 Phase inversion temperature (PIT) technique. (Reproduced from Svilenov and Tzachev, 2014.)

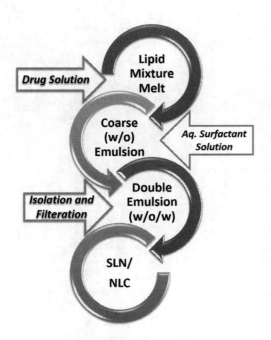

FIGURE 4.7 Double emulsion technique.

molecules. Table 4.2 summarizes a brief account of some modifiers (hydrophilic agents) explored for coating the LNs, such as chitosan (Garcia-Fuentes et al., 2005), PEG (Kang et al., 2018), cysteine (Mazzaferro et al., 2013), human serum albumin (Iqbal et al., 2011), and mannose (Fang et al., 2015) etc., owing to their promising results. After the functionalization of LNs, improved stability of the particles with enhanced transmucosal transport has been showcased (Garcia-Fuentes et al., 2005; Kang et al., 2018; Mazzaferro et al., 2013; Iqbal et al., 2011; Fang et al., 2015). Their ability to reduce the phagocytic uptake with minimal nonspecific interactions is also improved.

The first proposition includes the use of positively charged polymers such as chitosan and its derivatives. These polymers can efficiently bind to the negatively charged sialic and sulfonic acids in the mucus layer through electrostatic interactions, thereby increasing the mucoadhesion. Garcia-Fuentes et al., (2005) fabricated chitosan-coated LNs for oral delivery of peptides. Preventing the initial burst release, the functionalized LNs exhibited slow release with no degradation of the lipid matrix.

The second approach involves surface coating with polyethylene glycol (PEG). The popularity of this polymer functionalization has been reported (Kang et al., 2018), due to its exceptional property to increase the blood circulation time of LNs. Enhanced cellular transport and improved oral bioavailability of active polyphenol-loaded LNs was reported by Kang and co-workers (Kang et al., 2018).

Yet another approach is the sulfur-containing polymers, where thiolated polymers have been crafted to ameliorate the mucous binding properties of LNs. The polymers containing sulfhydryl groups demonstrate the phenomenal ability of forming covalent disulfide bonds with cysteine-rich subdomains of the mucus layer. Compared to the electrostatic interactions in the case of hydrophilic polymers, the mucoadhesion of thiomers is governed by the chemical covalent bonding. This increases the residence time of the nanocarrier in the gut causing an increase in the drug concentration and reducing the gastrointestinal vermiculation. Thus, the absorption of the API increases, along with an appreciable increase in the permeation rate and inhibits the first pass metabolism. The use of thiomers for improving the oral administration has been actively reported by several groups recently. Mazzaferro et al. (2013) used thiolated chitosan-coated nanoparticles and reported a significant increase in the permeability of docetaxel in the intestinal region. Iqbal et al. (2011) designed a reduced glutathione-combined poly(acrylic acid)–cysteine for increasing the oral bioavailability of paclitaxel. Cysteine-functionalized LNs were developed by Fang et al. (2015) for synergistic enhancement docetaxel in the oral mode of administration. Bioadhesive amphiphilic thiomers (Cy-PEG-MSA) are obtained by conjugating cysteine (Cy) with the amphiphilic polymer polyethylene glycol monostearate (PEG-MSA). In-vitro bioadhesion studies are conducted on unmodified NLCs and cysteine-functionalized NLCs. The introduction of cysteine to the NLCs exuberates the bioavailability of NLCs, thereby showing the great potential of cysteine as a modifier for LNs, as shown in Figure 4.13.

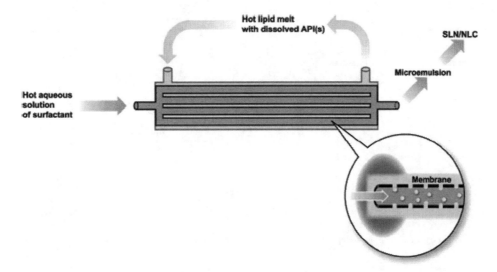

FIGURE 4.8 Membrane contractor technique. (Reproduced from Svilenov and Tzachev, 2014.)

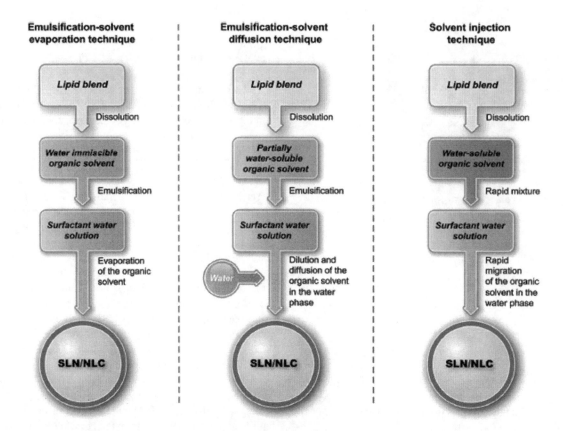

FIGURE 4.9   Emulsification-solvent evaporation technique. (Reproduced from Svilenov and Tzachev, 2014.)

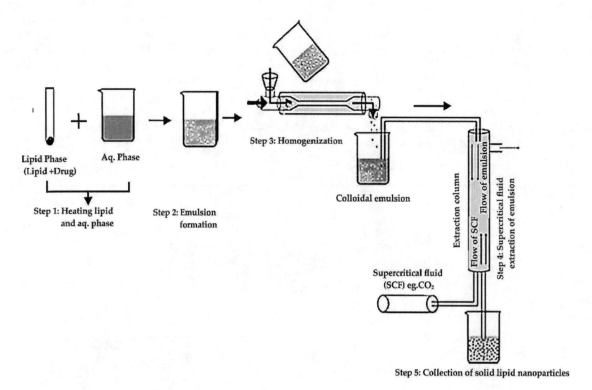

FIGURE 4.10   Supercritical fluid (SCF) technique. (Reproduced from Ganesan and Narayanasamy, 2017. Copyright 2020 Elsevier.)

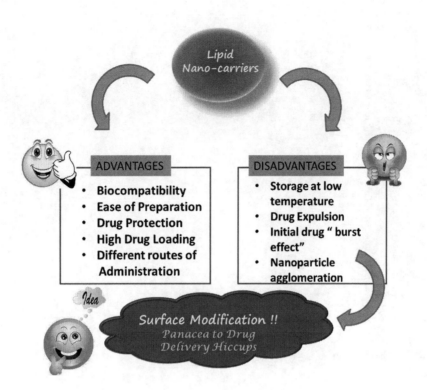

FIGURE 4.11 Advantages and disadvantages of lipid nanocarriers.

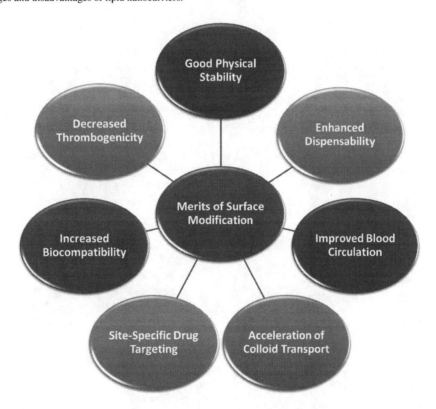

FIGURE 4.12 Merits of surface modifications.

Biodegradability, nontoxicity, and non-immunogenic characteristics make human serum albumin (HSA) a promising biomacromolecule and it draws great attention in modifying and stabilizing the drug carriers. The ligand-binding capability of HSA makes it a potential candidate for the construction of innovative structures for therapeutic carriers. Kuo and Chung (2011) investigated the physicochemical properties of HSA-grafted SLNs and NLCs loaded with nevirapine (NVP) (Figure 4.14).

HSA grafted on NVP-SLNs increased the viability of human brain-microvascular endothelial cells (HBMECs). HSA-grafted SLNs and NLCs have been proven to be effective formulations in the delivery of NVP for viral therapy. Table 4.2 gives a brief

TABLE 4.2

Modifier, Method of Preparation, and Advantages of Modification

Lipid Nanocarrier	Modifier	Method of Preparation	Advantage of Modification	Reference
SLN	Chitosan and its derivatives	Double emulsion-solvent emulsification method	Decrease in the burst release effect	Garcia-Fuentes et al. (2005)
NLC		Microemulsion method followed by sonication	Effective delivery to the brain and prolonging the retention time in the nasal epithelium with decreased dose and dosage frequency	Gartziandia et al. (2015)
SLN		Modified hot homogenization method followed by sonication	Bioavailability enhancement via. improved uptake through body fluids and cells	Baek et al. (2017)
SLN		Ultrasonication process	Improved SiRNA complexation, NIH3T3 cell transfection, and ERK1 downregulation	Tezgel et al. (2018)
SLN	Polyethylene glycol and its derivatives	Solvent diffusion method	Enhanced the oral absorption, transport efficacy and bioavailability	Kang et al. (2018)
SLN		Microemulsion method followed by sonication	The digestion fate of SLN in the gastro intestinal tract can be controlled depending on the length and concentration of PEGylated emulsifiers	Ban et al. (2018)
NLC	Cysteine	Microemulsion method followed by ultrasonication	Enhanced mucoadhesion and adsorption across the intestinal membrane	Fang et al. (2015)
SLN & NLC	Human serum albumin	Modified microemulsion method	Human brain micro vascular endothelial cells viability enhancement	Costa et al. (2018)
SLN	D-Mannose	Modified solvent emulsification-evaporation method	Effective in targeting alveolar macrophages	Jain et al. (2010)
SLN		Solvent injection method	Selectively ferry the bioactives and specifically to the tumor sites with the minimal side effects	Oliveira et al. (2016)

(Continued)

TABLE 4.2 (CONTINUED)

Modifier, Method of Preparation, and Advantages of Modification

Lipid Nanocarrier	Modifier	Modifier	Method of Preparation	Advantage of Modification	Reference
SLN	α-tocopheryl succinate		Hot melting homogenization method	Improved *in-vitro* anti-cancer efficacy	Mehrad et al. (2018)
SLN	Whey protein isolate		Microemulsion dilution method	Enhanced physico-chemical stability of β-carotene	Dal et al. (2017)
SLN	Apolipoprotein E-derived peptide		Microemulsion method	Improved brain delivery of therapeutics	Geske-Moritz et al. (2016)

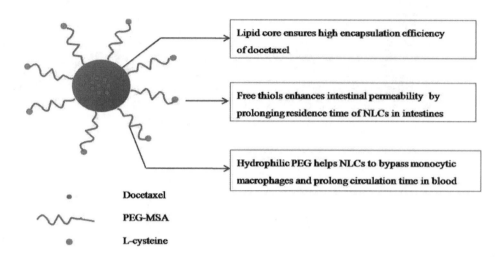

FIGURE 4.13 Diagrammatic representation of L-cysteine linked with PEG structure of LNs. (Reproduced from Fang et al., 2015. Copyright 2020 ACS.)

account of the type of hydrophilic modifiers with their positive outcomes.

4.4 Potential Applications of Lipid Nanocarriers

To combat the ever-rising cost of research and development, the efficacy and safety of the product are the essential parameters that governs its sustainability and long-lasting demand in the market. This has created increased interest in the lipid-based delivery systems because of their innumerable positive attributes, like biocompatibility, biodegradibilty, low chronic and acute toxicity, high drug pay-load, stability against coalescence under harsh conditions, etc. This channelizes LNs to a wide array of applications such as in-delivery of API, nutraceuticals, cosmeceuticals; gene therapy; food packaging and enhancement of shelf life,

thereby covering three major industries i.e. medicine, food, and cosmetics. A few commercialized products of SLNs and NLCs are: Capture-Dior, Mucosolvan Retard Capsules, Rifamsolin; and Cream Nanorepair Q10, IOPE Super Vital cream, respectively.

Delivery of Pharmaceuticals. SLNs and NLCs are smart drug carrier modules comprising of physiological lipid (biodegradable and biocompatible) materials dispersed in aqueous surfactant solutions. The success story of LNs can be easily assessed from a number of commercial products available in the market. LNs possess a variety of pharmaceutical applications owing to numerous routes of drug delivery viz., topical, oral, pulmonary, ocular, and parenteral forms. This highlights its ease of administration and its future perspective as a pharmaceutical carrier (Waghmare et al., 2012).

As Phytochemical Carriers. Owing to the stunted solubility, hindered bioavailability, low stability, and poor target specificity of phytochemicals (the natural health-promoting agents), LNs have been exploited to enhance the pharmacokinetic profile of phytochemicals, by improving the absorption ability, protecting the bioactive from premature degradation, and thereby prolongating the circulation time in the body. Moreover, LNs display selective uptake efficiency in the target cells (or tissue) over normal cells (or tissue), preventing off-target side-effects by interaction with the healthy biological environment. They have been widely employed as carriers for administration of phytochemicals like epigallocatechin gallate, quercetin, resveratrol, and curcumin to augment their bioactivities (Wang et al., 2014).

Gene Delivery. Gene therapy embodies a novel paradigm in the inhibition and cure of numerous inherited and acquired ailments. Cationic LNs efficiently act as a substitute mode of DNA delivery, due to their ease of scale-up production, virtuous storage stability and steam sterilization, and lyophilization-friendly characteristics (Bondi et al., 2010).

Delivery of Cosmeceuticals. The capability of protecting the labile compounds against chemical degradation (e.g. for tocopherols and retinol), makes LNs a promising carrier for cosmetic applications. The ability of LNs to show controlled release, occlusivity, and the potential for UV-blocking plays a vital role in increasing the efficacy of cosmetics. LNs in combination with molecular sunscreens act as a suitable candidate to achieve

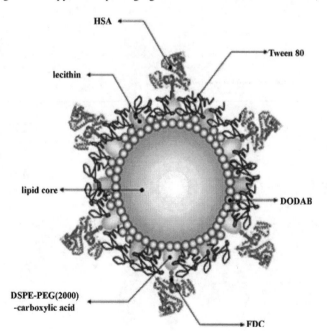

FIGURE 4.14 Schematic illustration of HSA coated SLNs and NLCs. (Reproduced from Kuo and Chung, 2011. Copyright 2020 Elsevier.)

improved photo-protection. Also, LNs are extremely successful for preventing and treating several skin diseases (Patravale et al., 2008).

4.5 Conclusions

The unique characteristics of LNs have made them the most effective and advanced delivery vehicles to administer API in a site-specific and sustained manner. LNs possess an ability of incorporating both hydrophilic and hydrophobic molecules from different classes including antibacterial, antioxidant, antiviral, anticancer, anti-inflammatory, antigens, proteins, nucleotides, genes, vitamins, plant extracts, flavonoids, carotenoids, food bioactives, etc. An in-depth elucidation of SLNs and NLCs, along with the role of different fabrication methodologies on their properties showcasing their advantages and disadvantages, have been highlighted. Since they are composed of physiological excipients, metabolic pathways are already in place within the body, which facilitate the transportation and metabolism of the encapsulated drugs. A controlled drug release rate, protection of the encapsulated drug, and the large-scale-production-like attributes of SLNs/NLCs have actually revolutionized the field of drug delivery. Surface modification as a panacea for drug delivery hiccups has also been elucidated in great detail. Also a brief account of various modifiers and their unique properties has been listed. Further work is going on at an accelerating pace to exploit the untamed potential of these exuberant nanocarriers.

Acknowledgments

Rohini Kanwar gratefully acknowledges CSIR, New Delhi. Shivani Uppal thanks the UGC Centre of Excellence in Applications of Nanomaterials, Nanoparticles & Nanocomposites. Surinder Kumar Mehta is thankful to DST–PURSE II for financial assistance.

REFERENCES

Aditya, N.P., Shim, M., Lee, I., Lee, Y., Im, M.H., & Ko, S. (2013). Curcumin and genistein coloaded nanostructured lipid carriers: In vitro digestion and antiprostate cancer activity. *Journal of Agricultural and Food Chemistry, 61*, 1878–1883.

Ahmed El-Harati, A., Charcosset, C., and Fessi, H. (2006). Influence of the formulation for solid lipid nanoparticles prepared with a membrane contactor. *Pharmaceutical Development and Technology, 11*, 153–157.

Baek, J.-S., and Cho, C.-W. (2017). Surface modification of solid lipid nanoparticles for oral delivery of curcumin: Improvement of bioavailability through enhanced cellular uptake, and lymphatic uptake. *European Journal of Pharmaceutics and Biopharmaceutics, 117*, 132–140.

Ban, C., Jo, M., Lim, S., and Choi, Y.J. (2018). Control of the gastrointestinal digestion of solid lipid nanoparticles using PEGylated emulsifiers. *Food Chemistry, 239*, 442–452.

Bevilacqua, A., Cibelli, F., Corbo, M.R., and Sinigaglia, M. (2007). Effects of high-pressure homogenization on the survival of Alicyclobacillus acidoterrestris in a laboratory medium. *Letters in Applied Microbiology, 45*, 382–386.

Bondì, M.L., and Craparo, E.F. (2010). Solid lipid nanoparticles for applications in gene therapy: A review of the state of the art. *Expert Opinion on Drug Delivery, 7*, 7–18.

Bunjes, H. (2011). Structure properties of solid lipid based colloidal drug delivery systems. *Current Opinion in Colloid & Interface Science, 16*, 405–411.

Chakraborty, S., Shukla, D., Mishra, B., and Singh, S. (2009). Lipid: An emerging platform for oral delivery of drugs with poor bioavailability. *European Journal of Pharmaceutics and Biopharmaceutics, 73*, 1–15.

Chattopadhyay, P., Shekunov, B.Y., Yim, D., Cipolla, D., Boyd, B., and Farr, S. (2007). Production of solid lipid nanoparticle suspensions using supercritical fluid extraction of emulsions (SFEE) for pulmonary delivery using the AERx system. *Advanced Drug Delivery Reviews, 59*, 444–453.

Costa, A., Sarmento, B., and Seabra, V. (2018). Mannose-functionalized solid lipid nanoparticles are effective in targeting alveolar macrophages. *European Journal of Pharmaceutical Sciences, 114*, 103–113.

Dal Magro, R., Ornaghi, F., Cambianica, I., Beretta, S., Re, F., Musicanti, C., Rigolio, R., Donzelli, E., Canta, A., Ballarini, E., Cavaletti, G., Gasco, P., and Sancini, G. (2017). ApoE-modified solid lipid nanoparticles: A feasible strategy to cross the blood-brain barrier. *Journal of Controlled Release, 249*, 103–110.

Ekambaram, P., Sathali, A.A.H., and Priyanka, K. (2011). Solid lipid nanoparticles: A review. *Scientific Reviews and Chemical Communication, 2*, 80–102.

Fang, G., Tang, B., Chao, Y., Xu, H., Gou, J., Zhang, Y., Xu, H., and Tang, X. (2015). Cysteine-functionalized nanostructured lipid carriers for oral delivery of docetaxel: A permeability and pharmacokinetic study. *Molecular Pharmaceutics, 12*, 2384–2395.

Ganesan, P., and Narayanasamy, D. (2017). Lipid nanoparticles: Different preparation techniques, characterization, hurdles, and strategies for the production of solid lipid nanoparticles and nanostructured lipid carriers for oral drug delivery. *Sustainable Chemistry and Pharmacy 6*, 37–56.

Garcia-Fuentes, M., Torres, D., and Alonso, M.J. (2005). New surface-modified lipid nanoparticles as delivery vehicles for salmon calcitonin. *International Journal of Pharmaceutics, 296*, 122–132.

Gartziandia, O., Herran, E., Pedraz, J.L., Carro, E., Igartua, M. and Hernandez, R.M. (2015). Chitosan coated nanostructured lipid carriers for brain delivery of proteins by intranasal administration. *Colloids and Surfaces B: Biointerfaces, 134*, 304–213.

Gaspar, D.P., Faria, V., Quintas, J.P., and Almeida, A.J. (2017). Targeted delivery of lipid nanoparticles by means of surface chemical modification. *Current Organic Chemistry, 21*, 2360–2375.

Geszke-Moritz, M., and Moritz, M. (2016). Solid lipid nanoparticles as attractive drug vehicles: Composition, properties and therapeutic strategies. *Materials Science and Engineering: C 68*, 982–994.

Heurtault, B., Saulnier, P., Pech, B., Proust, J.-E., and Benoit, J.-P. (2002). A novel phase inversion-based process for the preparation of lipid nanocarriers. *Pharmaceutical Research, 19*, 875–880.

Iqbal, J., Sarti, F., and Perera, G. (2011). Development and *in vivo* evaluation of an oral drug delivery system for paclitaxel. *Biomaterials, 32*, 170–175.

Jain, A., Agarwal, A., Majumder, S., Lariya, N., Khaya, A., Agrawal, H., Majumdar, S., and Agrawal, G.P. (2010). Mannosylated solid lipid nanoparticles as vectors for site-specific delivery of an anti-cancer drug. *Journal of Controlled Release 148*, 359–367.

Kang, X., Chen, H., Li, S., Jie, L., Hu, J., Wang, X., Qi, J., Ying, X. and Du, Y. (2018). Magnesium lithospermate B loaded PEGylated solid lipid nanoparticles for improved oral bioavailability. *Colloids and Surfaces B: Biointerfaces, 161*, 597–605.

Kanwar, R., Kaur, G., and Mehta S.K. Revealing the potential of didodecyldimethyl ammonium bromide as efficient scaffold for fabrication of nano liquid crystalline structures. (2016). *Chemistry and Physics of Lipids, 196*, 61–68.

Kuo, Y.-C., and Chung, J.-F. (2011). Physicochemical properties of nevirapine-loaded solid lipid nanoparticles and nanostructured lipid carriers. *Colloids and Surfaces B: Biointerfaces, 83*, 299–306.

Lahkar, S., and Das, M.K. (2013). Surface modified polymeric nanoparticles for brain targeted drug delivery. *Current Trends in Biotechnology and Pharmacy, 7*, 914–931.

Li, H., Zhao, X., Ma, Y., Zhai, G., Li, L. and Lou, H., (2009). Enhancement of gastrointestinal absorption of quercetin by solid lipid nanoparticles. *Journal of Controlled Release, 133*, 238–244.

Mahapatro, A., and Singh, D.K. (2011). Biodegradable nanoparticles are excellent vehicle for site directed in-vivo delivery of drugs and vaccines. *Journal of Nanobiotechnology, 9*, 55–61.

Mazzaferro, S., Bouchemal, K., and Ponchel, G. (2013). Oral delivery of anticancer drugs III: Formulation using drug delivery system. *Drug Discovery Today, 18*, 99–104.

Mehrad, B., Ravanfar, R., Licker, J., Regenstein, J.M., and Abbaspourrad, A. (2018). Enhancing the physicochemical stability of β-carotene solid lipid nanoparticle (SLNP) using whey protein isolate. *Food Research International, 105*, 962–969.

Mishra, B.B.T.S., Patel, B.B., and Tiwari, S. (2010). Colloidal nanocarriers: A review on formulation technology, types and applications toward targeted drug delivery. *Nanomedicine, 6*, 9–24.

Montenegro, L., Lai, F., Offerta, A., Sarpietro, M.G., Micicchè, L., Maccioni, A.M., Valenti, D., and Fadda, A.M. (2016). From nanoemulsions to nanostructured lipid carriers: A relevant development in dermal delivery of drugs and cosmetics. *Journal of Drug Delivery Science and Technology, 32*, 100–112.

Müller, R.H., Radtke, M., and Wissing, S.A. (2002). Solid lipid nanoparticles (SLN) and nanostructured lipid carriers (NLC) in cosmetic and dermatological preparations. *Advanced Drug Delivery Reviews, 54*, S131–515.

Negi, L.M., Jaggi, M., and Talegaonkar, S. (2014). Development of protocol for screening the formulation components and the assessment of common quality problems of nano-structured lipid carriers. *International Journal of Pharmaceutics, 461*, 403–410.

Oliveira, M.S., Mussi, S.V., Gomes, D.A., Yoshida, M.I., Frezard, F., Carregal, V.M., and Ferreira, L.A.M. (2016). α-tocopherol succinate improves encapsulation and anticancer activity of doxorubicin loaded in solid lipid nanoparticles. *Colloids and Surfaces B: Biointerfaces, 140*, 246–253.

Pardeike, J., Hommoss, A., and Müller, R.H. (2009). Lipid nanoparticles (SLN, NLC) in cosmetic and pharmaceutical dermal products. *International Journal of Pharmaceutics, 366*, 170–184.

Patel, D., Kesharwani, R. and Gupta, S. (2013). Development & screening approach for lipid nanoparticle: A review. *International Journal of Innovative Pharmaceutical Sciences and Research, 2*, 27–32.

Patravale, V.B., and Mandawgade, S.D. (2008). Novel cosmetic delivery systems: An application update. *International Journal of Cosmetic Science, 30*, 19–33.

Ramteke, K.H., Joshi, S.A., and Dhole, S.N. (2012). Solid lipid nanoparticle : A review. *IOSR Journal of Pharmacy, 2*, 34–44.

Redhead, H.M., Davis, S.S., and Illum, L. (2001). Drug delivery in poly (lactide-co-glycolide) nanoparticles surface modified with poloxamer 407 and poloxamine 908: In vitro characterisation and in vivo evaluation. *Journal of Controlled Release, 70*, 353–363.

Sanchez, F., and Sobolev, K. (2010). Nanotechnology in concrete: A review. Construction and Building Materials, 24, 2060–2071.

Schubert, M., and Müller-Goymann, C. (2003). Solvent injection as a new approach for manufacturing lipid nanoparticles–evaluation of the method and process parameters. *European Journal of Pharmaceutics and Biopharmaceutics, 55*, 125–131.

Svilenov, H., and Tzachev, C. (2014). Solid lipid nanoparticles: A promising drug delivery system. *Nanomedicine*, 187–237.

Tezgel, Ö., Szarpak-Jankowska, A., Arnould, A., Auzély-Velty, R., and Texier, I. (2018). Chitosan-lipid nanoparticles (CS-LNPs): Application to siRNA delivery. *Journal of Colloid and Interface Science, 510*, 45–56.

Trotta, M., Debernardi, F., and Caputo, O. (2003). Preparation of solid lipid nanoparticles by a solvent emulsification–diffusion technique. *International Journal of Pharmaceutics, 257*, 153–160.

Waghmare, A.S., Grampurohit, N.D., Gadhave, M.V, Gaikwad, D.D., and Jadhav, S.L. (2012). Solid lipid nanoparticles: A promising drug delivery system. *International Research Journal of Pharmacy, 3*, 100–107.

Wang, S., Su, R., Nie, S., Sun, M., Zhang, J., Wu, D., and Moustaid-Moussa, N. (2014). Application of nanotechnology in improving bioavailability and bioactivity of diet-derived phytochemicals. *Journal of Nutritional Biochemistry, 25*, 363–376.

Weber, S., Zimmer, A., and Pardeike, J. (2014). Solid lipid nanoparticles (SLN) and nanostructured lipid carriers (NLC) for pulmonary application: A review of the state of the art. *European Journal of Pharmaceutics and Biopharmaceutics, 86*, 7–22.

Yang, R., Gao, R., Li, F., He, H., and Tang, X. (2011). The influence of lipid characteristics on the formation, in vitro release, and in vivo absorption of protein-loaded SLN prepared by the double emulsion process. *Drug Development and Industrial Pharmacy, 37*, 139–148.

5

Biodegradable Nanomaterials for Cosmetic and Medical Use

Pierfrancesco Morganti, Gianluca Morganti, Serena Danti, Maria-Beatrice Coltelli, and Giovanna Donnarumma

CONTENTS

5.1 Introduction

In a world where non-biodegradable waste is invading land and oceans, it is time to shift towards a circular economy whereby the biological and technical cycles are linked and driven by innovative products respectful of the Planet's environment [1]. Nanobiotechnology may be of great help to make innovative and smart goods realized by polyglucoside polymers obtained from waste materials of industrial and agricultural origin [2]. It is estimated, for example, that the fishery industry's by-products and agro-food biomass, rich of chitoolisaccharides and oligosaccharides, are produced in a global quantity of about 300 billion tonnes/year, of which less than 20% is currently used for energy and goods [3].

While each year US$80–120 billion of plastic packaging material is lost to the economy, the amount of food that is wasted each year (66 tonnes of food thrown away every second) represents one third of its global production [1, 4]. Thus, on the one hand, it has been estimated that oceans could contain more plastic than fish (by weight) by 2050 (Figure 5.1). On the other hand, the United Nations' Food and Agriculture Organization (FAO) estimates that about 815 million people (10.7% of the 7.6 billion people in the world), suffered from chronic undernourishment in 2016 [5]. Additionally, the plastics industry could be consuming 20% of total petrol production and 15% of the annual carbon budget [1], while food waste accounts for 8% of global greenhouse gas emissions [5].

Thus, we have to rethink our way of living, basing our life on new economic, societal, and environmental values. Technological innovation should be much addressed in finding integrated solutions for producing goods using renewable sources compatible with health and the environment, through design for an 'end of life' approach, involving scientists, institutions, companies, students, citizens, consumers, and stakeholders. In this paper we try to propose the use of nanostructured biomaterials to make specific medical tissues and smart cosmetic products, considered of great help in reducing both food and plastic waste for a better and more pleasant world.

5.2 Polysaccharides and Biobased and Biodegradable Polyesters

Pollution in the land and oceans, generated by plastic goods made by fossil raw materials, is stimulating the increase of a sustainable circular economy based on the use of renewable resources as feedstock for the production of natural sugar-like polymers, i.e. polysaccharides and chitoolisaccharides [6, 7].

The most abundant natural polymers are cellulose and lignin, from plant biomass, and chitin and its derivatives from the fishery industry's by-products. Cellulose is a polymer made from glucose while lignin has a complex structure deriving from phenol-based fragments. Chitin is a chitoolisaccharide rich in nitrogen. All these polymers, plant and animal-derived, exhibit high hydrophilicity, interesting bio-functionality, biocompatibility and eco-compatibility, tunable size from sub-microns to nanometers, large surface area for multivalent bio-conjugation, and an interior network for the incorporation of drugs and cosmetic ingredients [7].

Polysaccharides are natural inexpensive and easily degradable biopolymers, which represent about 75% of all organic material on earth [8]. It is worth mentioning that a polysaccharide is a high-molecular-weight compound made up of a small repeating unit of glucose, termed monomer, which is a low-molecular-weight compound connected together by glycosidic linkages to give a polymer (Figure 5.2).

These polymers may be of microbial origin (xanthan gum), algal origin (alginate), and animal origin (chitin), being classified into positively charged polysaccharides (chitin) and negatively charged ones (lignin, hyaluronic acid, etc). When positively charged, they possess peculiar characteristics such as mucoadhesiveness, a permeability enhancing effect, and antimicrobial activity, rendering them particularly useful to treat both skin and hair, negatively charged [9]. Skin cellular surface, for example, dominated by negatively charged sulphated proteoglycan molecules, anchored to cell membrane and linked to glycosaminoglycans, plays pivotal roles in cellular proliferation, migration, and motility [10]. On the other hand, positively charged block

FIGURE 5.1   Slide from Ellen McArthur Foundation about ocean pollution.

polymeric nanoparticles, bound strongly to the cell membrane, facilitate a higher cellular uptake and intercellular distribution, affecting the skin permeation of the encapsulated active ingredients. At the same time the negative charge of the hair repels, for example, the negative charge of the nanoparticles (NPs) so that an alkaline pH may increase the negative electrical charge of a hair fiber surface, increasing friction between the fibers, and leading to cuticle damage and fiber breakage (Figure 5.3).

Thus, the formulation of NPs with different surface properties and charge (zeta potential) may influence their cellular uptake. As a consequence, the degree of skin and hair binding is probably more important when treated by the positively charged particles than with the negatively ones, being skin and hair negatively charged at neutral pH, as previously reported.

However, nanosize, crystallinity, and zeta potential of polysaccharides are important structural characteristics able to influence the various properties of the polymers. These parameters, in fact, affecting pharmacokinetic, safeness, and effectiveness of the NPs with regards to skin and hair, may impact any pharmaceutical and cosmetic treatment.

Thus, the shape, size, and surface properties of any carrier, based on the use of nanoparticles, contribute in general to its ability to penetrate the tissues. In particular, some studies have demonstrated that because of the existing channels within the natural extracellular matrix (ECM), these NPs may easily diffuse into the skin [11]. This is also the reason why some polysaccharides,

under the form of micro-/nanofibers, are often used as delivery systems for both skin and hair.

Within the framework of nanostructured carrier for biomedical and cosmetic applications, a nanofiber generally refers to a fiber with a diameter less than 100 nanometers (nm). But it may be referred also to a micrometer range of around 1000nm, generally manufactured, for example, by electrospinning technology (Figure 5.4). Thanks to this technique, it is possible to produce large volumes of nanofibers with diverse molecules [6].

This methodology consists of a simple technique by which, controlling their physicochemical and biological characteristics, it is possible to make smart non-woven tissues. These tissues may be realized indifferently by nanofibers of pure polymers and/ or composite blends with synthetic polymers, to obtain tissues which have the properties that fulfill the requirement of specific biological applications. They, in fact, possess characteristics of safeness and effectiveness, because of their morphological similarities to the natural ECM (Figure 5.5). However, the nanofibers, used for medical scaffolds or cosmetic carriers, can be obtained from natural polymers, synthesized in biological systems by specific metabolic pathways. The fibers, therefore, can be synthesized from biopolyesters formed by or industrial processes with natural ingredients, such as polylactic acid (PLA) or poly(butylene succinate) (PBS) or produced by microorganisms like polyhydroxyalkanoates (PHAs) [6].

Unlike electrospinning, these natural polymers may be processed by melt extrusion to obtain traditional fibers through spinning by small orifices by high velocity stream of heated air (Figure 5.6). Grancaric et al. evidenced that the materials primarily used for textiles in the near future will be: polylactides (PLA), poly(hydroxy-alkanoates) (PHA), poly(hydroxybutyrate) (PHB), poly(glycolide) (PGA) and its blends, bio-polyester (bio-PES), bio-polyamide (bio-PA), thermoplastics based on casein (milk protein), and planted products: soy, kenaf, jute, silk, etc. [12]. In this sector, the complex between CN and phosphate ions was investigated as a highly efficient flame-retardant that could be used in technical textile applications [13].

In conclusion, biopolymers can provide an alternative to many petroleum-derived polymers, enabling the development of novel

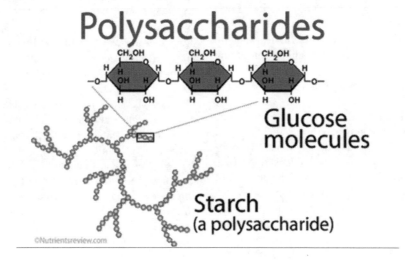

FIGURE 5.2   Polysaccharide structure consisting of glucose units (Courtesy of nutrientsreview.com).

FIGURE 5.3   Hair damage due to excessive combing.

applications sometimes with properties that exceed those of synthetic non-biodegradable polymers.

As previously reported, since the realization of global sustainability depends on producing materials and energy from renewable sources, there is an increasing need to develop bio-based polymers able to replace petroleum-based ones. Plant dry matter, known as ligno-cellulosic biomass, provides an ample renewable resource that can be processed to obtain added-value products, such as textiles, cosmetics, and commodity chemicals. Lignin, abundantly available from the large-scale renew-able biomass, is a 'natural thermoset polymer' consisting of phenyl propane basic units derived primarily from methoxylated hydroxycinnamyl alcohol building blocks of p-coumaryl alcohol, coniferyl alcohol, and sinaptyl alcohol (Figure 5.7). It therefore represents interesting bio-composite compounds with physicochemical properties that match or exceed those of petroleum-based ones, because it is obtained from cellulosic fibers as waste from the paper and pulp industries [14, 15]. Although structurally more complex than cellulose, the higher carbon content and lower oxygen content of lignin renders this macromolecule an attractive feedstock for the production of biofuels and chemicals [15]. Thus, for its branched and cross-linked poly-phenolic chains and its antioxidant, antibacterial and absorbing UV radiation properties, lignin is used as support material for pharmaceutical, food, and cosmetic applications [16–18].

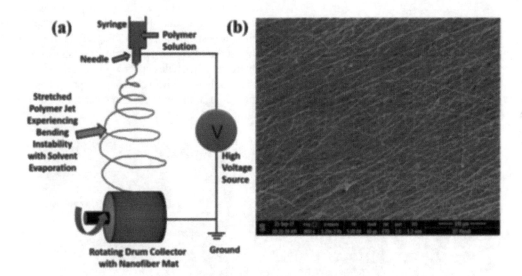

FIGURE 5.4   Electrospinning technology: (a) scheme of the instrumentation; (b) electrospun tissue.

FIGURE 5.5   Comparison between chitin nanofiber electrospun non-woven (left) and ECM (right).

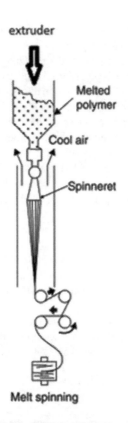

FIGURE 5.6   Melt spinning technology to obtain polymeric fibers.

the largest lignin stream by volume, it is not available commercially in isolated form in the same abundance, because black lignin liquor plays a key role as an internal energy recovery and for recovering the inorganic chemicals used in the pulp process [15]. However, its market share, dominated by the North America region followed the by central and eastern EU and China, is considerably increasing, owing to its organic nature [19–21]. Among the regions, Asia Pacific, excluding Japan, is estimated to grow chiefly in the key application of the lignin market, covering a concrete segment of lignin product varieties. Until today, the major consumption of lignin has been as a dispersant ingredient, followed by binder and adhesive applications, used principally for construction, mining, and animal feed. The availability of lignin-like solid is going to increase also because of the diffusion of biorefinery plants. The solid waste related to the production of levulinic acid from the well-fissed Arundo donax plant, shows similarities with some lignins [22], and was also considered as a reinforcement material for biopolyurethanes foam [23–25].

Additionally, a growing demand of centralized systems for dust removal, driven by the worldwide growth in air pollution, is likely to emerge as the major growth driver over the coming years [19]. Moreover, due to the development of advanced and sustainable technologies realized by significant investment in R&D by major manufacturers, it is increasing the extraction and use of sulphur-free lignin for producing lignin-based polymers and dust controllers, necessary to reduce air pollution and prevent pesticide pollution in agricultural systems [26]. Finally, due to the interesting characteristics shown from lignin-derived micro-/nanoparticles covered in their surface by electronegative charges, new innovative and widespread applications have been recovered in the medical and cosmetic field, as will be reported subsequently.

Lignin is a natural and renewable material available at an affordable price with a reported worldwide production of approximately 500 million tons/year as ligno-sulphonated lignin from paper production and 100 million tonnes/year of Kraft lignin [19, 20]. Although the Kraft method currently constitutes

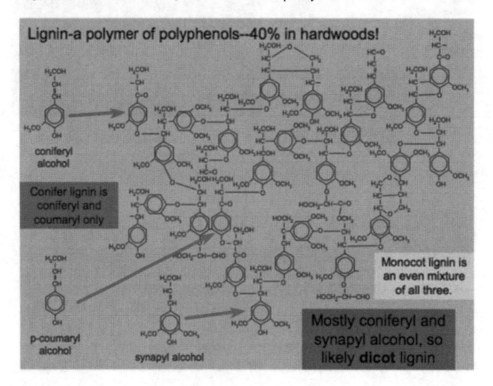

FIGURE 5.7   Complex lignin chemical structure.

5.3 Chitin and Chitosan from Fishery's By-Products

On the one hand, the synthetic polymers used in abundance from the past couple of decades have enhanced the comfort and quality of life, while on the other hand they have resulted in great waste problems of accumulation, because of their non-biodegradable nature and difficulty to recycle. Thus, the development of environmental-friendly and degradable biopolymers, derived from natural sources, has resulted in a worldwide production and application in many areas, generating a more sustainable growth and the transition to a green bio-economy [27].

Among the natural ingredients, chitin and its derived compounds may be classified as the most common and promising biopolymers, because they are obtained from the abundant and inexpensive renewable raw material, recovered by fishery's by-products, especially crustacean waste.

Chitin is a long-chain polymer of N-acetyl-glucosamine. It is a polysaccharide which, similar to cellulose and resembling the structural function of keratin in humans (Figure 5.8), renders a tough, protective covering or structural support of crustaceans, making up the cell wall of fungi and the exoskeleton of insects.

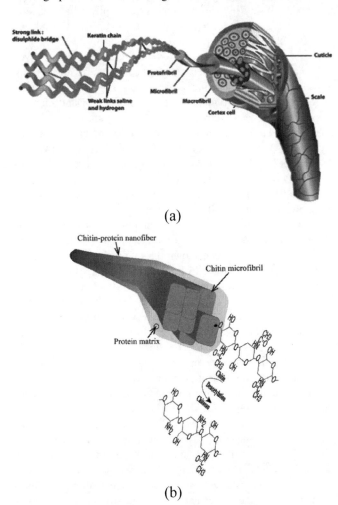

(a)

(b)

FIGURE 5.8 Similarity between (a) keratin, representing in humans the structural function, and (b) chitin, representing the structural function in a crustacean exoskeleton.

Unlike cellulose, chitin has a great commercial interest for its high content of nitrogen (around 7%). Moreover, it is a natural polymer which is nontoxic, biodegradable, biocompatible, and with moisturizing properties. For its hydrophilic nature, therefore, it resembles the hyaluronic acid activity, considering that it has adsorbing properties higher than those of synthetically substituted cellulose [28].

Chitosan, as a linear copolymer composed of N-acetyl-D-glucosamine and D-glucosamine units, is a chitin-derived compound obtained by alkaline treatment necessary to partially remove the acetyl groups. When the deacetylation is less than 60%, the polymer is considered chitin, while from 60% to 98% it is classified as chitosan. Both chitin and chitosan are naturally occurring substances which, as the cellulose fiber in plants, possess many of the same fiber properties. But, unlike plant fibers, they have the ability to significantly bind fat, acting like a 'sponge' at the level of skin and in the digestive tract, especially when present in their micro/nano size. Thus, size dimension, crystallinity, shape, and purity make both these polymers biochemically and bio-pharmacologically more effective when used in the medical and cosmetic field [29]. Just as an example, chitin at a dimension less than 40 microns has shown an evident anti-inflammatory effectiveness, releasing IL-10 cytokines, while with a dimension of 70 microns has a pro-inflammatory activity due to the release of the TNF-alpha and IL17 intra-cellular cascade. (Figure 5.9) [30].

Chitin and chitosan are linear semicrystalline polymers composed by macromolecules which, joined by hydrogen bonds, form their polymeric macro-structures. These nanofibrillar bundles are called micro-fibrils, and resemble the human muscle structure (Figure 5.10).

However, chitin is easily degraded by human and environmental enzymes, resembling also the keratin polymer, that represents

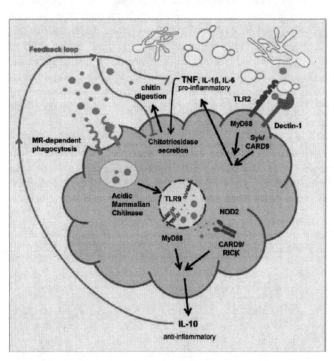

FIGURE 5.9 Pro-inflammatory and anti-inflammatory activity of chitin depends on its size (Courtesy of Wagener [30]).

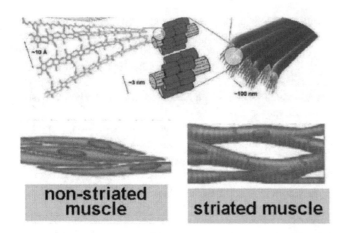

FIGURE 5.10 Complex chitin nanofibrils assembly resembling the human muscle structure.

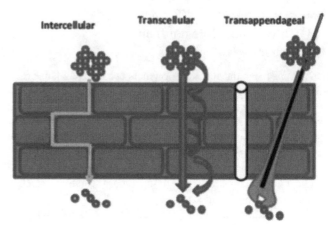

FIGURE 5.11 Intercellular, transcellular, and transappendageal skin penetration routes.

the key structural material making up hair, nails, and the outer layer of the outer human skin. In conclusion, each chitin microfibril is made by ordered crystallites and amorphous domains alternating along the fibril (Figure 5.10, top), so that its grade of crystallinity seems to represent an important structural characteristic, able to influence the final polymer properties [31]. It has a strong affinity with human cells, is nontoxic and its micro-/nanodimension reveals a large surface area/weight ratio with a consequent high porosity. Moreover, the capacity and ability of adsorbing sebum, heavy metals, dyes [32], and low molecular weight compounds, more than starch and other active substances, renders chitin useful to formulate frequent washing shampoos or emulsion for greasy hair or skin [33], and/or to realize protective formulations against the side effects of some precursors of hair color cosmetics (data not reported). Additionally, chitin and its derived compounds are found to significantly change the secondary structure of keratin in the stratum corneum, to increase its water content, enhance the cell membrane fluidity to various degrees, interacting with both skin lipids and proteins. In so doing, when used as carriers, these polymers disorganizing the skin lipid lamellae by the formation of larger aqueous pores and fluidization of the lipids' membrane, increase the penetration of the loaded active ingredients by intercellular and/or transcellular routes (Figure 5.11) [34].

Thus, for their multifunction activity, chitin and its derivatives have been used as biological carriers to stimulate the hair regrowth exercising an interesting stimulus for CD34-positive radicular-follicular (i.e. stem) cells [35], in tissue engineering and drug delivery applications, as well as for its wound healing and other elected biomedical applications [36].

5.4 Chitin Nanofibrils, Nano-Lignin, and Their Complexes

As previously reported, chitin nanofibrils showed interesting properties because of their high surface/volume ratio that renders the micro-/nano-molecule more efficacious than large size ones. The alpha- CN, obtained by a patented methodology [37], in fact, have a mean dimension of 240×7×5 nm with an aspect of needle-like particles, hierarchically organized like nanosized

fibers arranged in an antiparallel fashion (Figure 5.12) [38]. It is also characterized by a well-ordered crystalline phase with strong hydrogen bonding, having high-binding energy.

As a molecule it is insoluble in water, but it shows a high dispersing ability in water so that, for example, it induces both radical oxygen species (ROS) production and chitin-induced defense-related gene expression in some plants [39]. Also, in humans, the administration in vivo of chitin and chitin-derived micro/nano polymers, used by emulsions or dressing tissues, shows an mmune-potential effectiveness by the activation of macrophages and stimulation of NO production and chemiotaxis. Macrophage activation, in fact, results in increased metabolic activity, stimulating the secretion of growth factors, cytokines, and inflammatory mediators [40, 41], as shown also by a recent study of our group [42]. For this purpose, the anti-inflammatory and wound repair activity of chitin-lignin nanofibers have been analyzed on human keratinocytes treated by lipopolysaccharide(LPS) of P Aeruginosa to evaluate the expression of the beta-defensin 2 (hBD-2) and metalloproteinase 2 and 9 (MMP-2 and -9), involved in the mechanism of tissue regeneration. Thus, it has been shown that this polymer significantly reduced the pro-inflammatory cytokines induced by LPS, also modulating the expression of MMPs and hBD-2. It is interesting to underline that a previous study has been shown how CNs seem more effective as a cicatrizing agent when embedded into the skin-tissue and kept in situ for at least four days, than applied as a spray suspension or gel [43]. For this clinical study in traumatic wounds, the used non-woven tissue induced better epithelial differentiation and keratinization with a better trophism of the basal lamina. In conclusion, it has been reported that CN could be used by the spray form as a first-aid tool on bleeding abrasion; by gel to enhance physiological repair for areas with thin epidermal layer; and by tissue or other smart dressing to avoid scars appearing [43]. These results were successively confirmed by a recent clinical study that, realized by our group on first and second grade burned scars, used a non-woven tissue made by CN fibers bound to low quantity of nanostructured silver [44]. The study has shown that this biodegradable tissue was more effective to regenerate the skin affected by burns in a shorter time, compared to the traditional non-woven tissues, used normally in hospitals.

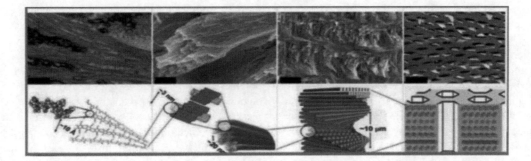

FIGURE 5.12 Hierarchical organization of chitin in exoskeleton of crustaceans (Courtesy of Nikolov et al. [38].

All these studies suggest that CNs, as well as its association with lignin, could improve their ability in their scaffold function.

It is also necessary to emphasize that chitin, sensed by the immune system of both plants and humans as a pathogen-associated molecular pattern (PAMP) through specific membrane-bound receptors, plays a key role in defense against pathogens [45]. Thus, since this polymer is expressed by microorganisms that are involved in many skin allergies, the keratinocyte-chitin interactions are considered also important in the regulation of epidermal immunity.

Regarding the physicochemical characteristics, it has been calculated that while chitin has a weight of 203.192,5 g/mole, CN have an average weight of $0,074 \cdot 10^6$ ng [46]. These insoluble nanocrystals also have the capacity to remain suspended in water solution for long periods. Moreover, each crystal having a high number of NH_2- groups on its high surface – and being consequently positively charged – CNs are able to complex different electronegative compounds, forming block polymeric micro/nanoparticles, for example, with hyaluronic acid and nano-lignin [47–49]. Thus, for its nanometric dimension leading to a high-surface-area-to volume ratio, CN interacts closely with a microbial membrane, displaying more significant antibacterial and antifungal effects than those of bulk counterparts. Moreover, these micro-/nanoparticles, capable also of entrapping active ingredients, enable a better diffusion through the skin layers, when covered by positive charges on their surface. On the other hand, when the nanoparticles are negatively charged, the active ingredients remain at the level of the outermost skin, acting as a deposit, as previously reported.

The micro-/nanoparticles of CN, lignin, and chitin-lignin complex, after deposition from diluted water suspensions, were investigated by Scanning Electron Microscopy (SEM) (Figure 5.13).

CN appeared as 'whiskers', having a nanometric thickness and a micrometric length. The spray-dried complex CN–lignin showed a completely different morphology, consisting of micrometric disks having a round or ellipsoidal shape (Figure 5.13), but the presence of a nanostructured system was evident in which the presence of both CN and lignin particles could be observed. This morphology can enable the spray-dried CN–lignin complex

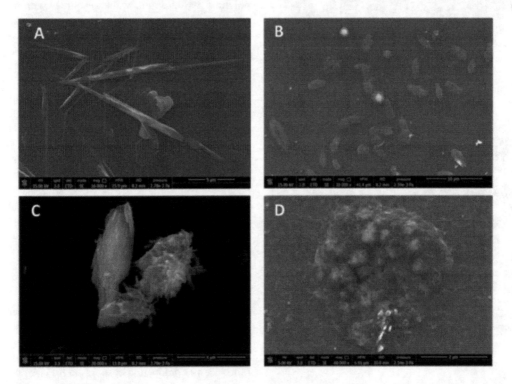

FIGURE 5.13 FE-SEM micrographs of (A) pure CN; (B) CN–lignin complexes; (C) and (D) CN–lignin complex at higher magnifications [49].

to be easily suspended in water to obtain flat micrometric nano-structured agglomerates that can deposit onto a surface and be used to modify its properties thanks to CN and lignin functionalities (anti-microbial and anti-oxidant respectively).

The original spray-dried CN–lignin powder consisted of almost spherical particles of micrometric dimensions, as can be observed in Figure 5.14.

The infrared ATR spectra (Figure 5.15) performed onto pure CN and lignin, as well as onto the spray-dried CN–lignin complex powder evidenced that the obtained spectrum of the latter showed the bands already observed in the previous spectra of CN and lignin, but slightly shifted because of the interactions occurring between CN and lignin.

Regarding the thermogravimetric behavior, this characterization can be interesting to investigate the thermal stability of compounds to predict their behavior in processing with polymers to obtain functional composites [50–52]. The behavior of CN and lignin was different, but they formed a high amount of carbonaceous residue during thermal decomposition. The amount of residue decreased significantly for CN–lignin complex. Moreover, the thermogravimetric trend of the complex did not include a mass loss at 40°C that was evident in the TGA thermogram of the pure lignin. Hence, lignin appeared more stable thanks to the formation of the complex with CN. However, the thermal stability of CN was slightly reduced due to the presence of lignin.

The performed characterization gave indications for a suitable application of CN–NL complexes as coating for material surfaces when deposited from water suspensions. All the investigated biopolymeric components (CN, lignin, and CN–lignin complexes) demonstrated good cytocompatibility with keratinocyte cells and mesenchymal stromal cells, and optimal concentrations were selected for each of them. These findings demonstrate that CN–lignin complexes are very promising for skin contact application, such as in biomedical, personal care, and cosmetic products.

5.5 Conclusion

In conclusion, bio-nanotechnology and the use of natural polysaccharides offer the potential for new eco-friendly industrial processes that require less energy and are based on renewable raw materials. With these natural raw materials, obtainable from industrial and agricultural biomass, such as Chitin and Lignin, it is possible to produce biodegradable nanocomposites which, being also characterized for their thermoplastic processing capability [49–53], help to preserve non-renewable resources, contributing to a sustainable development. These bio-nanocomposites, in fact, may be used to produce bio-plastic containers and many goods, such as non-woven tissues for bio-medical and cosmetic use, necessary to reduce one of the major environmental problems, as seen in the non-biodegradability of many fossil-derived products. Thus, preference is reserved for bio-based polymers [54], indicating them to be the most preferable compounds because of their biodegradability and compostability. According also to our in-progress studies [55], the best examples of biopolymers based on renewable resources seem to be represented by cellulose, lignin, chitin derivations, starch, polylactide (PLA), and polyhydroxyalkanoates (PHAs) used to make micro-/nanocomposites. However, it is necessary to underline that most biopolyester-based bio-composites, obtained from corn and sugarcane by microorganisms and used to make, for example, bioplastics, result in a biodegradable product but are not durable enough for long-term application. In any case, the durability of these compounds and composites would deserve more deepening. PLA has disadvantages such as an inherent brittleness, poor heat resistance, and low-melt strength, having on the one hand advantage of being available from renewable agriculture resources with a reduction in carbon dioxide emissions. On the other hand, PHAs, synthesized by microorganisms, currently result in relatively high costs, but with interesting mechanical and barrier properties that, together with their high compatibility with human blood and tissue, and interesting biodegradability also in soil and marine water, make them potentially useful for packaging, personal care, cosmetic, and biomedical applications. For this purpose, it is worth remembering that the monomer of poly(hydroxybutyrate) (PHB) is a normal metabolite of human blood which is easily absorbable, so that it can be used as a surgical implant [55].

Due to all these considerations, it is evident there is a need to reduce the production costs of both PHA and PHB polymers and to use reinforcing agents as nanocellulose or nanochitin for

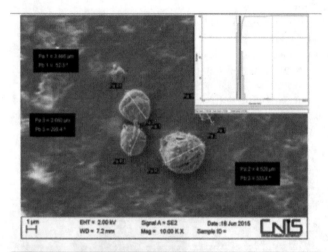

FIGURE 5.14 Morphology of the original spray-dried CN–lignin powder.

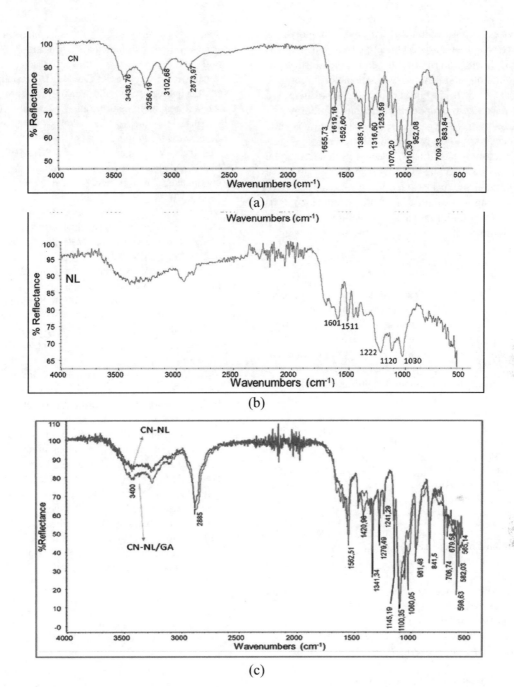

FIGURE 5.15 Infrared ATR spectra of (a) pure CN; (b) lignin; (c) CN–lignin complex and the same with entrapped glycyrrhetinic acid (GA) [49].

increasing the durability and sustainability of PLA during all the stages of its life cycle [56–59].

However, it is important to remember that due to their nontoxicity, high biodegradability, and hydrophilic nature, both chitin and chitosan have significant skin moisturizing activity, finding interesting position as active ingredients and smart carriers in cosmetic products [60, 61]. Moreover, since the bioadhesive properties that make them adhere to hard and soft tissues, these chitooligosaccharides are used not only for the skin but also in dentistry, orthopedics, ophthalmology, and surgical procedures [61]. Finally, due to the possibility of chemically modifying the amino and hydroxyl groups of their molecules and binding together selected electropositive with electronegative natural polymers, this could promote new biological activities,

modifying the mechanical properties of the designed nanocomposites and complexes. These innovative compounds could be useful for realizing the development of innovative and smart chitin and chitosan block polymeric nanoparticles able to load, carry, and deliver hydrophilic and hydrophobic drug and cosmetic active and skin-friendly ingredients into the different skin layers, meanwhile safeguarding the environmental eco-system [62–64].

Moreover, it is interesting to underline that, according to our group, recent technology, binding the active ingredients directly into fibers of non-woven tissues (as scaffolds), it has been possible to locally control the release of the active ingredients for the designed period (unpublished data). It is, in fact, possible to control the time-release of the selected bioactive molecules, by the adopted process of fabrication and the type of biopolymers used, the surface characteristics of which (i.e. wettability, chemistry, charge, roughness, and rigidity) depend upon the survival, adhesion, proliferation, and differentiation of the different type of cells. In conclusion, to design and produce innovative and smart products, such as the reported biodegradable non-woven tissues to be used as active carriers for medical and cosmetic use, it will be necessary to study in depth the synergistic effects between the natural polymers (fibers) selected (which have to be able to make a non-woven tissue ECM-like), and the cargo active ingredients. The final product, in fact, to be safe and effective has to maintain strength, flexibility, bio-stability, and biocompatibility of a natural scaffold, mimicking the natural human skin structure and environment.

REFERENCES

1. United Nations. (2019). *New Plastics Economy Global Commitment*, March 13. Ellen MacArthur Foundation.
2. Phoenix, D.A., and Ahmed, W., (eds.) (2014). *Nanobiotechnology*. Manchester, UK: One Central Press.
3. Morganti, P., Carezzi, F., Del Ciotto, P., Morganti, G., Nunziata, M.L., Gao, X.H., Chen, H.D., Tishenko, G., and Yudin, V.E. (2014). Chitin nanofibrils: A natural multifunctional polymer. In D.A. Phoenix, and W. Ahmed, *Nanobiotechnology* (pp. 1–22). Manchester,UK: One Central Press.
4. MacArthur, D.E., Samans, R., Waughray, D., Stuchrey, M.R. (2017). *The New Plastics Economy*. New York: Ellen MacArthur Foundation.
5. FAO (Food and Agriculture Organization of the United Nations). (2011). Global Initiative on Food Losses and Waste Reduction, Save Food. FAO Report, Rome, Italy.
6. Kamble, P., Sedarani, B., Majumdar, A., and Bhulkar S.C. (2017). Nanofiber based drug delivery systems for skin: A promising theraplastic approach. *Journal of Drug Delivery Science and Technology*, *41*, 124–133. doi:1016/J. jddst.2017.07.003.
7. Bathia, S. (2016). *Systems for Drug Delivery: Safety, Animal and Microbial Polysaccharides*. Basel, Switzerland: Springer. doi:10.1007/978-3-319-41926-8_3.
8. Atala, A., and Moony, D. (Eds.). (1997). *Synthetic Biodegradable Polymer Scaffolds*. New York: Birkhauser Bioscience.
9. Ruel-Gariepy, E., and Leroux, J.C. (2006). Chitosan: A natural polycationic with multiple applications. *ACS Symposium Series*, *934*, 243–259. doi:10.1021/bk-2006-0934-ch012.
10. Honary, S., and Zahir, F. (2013). Effect of zeta potential on the properties of nano-drug delivery systems: A review (part 1). *Tropical Journal of Pharmaceutical Research*, *12*(2), 255–264.
11. Barua, S., and Mitragori, S. 2014. Challenges associated with penetration of nanoparticles across cell and tissue barriers: A review of current status and future prospects. *Nano Today*, *9*(2), 223–243. doi:10.1016/jnantod2014.04.008.
12. Grancaric, A.M., and Jerkovic, I., and Tarbuk, A. (2013). Bioplastics in textile. *Polimeri*, *34*(1), 9–14.
13. Riehle, F., Hoenders, D., Guo, J., Eckert, A., Ifuku, S., and Walther, A. (2019). Sustainable chitin nanofibrils provide outstanding flame-retardant nanopapers. *Biomacromolecules*, *20*, 1098–1108.
14. Ten, E., and Vermerric, W. (2013). Functionalized polymers from lignocellulosic biomass: State of the art. *Polymers*, *25*(2), 600–642.
15. Rinaldi, R., Jastrzebski, R., Clough, M.T., Ralph, M., Kennema, M., Bruijnincx, P.C.A., and Weckhuysen, B.M. (2016). Paving the way for lignin valorisation: Recent advances in bioengineering, biorefinering and catalysis. *Angewandte Chemie International Edition*, *55*, 2–54.
16. Martinez, V., Mitjans, M., and Vinardell, M.P. (2012). Pharmacological applications of lignins and lignins related compounds: An overview. *Current Organic Chemistry*, *16*, 1863–1870.
17. Vinardell, M.P., and Mitjians, M. (2017). Lignins and their derivatives with beneficial effects on human health. *International Journal of Molecular Sciences*, *18*, 1–15. doi:10.3399/ijms18061219.
18. Beisl, S., Friedl, A., and Miltner, A. (2017). Lignin from micro-to nanosize: Applications. *International Journal of Molecular Sciences*, *18*, 1–24. doi:10.3390/ijms18112367.
19. Lignin Institute. (2019). Promoting future technologies for a multi-product conversion in environ-mentally cyclic processes where lignin is a major component. www.ili-lignin.com/aboutlignin.
20. FMI. (2019). Lignin: Waste market: Global industry analysis and opportunity assessment 2016–2026. Future Market Insight, London, UK. www.futuremarmetinsights.com/reports/lignin-waste-market.
21. Orbis Company. (2017). Lignin products global market size, sales data 2017–2028. Applications in animal feed industry, orbis research. April 20, Dallas, USA.
22. Licursi, D., Antonetti, C., Bernardini, J., Cinelli, P., Coltelli, M.B., Lazzeri, A., Martnelli, M., and Raspolli Galletti, A.M. (2015). *Industrial Crops and Products*, *76*, 1008–1024.
23. Bernardini, J., Licursi, D., Anguillesi, I., Cinelli, P., Coltelli, M.B., Antonetti, C., Galletti, A.M., and Lazzeri, A. (2017). Exploitation of arundo donax L. Hydrolysis residue for the green synthesis of flexible polyurethane foams. *Bioresources*, *12*(2), 3630–3655.
24. Bernardini, J., Cinelli, P., Anguillesi, I., Coltelli, M.B., and Lazzeri, A. (2015). Flexible polyurethane foams green production employing lignin or oxypropylated lignin. *European Polymer Journal*, *64*, 147–156.
25. Bernardini, J., Anguillesi, I., Coltelli, M.B., Cinelli, P., and Lazzeri, A. (2015). Optimizing the lignin based synthesis of flexible polyurethane foams employing reactive liquefying agents. *Polymer International*, *64*, 1235–1264. doi:10.1002/pi.4905.
26. Garrido-Herrera, F.J., Daza-Fernandez, I., Gonzales-Pradas, E., and Fernandez-Perez, M. (2009). Lignin-based formulations to prevent pesticides pollution. *Journal of Hazardous Materials*, *168*, 220–225.
27. Morganti, P., Chen, H.D., and Hong, L.Y. (2019). Green-bio-economy and bio-nanotechnology for a more sustainable environment. In P. Morganti (Ed,), *Bionanotechnology to Save the Environment. Plant and Fishery's Biomass as Alternative to Petrol* (pp. 29–59). Basel, Switzerland: Multidisciplinary Digital Publishing Institute (MDPI). doi:10.3390/books978-3-03842-693-6.
28. Kim, S.K. (2014). *Chitin and Chitosan Derivatives*. Boca Raton, FL: CRC Press/Taylor & Francis.

29. Alvarez, F.J. (2014). The effect of chitin size, shape, source and purification method on immune recognition. *Molecules*, *19*, 4433–4451. doi:10.3390/molecules19044433.

30. Wagener, J., Subbarao-Malireddi, R.K., Lenardon, M.D., Koberle, M., Vautier, S., McCallum, D.M., Biedermann, T., Shaller, M., Netea, M.G., Thirumala-Devi Kanneganti,Brown, G.D., and Brown A.J. (2014). Fungal chitin dampens inflammation through IL-10 induction mediated by NOD2 and TLR9 activation. *PLOS Pathogens*, *10*(4), e1004050. doi:10.1371/Journal.ppet.1004050.

31. Whathanaphanit, A., and Rujiravanit, R. (2010). Structural and biological activity of chitin nanofibrilss. In R. Ito, and Y Matsuo (Eds.), *Handbook of Carbohydrate Polymers: Development Properties and Applications* (pp. 535–554). New York: NOVA Publisher Inc.

32. Wan Ngah, W.S., Teong, L.C., and Hanafiah, M.A.K.M. (2011). Adsorbtion of dyes and heavy metal ions by chitosan composites: A review. *Carbohydrate Polymers*, *83*, 1446–1456.

33. Liaqat, F., and Eltem, R. (2018). Chitooligosaccharides: A comprehensive review. *Carbohydrate Polymers*, *184*, 243–259. doi:10.1016/jcarbpol.2017.12067.

34. Nawab, A., Rahim Khan, N., and Wong, T.W. (2016). Chitosan and its roles in transdermal drug delivery. In V.K. Thakur, and M.K. Thakur (Eds.), *Handbook of Sustainable Polymers. Processing and Applications* (pp. 557–586). Singapore: Pan Publishing Pte Ltd.

35. Biagini, G., Zizzi, A., Giantomassi, F., Orlando, F., Lucorini, G., Mattioli-Belmonte, M., Tucci, M.G., and Morganti, P. (2008). Cutaneous absorption of nanostructured chitin associated with natural synergistic molecules (lutein). *Journal of Applied Cosmetology*, *26*, 69–80.

36. Anitha, A., Sowmya, S., Sudheesh, K., Deepthi, S., Chennazhi, K.P., Ehrlich, H., Tsurkan, M., Jayakumar, R. (2014). Chitin and chitosan in selected biomedical applications. *Progress in Polymer Science*, *39*, 1644–1667. doi:10.1016/j.progpolymsci.2014.02.008.

37. Morganti, P. (2017). Preparation of chitin and dervatives thereof for cosmetc and therapeutic use. US 8,383,157 B2 (2013); EP 2 995 321B1.

38. Nicolov, S., Petrov, M., Lymperakis, L., Friak, M., Sachs, S., Fabritius, H., Raabe, D., and Neugebaur, J. (2010). Revealing the design principles of high-performance biological composites ab initio and multiscale simulations: The example of Lobster cuticle. *Advanced Materials*, *22*, 519–526.

39. Egusa, M., Matsui, H., Urakami, T., Okuda, S., Ifuku S., Nakagami, H., and Kaminaka, H. (2015). Chitin nanofiber elucidater the elicitor activity of polymeric chitin in plants. *Frontiers in Plant Sciences*, *6*, 1098. doi:103389/fpls.2015.01098.

40. Mori, T., Murakami, M., Okumura, M., Kadosawa, T., Uede, T., and Fujinaga, T. (2005). Mechanism of macrophage activation by chitin derivatives. *Journal of Veterinary Medical Science*, *67*(1), 51–56.

41. Bueter, C.L., Lee, C.K., Rathinam, V.A.K., Healy, G.J., Taron, C.H., Specht, C.A., and Levitz S.M. (2011). Chitosan but not chitin activates the inflammation by a mechanism dependent upon phagocytosis. *Journal of Biological Chemistry*, *286*(41), 33447–33455. doi:10.1074/jbcM111.274936.

42. Morganti, P., Fusco, A., Paoletti, I., Perfetto, B., Del Ciotto, P., Palombo, M., Chianese, A., Baroni, A., and Donnarumma, G. (2017). Antiinflammatory, immunomodulatory and tissue repair activity of human keratinocytes by green innovative nanocomposites. *Materials*, *10*, 843.

43. Muzzarelli, R.A.A., Morganti, P., Morganti, G., Palombo, P., Palombo, M., Biagini, G., Mattioli-Belmonte, M., Giantomassi, F., Orlandi, F., and Muzzarelli C. (2007). Chitin nanofibrils/chitosan glycolate composites as wound medicaments. *Carbohydrate Polymers*, *7*, 274–284. doi:10.1016/jcarbpol.2007.04.008.

44. Anniboletti, A., Palombo, M., Moroni, S., Bruno, A., Palombo, P., and Morganti P. (2019). Clinical activity of innovative non-woven tissues. In P. Morganti (Ed.), *Bionanotechnology to Save the Environment Plant and Fishery's Biomass as Alternative to Petrol* (pp. 340–360), Basel, Switzerland: MDPI. doi:10.3399/books978-3-03842-693-6.

45. Komi, D.E.A., Sharma, L., and De la Cruz, C.S. (2018). Chitin and its effects on inflammatory and immune responses. *Clinical Reviews in Allergy & Immunology*, *54*(2), 213–223. doi:10.1007/s12016-017-8600-0.

46. Muzzarelli, R.R.A. (2014). Private communication.

47. Morganti, P., Morganti, G., and Nunziata, M.L. (2019). Nanofibrils, a natural polymer from fishery waste: Nanoparticle and nanocomposite characteristics. In P. Morganti (Ed.), *Bionanotechnology to Save tihe Environment Plant and Fishery's Biomass as Alternative to Petrol* (pp. 60–81). Basel, Switzerland: MDPI.

48. Donnarumma, G., Perfetto, B., Baroni, A., Paoletti, I., Tufano, A.M., Del Ciotto, P., and Morganti, P. (2019). Biological activity of innovative polymeric nanoparticles and non-wooven tussue. In P. Morganti (Ed.), *Bionanotechnology to Save the Environment Plant and Fishery's Biomass as Alternative to Petrol* (pp. 321–339). Basel, Switzerland: MDPI.

49. Danti, S., Trombi, L., Fusco, A., Azimi, B., Lazzeri, A., Morganti, P., Coltelli, M.B., and Donnarumma, G. (2019). Chitin nanofibrils and nanolignin as functional agents in skin regeneration. *International Journal of Molecular Sciences*, *20*, 2669.

50. Coltelli, M.B., Cinelli. P., Gigante, V., Aliotta, L., Morganti, P., Panariello, L., and Lazzeri, A. (2019). Chitin nanofibrils in poly(lactic acid) (PLA) nanocomposites: Dispersion and thermo-mechanical properties. *International Journal of Molecular Sciences*, *20*, 504. doi:10.3390/ijms 20030504.

51. Coltelli, M.B., Gigante, V., Panariello, L., Aliotta, L., Morganti, P., Danti, S., Cinelli, P., and Lazzeri, A. (2019). Chitin nanofibrils in renewable materials for packaging and personal care applications. *Advanced Materials Letters*, *10*(6), 425–430.

52. Panariello, L., Coltelli, M.B., Buchignani, M., and Lazzeri, A. (2019). Chitosan and nano-structured chitin for biobased anti-microbial treatments onto cellulose based materials, *European Polymer Journal 113*, 328–339.

53. Muller, K., Zollfrank, C., Schmid, M. (2019). Natural polymers from biomass resources as feedstocks for thermoplastic materials. *Macromolecular Materials and Engineering*, *304*, 1800760. doi:10.1002/mame.201800760.

54. Alvarez-Chavez, C.R., Edwards, S., Rafael, M.E., and Geiser K. (2012). Sustainability of bio-based plastics: General comparative analysis and recommendations for improvement. *Journal of Cleaner Production*, *23*, 47–56.

55. Morganti, P., Morganti, G., Gagliardini, A. (2019). Repairing the skin as a fabric. *Goap Dermatology*, *1*(1), 1–9.

56. Bugnicourt, E., Cinelli, P., Lazzeri, A. and Alvarez, V. (2016). The main characteristics, properties, improvements and market data of polyhydroxyalvankanoates. In M. Kumar Thakur, and M. Kumari Thakur (Eds.), *Sustainable Polymers. Processing and Applications* (pp. 899–927). Singapore: Pan Stanford Publishing Ltd.

57. Nakagaiyo, A.N., Kanzawa, S., and Takagi, H. (2018). Polylactic acid reinforced with mixed cellulose and chitin nanofibers-effct of mixture ratio on the mechanical properties of composites. *Journal of Composiste Science*, 2, 36. doi:10.3390/jcs2020036.

58. Cinelli, P., Seggiani. M., Mallegni, M., Gigante, V., and Lazzeri, A. (2019). Processability and degradability of PHA-based composites in terrestrial environments. *International Journal of Molecular Sciences*, 20, 284. doi:10.3390/ijms20020284.

59. Aliotta, L., Gigante, V., Coltelli, M.B., Cinelli, P., and Lazzeri, A. (2019). Evaluation of mechanical and interfacial properties of bio-composites based on poly(lactic acid)with natural cellulose fibers. *International Journal of Molecular Sciences*, 20, 960. doi:10.3390/ijms20040960.

60. Morganti, P., Palombo, M., Tishchenko, G., Yudin, V.E., Guarneri, F., et al. (2014). Chitin-hyaluronan nanoparticles: A multifunctional carrier to deliver anti-aging active ingredients through the skin. *Cosmetics*, *1*(3), 140–158.

61. Morganti, P., and Coltelli, M.B. (2019). A new carrier for advanced cosmeceuticals. *Cosmetics*, *6*, 10. doi:10.3390/cosmetics6010010.

62. Sudha, P.N., Gomathi, T., Nasreen, K., Jayachandran Venkatesan, J., and Kim S.K. (2014). Recent advancements in research on chitin and chitosan derivatives for drug delivery application. In S.K. Kim (ed.), *Chitin and Chitosan Derivatives* (pp. 463–479). Boca Raton, FL: CRC Press.

63. Morganti, P. (2016). Use of chitin nanofibrils from bioass for an innovative bioeconomy. In J. Ebothe', and W. Ahmed (Eds.), *Nanofabrication Using Nanomaterials* (pp. 1–22). Manchester UK: One Central Press.

64. Morganti, P. (Ed.). (2019). *Bionanotechnology to Save the Environment. Plant and Fishery's Bio-mass as Alternative to Petrol*. Basel, Switzerland: MDPI Editorial Office.

6

Hydrogel Nanocomposites as an Advanced Material

Ankur H. Gor and Pragnesh N. Dave

CONTENTS

6.1 Introduction

After the end of the eighteenth century, a process of development was promoted in research and advancement of materials. The main aim of the development was centered toward adapting existing technology to find a new way of producing advanced materials (Fajardo et al. 2015). Different approaches have been modified for the advancement of materials and researchers are still focusing on how to develop advanced materials. In a growing world, it is essential to develop advanced materials with an idea of the cost consumption and efficiency of materials. The efficiency of materials is one of the prime goals of researchers and different strategies have been endorsed to increase it. The hydrogel technology with science nanotechnology holds great interest which continues to grow. The different approaches applied to the introduction of nano-scale technology with conventional hydrogel methods towards the engineering of advanced materials. The formation of nanocomposite hydrogels leads to the formation of mechanically strong hydrogels with which they are able to show parent characteristics of hydrogels. Hydrogels are mainly defined as polymers with three-dimensional cross-link networks. They have characteristics such as insolubility in water and the intelligence to consume a large amount of water. Hydrogels have properties similar to solids and liquids (Gong et al. 2006). A variety of physical and chemical techniques are used to construct the hydrogels. In 1894, the term 'hydrogel' was published as a term in literature (Bemmelen 1894). Poly (HEMA) polymer was prepared by DuPont's scientists with brittle, glassy, and hard characteristics in 1936. Due to these characteristics, it was not recognized as being of importance. Poly (HEMA) cross-linked hydrogel with elastic gel property was developed by Lim and Wichterle in 1960. After that, a number of hydrogels were prepared and this continues to this today (Wichterle and Lim 1960, Dave and Gor 2018). Hydrogels have received greater interest from researchers due to their promising application in vast areas from the last decade. Hydrogels are able to retain a large quantity of water due to the existence of hydrophilic groups such as $-OH$, $-CONH$, $-CONH_2$, $-SO_3H$. In recent times, hydrogel technology has been applied in diverse fields such as drug delivery, metal ion adsorption, dye removal, removal of antibiotics, protein drug delivery, antibacterial drug release for wound dressings, tissue engineering, and degradation of dyes (Li et al. 1993, Guilherme et al. 2007, Coviello et al. 2007, Matricardi et al. 2009, Ferris et al. 2013, Harikumar et al. 2013, Singh et al. 2017, Gor and Dave 2019). 'Polymer' is derived from Greek, 'poly' describes 'many' while 'meros' describes a unit of high molecular mass. Polymer exists as a high molar mass of material that is formed by the repeating of a monomer unit. The process in which a monomer unit combines to form polymer is called polymerization (Kaushik et al. 2016). Synthetic polymers have an extensive variety of applications due to their specific design, flexibility, economical feature, mechanical properties, and thermal stability but they carry limitations which may pose a hindrance (Sun et al. 2016, Sionkowska 2011). To overcome these hindrances, hydrogel nanocomposites have become an interesting area of research. Nanotechnology has gained much consideration from researchers, along with their efforts in improving and modifying existing technologies from the last decade. Nanotechnology is a branch of material science. A variety of methods have been used in the preparation of nanomaterials. Nanomaterial has a size from 1 to 100nm. Nano-sized materials have unique characteristics as compared to their bulk materials (Zheng et al. 2014). In the early stage of nanotechnology, nano-size particles of metals, metal oxide, and carbon types of materials formed but now it is possible to develop nano-size materials of diverse types of synthetic monomer, and also natural polymers (Bhushan 2012, Dai 2006). Different types of organic/inorganic nanocomposites contain the specific organic/inorganic constituent and it is possible to fabricate complex nanometer-scale structures. Since the last decade, nanocomposites of the polymer and clay have been widely studied and successfully established for a wide range of applications (Okada and Usuki 2006). The evolution of an inorganic with organic network was achieved by Haraguchi in the form of 'nanocomposite hydrogel' (Haraguchi and Takehisa 2002). The nanocomposite hydrogel exhibited exceptional properties in terms of swelling/shrinking, and the optical and mechanical makes them one of the alternatives to traditional hydrogels. In nanocomposite hydrogels, nanoparticles do not just improve the physical characteristics of hydrogels but act as a multifunctional crosslinker. Due to its superior characteristics and hybrid composition, nanocomposite hydrogel has received more attention when compared to traditional hydrogels. Varieties of chemical or physical crosslinking approaches are used to construct hybrid nanocomposites with nanomaterials and polymers. The stiffness of nanoparticles and softness, the flexibility of an organic polymer matrix, therefore mean that a nanocomposite hydrogel can exhibit novel or improved properties (Haraguchi and Takehisa 2002, Paranhos et al. 2007, Wang et al. 2010, Gaharwar et al. 2014). The hydrogel has been exploited as a reaction chamber for the synthesis and encapsulation of metal nanoparticles. The adopting hydrogel network as a reaction vessel provides certain advantages: first, the hydrogel three-dimensional network averts the clump of nanoparticles and provides a flexible device for utilization as a catalyst for drug delivery (Sahiner 2013). The enclosed metal particle inside the hydrogel network offers a versatile material. The use of hydrogel networks for nanoparticle preparation and utilization as a reactor is a novel concept and provides combinatory advantages over conventional systems. More essentially, a network of hydrogel is able to encapsulate the toxic metal and resolve environmental concerns, while also contributing better activity, long life of the catalyst, utilization of the catalyst in multiple reactions and a better trait of catalyst (Sahiner and Ozay 2011; Sahiner et al. 2011a). The different types of nanoparticles with the type of hydrogels based on the synthetic as well as natural polymer is a tactical new way of creating advanced composite materials with superior properties as compared to the traditional material. This paper looks forward to reviewing and analyzing recent trends in nanoparticles containing hydrogels with an objective of highlighting the new technologies used in the preparation of nanoparticle-containing nanocomposite hydrogels. Another objective of this paper is to scrutinize the outcomes of nanoparticles as additives on hydrogel development, properties, and applications. We concentrated on the different synthetic- or natural polymer-based 4−hydrogel/nanomaterial combinations and desired properties which are features of the final composite materials. The final objective is to recognize the present challenges in developing the foundation for new composites.

6.2 Synthesis Approaches of Hydrogel-Nanocomposites

In recent years, a number of hydrogel-nanoparticle composites have been constructed in a manner to make them advanced materials. Scientists are still focusing on developing nanocomposites in wide ranges. Different approaches are described below:

6.2.1 Hydrogel-Nanocomposite Formation by Suspension of Nanoparticles

This is the elementary method to construct a nanocomposite hydrogel in a suspension of nanoparticles. The optical responsive hydrogel-nanocomposites can be constructed by this method (Thoniyot et al. 2015). This method is most effective because a possible variation can be made with nanoparticles, monomer, and other additives are possible to obtain hydrogel with high-efficiency materials. A. L. Daniel-da-Silva et al. (Silva et al. 2013) reported K-carrageenan nanocomposite hydrogels with gold nanoparticles (Au NPs). Hydrogel nanocomposites prepared by using the suspension of Au nanoparticles with K-carrageenan polysaccharide as a natural monomer and KCl. Hydrogel composite was formed with a different content of Au NPs. Sozeri et al. (2013) synthesized polyacrylamide and nickel ferrite ($NiFe_2O_4$) nanoparticle-composite hydrogels. Nanocomposite hydrogels were prepared in suspension of $NiFe_2O_4$ with 10^{-4} M solution of pyrine. After that acrylamide, N, N, N′, N′- tetramethyl-ethylenediamine (TEMED), ammonium persulphate (APS), N, N methylene bis-acrylamide (MBA) as an initiator-crosslinker system was added. Huang et al. (2012)developed the graphene oxide(GO)/P(Acrylic acid-co-Acrylamide) (P(AA-co-AM) nanocomposites by using an in situ free radical solution technique. GO/P(AA-co-AM) was prepared by using graphite oxide, acrylic acid, MBA, APS. GO was prepared by using a modified hummers' method. In the nanoparticle-hydrogel composite, the content of GO varied from 0.05 to 0.50 wt.%. Shi et al. (2013) developed hydrogel-nanoparticle systems based on GO/sodium alginate/polyacrylamide. The fabrication of polysaccharide hydrogels carried with grapheme oxide and polymerization was done using a free radical polymerization mechanism. The dispersion of GO was prepared in water; after that, sodium alginate and acrylamide was used as a polysaccharide-monomer pair with other additives such as crosslinker and initiator added in the GO dispersion.

Zhang et al. (2014) prepared GO-fabricated carboxylmethyl cellulose (CMC) sodium nanocomposite hydrogel. Hydrogel nanocomposites were constructed using GO, CMC sodium, acrylamide, and crosslinker-initiator systems. The mechanical robustness of hydrogel was improved due to the hydrogen bond formation between the grapheme oxide (GO) sheets and polyacrylamide chains. Sun et al. (2015) prepared a magnetic adsorbent from xylan, polyacrylic acid, Fe_3O_4 magnetic nanoparticles. Xylan-based magnetic nanocomposites were prepared with directly prepared Fe_3O_4 nanoparticle added in the xylan suspension and N, N methylenebisacrylamide (MBA) as a crosslinker; APS and sodium thiosulphate as an initiator-accelerator system was added. Shi et al. (2015) prepared near infrared responsive (NIR) nanosheet-hydrogel nanocomposites with a temperature responsive monomer NIPA (N-isopropylacrylamide). Hydrogel nanocomposites were prepared with a graphene oxide sheet as a fabricating additive and NIPA as a monomer with MBA as a crosslinker and APS-TEMED as an accelerator-initiator. The fabrication and NIR light-responsive property behavior of hydrogel is schematically illustrated in Figure 6.1.

6.2.2 Hydrogel-Nanocomposite Formation by Nanoparticles as a Crosslinker

This interesting method involves the formation of nanocomposite hydrogel by using the specific function group present on the surface of the nanoparticles as crosslinking points. Thomas et al. (2011) prepared clay nanoparticle hydrogel by using a laponite-RD, NIPA, KPS and TEMED composition. The amount of

clay nanoparticle used 0.2 gm and 0.4 gm in 10 ml of deionized water. Laponite nanoparticles provide the crosslinking points with the NIPA and are effective at extracting cationic from an aqueous solution. Campbell et al. (2013) reported hydrogel based on the dextran functionalized with aldehyde and surface of superparamagnetic nanoparticles functionalized with hydrazine-functionalized poly(N-isopropylacrylamide) (pNIPA). The functionalization of pNIPA with hydrazine play a crucial role because after that it peptized and forms a strong interaction with a supermagentic iron oxide nanoparticle. The composite hydrogels were fabricated via condensation of both additives in a different assembly and therapeutic agents added in both reactive phases. The resulting nanocomposites displayed special characteristics with their degradability. Wang et al. (2010) reported nanocomposites by using the semiconductor metal nanoparticles as a photo initiator and as a cross-linker to form nanocomposites with the utility of clay nano sheets and polymerization carried out under sunlight in place of the photon source. Nanocomposites with zinc oxide (Zno) were prepared in sunlight while composite based on the other semiconductor was prepared in artificial

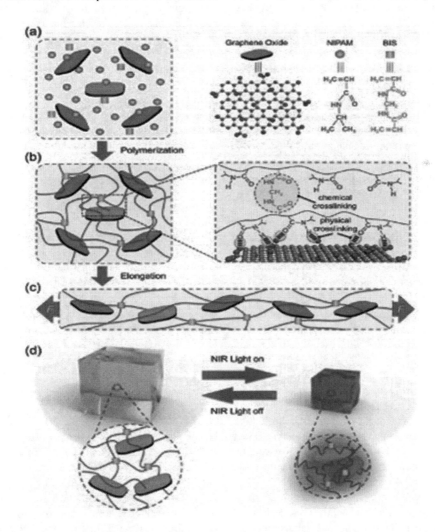

FIGURE 6.1 Schematic illustration of the fabrication process and performance mechanism of the proposed PNIPAM-GO nanocomposite hydrogels. (a) GO nanosheets are homogeneously dispersed in the monomer solution. (b) The PNIPAM-GO nanocomposite hydrogels are formed by both chemical and physical cross-linking, in which the PNIPAM chains are chemically cross-linked by BIS, and the hydrogen bond interactions between GO nanosheets and PNIPAM chains result in the physical cross-linking (c, d) The PNIPAM-GO hydrogels exhibit ultrahigh tensibility (c) and reversible NIR lightresponsive property (d). (Adapted with permission from Shi et al. 2015. Copyright (2015) American Chemical Society.)

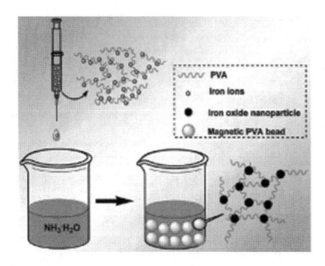

FIGURE 6.2 Schematic Illustration of Preparing Magnetic PVA (mPVA) Gel Beads by OnePot Strategy. (Adapted with permission from Zhou et al. 2012).Copyright (2012) American Chemical Society.)

sunlight. All semiconductor nano-composite hydrogel exhibited better mechanical properties.

Zhou et al. developed iron oxide cross-linked hydrogels magnetic poly(vinyl alcohol) (PVA) gel beads. Gel beads were formed by the drop wise introduction of a mixture of iron salts and PVA solution in alkaline medium. Mainly iron oxide nanoparticles (Fe_3O_4) were fabricated material with a uniform size and excellent magnetic response. After the freeze thawing treatment, the drug-loading level of the beads was high. In Figure 6.2, a schematic illustration of the preparation of magnetic PVA (mPVA) gel beads by the one-pot strategy described. The drug release from magnetic beads can be modified by adjusting the applied magnetic field, the amount of Fe_3O_4 nanoparticle and temperature (Zhou et al. 2012).

Erne et al. (Berkum et al. 2015) prepared remnant magnetic nanocomposites. Structured hydrogel by using acrylic acid, hydroxylethyl acrylate, diethyleneglycol diacrylate, cobalt ferrite nanoparticle dispersion and V-50 initiator. The polymerization of gels was carried out at 80°C. The amount of cobalt ferrite nanoparticles wass added at 28 mg per ml. Formed nanocomposites partly crosslinked with cobalt ferrite nanoparticles and remnant magnetization performance attributed to nanoparticles. They developed the nanocomposites with exceptional remnant magnetization during swelling and shrinking. The remnant magnetization of gels changed during swelling. First, there was change in the average orientation of magnetic dipolar structures and it enhanced the remnant field but more swelling resulted in a lower amount of magnetic nanoparticles. Thus, the diminution of the remanent field was observed. The change in thickness was observed during swelling and shrinking behavior. In swelling, gels were less thick than shrinking due to enhancement in the remnant field.

G. De Filpo et al. reported hydrogels based on natural polysaccharide and titanium dioxide for the application of cleaning and disinfection of parchment. Development of nanocomposites was carried out by using titanium dioxide, surfactant, and gellan gum. TiO_2 dispersion was prepared in water and placed in ultrasonic probe. Surfactant was added in the TiO_2 dispersion for better dispersion. Gellan gum and TiO_2 nanocomposites were

prepared by keeping the same amount of solvent (water) with a variation of content of gellan gum /TiO_2 dispersion (Filpo et al. 2015) Zhao et al. 2012 developed nanoparticle-crosslinked hydrogels by using AA and N, N dimethylacrylamide (DMAA). Titanium crosslinked gels prepared in this way of changing the concentration of acrylic acid in the total amount of monomer present in the hydrogels and the amount of TiO_2 constant. The crosslinking formation in hydrogel network is due to hydrogen bonds between the carboxyl group and hydroxyl present in the hydrogel systems (Zhao et al. 2012).

Makoto et al. reported nano silica-crosslinked nanocomposite gels. Nanocomposites were formed by the reaction between functionalized silanol and copolymer and copolymer poly[N-hydroxyethyl acrylamide-co-(3-methacryloxy propyltrimethoxy silane)] by free radical polymerization. The fabrication of hydrogels with inorganic nanoparticles as a crosslinker provides the mechanical robustness of hydrogels (Makoto et al. 2015) (Figure 6.3).

P. Liu et al. prepared covalently crosslinked clay nanoparticles with a copolymer of acrylamide and acrylic acid. Selective adsorption of metal from the mixture of metal ion solution displayed by clay nanoparticles incorporated in a hydrogel composite. The modified clay nanoparticles were used as a sole crosslinker. Modification of clay nanoparticles were done by the acidification method and obtained nanorods employed in the co-polymerization of polymer. The application of nanocomposite hydrogel as an adsorbent, and the component group present in the network structure, play a crucial role. The component groups present in the hydrogels were adjusted by variation in the concentration of acrylic acid and acrylamide. Nanocomposite hydrogels have an excellent selective adsorption capacity of Cu^{2+} and Pb^{2+} in the mixture of the heavy metal ions solutions including Pb^{2+}, Cd^{2+}, Zn^{2+}, Cu^{2+}, and Ni^{2+} and excellent reusability (Liu et al. 2015). Nanoparticle as a crosslinker in the hydrogel formation has been used in the last few years but now researchers are focusing on the modification of the component of the nanoparticle surface then using it in the formation of hydrogels. Garcia-Astrain et al. (2015) reported the formation nanocomposites using TiO_2 as a crosslinker and crosslinking took place between furan-modified pigskin gelatin and maleimide-coated TiO_2. The functionalization of TiO_2 by using a bifunctional dopamine-maleimide linker was employed. Furan-modified gelatin and functionalized TiO_2 mixed and hydrogel were formed by the Diels–Alder 'click' reaction (Garcia-Astrain et al. (2015)).

6.2.3 Hydrogel-Nanocomposite Formation by the Freeze-Thawing Method

Nanocomposites can be formed by using the chemical and physical crosslinked methods. In physical crosslinking methods, a nanocomposite network was constructed by the entanglement of polymer chains with nanoparticles in the hydrogels (Song et al. 2010). Kokabi et al. reported the nanocomposites of clay and synthetic polymer: A series of nanocomposites of polyvinyl alcohol containing 0–10 wt % of the organically modified montmorillonite clay was prepared. Modification of clay was carried out by using CTAB. Nanocomposites contain the amount of PVA 15 wt.% and 0, 2, 5, 7, and 10 wt % of modified clay. Clay nanocomposites were prepared by subsequent freeze-drying processes.

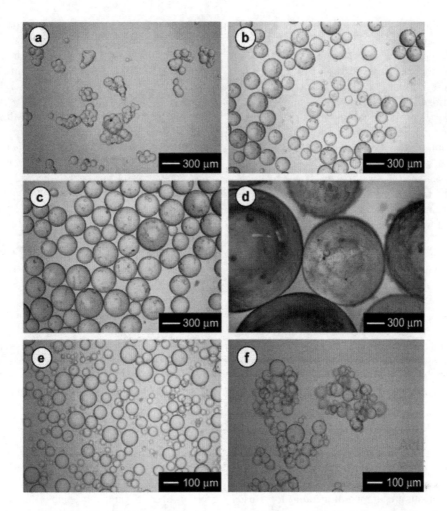

FIGURE 6.3 Optical microscopic images of the hybrid hydrogel particles prepared in various suspension media. (a) $[pSiHm_{100}] = 5$ wt%, $[SiNP] = 5$ wt%, Silicone oil-100 CS without APS, (b) $[pSiHm_{100}] = 5$ wt%, $[SiNP] = 2.5$ wt%, Silicone oil-100 CS with APS, (c) $[pSiHm_{100}] = 5$ wt%, $[SiNP] = 5$ wt%, Silicone oil-100 CS with APS, (d) $[pSiHm_{100}] = 5$ wt%, $[SiNP] = 5$ wt%, Silicone oil-50 CS with APS. (e) $[pSiHm_{100}] = 5$ wt%, $[SiNP] = 5$ wt%, Silicone oil-300 CS with APS, (f) $[pSiHm_{100}] = 5$ wt%, $[SiNP] = 5$ wt%, Silicone oil-100 CS with MAPTS. (Adapted from Makoto et al. 2015. Copyright (2015), with permission from Elsevier.)

In clay hydrogel nanocomposites, it is observed that swelling is inversely proportional to the amount of clay nanoparticles. Clay hydrogel nanocomposites obey the diffusion-controlled mechanism (Kokabi et al. 2007).

Ibrahim and El-Naggar et al. reported that freezing-thawing cycles assisted the entanglement of polymer chains and crosslinking through weak bonds between the hydroxyl group of PVA chains and oxyanions on the surface of montmorillonite. Typically, these types of nanocomposites are able to undergo sol-gel transition, so they exhibit thermo-responsive characteristics. Construction of nanocomposites with clay nanoparticles leads to thermal stability and it can be also increased by increasing the amount of clay nanoparticles. In short, the freeze-thaw process is a green method and does not allow the use of any chemical additives such as crosslinkers. The freezing- thawing method can be used for nanocomposite hydrogel with electron beam irradiation (Ibrahim and El-Naggar 2012).

Nanocomposites prepared by physical methods such as freezing-thawing cycles have certain advantages over conventional methods. Nanocomposites prepared by freezing-thaw nanocomposites have a physical crosslinked structure. Therefore, there

is no need to use unhygienic chemicals. Due to that, the biocompatibility of materials can be achieved and nanocomposites formed by this method are mechanically strong (Sirousazar et al. 2012b). While in another strategies, M. Kokabi et al. reported physical crosslinked hydrogels based on the poly vinyl alcohol and Na-montmorillonite. A series of samples consisting of 15 wt.% PVA with Na-montmorillonite 0, 3, 6, 9, 12, and 15 wt. %. Na-MMT content in the nanocomposite samples is proportional to the hardness of the samples. The hardness increased with the increasing amount of Na-MMT due to the clay nanoparticles acting as a crosslinker that enhances the hardness of the material (Kokabi et al. 2007).

Parparițaa et al. (2014) reported on the nanocomposite hydrogels based on the chitosan/PVA/MMT. Two types of hydrogels formed, firstly prepared PVA, chitosan-based gels and clay dispersion prepared in acid acetic solution; after that, the dispersion was added in the chitosan solution. The blend was obtained by the mixing of PVA solution and chitosan/MMT clay dispersion. Finally, the mixtures of PVA/CS and PVA/MMT were cast onto glass petri dishes. Three repeated cycles of freeze and thaw were carried out to keep the solutions of the blends at -20°C for 12 hours and then maintaining them at room temperature for the same time (Parpariţaa et al. 2014).

Shi et al. 2016 prepared graphene-based nanocomposites by a freeze-thaw method. Inorganic nanoparticles such as clay and CNT act as crosslinkers and enhance the mechanical robustness of nanocomposite hydrogels. The PVA/GO nanocomposites were synthesized by a freeze-thaw technique and after γ-ray irradiation. Irradiation leads to either crosslinking or a random chain scission of polymer. Homogenous with denser hydrogel based on PVA and graphene was formed. The irradiation dose also effected the structure of hydrogel and the water content in the hydrogel. The hydrogel was prepared with a high irradiation dose, and the pore of hydrogels became smaller. The irradiation dose had an effect on the amount of water in the PVA/GO hydrogel. It is clear that the water of PVA/GO hydrogel significantly decreases with the cumulative irradiation dose. The γ-ray irradiation dose also had an effect on the mechanical robustness of hydrogel that leads to improvement in the mechanical robustness of the hydrogel observed. The reduced GO has a larger surface area and surface activity that provides the additional crosslinking points in the network. Increasing the crosslink density enhances the mechanical robustness of hydrogels. The γ-ray irradiation dose also has an effect on the tribological characteristic of hydrogel nanocomposites (Shi et al. 2016) (Figure 6.4).

6.2.4 Hydrogel-Nanoparticle Formation by the in situ Method

Forming metal nanoparticles by the use of the three-dimensional structure of hydrogel is a new technique. The pre-shaped dried hydrogels are kept in a metal ion solution to load hydrogel with metal ions. Regarding n situ preparation of metal nanoparticles in the hydrogel network, the size of a hydrogel network does not influence the metal particle size but the capacity of a hydrogel network to retain the metal in the network is important to obtain the specific size of the metal nanoparticle.

The metal ion retaining-capacity of hydrogels is dependent upon a number of factors such as immersing time, the presence of functional component in it, and characteristic hydrogels. Crosslinking in the hydrogel network and the medium of metal ion solution also affect it. The different composition of hydrogels with their metal nanocomposites is illustrated in Figure 6.5 (A) & (B) (Mekewi and Darwish 2015). It can be observed that the color of the metal nanoparticle-hydrogel complex varies for different metals and different composites of hydrogels. Once metal ions are loaded into the hydrogel network, it is necessary to remove the surface embedded or unbound metal ions. Therefore, hydrogels are rinsed to eliminate released or physiosorbed metal ions. The cleaned metal ion-filled hydrogels are immersed in media of a reducing agent. Different reducing agents such NaBH$_4$, NH$_3$, H$_2$, N$_2$H$_4$, NaOH, Na$_2$S, and citrate are used to reduce the size of metal ions.

The scheme of silver metal nanoparticle-hydrogel preparation is illustrated in Figure 6.6. First, bare hydrogels are prepared by using the free radical mechanism. Prepared hydrogel is immersed into the AgNo$_3$ solution. While immersing bare hydrogel in the silver nitrate solution, silver ion is adsorbed into the hydrogel network. Silver ion filled hydrogel acts as a vessel for the synthesis of nanoparticles and the silver-loaded hydrogel is placed in the aqueous extract of A. indica. Extract of A. Indica contains a number of reducing components, therefore it reduces silver ion particles to silver nanoparticles. After some time, silver ions start to reduce and convert into the nanosize of particles. Different types of reducing agents are available in nature. In the last few years, nanoparticles have been prepared by using leaf extracts, polysaccharides, and extract of resins. They have become an alternative to the chemical reducing agent (Jayaramudu et al. 2013).

S. Pourbeyram et al. prepared copper nanocomposites by using acrylamide and acrylic acid as monomers. The free radical

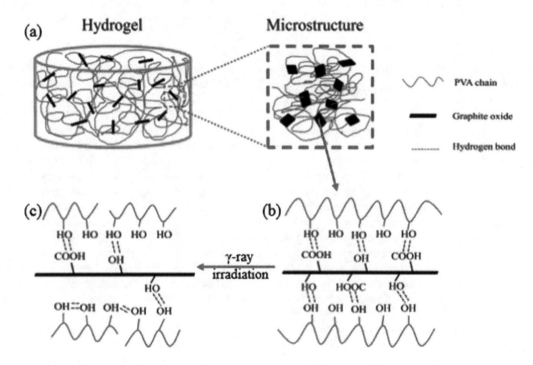

FIGURE 6.4 Proposed structure of PVA/GO hydrogels. (Adapted with permission from Shi et al. 2016. Copyright (2016) American Chemical Society.)

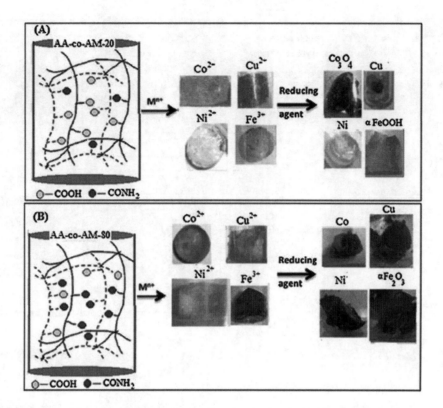

FIGURE 6.5 Schematic representation of metal nanoparticles (M: Co, Cu, Ni, and Fe) embedded in (A) AA-co-AM-20 and (B) AA-co-AM-80 hydrogels with digital camera images and (C) XRD patterns of M@P(AA-co-AM) nanocomposites: (a) Co@AA-co-AM-20, (b) Co@AA-co-AM80, (c) Cu@AA-co-AM-20, (d) Cu@AA-co-AM-80, (e) Ni@AA-co-AM-80, (f) Fe@AA-coAM-20 and (g) Fe@AA-co-AM-80. (Adapted from Mekewi and Darwish 2015. Copyright (2015), with permission from Elsevier.)

polymerization mechanism used to construct nanocomposites and hydrogels were placed in the solution of copper chloride. The color of nanocomposites mainly depends on the nature of the metal ion filled in the hydrogel network and temperature plays an important role. As described in Figure 6.7 (A), digital photographs of copper filled hydrogels with blue color. These metal ions filled hydrogels reducing into their nanoparticles by using a hydrazine solution as a reducing agent. As described in Figure 6.7(B), we can see metal nanoparticles in a hydrogel network. The color of nanocomposites altered from a blue to wine red color after treatment with a reducing agent (Pourbeyram et al. 2014).

In Figure 6.8(A), while the size distribution profile of nanocomposite hydrogel is obtained by DLS, it is indicated that copper nanoparticles were almost homogeneous, with an average particle size of approximately 20nm. In Figure 6.8(B), the SEM image exhibited that copper nanocomposite hydrogel homogenous structure and copper nanoparticles were well distributed over the porous surface of the hydrogel network. H. Ahmad et al. developed nanocomposites by using P(NIPA), poly(methacrylic acid), and Fe_3O_4. Sequential polymerization was used to form P(NIPA) and pH-sensitive crosslinked poly(methacrylic acid). Meanwhile, microspheres P(NIPA-MBA)/P(methacrylic acid-ethylene glycol dimethacrylate) IPN hydrogel microspheres were prepared by sequential emulsion copolymerization. Microsphere hydrogel was placed in the solution of $FeCl_3 \cdot 6H_2O$ and $FeSO_4$. Metal iron oxide reduced by the ammonium hydroxide solution was used as a reducing agent. Then the mixture was cooled at 273 K and iron doped IPN hydrogel microspheres frequently washed by applying a magnetic field (Ahmad et al. 2014).

Spasojevic et al et al. formed the hydrogel based on poly(N-isopropylacrylamide/itaconic acid) with silver nanoparticles. The base monomer was N-isopropylacrylamide, comonomer was itaconic acid, and hydrogels were prepared by using free radical polymerization. The prepared hydrogels were placed in the 1.0×10^{-2} mol/dm³ of $AgNO_3$ solution. The reduction of metal ion-loaded hydrogel composites was carried out by using gamma irradiation Spasojevic et al. 2015). Bajpai et al. prepared polysaccharide and vinyl monomer-based nanocomposite hydrogels. A semi-ipn network of sodium polyacylate was prepared in the existence of gum acacia by using a radical polymerization mechanism. Radical polymerization was carried out with a crosslinker and initiator. The prepared hydrogel equilibrated in pre-concentrated solution silver nitrate solution for a period of 12 h. The reduction of silver metal ion was carried out by using the clove extract for a period of 12 h. The nanocomposite hydrogel become brownish after the complete reduction of nanoparticles (Bajpai and Kumari 2015). Namazi et al. reported hydrogels based on the carboxylmethyl cellulose and double layered hydroxides prepared by an intercalate co-precipitation method. In this method, the aqueous solution of mixed metal ions solution is dropped into an alkali-CMC solution. Then, the precipitation of LDH layers is observed and hydrogel formation takes place by the intercalating of CMC chains into the LDH. LDH mainly acts as an inorganic crosslinker and anionic CMC transforms into a gel form. Silver nanoparticles were synthesized in the hydrogel network using in situ method. 1.0 g of dried hydrogel swelled in distilled water for 48 hrs. After that it is transferred into the solution of silver nitrate solution and allowed to reach an equilibrium state. Silver

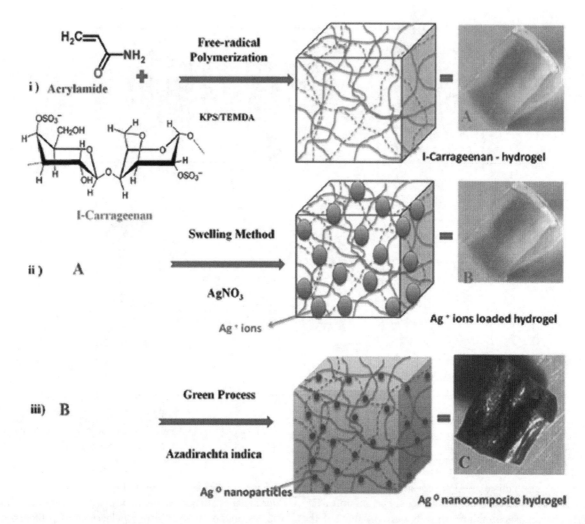

FIGURE 6.6 Biodegradable P(IC-AM) silver nanocomposite hydrogel preparative schematic illustration in three steps. (i) The fabrication of P(IC-AM) hydrogels via free-radical reaction; (ii) the preparation of Ag+ ions-loaded hydrogels via swelling method and (iii) the synthesis of Ag0 nanocomposite P(IC-AM) hydrogels via green process (The Ag0 nanoparticles were prepared by reducing AgNO₃ with A. indica in the P(IC-AM) hydrogels network). (Adapted from Jayaramudu et al. 2013. Copyright (2013), with permission from Elsevier.)

ion-loaded hydrogel composite centrifuged and critical washing is required to remove any free silver ions. The immobilization of Ag NPs in network of hydrogel is as result of strong localization of particles within existing groups in a network. CMC contain a number of carboxyl and hydroxyl groups, therefore it is used as a capping and reducing agent in the synthesis of silver NPs (Yadollahi et al. 2015). Narayanan et al. (2014) edeveloped silver

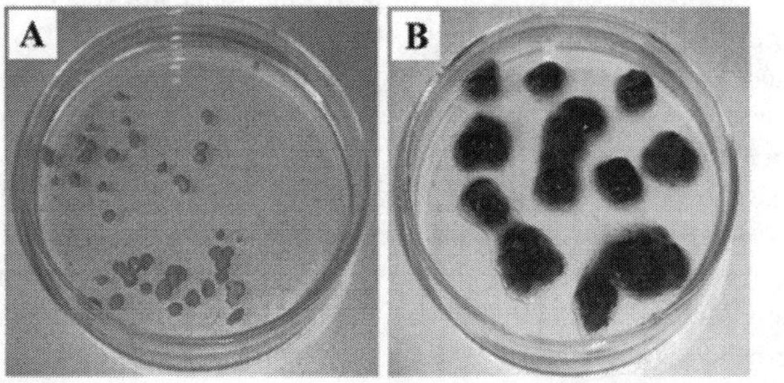

(Ag) nanoparticle-hydrogel nanocomposites by using amidodiol. The role of amidodiol is versatile in the synthesis, it can act as a physical crosslinker and reducing agent. Amidodiol was formed by the reaction amine with butyrolactone using hexamethylenediamine. Hydrogel nanocomposites were prepared by using AA, silver nitrate, ammoniumper sulphate, and amidodiol acts as a cross-linking-cum-reducing agent. The dual property of amidodiol in the formation of hydrogels is due to the existence of hydroxyl groups and two amide groups in it. These components are responsible for the comprehensive hydrogen bonding that enhances the crosslinking density and reducing agent characteristic Narayanan et al. (2014).

6.3 Applications of Nanocomposite Hydrogel

6.3.1 Nanocomposite Hydrogel for Wastewater Treatment

The availability of pure water is a severe environmental problem. Hydrogels as an adsorbent have a significant impact on it but the incorporation of nanoparticle in the hydrogels network

FIGURE 6.7 Digital photos of (A) hydrogel loaded by copper ions and (B) hydrogel–copper nanocomposite. (Adapted from Pourbeyram et al. 2014. Copyright (2014), with permission from Elsevier.)

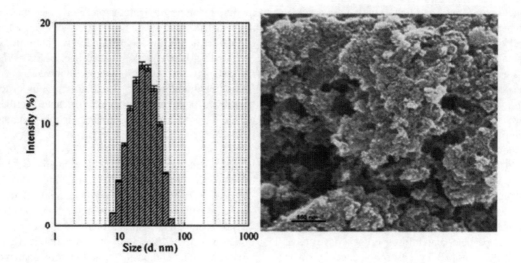

FIGURE 6.8 (A) Size distribution profile and (B) scanning electron microscopy of hydrogel–copper nanocomposite. (Adapted from Pourbeyram et al. 2014. Copyright (2014), with permission from Elsevier.)

has extraordinary properties. Thus, nanocomposite hydrogels have shown good promise as a character for water treatment. Different types of magnetic hydrogels have been reported for the treatment of wastewater in literature (Zhou et al. 2014, Tang et al. 2014). Zheng et al. (2014) reported an effective approach to form nanocomposite hydrogels by using metal oxide as photo initiator. A magnetic nanocomposite hydrogels was synthesized for the abatement of Lanthanum (La)(III). Nanocomposites were prepared from ZnO nanoparticles, Fe_3O_4 magnetic nanoparticles, Clay-NS, N,N-Dimethylacrylamide (DMAA), 2-Acrylamido-2-methylpropane sulfonic acid (AMPSNa), and photo polymerisation took place when mixture was subjected to the UV radiation, as illustrated in Figure 6.9 (A). Under UV radiation, electrons and holes were generated by zinc oxide nanoparticles. In Figure

6.9(B) & (C) it is shown that nanocomposite hydrogels have a porous structure and monomers polymerized by a photo initiator. This hydrogel is able to abate La(III) ions selectively from the mixtures of other ions. Selective removal of La(III) ions occurred because La(III) ions have an affinity towards carboxylic groups The hydrogel has good reuse ability because its adsorption ability remains after six cycles of adsorption–desorption. The magnetic properties of hydrogel make its collection easy after completion of the adsorption operation (Zheng et al. 2014).

Sahiner et al. 2010 have reported amidoximated microgels by using poly(methacrylic-co-acrylonitrile microgels. First, poly(methacrylic-co-acrylonitrile) microgels were prepared and amidoximation carried out to convert the nitrile groups into the amidoxime groups. Cobalt-iron magnetic NPs were prepared by

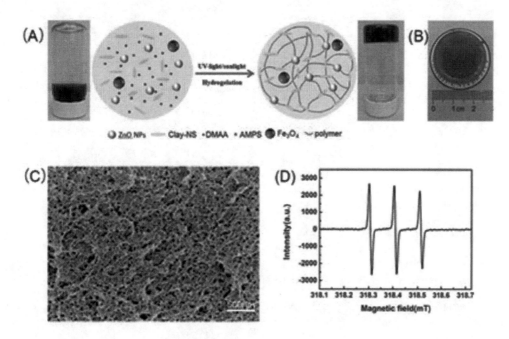

FIGURE 6.9 (A) Schematic preparation of magnetic nanocomposite hydrogel via ZnO-initiated polymerization. (B) Photograph of a self-standing hydrogel. (C) SEM image of a magnetic nanocomposite hydrogel after supercritical drying. (D) The EPR spectrum of the precursor comprising ZnO and DMAA under UV irradiation. (Adapted with permission from Wang et al. 2014. Copyright (2014) American Chemical Society.)

the in situ reduction of Co (II) and Fe (II) with a microgel technique. Amoxidated-nanocomposite hydrogels were established to be a good adsorbent for the elimination of metal ions, herbicides, and organic dyes (Ajmal et al. 2015a). Mittal et al. have reported the fabrication of hydrogel-based polysaccharide with nano SiO_2. The development of poly (acrylic acid-co-acrylamide) grafted onto gum karaya hydrogel and was fabricated with nano silica for remediation of methylene blue (MB). The results of their finding indicated that after addition of nano SiO_2 into the hydrogel, the monolayer abatement capacity of nanocomposites improved the hydrogels. By using the biodegradable polysaccharide in graft-polymer, hydrogel could be an efficient adsorbent of methylene (Mittal et al. 2015). Wheat xylan/poly(acrylic acid) nanocomposites containing Fe_3O_4NPs have been prepared from wheat straw xylan and Fe_3O_4NPs, and it has been reported that magnetic nanoparticle-containing hydrogels have the ability to eliminate 90% of MB by adsorption mechanism. The nanoparticle-containing hydrogel showed a good adsorption property due to the characteristics of the network and macroporous structure (Boruah et al. 2014). Samandari et al. prepared magnetic gelatin beads by using gelatin and it is entrapped in the carboxylic acid functionalized MWCNTs (multi-walled carbon nanotube) and physically bonded Fe_3O_4NPs nanoparticles. A simple co-precipitation method was employed to form iron oxide nanoparticles and gelatin beads were formed by an emulsification method. Magnetic hydrogel is intelligent at eliminating both cationic and anionic organic dyes. Magnetic beads showed slightly higher removal efficiency of anionic dye (96%) as compared to cationic dye (76%) (Samandari et al. 2017). M. Ghaemy et al. prepared the novel magnetic hydrogel beads by using

modified starch, vinyl imidazole, vinyl alcohol, and iron oxide nanoparticle. Modification of starch was carried out and graft copolymer of modified starch and polyvinyl imidazole prepared then CMS-g-PVI/PVA/Fe_3O_4 hydrogel beads were prepared by the instantaneous gelation method, followed by chemical cross-linking by glutaraldehyde. The highly porous structure and distribution of similar pores means that magnetic beads could be a good absorbent for dyes and organic dyes. Magnetic hydrogel beads showed a maximum sorption of Cu^{2+} ions is 83.60 mgg^{-1} and for CV dye is 91.58 mgg^{-1}. The experimental adsorption isotherm data were fitted with a Langmuir model and magnetic hydrogel showed extraordinary reusability (Pour and Ghaemy 2015). M.M.E. Breky et al. used in situ intercalative polymerization to form TiO_2/Poly (acrylamide–styrene sodium sulfonate) in the presence of TiO_2 nanoparticles as an inorganic filler, cross-linked by using MBA, and polymerization carried out by using γ-radiation. The composite exhibited good abatement efficiency for the radioactive elements. The maximum experimental abatement capacities for Cs^+, Co^{2+}, and Eu^{3+}were found to be 120, 100.9, and 85.7 mg/g respectively (Borai et al. 2015). In Table 6.1, nanocomposite hydrogel and their application in waste-water treatment are given.

6.3.2 Nanocomposite Hydrogels as Catalyst

Metal nanoparticles can be employed as a catalyst in different chemical paths, while nanocomposite hydrogel can be used as a catalyst. Hydrogels have their own characteristics like soft nature, flexibility, and presence of metal nanoparticle that makes their use as a catalyst in a different reaction. In addition, the

TABLE 6.1

Nanocomposite Hydrogels and Their Application for the Removal of Pollutants with Capacity and Isotherm

Nanocomposites	Pollutant	Removal capacity (mg/g)	Isotherm	References
TiO_2 nanotube/reduced graphene oxide	Ciprofloxacin	181.8	L	Zhuang et al. (2015)
Graphene oxide – Fe_3O_4	Ciprofloxacin	769.23	L	Pourjavadi et al. (2015)
Starch/poly vinylalcohol/fumarate-alumoxane	Ammonium	19.01	L	Shahrooie et al. (2015)
TETA-NBC	Hg$^+$	407.9	F	Varghese et al. (2015)
CD-NBC		292.1	F	
Graphene oxide/ APTMACl	AR88	1140.2	L	Dong and Wang (2016b)
APTMACl / Fe_3O_4La	Fluoride	136.78	S	Dong and Wang (2016a)
Cellulose/clay nanocomposite	Methylene blue	782.9	L	Peng et al. (2016)
Gum karaya/Poly acrylamide/ Nickel sulphide NPs	Rhodamine 6G	1244.71	L	Kumar et al. (2016)
Nitrogen doped carbon hydrogel/FeMg layered double hydroxide	Lead	344.8	L	Ling et al. (2016)
Graphene oxide/ Chitosan/ Poly(acrylic acid)	Lead	138.89	L	Medina et al. (2016)
Gum xanthan/Fe_3O_4	Methyl violet	642	L	Mittal et al. (2016)
Poly(acrylamide-co-itaconic acid)/MWCNTs	Lead	101.01	L	Mohammadinezhad et al. (2017)
Alginate/graphene double-network	Cu2+ and $Cr_2O_7^{2-}$	169.5 and 72.5	L	Zhuang et al. (2015)
Graphene oxide/AMPS/Poly acrylamide	Methylene blue	714.29	L	Pourjavadi et al. (2016)
Modified Gum Tragacanth/Graphene Oxide	Pb(II)	142.50	L	Sahraei and Ghaemy (2017)
	Cd(II)	112.50		
	Ag(I)	132.12		

L = Langmuir, S = Sips, F = Freundlich.

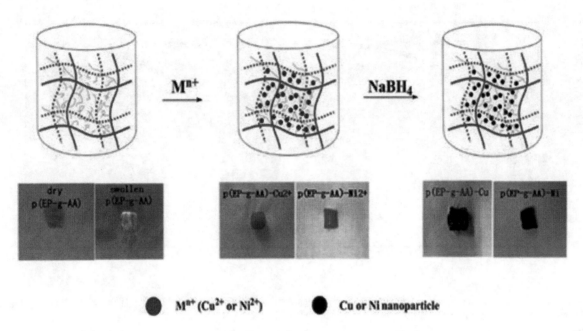

FIGURE 6.10 The preparation process of p(EP-g-AA)-M. (Adapted from Li et al. 2017. Copyright (2017), with permission from Elsevier.)

stimuli responsive property of hydrogels also provides benefits in the catalyst system. For example, certain reactions intervene at a particular temperature – there is no need to change the temperature of the whole reactor. Therefore, by just designing a smart temperature-responsive hydrogel–M catalyst system, all the necessary conditions can be delivered. This may be very significant for industrial application. In addition to its versatility, hydrogel–M catalyst systems are also superior in terms of their catalytic activity, storage capability, and the ability to host even toxic metal catalysts. For in the case of environmental interest, hydrogel adopted to form toxic catalysts can be safely utilized because of the polymeric network contributing as a flexible shelter (Seven and Sahiner 2013).

The production of hydrogen is a very crucial reaction because it has been considered as environmentally friendly and a clean energy source. Hydrogen gas is recognized as a substitute to traditional energy sources. Therefore, nowaday's research has been assigned to hydrogen production by using different metal hydrides. Among them sodium boron hydride is more frequently used because of its non-toxicity, high stability, hydrogen storage capacity, and non-flammable characteristic (Seven and Sahiner 2013, Sahiner et al. 2011b, 2011c, Sahiner and Ozay 2011a). Generally, the hydrolysis reaction is accomplished in water and a catalyst is used. Sahiner et al. have reported poly(acrylamide-co-vinyl sulfonic acid)/Co and Ni metal nanoparticle composite and they evolved their application in the hydrogen production. The consequence of both nanoparticles on hydrogen production evolved with sodium borohydride (Seven and Sahiner 2013). Sahiner et al. have reported the micro poly(3-sulfopropyl methacrylate) hydrogel for in situ preparation of Co(II) and Ni(II) metal nanoparticles and hydrogen generation from sodium borohydride by a hydrolysis process. The results of their finding indicated that in nanocomposites, the production rate of hydrogen is faster than that of macro hydrogel. The reuse results indicated that the micro nanocomposite hydrogel catalyst could be used in five cycles with a 100 % conversion rate (Turhan et al. 2013). Q.

Li have reported novel ploy (Enteromorpha-g-acrylic acid) synthesized and used nanocomposites for hydrogen production in the hydrogel network by utilizing sodium borohydride as reducing agents. The schematic presentation of the preparation processes of hydrogel and their metal complexes are illustrated in Figure 6.10. It was observed that the difference in the color of hydrogel metal complex and nanoparticle contain hydrogels of both nickel and cobalt. Differences in color of both hydrogels are due to the composition and metal ion properties of nanocomposite hydrogel. Nanoparticle-containing hydrogels have been utilized in the formation of hydrogen by hydrolysis of sodium borohydride (Li et al. 2017).

The mechanism of hydrogen formation by adopting sodium borohydride is shown in Figure 6.11. The studies performed to evaluate the effect of the type of metal ion and concentration of catalyst on hydrogen production and the produced volumes of hydrogen versus time for different catalysts are demonstrated in Figure 6.12.

It was found that copper containing a nanocomposite catalyst exhibited lower catalytic efficiency as compared to nickel containing a nanocomposite catalyst (Zhou et al. 2017). From an environmental point of view, nitro compound is seen as the most toxic pollutant and increased worries in relation to the plant and animal kingdom. Nitro compounds are formed in many industrial operations and are expelled into the environment. Therefore, converting hazardous pollutants into their less harmful or harmless

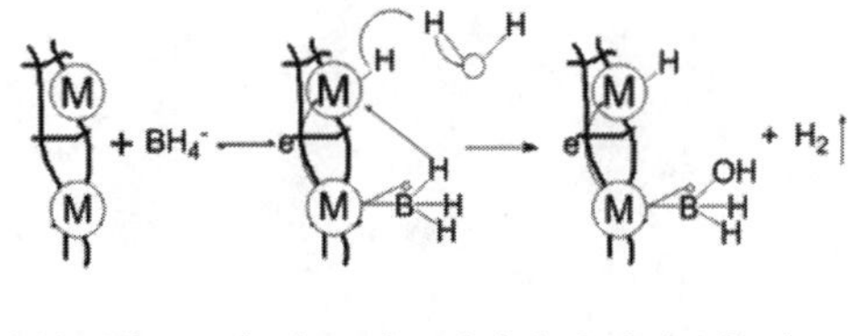

FIGURE 6.11 The mechanism of catalytic hydrolysis of hydrogen production for NaBH4. (Adapted from Li et al. 2017. Copyright (2017), with permission from Elsevier.)

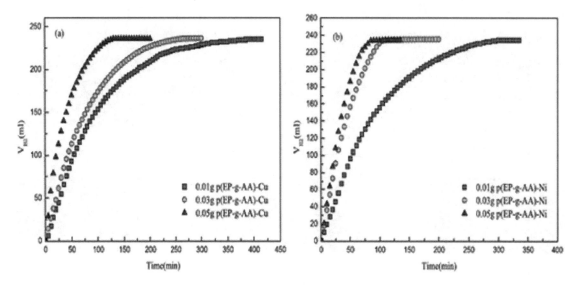

FIGURE 6.12 The produced volume of hydrogen versus time for different amounts of metal nanoparticles ((a) for p(EP-g-AA)-Cu,(b) for p(EP-g-AA)-Ni). (Adapted from Li et al. 2017. Copyright (2017), with permission from Elsevier.)

species is a challenge. Converting nitro compound containing an organic component into amine-substituted aromatic compounds is more useful. Therefore, to obtain amino compounds from nitro compounds is vital, and different types of metal nanocatalysts such as Co (Cobalt), Ni (Nickel), Au (Gold), Ru (Ruthenium), Ag (Silver), Cu (Copper), Pt (platinum), Fe (Iron), TiO_2 (Titanium dioxide), and so on, are used in the presence of a reducing agent (Sahiner 2013). Sarkar et al. reported how they crosslinked pectin-stabilized exfoliated titanium dioxidenano sheet-reinforced silver nanoparticles for the reduction of p-nitrophenol. In the hydrogen gaining of the nitro aromatic compound, different parameters have their own influence on it. The full hydrogen gaining of 4-NP could be completed within 16 s using a 5 mg nanocomposite catalyst. The excellent catalyst efficiency is due to the cumulative effect of crosslinked amylopectin and the in situ fabrication of Ag NPs.

Reduction of niro phenol was completed with 16 s with freshly prepared sodium borohydride solution. In Figure 6.13(a), it is shown that the quick color variation of nitro compound changed and the shift of wavelength was observed. The stability of catalyst

studies is seen by the repeated use after each cycle: the catalyst was easily reused more than 15 times (Mandal et al. 2017).

Sharmaet al. have reported biopolymer-containing hydrogel nanocomposites. Chitosan-based non-toxic biocompatible copper nanoparticles containing nanocomposite hydrogels were prepared by an in situ free radical polymerization using acrylamide as a monomer unit. Chitosan polysaccharide-based hydrogel nanocomposites were used as a catalyst in the reduction of the 4-nitrophenol into 4-aminophenol (Sharma et al. 2017). El Fadl et al. prepared chitosan–poly(AA–co-AA) based hydrogels formed by γ-radiation and used as reaction vessels to form iron oxide metal nanoparticles. Finally, formed magnetic nanocomposite hydrogel have been used to convert 4-nitrophenol into 4-aminophenol (El Fadl 2016).

Ajmal et al. prepared micro-sized hydrogels by using suspension polymerization of a zwitterionic monomer 2–(methacryloyloxy) ethyl]dimethyl (3–sulfopropyl) ammonium hydroxide (SBMA). The developed micro-gels used as reactor for the formation of Ni NPS. Nanoparticle containing hydrogel is used as catalytic agent for the reduction of 4–nitroaniline, 2–nitrophenol

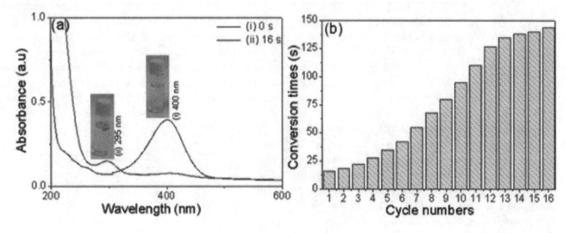

FIGURE 6.13 (a) UV-Vis spectra of p-nitro phenol reduction (i) before treatment (ii) 16 s after treatment of cl-AP/exf.LT-AgNPs, (b) Stability of the cl-AP/exf.LT-AgNPs during 16 cycles for p-nitro phenol reduction, (Adapted with permission from Sarkar et al. 2017. Copyright (2017) American Chemical Society.)

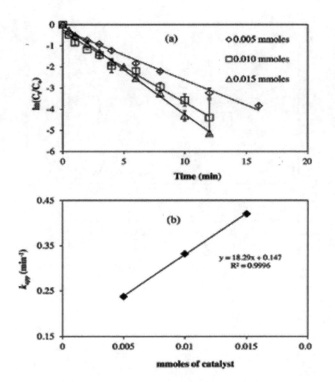

FIGURE 6.14 (a) Plots of ln (Ct/Co) against time with different amounts of p(SBMA)-Ni cata-lyst (b) Increase in k_{app} with the increase in amount of catalyst. Reaction conditions:0.01 M 4-NP = 50 ml, 0.756 g NaBH4, 30°C, 750 rpm. (Adapted from Ajmal et al. 2015a). Copyright (2017), with permission from Elsevier.)

(2-NP) and 4–nitrophenol (4–NP). Nickel nanoparticle-hydrogel nanocomposites exhibited an excellent catalytic trait. The dose of the catalyst is a significant variable that has an effect on the reduction of the reductive compound. The effect of the amount of catalyst on the rate of reaction and apparent rate constants are presented in Figure 6.14 (Ajmal et al. 2015a).

The amount of catalyst also affects the reduction rate of 4-nitri phenol assessed by inspecting the rate of reaction with three different amount of catalyst (.005, 0.010 and 0.015 mmoles of Ni). The rate of reaction increased with increasing the amount of catalyst. This provides additional information for controlling the rate of reduction (Ajmal et al. 2015b).

6.3.3 Nanocomposite Hydrogel for Drug Delivery

Nanocomposite hydrogels are a new strategic product, which have a structured network and contain nanoparticles with other additives. In the last decade, varieties of nanocomposites have been applied in the pharmaceutical and biomedical fields. The broad application range of nanocomposite hydrogels is due to their biodegradable and biocompatible characteristics. Nanocomposite hydrogels are known as sustainable drug delivery agents because they allow the release of a drug at a specific site and in a reduced sustainable manner (Hamidi et al. 2008). Nanocomposites could be a suitable vehicle for the load and transportation of a biologically active component. There are many aspects such as the component of nanocomposite hydrogel, and the swelling ratio that has an effect on the release of the biological active molecule. Nanocomposite hydrogels are able to deliver local drug delivery and stimuli-responsive drug delivery.

G. R. Bardajee et al. used iron oxide nanoparticles to form novel super-magnetic nanocomposite hydrogel of poly((2-dimethylamino)ethyl methacrylate) onto salep by employing a graft copolymerization technique. Fe_3O_4NPS prepared in situ methods. Nanocomposite hydrogel exhibited the stimuli-responsive swelling behavior with respect to pH, temperature, and the applied magnetic field. In vitro drug discharge examined at different pH, the highest amount of discharge of deferasirox was found at neutral pH. Nanocomposite gel exhibited a higher amount of drug delivery in the nonappearance of the outside magnetic field due to the presence of magnetic field nanoparticles forming a compact structure and hence the delivery of the drug was reduced (Bardajee et al. 2014).

T. Hoare et al. prepared injectable combined magnetic hydrogels with thermosensitive microgels. NIPAM-Hzd was produced via copolymerization of NIPAM and acrylic acid, followed by EDC-mediated coupling of a large overabundance of adipic acid dihydrazide. Injectable magnetic gel was produced by using PNIPAM-Hzd, PEG-SPIONs, FITC-dextran, and 8 wt % microgels. Magnetic particles were covalently linked to the hydrogel network structure, and fabricated by reacting aldehyde-functionalized dextran with hydrazide-functionalized PNIPAM-coated SPIONs. The resulting hydrogel nanocomposite exhibited enhanced a mechanical strength and an ability to release deliver pulsatile releases of a drug in an alternating magnetic field (Campbell et al. 2015).

Singh et al. formed the injectable hydrogel prepared from PAEU,LDH-hGH. To avoid the diffusion of growth hormone, new protein delivery systems were developed by the self-assembly and intercalation of a new protein delivery system by the self-assembly and intercalation of 2D-LDHNPs and hGH. The injectable nanocomposite hydrogel was prepared by dispersing PAEU and LDH–hGH. The hydrogel pH was controlled at 7 and it was stabilized at 2°C overnight. The LDH–hGH-filled hydrogels exhibited a disintegrate release and a controlled release for 120 hrs in vivo and 312 hrs in vitro (Singh et al. 2015). F. Song et al. reported how hybrid nanocomposite hydrogels were developed by using tyramine-modified HA/graphene oxide. Nanocomposite hydrogels were prepared through an in-situ crosslinking initiated by horseradish peroxidase. Mainly GO act as a physical crosslinker that enhance the mechanical robustness of hydrogel and displayed the pH swelling behavior. The drug release property of hydrogel is carried out with rhodamine B as a model drug. The hydrogel exhibited the prolonged release of rhodamine B at a lower pH due to GO acting as a crosslinker and the increase in crosslinking density reducing the release of the drug, as compared to parent HA gels (Song et al. 2015). Tao et al. developed paclitaxel (PTX)/MPEG-PDLLA nanoparticles in a hydrogel nanocomposite. PTX-loaded NPs were formed with MPEG-PDLLA through an emulsion/solvent evaporation approach. Nanoparticle-containing hydrogel nanocomposite was prepared through a self-assembly route. Nanocomposite hydrogel was released through a sustained and controllable manner. Therefore, this drug delivery device has potential application in dealing with residual glioma (Tao et al. 2017). Guilherme et al. reported nanocomposites of modified starch and Co-doped zinc ferrite nanoparticles ($CoZnFeO_4$) for the application of a magnetic field-responsive drug delivery. Hydrogel nanocomposite was prepared in the suspension of

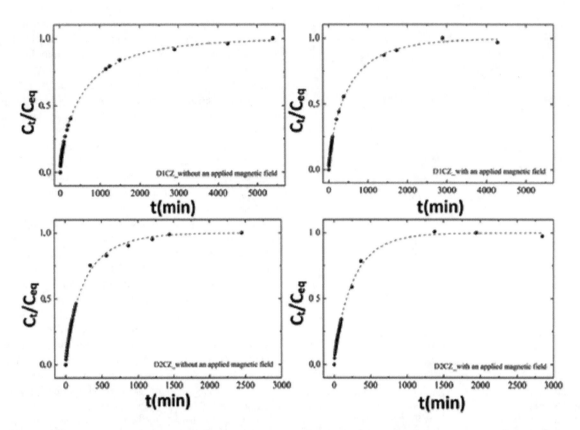

FIGURE 6.15 Time-dependent release curves of prednisolone from D2CZ and D1CZ hydrogel nanocomposites into a PBS buffer solution of pH 7.4 at a temperature of 37°C, without and with an applied magnetic field. (Adapted from Guilherme et al. 2015. Copyright (2014), with permission from Elsevier.)

pre-prepared $CoZnFeO_4NPs$ with vinylated starch and N',N'-dimethylacrylamide via an ultrasound-assisted radical cross-linking/polymerization reaction. Prednisolone was selected as a model drug for a drug release study. In Figure 6.15, the time-dependent release curves of a prednisolone applied magnetic field are shown (Guilherme et al. 2015).

The results of their finding indicated that an applied magnetic field triggered the release of prednisolone from both composite of hydrogels (Guilherme et al. 2015). Yadollahi et al. established the use of chitosan/Zno-based beads for the drug delivery device. Chitosan/ZnO-nanocomposite hydrogel was prepared by using chitosan, Zinc nitrate, and sodium tripolyphosphate as an ionic crosslinker. Nanocomposite hydrogel was prepared with a different content of nanoparticles. Ibuprofen was selected as a model drug for drug delivery. They conducted studies to evaluate the amount of nanoparticle on the release behavior of gels (Yadollahi et al. 2016). The drug release behavior of chito/Zno with different content of nanoparticle is illustrated in Figure 6.16.

It was observed that the amount of Zno increased the release of the drug and became prolonged because the incorporation of zinc oxide into beads leads to a smooth network formation (Yadollahi et al. 2016). The release of drugs as well as the loading of the drug is also important for drug delivery properties. Nanoparticles also have an impact on the drug delivery. H. Hamidian et al. reported how starch-g-poly(ethylene phthalate)/Fe_3O_4 hydrogel nanocomposite was prepared using poly(ethylene phthalate) grafted onto starch. To inspect the drug delivery property, the experiment goes throughby using tungstophosphoric acid. The loading of tungstenphosphoric can be changed by a different nanoparticle

amount. In this case, increasing the amount of Fe_3O_4 nanoparticle drops out drug loading (Hamidian and Tavakoli 2016). Mandal et al. reported how gold nanoparticle incorporated polysaccharide-based hydrogel formed by using CMC crosslinked with poly (methacrylic acid). Nanocomposite hydrogels have been used in the drug delivery of diclofenac sodium and diltiazem hydrochloride. Nanocomposite hydrogel creates the interaction site for the drug to load in the network character. Nanocomposite

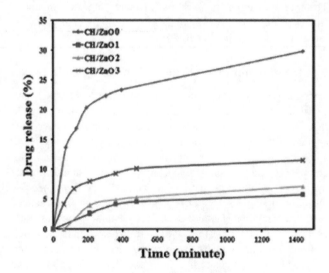

FIGURE 6.16 Drug release behavior of CH/ZnO nanocomposite beads at pH 7.4. (Adapted from Yadollahi et al. 2016. Copyright (2015), with permission from Elsevier.)

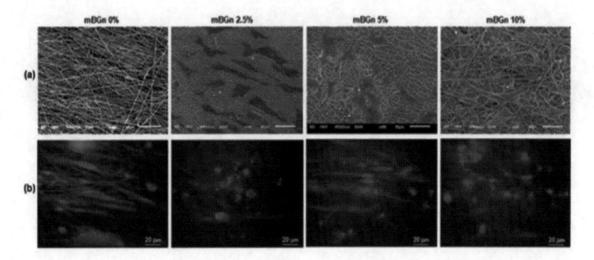

FIGURE 6.17 Cell compatibility of the fiber scaffolds generated with different compositions, briefly assessed using rat periodontal ligament stem cells (rPDSCs). After 10 days of culture on the scaffolds, (a) SEM and (b) fluorescence images of cells were observed. (Adapted with permission from El-Fiqi et al. 2015. Copyright (2015) American Chemical Society.)

hydrogel showed the most sustained release behavior for both two drugs because nanocomposites presumed a high mechanical strength and lower swelling profile and physical interaction of gold nanoparticle; the hydrogel network also has an impact on the release behavior of drugs (Mandal et al. 2017).

6.3.4 Hydrogel Nanocomposites for Tissue Engineering

Recent years have seen a remarkable interest in, and a movement of development of nanocomposite hydrogels in the biomedical and pharmaceutical field. Nanocomposite hydrogels can be pitched to mimic native tissues. Extended nanocomposite hydrogels have been reported for use in the application in tissue engineering. Nanocomposite hydrogels have been constructed with the required properties for comprehensive applications in the biomedical field. Typically, nanocomposite hydrogels can be formed by using numerous nanomaterials such as silicate, carbon nanotube, graphene oxide, silver, gold, iron oxide, polymeric nanoparticles. Paul et al. reported the photo crosslinked gelatin methacryloyl and Laponite-biocompatible hydrogel. This project demonstrated that the Laponite nanosilicates enhance the differentiation of hMSCs into osteogenic in the nanocomposite network. Laponite nanosilicates act as an initiator for the osteogenic differentiation of (hMSCs) in the nanocomposite network. Nanocomposites exhibited the local non-inflammatory, biocompatibility, and nanocomposite hydrogel indorsed the migration and proliferation of the hMSCs. Therefore, nanocomposite hydrogel could potentially be used for bone tissue engineering (Paul et al. 2016). H. Liu et al. prepared a hybrid hydrogel based on type II collagen, hyaluronic acid (HA), and polyethylene glycol (PEG), and magnetic hybrid hydrogel prepared by incorporation of I magnetic nanoparticle. Due to the magnetic field responsive characteristic, magnetic gel provides insight into a tissue-defective site under magnetic guidance. Magnetic hydrogel showed a similar cell density to the glass. Therefore, hybrid magnetic hydrogel has potential for use in cartilage tissue engineering applications (Zhang et al. 2015).

Samandari et al. reported the use of a bioactive glass nanoparticle containing hybrid hydrogels. Nanocomposite hydrogel consists of a gelatin collagen bioactive glass nanoparticle. Gel/Col/BG scaffold exhibited a significant growth of cell numbers and viability, in comparison with parent moiety (Samandari et al. 2017).

Bioactive nanoparticle-incorporated hydrogel increased the discharge of VEGF from cells on scaffolds, therefore EnSCs can efficiently differentiate into cardiomyocytes. Their finding indicated that the bioactive nanoparticle-incorporated hydrogel has potential use for angiogenesis in myocardial tissue engineering (Barabadi et al. 2016). Ahmed El-Fiqi et al. prepared the bioactive glass nanoparticle incorporated by the prepared hydrogel. Scaffolds were prepared by using olycaprolactone-gelatin inclusion mesoporous bioactive glass nanoparticles. After the addition of bioactive glass nanoparticle into the hydrogel network, it improved their mechanical robustness, elasticity, and hydrophilicity. Incorporation of bioactive glass nanoparticle enhanced the drug-loading capacity and the proliferation and osteogenic differentiation of stem cells derived from periodontal ligament was improved. The cell growth morphologies were taken by SEM and CLSM and are described in Figure 6.17. The scaffolds could release silicate and calcium ions (El-Fiqi et al. 2015). Later, Zhai et al. reported a fabrication of the effect of different nanoparticles on the mechanical robustness and biocompatibility of alginate hydrogels. Alginate hydrogel fabricated by titanium oxide, silica, aluminum oxide, and hectorite. Nanocomposite hydrogels exhibited strengthened mechanical robustness and extensibility. The negative-charged surface of hectorite is responsible for its elasticity. Furthermore, cell adhesion experiments exhibited good cell adhesion (Zhai et al. 2015).

Tentor et al. reported how CS/Pec/gold nanoparticles were prepared with different amounts of gold nanoparticles by using a tilting method. When the content of pectin was decreased and the amount of nanoparticle increased, then the gelation temperature decreased. Nanocomposites showed the potential to substitute proliferation and growth of bone cells and make them possible stimulators for the rebuilding of bone tissues.

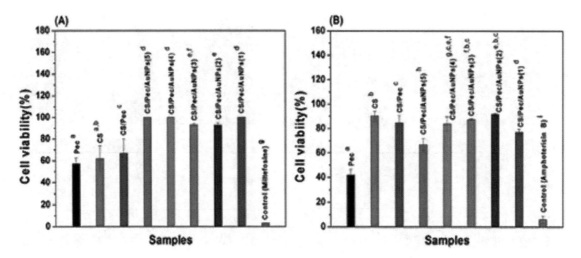

FIGURE 6.18 (A) HT-29 cell viabilities against Pec, Cs, Cs/Pec, and Cs/Pec/AuNPs composites at a concentration of 1000 μg mL−1, after 72 h of incubation. Cell viability toward the positive control (miltefosine) was analysed at 50 μg mL−1. Data shown are expressed as mean ± SD (n = 3). Based on a t-test, different underlined letters in Figure 6.5A imply that results were significantly distinct (p < 0.05). (B) VERO cell viabilities against Pec, Cs, Cs/Pec, and Cs/Pec/AuNPs composites at a concentration of 1000 μg mL⁻¹, after 72 h of incubation. Cell viability toward the positive control (amphotericin B) was analysed at 50 μg mL⁻¹. Data shown are expressed as mean ± SD (n = 3). (Adapted from Tentor et al. 2017. Copyright (2017), with permission from Elsevier.)

Gold nanoparticle-fabricated hydrogels showed the potential for the foster proliferation and are able to reconstruct the bone tissue with the growth of bone cells. The cytotoxic assay result is described in Figure 6.18. Nanocomposites showed cell viability performances for HT–29 and VERO cells (Tentor et al. 2017).

Bonifacio et al. reported on the use of nanocomposites derived from glycerol (Gly), gellan gum (GG), and halloysite nanotubes (HNT) for tissue engineering purposes. Nanocomposites have a biocompatibility and a mechanical robustness. Halloysite nanotube-incorporated hydrogels exhibited tunable physical properties and cell viability. The cells encapsulated within the different systems exhibited a comparable feasibility retort to those observed when seeded on the hydrogels' surface, (Bonifacio et al. 2017).

6.4 Conclusion

This overview enlightens us as to the variety of hydrogel nanocomposites and their compositions, characteristics of structures, and properties that can be delivered by combining various inorganic and polymer nanoparticle within hydrogels. These types of associations may be useful for polymeric matrices and nanoparticle material. Typically, hydrogels' structure prevents them from agglomeration, controls their size, and provides the final composite material with biocompatible and biodegradable properties. The incorporation of metal nanoparticles in a hydrogel network improves the mechanical, chemical, and physical properties with tunable functions.

In this paper, we have reviewed nanoparticle-hydrogel composites as an advanced material and suitable for diverse applications. Nanoparticle-hydrogel composites exhibited stimuli responsive and multi-functional characteristics. We predicted that the different synthetic techniques and applications of nanocomposite hydrogel described in this paper provide for a better understanding of systems and empower the reader to develop advanced combinations for new applications. The development

of new composite material and the prophecies of the resultant properties are major areas to be developed in this emerging field. Such on the basis of prophecies by taking into account research findings will form a floor of construction to develop new nanocomposite hydrogel with most favorable properties require in fascinated applications.

ABBREVIATIONS

N,N,MBA	N, N, Methylene bis acrylamide
APS	Ammonium persulphate
AA	Acrylic acid
NIPM	N-Isopropylacrylamide
TEMED	N, N, N′, N′- tetramethyl-ethylenediamine
CTAB	Cetyltrimethylammonium bromide
A. indica	Azadirachta indica
DMAA	N, N-dimethylacrylamide
AMPSNa	2-methylpropanesulfonic acid sodium.
MPEG-PDLLA	Monomethoxy poly(ethylene glycol)-block-poly(Ò D,L-lactide)
PTX	Paclitaxel

REFERENCES

Abou El Fadl, F.I. (2016). Synthesis and characterization of chitosan–poly (acrylamide–co-acrylic acid) magnetic nanocomposite hydrogels for use in catalysis. *Russian Journal of Applied Chemistry*, 89(10), 1673–80.

Ahmad, H., M. Nurunnabi, M.M. Rahman, K. Kumar, K. Tauerb, H. Minamic, and M.A. Gafur. (2014). Magnetically doped multi stimuli-responsive hydrogel microsphereswith IPN structure and application in dye. *Colloids and Surfaces A: Physicochemical and Engineering Aspects*, 459, 39–47.

Ajmal, M., S. Demirci, M. Siddiq, N. Aktas, and N. Sahiner. (2015a). Betaine microgel preparation from 2-(methacryloyloxy) ethyl] dimethyl (3-sulfopropyl) ammonium hydroxide and its use as acatalyst system nanocomposite hydrogels for drug delivery. *Colloids and Surfaces A: Physicochemical and Engineering Aspects, 486*: 29–37.

Ajmal, M., M. Siddiq, N. Aktas, and N. Sahiner. (2015b). Magnetic Co-Fe bimetallic nanoparticle containing modifiable microgels for the removal of heavy metal ions, organic dyes and herbicides from aqueous media. *RSC Advances, 5*(54), 43873–84.

Bajpai, K.S., and M. Kumari. (2015). A green approach to prepare silver nanoparticles loaded gumacacia/poly(acrylate) hydrogels. *International Journal of Biological Macromolecules, 80*, 177–88.

Barabadi, Z., M. Azami, E. Sharifi, R. Karimi, N. Lotfibakhshaiesh, R. Roozafzoon, M.T. Joghataei, and J. Ai. (2016). Fabrication of hydrogel based nanocomposite scaffold containing bioactive glass nanoparticles for myocardial tissue engineering. *Materials Science & Engineering C, 69*, 1137–46.

Bardajee G.R., Z. Hooshyar, M.J. Asli, F.E. Shahidi, N. Dianatnejad. (2014). Synthesis of a novel supermagnetic iron oxide nanocomposite hydrogel based on graft copolymerization of poly((2-dimethylamino)ethyl methacrylate) onto salep for controlled release of drug. *Materials Science and Engineering C, 36*, 277–86.

Bhushan, B. (2012). *Encyclopedia of Nanotechnology*. Springer.

Bonifacio, M.A., P. Gentile, P., Ferreira, S. Cometa, E.D. Giglio. (2017). Insight into halloysite nanotubes-loaded gellan gum hydrogels for soft tissue engineering applications. *Carbohydrate Polymers, 163*, 280–91.

Borai, E.H., M.M.E. Breky, M.S. Sayed, and M.M. Abo-Aly. (2015). Synthesis, characterization and application of titanium oxide nanocomposites for removal of radioactive cesium, cobalt and europium ions. *Journal of Colloid and Interface Science, 450*, 17–25.

Boruah, M., P. Gogoi, A.K. Manhar, M. Khannam, M. Mandal, and S.K. Dolui. (2014). Biocompatible carboxymethylcellulose-g-poly(acrylic acid)/OMMT nanocomposite hydrogel for in vitro release of vitamin B12. *RSC Advances, 4*(83), 43865–73.

Campbell, S., D. Maitland, and T. Hoare. (2015). Enhanced pulsatile drug release from injectable magnetic hydrogels with embedded thermosensitive microgels. *ACS Macro Letters, 4*(3), 312–16.

Campbell, S.B., M. Patenaude, and T. Hoare. (2013). Injectable superparamagnets: Highly elastic and degradable poly(n-isopropylacrylamide)–superparamagnetic iron oxide nanoparticle (SPION) composite hydrogels. *Biomacromolecules, 14*(3), 644–53.

Chen, J., S. Jo, and K. Park. (1995). Polysaccharide hydrogels for protein drug delivery. *Carhohydrare Polymers, 28*, 69–76.

Coviello, T., P. Matricardi, C. Marianecci, and F. Alhaique. (2007). Polysaccharide hydrogels for modified release formulations. *Journal of Controlled Release, 119*(1), 5–24.

Dai, L. (2006). *Carbon Nanotechnology*.1st ed. Elsevier Science.

Dave, Pragnesh N., and Ankur Gor. (2018). Natural polysaccharide-based hydrogels and nanomaterials. In *Handbook of Nanomaterials for Industrial Applications* (Chap. 3, pp. 33–66).

De Filpo G., A.M. Palermo, R. Munno, L. Molinaro, P. Formoso, and F.P. Nicoletta. (2015). Gellan gum/titanium dioxide nanoparticle hybrid hydrogels for the cleaning and disinfection of parchment. *International Biodeterioration & Biodegradation, 103*, 51–58.

Dong, S., and Y. Wang. (2016a). Characterization and adsorption properties of a lanthanum-loaded magnetic cationic hydrogel composite for fluoride removal. Water Research, *88*, 852–60.

Dong, S., and Y. Wang. (2016b). Removal of acid red 88 by a magnetic graphene oxide/cationic hydrogel nanocomposite from aqueous solutions: Adsorption behavior and mechanism. *RSC Advances, 6*(68), 63922–32.

El-Fiqi, A., J.H. Kim, and H.W. Kim. (2015). Osteoinductive fibrous scaffolds of biopolymer/mesoporous bioactive glass nanocarriers with excellent bioactivity and long-term delivery of osteogenic drug. *ACS Applied Materials & Interfaces, 7*(2), 1140–52.

Fajardo, A.R., A.G.B. Pereira, A.D. Rubira and E.C. Muniz. (2015). Ch. 9: Stimuli-Responsive Polysaccharide-Based Hydrogels. In: *Book Polysaccharide Hydrogels: Characterization and Biomedical Applications*. Edited by Pietro Matricardi, Franco Alhaique, and Tommasina Coviello. Copyright (c) 2015 Pan Stanford Publishing Pte. Ltd.

Ferris, C.J., K.J. Gilmore, G.G. Wallace and M.I.H. Panhuis. (2013). Modified gellan gum hydrogels for tissue engineering applications. *Soft Matter, 9*(14), 3705–11.

Gaharwar, A.K., N.A. Peppas, and A. Khademhosseini. (2014). Nanocomposite hydrogels for biomedical applications. *Biotechnology and Bioengineering, 111*(3), 441–53.

Garcia-Astrain, C., Ahmed, I., Kendziora, D., Guaresti, O., Eceiza, A., Fruk, L., Corcuera, M.A., Gabilondo, N. (2015). Effect of maleimide-functionalized gold nanoparticles on hybrid biohydrogels properties. RSC Advances, *5*, 50268–77.

Gong, J.P. (2006). Friction and lubrication of hydrogels: Its richness and complexity. *Soft Matter, 2*, 544–52.

Gor, A.H., and Dave, P.N. (2019). Adsorptive abatement of ciprofloxacin using $NiFe_2O_4$ nanoparticles incorporated into G. ghatti-cl-P(AAm) nanocomposites hydrogel: Isotherm, kinetic, and thermodynamic studies. *Polymer Bulletin*. https://doi.org/10.1007/s00289-019-03032-2.

Guilherme, M.R., E.C. Muniz, E.A.G. Pineda, and A.F. Rubira. (2015). Hydrogel nanocomposite based on starch and a Co-doped zinc ferrite nanoparticle that shows magnetic field-responsive drug release changes. *Journal of Molecular Liquids, 210*, 100–05.

Guilherme, M.R., A.V. Reis, A.T. Paulino, A.R. Fajardo, E.C. Muniz, and E.B. Tambourgi. (2007). Superabsorbent hydrogel based on modified polysaccharide for removal of Pb^{+2} and Cu^{+2} from water with excellent performance. *Journal of Applied Polymer Science, 105*, 2903–09.

Hamidi, M., A. Azadi, and P. Rafiei. (2008). Hydrogel nanoparticles in drug delivery. *Advanced Drug Delivery Reviews, 60*, 1638–49.

Hamidian, H., and T. Tavakoli. (2016). Preparation of a new Fe_3O_4/starch-g-polyester nanocomposite hydrogel and a study on swelling and drug delivery properties. *Carbohydrate Polymers, 144*, 140–48.

Haraguchi, K., and T. Takehisa. (2002). Nanocomposite hydrogels: A unique organic–inorganic network structure with extraordinary mechanical, optical, and swelling/de-swelling properties. *Advanced Materials, 14*(16), 1120–24.

Harikumar, P.S., L. Joseph, and A. Dhanya. (2013). Photocatalytic degradation of textile dyes by hydrogel supported titanium dioxide nanoparticles. *Journal of Environmental Engineering & Ecological Science, 2*(1), 1–9.

Huang, Y., M. Zeng, J. Ren, J. Wang, L. Fan, and Q. Xu. (2012). Preparation and swelling properties of graphene oxide/poly(acrylic acid-co-acrylamide) super-absorbent hydrogel nanocomposites. *Colloids and Surfaces A: Physicochemical and Engineering Aspects, 401,* 97–106.

Ibrahim, S. M., and A.A. El-Naggar. (2012). Preparation of poly(vinyl alcohol)/ clay hydrogel through freezing and thawing followed by electron beam irradiation for the treatment of wastewater. *Journal of Thermoplastic Composite Materials, 26,* 1332–48.

Jayaramudu, T., G.M. Raghavendra, K. Varaprasad, R. Sadiku, K. Ramam, and K.M. Raju. (2013). Iota-Carrageenan-based biodegradable Ag⁰ nanocomposite hydrogels for the inactivation of bacteria. *Carbohydrate Polymers, 95,* 188–94.

Kaushik, K., R.B. Sharma, and S. Agarwal. (2016). Natural polymers and their applications. *International Journal of Pharmaceutical Sciences Review and Research, 37*(2), 30–36.

Kokabi, M., M. Sirousazar, and Z. M. Hassan. (2007). PVA–clay nanocomposite hydrogels for wound dressing. *European Polymer Journal, 43,* 773–81.

Kumar, N., H. Mittal, V. Parashar, S.S. Ray, and J.C. Ngila. (2016). Efficient removal of rhodamine 6G dye from aqueous solution using nickel sulphide incorporated polyacrylamide grafted gum karaya bionanocomposite hydrogel. *RSC Advances, 6*(26), 21929–39.

Li, Y.B., X. Zhang and W. Chen et al. (1993). In: Trans 19th Annual Meeting of the Society for Biomaterials. Birmingham, AL, p.165.

Ling, L.L., W.J. Liu, S. Zhang, and H. Jiang. (2016). Achieving high-efficiency and ultrafast removal of Pb(ii) by one-pot incorporation of a N-doped carbon hydrogel into FeMg layered double hydroxides. *Journal of Materials Chemistry A, 4*(26), 10336–44.

Liu, P., L. Jiang, L. Zhu, J. Guo, and A. Wang. (2015). Synthesis of covalently crosslinked attapulgite/poly (acrylic acid-co-acrylamide) nanocomposite hydrogels and their evaluation as adsorbent for heavy metal ions. *Journal of Industrial and Engineering Chemistry, 23,* 188–93.

Makoto, T., A. Md. Ashraful, G. Hiroyuki, and I. Hirotaka. (2015). Microspherical hydrogel particles based on silica nanoparticle-webbed polymer networks. *Journal of Colloid and Interface Science, 455,* 32–38.

Mandal, B., A.P. Rameshbabu, S. Dhara, and S. Pal. (2017). Nanocomposite hydrogel derived from poly (methacrylic acid)/carboxymethyl cellulose/AuNPs: A potential transdermal drugs carrier. *Polymer, 120,* 9–19.

Matricardi, P., C. Cencetti, R. Ria, F. Alhaique, and T. Coviello. (2009). Preparation and characterization of novel gellan gum hydrogels suitable for modified drug release. *Molecules, 14,* 3376–91.

Medina, R.P., E.T. Nadres, F.C. Ballesteros, and D.F. Rodrigues. (2016). Incorporation of graphene oxide into a chitosan-poly(acrylic acid) porous polymer nanocomposite for enhanced lead adsorption. *Environmental Science: Nano, 3*(3), 638–46.

Mekewi, M.A., and A.S. Darwish. (2015). Elaboration of metal (M: Co, Cu, Ni, Fe) embedded poly(acrylic acid-co-acrylamide) hydrogel nanocomposites: An attempt to synthesize uncommon architectured "Auto-active" nanocatalysts for treatment of dyeing wastewater. *Materials Research Bulletin, 70,* 607–20.

Mittal, H., A. Maity, and S.S. Ray. (2015). Synthesis of co-polymer-grafted gum karaya and silica hybrid organic–inorganic hydrogel nanocomposite for the highly effective removal of methylene blue. *Chemical Engineering Journal, 279,* 166–79.

Mittal, H., V.K. Saruchi, and S.S. Ray. (2016). Adsorption of methyl violet from aqueous solution using gumxanthan/Fe₃O₄ based nanocomposite hydrogel. *International Journal of Biological Macromolecules, 89,* 1–11.

Mohammadinezhad, A., G.B. Marandi, M. Farsadrooh, and H. Javadian. (2017). Synthesis of poly(acrylamide-co-itaconic acid)/MWCNTs superabsorbent hydrogel nanocomposite by ultrasound-assisted technique: Swelling behavior and Pb (II) adsorption capacity. *Ultrasonics Sonochemistry, 49,* 1–12. doi: https://doi.org/ 10.1016/j.ultsonch.2017.12.028.

Narayanan, R.K., S.J. Devaki, and T.P. Rao. (2014). Robust fibrillar nanocatalysts based on silver nanoparticle-entrapped polymeric hydrogels. *Applied Catalysis A: General, 483,* 31–40.

Okada, K., and A. Usuki. (2006). Twenty years of polymer–clay nanocomposites. *Macromolecular Materials and Engineering, 291,* 1449–76.

Ozay, O., E. Inger, N. Aktas, and N. Sahiner. (2011a). Hydrogel assisted nickel nano particle synthesis and their use in hydrogen productionfrom sodium boron hydride. *International Journal of Hydrogen Energy, 36,* 1998–2006.

Ozay, O., E. Inger, N. Aktas, and N. Sahiner. (2011b). Hydrogen production from ammonia borane via hydrogel template synthesized Cu, Ni, Co com-posites. *International Journal of Hydrogen Energy, 36,* 8209–16.

Paranhos, C.M., B.G. Soares, R.N. Oliveira, and L.A. Pessan. (2007). Poly(vinyl alcohol)/clay-based nanocomposite hydrogels: Swelling behavior and characterization. *Macromolecular Materials and Engineering, 292*(5), 620–26.

Parpariţaa, E., C.N. Cheaburu, S.F. Paţachia, and C. Vasile. (2014). Polyvinyl alcohol/chitosan/montmorillonite nanocomposites preparation by freeze/thaw cycles and characterization. *Acta Chemica Iasi, 22*(2), 75–96.

Paul, A., V. Manoharan, D. Krafft, A. Assmann, J.A. Uquillas, S.R. Shin, A. Hasan, M.A. Hussain, A. Memic, A.K. Gaharwar, A. Khademhosseini. (2016). Nanoengineered biomimetic hydrogels for guiding human stem cell osteogenesis in three dimensional microenvironments. *Journal of Materials Chemisrty, 4*(20), 3544–55.

Peng, N., D. Hu, J. Zeng, Y. Li, L. Liang, and C. Chang. (2016). Superabsorbent cellulose–clay nanocomposite hydrogels for highly efficient removal of dye in water. *ACS Sustainable Chemistry & Engineering, 4*(12), 7217–24.

Pour, Z.S., and M. Ghaemy. (2015). Removal of dyes and heavy metal ions from water by magnetic hydrogel beads based on poly(vinyl alcohol)/carboxymethyl starch-g-poly(vinyl imidazole). *RSC Advances, 5*(79), 64106–18.

Pourbeyram, S., and S. Mohammadi. (2014). Synthesis and characterization of highly stable and water dispersible hydrogel–copper nanocomposites. *Journal of Non-Crystalline Solids, 402,* 58–63.

Pourjavadi, A., M. Nazari and S.H. Hosseini. (2015). Synthesis of magnetic graphene oxide-containing nanocomposite hydrogels for adsorption of crystal violet from aqueous solution. *RSC Advances, 5*(41), 32263–71.

Pourjavadi, A., M. Nazari, B. Kabiri, S.H. Hosseini, and C. Bennett. (2016). Preparation of porous graphene oxide/hydrogel nanocomposites and their ability for efficient adsorption of methylene blue. *RSC Advances*, 6(13), 10430–37.

Sahiner, N. (2013). Soft and flexible hydrogel templates of different sizes and various functionalities for metal nanoparticle preparation and their use in catalysis. *Progress in Polymer Science*, 38, 1329–56.

Sahiner, N., S. Butun, and P. Ilgin. (2011a). Hydrogel particles with core shell morphology for versatile applications: Environmental, biomedical and catalysis. *Colloids and Surfaces A: Physicochemical and Engineering Aspects*, 386, 16–24.

Sahiner, N., and O. Ozay. (2011). Responsive tunable colloidal soft materials basedon p(4-VP) for potential biomedical and environmental applications. *Colloids and Surfaces A: Physicochemical and Engineering Aspects*, 378, 50–59.

Sahiner, N., H. Ozay, O. Ozay, and N. Aktas. (2010). New catalytic route: hydro-gels as templates and reactors for in situ Ni nanoparticle synthesis and usage in the reduction of 2- and 4-nitrophenols. *Applied Catalysis A: General*, 385, 201–07.

Sahiner, N., O. Ozay, N. Aktas, E. Inger, and J.B. He. (2011b). The on demand generation of hydrogen from Co–Ni bimetallic nano catalyst prepared by dual use of hydrogel: As template and as reactor. *International Journal of Hydrogen Energy*, 36, 15250–58.

Sahiner, N., O. Ozay, E. Inger, and N. Aktas. (2011c). Controllable hydrogen generation by use smart hydrogel reactor containing Ru nano catalystand magnetic iron nanoparticles. *Journal of Power Sources*, 196, 10105–11.

Sahraei, R., and M. Ghaemy. (2017). Synthesis of modified gum tragacanth/graphene oxide composite hydrogel for heavy metal ions removal and preparation of silver nanocomposite for antibacterial activity. *Carbohydrate Polymers* 157, 823–833.

Samandari, S.S., H.J. Yekta, and M. Mohseni. (2017). Adsorption of anionic and cationic dyes from aqueous solution using gelatin based magnetic nanocomposite beads comprising carboxylic acid functionalized carbon nanotube. *Chemical Engineering Journal*, 308, 1133–44.

Sarkar, A.K., A. Saha, L. Midya, C. Banerjee, N.R. Mandre, A.B. Panda, and S. Pal. (2017). Cross-linked biopolymer stabilized exfoliated titanate nanosheet-supported AgNPs: A green sustainable ternary nanocomposite hydrogel for catalytic and antimicrobial activity. *ACS Sustainable Chemistry & Engineering*, 5(2), 1881–91.

Seven, F., and N. Sahiner. (2013). Poly(acrylamide-co-vinyl sulfonic acid) p(AAm-co-VSA) hydrogel templates for Co and Ni metal nanoparticle preparation and their use in hydrogen production. *International Journal of Hydrogen Energy*, 38, 777–84.

Shahrooie, B., L. Rajabi, A.A. Derakhshan, and M. Keyhani. (2015). Fabrication, characterization and statistical investigation of a new starch-based hydrogel nanocomposite for ammonium adsorption. *Journal of the Taiwan Institute of Chemical Engineers*, 1–15.

Sharma, S., Deepak, A. Kumar, S. Afgan, and R. Kumar. (2017). Stimuli-responsive polymeric hydrogel-copper nanocomposite material for biomedical application and its alternative application to catalytic field. *Chemistry Select*, 2, 11281–87.

Shi, K., Z. Liu, Y.Y. Wei, W. Wang, X.J. Ju, R. Xie, and L.Y. Chu. (2015). Near-infrared light-responsive poly(Nisopropylacrylamide)/graphene oxide nanocomposite hydrogels with ultrahigh tensibility. *ACS Applied Materials & Interfaces*, 7(49), 27289–98.

Shi, Y., D. Xiong, J. Li, and N. Wang. (2016). In situ reduction of graphene oxide nanosheets in poly(vinyl alcohol) hydrogel by γ-ray irradiation and its influence on mechanical and tribological properties. *Journal of Physical Chemistry C*, 120(4), 19442–53.

Silva, S.S., A.R.C. Duarte, J.M. Oliveira, J.F. Mano, and R.L. Reis. (2013). Alternative methodology for chitin–hydroxyapatite composites using ionic liquids and supercritical fluid technology. *Journal of Bioactive and Compatible Polymers* 28(5), 481–491.

Singh, B., S. Sharma, and A. Dhiman. (2017). Acacia gum polysaccharide based hydrogel wound dressings: Synthesis, characterization, drug delivery and biomedical properties. *Carbohydrate Polymers*, 165, 294–303.

Singh, N.K., Q.V. Nguyen, B.S. Kim and D.S. Lee. (2015). Nanostructure controlled sustain delivery of human growth hormone using injectable, biodegradable, pH / temperature responsive nanobiohybrid hydrogel. *Nanoscale*, 7(7), 3043–54.

Sionkowska, A. (2011). Current research on the blends of natural and synthetic polymers as new biomaterials: Review. *Progress in Polymer Science*, 36, 1254–76.

Sirousazar, M., M. Kokabi, and Z.M. Hassan. (2012a). Swelling behavior and structural characteristics of polyvinyl alcohol/montmorillonite nanocomposite hydrogels. *Journal of Applied Polymer Science*, 123, 50–58.

Sirousazar, M., M. Kokabi, Z.M. Hassan, and A.R. Bahramian. (2012b). Polyvinyl alcohol/na-montmorillonite nanocomposite hydrogels prepared by freezing–thawing method: Structural, mechanical, thermal, and swelling properties. *Journal of Macromolecular Science R, Part B: Physics*, 51, 1335–50.

Song, F., W. Hu, L. Xiao, Z. Cao, X. Li, C. Zhang, L. Liao, and L. Liu. (2015). Enzymatically crosslinked hyaluronic acid/graphene oxide nanocomposite hydrogel with pH responsive release. *Journal of Biomaterials Science, Polymer Edition*, 26(6), 339–52.

Song, F., L.M. Zhang, J.F. Shi, and N.N. Li. (2010). Viscoelastic and fractal characteristics of a supramolecular hydrogel hybridized with clay nanoparticles. *Colloids and Surfaces B: Biointerfaces*, 81(2), 486–91.

Sozeri, H., E. Alvero glu, U. Kurtan, M. Senel, and A. Baykal. (2013). Synthesis, electrical and magnetic characterization of polyacrylamide hydrogels including $NiFe_2O_4$ nanoparticles. *Journal of Superconductivity and Novel Magnetism*, 26(1), 213–18.

Spasojevic, J., A. Radosavljevic, J. Krstic, D. Jovanovic, V. Spasojevic, M.K. Krusic, and Z.K. Arevic-Popovic. (2015). Dual responsive antibacterial Ag-poly(N-isopropylacrylamide/ itaconic acid) hydrogel nanocomposites synthesized by gamma irradiation. *European Polymer Journal*, 69, 168–85.

Sun, H.S., Y.C. Chiu, and W.C. Chen. (2016). Renewable polymeric materials for electronic applications. *Polymer Journal*, 49(1), 1–13.

Sun, X.F., B. Liu, Z. Jing, and H. Wang. (2015). Preparation and adsorption property of xylan/poly(acrylic acid) magnetic nanocomposite hydrogel adsorbent. *Carbohydrate Polymers*, 118, 16–23.

Tang, S.C.N., D.Y.S. Yan, and I.M.C. Lo. (2014). Sustainable wastewater treatment using microsized magnetic hydrogel with magnetic separation technology. *Industrial & Engineering Chemistry Research*, *53*(40): 15718–24.

Tao, J., J. Zhang, Y. Hu, Y. Yang, Z. Gou, T. Du, J. Mao, and M. Gou. (2017). A conformal hydrogel nanocomposite for local delivery of paclitaxel. *Journal of Biomaterials Science, Polymer Edition*, *28*(1), 107–18.

Tentor, F.R., J.H. de Oliveira, D.B. Scariot, D.L. Bidoia, E.G. Bonafe, C.V. Nakamura, S.A.S. Venter, J.P. Monteiro, E.C. Muniz, and A.F. Martins. (2017). Scaffolds based on chitosan/pectin thermosensitive hydrogelscontaining gold nanoparticles. *International Journal of Biological Macromolecules*, *102*, 1186–94.

Thomas, P.C., B.H. Cipriano, and S.R. Raghavan. (2011). Nanoparticle-crosslinked hydrogels as a class of efficient materials for separation and ion exchange. *Soft Matter*, *7*(18), 8192–97.

Thoniyot, P., M.J. Tan, A.A. Karim, D.J. Young, and X.J. Loh. (2015). Nanoparticle-hydrogel composites: Concept, design, and applications of these promising, multi-functional materials. *Advanced Science*, *2*(1–2), 1400010.

Turhan, T., Y.A. Guvenilir, and N. Sahiner. (2013). Micro poly(3-sulfopropyl methacrylate) hydrogel synthesis for in situ metal nanoparticle preparation and hydrogen generation from hydrolysis of NaBH₄. *Energy*, *55*, 511–18.

Van Bemmelen, J.M. (1894). Des Hydrogel und das kristallinische Hydrat des Kupferoxydes. *Zeitschrift für anorganische Chemie*, *5*, 466–83.

Van Berkum, S., P.D. Biewenga, S.P. Verkleij, K.W Boere, A. Pal, A.P. Philipse, and B.H. Erne. (2015). Swelling enhanced remanent magnetization of hydrogels cross-linked with magnetic nanoparticles. *Langmuir*, *31*: 442–50.

Varghese, L.R., and N. Das. (2015). Removal of Hg (II) ions from aqueous environment using glutaraldehyde crosslinked nanobiocomposite hydrogel modified byTETA and β-cyclodextrin: Optimization, equilibrium, kinetic and ex situ studies. *Ecological Engineering*, *85*, 201–11.

Wang, Q., J.L. Mynar, M. Yoshida, E. Lee, M. Lee, K. Okuro, K. Kinbara, and T. Aida. (2010). High-water-content mouldable hydrogels by mixing clay and a dendritic molecular binder. *Nature*, *463*(7279), 339–43.

Wichterle, O., and D. Lim. (1960). Hydrophilic gels for biological use. *Nature*, *185*, 117–18.

Yadollahi, M., S. Farhoudian, S. Barkhordari, I. Gholamali,H. Farhadnejad, and H. Motasadizadeh. (2016). Facile synthesis of chitosan/ZnO bio-nanocomposite hydrogel beads as drug delivery systems. *International Journal of Biological Macromolecules*, *82*, 273–78.

Yadollahi, M., H. Namazi, and M. Aghazadeh. (2015). Antibacterial carboxymethyl cellulose/Ag nanocomposite hydrogelscross-linked with layered double hydroxides. *International Journal of Biological Macromolecules*, *79*, 269–77.

Zhai, Y., H. Duan, X. Meng, K. Cai, Y. Liu, and L. Lucia. (2015). Reinforcement effects of inorganic nanoparticles for double-network hydrogels. Macromolecular Materials and Engineering, *300*(12), 1290–99.

Zhang, H., D. Zhai, and Y. He. 2014. Graphene oxide/polyacrylamide/carboxymethyl cellulosesodium nanocomposite hydrogel with enhanced mechanical strength: Preparation, characterization and the swelling behavior. *RSC Advances*, *4*(84), 44600–09.

Zhang, N., J. Lock, A. Sallee, and H. Liu. (2015). Magnetic nanocomposite hydrogel for potential cartilage tissue engineering: Synthesis, characterization, and cytocompatibility with bone marrow derived mesenchymal stem cells. *ACS Applied Materials & Interfaces*, *7*(37), 20987–98.

Zhao, L., H. Benzhao, and Z. Faai. (2012). Facile one-pot synthesis of iron oxide nanoparticles cross-linked magnetic poly(vinyl alcohol) gel beads for drug delivery. *ACS Applied Materials & Interfaces*, *4*(1), 192–99.

Zheng, X., D. Wu, T. Su, S. Bao, C. Liao, and Qigang Wang. (2014). Magnetic nanocomposite hydrogel prepared by ZnO-initiated photopolymerization for La (III) adsorption. *ACS Applied Materials & Interfaces*, *6*(22), 19840–49.

Zhou, Y., S. Fu, L. Zhang, H. Zhan, and M.V. Levit. (2014). Use of carboxylated cellulose nanofibrils-filled magnetic chitosan hydrogel beads as adsorbents for Pb(II). *Carbohydrate Polymers*,*101*, 75–82.

Zhuang, Y., F. Yu, and J. Ma. (2015). Enhanced adsorption and removal of ciprofloxacin on regenerable long TiO₂ nanotube/graphene oxide hydrogel adsorbents. *Journal of Nanomaterials*, *2015*, 1–8.

7

Remediation of Wastewater

Shalini Chaturvedi and Pragnesh N. Dave

CONTENTS

7.1 Introduction

Water is the most abundant natural resource on the earth and essential for the ecosystem. Water available for human consumption is nearly about 1% of that resource which is available (Adeleye et al., 2016; Grey et al., 2013). Due to the expensive cost of potable water, increasing populations, several environmental concerns, and climatic changes, it was estimated that 1.3 billion people have a lack of an adequate drinking water supply (Adeleye et al., 2016; WHO, 2015). Continuous contamination of fresh water resources by pollutants from a type of organic and inorganic compounds make the condition more complicated (Schwarzenbach et al., 2006). Setting up treatment plants for the water remediation of wastewater can reduce the concern for drinking water (Ferroudi et al., 2013). Treatment by old traditional methods are not so efficient at completely removing the new types of contaminants (Qu et al., 2012, 2013, 2014). These wastewater treatment technologies have several disadvantages, such as incomplete removal of pollutants, a high-energy requirement, and the formation of toxic sludge (Ferroudi et al., 2013). The biological wastewater treatment is usually very slow and some toxic contaminants are present, which create toxicity for microorganisms (Zelmanov et al., 2008). So, in the present scenario there is a requirement for more efficient potential technologies for the treatment of wastewater, generated from municipal and industrial wastages (Burkhard et al., 2000; Crini and Badot, 2007; Ferroudi et al., 2013; Parsons and Jefferson, 2006).

For water remediation there is need to develop and improve the existing methods through some new interventions. Currently, advances in nanotechnology have shown a potential for the remediation of wastewater (Sadegh et al., 2014; Zare et al., 2013; Gupta et al., 2015). Nanotechnology is the science which is applied on a nanometer scale level. Nanomaterials are the tiny structures that humans develop which have the size of a few nanometers (1–100 nm) (Amin et al., 2014; Chaturvedi et al., 2012). Several types of nanomaterials have been developed such as nanowires, nanotubes, films, particles, quantum dots, and colloids (Edelstein and Cammaratra, 1998; Lubick and Betts, 2008). Lots of nanomaterials are now synthesized which are very efficient, eco-friendly, low-cost and have potential abilities for the removal of contaminants from industrial effluents, surface water, ground water, and drinking water (Gupta et al., 2015; Theron et al., 2004, 2008; Brumfiel, 2003).

Wastewater remediation based on nanomaterials can be classified on the basis of their nature like as such nano-adsorbents, nanocatalysts and nano-membranes. In this technology, nano-absorbents are used for the removal of contamination (Kyzas and Matis, 2015; Shamsizadeh, 2014; Tang et al., 2005, 2012, 2014; Zhang et al., 2014a, b). Nano-adsorbent produces atoms of the chemically active with a high adsorption capacity on the surface of the nanomaterial (Kyzas and Matis, 2015). Mostly, applied nano-absorbents are like activated carbon, metal oxides, silica, clay materials, and modified compounds in the form of composites (El Saliby et al., 2008). Nanocatalysts are second form of the nanomaterials used for wastewater remediation like metal oxides and semiconductors. Nanocatalysts now receive considerable attention for developing wastewater treatment technologies for the degradation of pollutants, like electrocatalysts (Dutta et al., 2014), Fenton-based catalysts (Kurian and Nair, 2015 which enhance oxidation of organic pollutants and have antimicrobial properties (Chaturvedi et al., 2012). Nano-membranes are also used in the wastewater remediation processes. The pressure-based treatment process of wastewater also improves water quality (Ríos et al., 2012). Nano-filtration is mostly applied in industries of the remediation of waste water. Nanosize filters are low in price, highly efficient, and user friendly (Petrinic et al., 2007; Hital et al., 2004; Babursah et al., 2006; Rashidi et al., 2015). Nano-membranes are formed by the use of nanomaterials such as nano metal, non-metal particles, and nano-carbon, etc. (El Saliby et al., 2008). In this chapter we mainly focused on nanomaterials which are used in wastewater remediation and how they work.

7.2 Water Nanotechnology

In the last few decades, several literatures have become available on the use of nanotechnology in wastewater treatment. Here we discuss a few applications of nanotechnology in the remediation of water and wastewater (Figure 7.1).

7.2.1 Adsorption and Separation

In the remediation of water and wastewater, there are adsorbents or the membrane-based separation process is mostly used. Traditional adsorbents have some disadvantages such as low capacity, low selectivity, and a short adsorption-regeneration cycle, and a high cost of the adsorbents. Nano-adsorbents, i.e., carbon nanotubes (CNTs), grapheme, nanosized metal or

metal oxides, and nanocomposites, have a large specific area, fast kinetics, high reactivity, and a specific affinity to various contaminants. Their effectiveness towards adsorption of certain contaminants is higher than that of conventional adsorbents (Ali, 2012; Khajeh et al., 2013).

Water remediation with the use of membrane separation is also a key module for the treatment of municipal wastewater. Size exclusion is the main concept of removal of contaminants by membrane separation. However, there are complications that come in this process, like the inherent trade-off between membrane selectivity and permeability, fouling, high-energy consumption, and operational complexity. These issues are can be resolved by the use of advanced nanocomposite membranes which are made by the incorporation of functional nanoparticles into the membrane. These advanced membranes have enhanced physiochemical properties such as enhanced mechanical and thermal stability, hydrophilicity, and porosity. Also, they show enhanced permeability, or anti-fouling, antimicrobial, adsorptive, or photocatalysis capabilities (Pendergast et al., 2011; Yin and Deng, 2015).

7.2.2 In Catalysis

For the removal of trace contaminants and microbial pathogens, advanced catalytic or photocatalytic oxidation processes are now in use. This method enhances the biodegradability for hazardous and non-biodegradable contaminants (Reddy et al., 2016). Nanocatalysts show significantly enhanced catalysis performance over their bulky counterparts. The electron hole redox potential and photo-generated charge distribution shows size-dependent behavior varied with varying sizes (Song et al., 2010; Turki et al., 2015). With the immobilization of nanoparticles onto various substrates, the stability of the nanocatalyst was improved and the resultant nanocomposites were compatible with an existing photo-reactor (Petronella et al., 2016).

7.2.3 Disinfection

Stopping the spread of waterborne disease disinfection is the most critical step in water treatment. A disinfectant process is requires a broad antimicrobial spectrum within a short time, and also does not generate any harmful by-products. It is of low toxicity for health and ecosystems; has low energy costs and ease

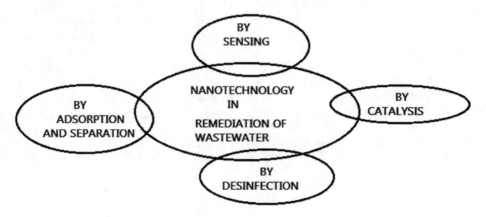

FIGURE 7.1   Application of nanotechnology in remediation of water and wastewater.

of operation; it is easily stored and must not be corrosive; and is capable of safe disposal. Now nanomaterials possess strong antimicrobial properties, including chitosan nanoparticles (Higazy et al., 2010), nano silver (nAg) (Rai et al., 2009), photocatalytic TiO2 (Hebeish et al., 2013), and carbon-based nanomaterials (Martynkova and Valaskova, 2014). These nanomaterials release toxic metal ions (e.g., Ag+) to kill microorganisms by destroying cell membranes through direct contact (e.g., chitosan nanoparticles) or generating reactive oxygen species (ROS, e.g., TiO2). These antimicrobial nanomaterials inactivate the microorganisms and minimize the formation of harmful disinfection byproducts (DBPs) (Li et al., 2008b). Some nano-disinfectants can operate continuously with high efficiency and a low energy consumption, which makes them attractive for de-centralized water and wastewater treatment.

7.2.4 Sensing

Nowadays, a very low concentration of micro-pollutants cannot be detected by classical sensing and monitoring methods. Sometimes during an emergency, rapid and in-situ detection of pathogens and highly toxic pollutants is required. Nanomaterials like graphene, carbon nanotubes, quantum dots, and noble metals possess unique optical/magnetic and electrochemical properties. In the presence of these nanomaterials, sensors/electrodes may selectively collect trace pollutants for detection. Spectroscopic response can be enhanced with the use of nanomaterials (e.g. Raman shift or surface plasmon resonances) (Das et al., 2015; Pradeep, 2009). Also, nanocomposite sensors may be used for environmental monitoring and sensing (Srivastava et al., 2016).

7.3 Nanomaterial Use in Water Remediation

The use of nanotechnology in water remediation is based on the innovation of various nanomaterials. In this section we discuss a number of nanomaterials that are used in water and wastewater remediation (Figure 7.2 and Table 7.1).

7.3.1 Metal and Metal Oxide-Based Nanoparticles

Metal oxide-based nanoparticles are mainly used for the removal of hazardous pollutants from wastewater. These nanoparticles are titanium oxide/dendrimer composites (Barakat et al., 2013a, b), titanium oxides (Gao et al., 2008), zinc oxides (Tuzen and Soylak, 2007), manganese oxides, magnesium oxide (Gupta et al., 2011), and ferric oxides (Xu et al., 2008 a, b). Oxide-based nanoparticles have less solubility, minimal environmental impact, and no secondary pollutants (Gupta et al., 2015).

Ferric oxide nanoparticles have simple synthesis processes which make them a low-cost material for the adsorption of toxic metals. They are eco-friendly and can be used directly in a contaminated environment and reduce the chance of secondary contamination (Li et al., 2003). Adsorption of Fe_2O_3 nanoparticles depend on the temperature, adsorbent dose, pH, and incubation time (Gupta et al., 2015). Adsorption capacity of Fe2O3 can be improved by surface modification (Wang et al., 2015; Ozmen et al., 2010), with the addition of 3-aminopropyltrimethoxysilane (Palimi et al., 2014). Modification of these nano-adsorbents enhances the ability for the removal of different pollutants such as Cr^{3+}, Co^{2+}, Ni^{2+}, Cu^{2+}, Cd^{2+}, Pb^{2+}, and As^{3+} simultaneously (Gupta et al., 2015).

Nanosize manganese oxides (MnO) have an enhanced adsorption ability due to their large surface area and due to their polymorphic structure (Luo et al., 2010). They are mainly used in the removal of various heavy metals like arsenic from wastewater (Wang et al., 2011). Modified MnOs contain nanoporous/nanotunnel manganese oxides and hydrous manganese oxide (HMO) (Gupta et al., 2015). Removal of other heavy metals like Pb (II), Cd (II), and Zn (II) is done by adsorption on HMOs. Removal of heavy metals are achieved by the inner-sphere formation mechanism that can be defined by an ion-exchange process (Gupta et al., 2015). However, the adsorption of divalent metals is a two-step process on the surface of HMOs. In this two step-process, in the first step metal ions adsorb on the external surface of HMOs, and in the second step intra-particle diffusion oocurs (Parida et al., 1981).

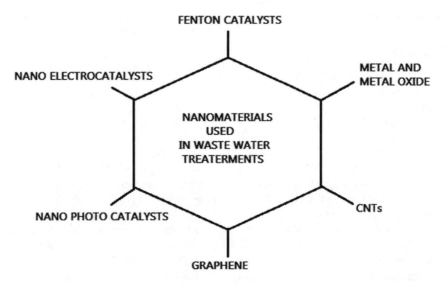

FIGURE 7.2 Nanomaterials used in water remediation from wastewater.

TABLE 7.1

Nanomaterials and Nano-Membranes Use in Water Remediation

S. No.	Nanomaterials
1.	**Metal and metal oxides** Ex: MnO, ZnO, Fe_2O_3, TiO_2, MgO
2.	**Carbon nanotubes: CNTs** modified with metal and metal oxide like MnO_2; Al_2O_3, Fe_2O_3
3.	**Graphene:** Graphene oxide; graphene oxide doped with TiO_2
4.	**Photocatalysts:** TiO2; ZnO; CdS:Eu, ZnO:Co, ZnSe:Mn, ZnS:Mn, ZnS:Cu, CdS:Mn, ZnS:Pb
5.	**Electrocatalysts:** Pt; Pd; carbon black XC72 supported Pt; Pd composites with nanomaterials like pristine multi-walled carbon nanotubes (pMWCNT), amine-modified multi-walled carbon nanotubes, carboxylated multi-walled carbon nanotubes, hydroxyl-modified pristine multi-walled carbon nanotubes, carboxylated graphene, and XC72 carbon black.
6.	**Fenton's catalysts:** Spinel ferrites capped with Ni, Zn, Co, Cu, Ba, Co, and Mn, carbonaceous materials

S. No.	Nano-Membranes
1.	**Carbon nanotube membranes:** Carboxyl multi-walled carbon nanotubes/calcium alginate composite membrane; Chitosan/Silica-coated carbon nanotubes composite membranes
2.	**Electrospun nanofiber membranes:** like poly (vinyl chloride), polyvinylidene fluoride, polybenzimidazole, polyurethanes, Nylon-6, polystyrene, poly(vinyl phenol), Kevlar (poly (p-phenylene terephthalamide), poly(vinyl alcohol), polycarbonates, poly(e-capro-lactone), polysulfones, poly(ethylene terephthalate)
3.	**Hybrid nano-membranes:** Zeolite-impregnated polysulfone (PSf) nanoparticle membrane; nanostructured polymer-based (styrene, divinyl benzene, potassium persulfate, sorbitan monooleate) membrane/sorbent

Zinc oxide has a porous nanostructure which is suitable for the adsorption of heavy metals. For the removal of heavy metals from wastewater nanoplates, microspheres with nano-sheets, nano-assemblies, and hierarchical ZnO nano-rods are mainly used (Ge et al., 2012; Kumar et al., 2013). Modified forms of ZnO nanoplates possess an enhanced removal efficiency for heavy metals like Cu (II) than commercial ZnO (Wang et al., 2010). Removal of heavy metals like Co^{2+}, Ni^{2+}, Cu^{2+}, Cd^{2+}, Pb^{2+}, Hg^{2+}, and As^{3+} were done by nano-assemblies (Singh et al., 2013). Pb^{2+}, Hg^{2+}, and As^{3+} have an electropositive nature due to which microporous nano-assemblies show more affinity for their adsorption (Gupta et al., 2015).

Magnesium oxide (MgO) is used for the removal of heavy metals from contaminated water. To enhance the adsorption capacity of MgO, some modifications were carried out in NPs morphology like nanobelts (Zhu et al., 2001; Yin et al., 2002), nanotubes (Yin et al., 2002), nanocubes (Li et al.; 2002), fishbone fractal nanostructures induced (Liang et al., 2004), nanorods (Engates et al, 2011), by nanowires (Mo et al., 2005), and three-dimensional entities (Klug and Dravid, 2002).

7.3.2 Carbon Nanotubes (CNTs)

Carbon nanotubes (CNTs) have the potential ability to adsorb and remove heavy metals and organic contaminants from wastewater (Ren et al., 2011). Carbon nanotubes' absorbents have some disadvantages like low dispersion, difficulty in separation, and their small particle size. To resolve these issues, modified CNT such as multi-wall carbon nanotubes (Tang et al., 2012; Tarigh et al., 2013) are used. Magnetized CNTs have a great ability to disperse and can remove heavy metals such as Pb(II) and Mn(II) (Tarigh et al., 2013), Cu(II) (Tang et al., 2012) from wastewater by using a magnet (Madrakian et al., 2011). Alumina-coated carbon nanotubes exhibited greater removal ability than the uncoated carbon nanotubes (Gupta et al., 2011). Modification on the surface of carbon nanotubes increases its overall adsorption activity. These modifications were done by techniques like acid treatment (Ren et al., 2011; Ihsanullah et al., 2015), metal

impregnation (Tawabini et al., 2011; Zhang et al., 2012), and functional molecule/group grafting (Shao et al., 2010 a, b; Chen et al., 2012). These techniques improve the characteristics of carbon nanotubes' surface like a high surface area, surface charge, dispersion, and hydrophobicity.

Acid treatment of carbon nanotubes was done by use of acids like HNO3, KMnO4, H2O2, H2SO4, and HCl (Ren et al., 2011; Fu and Wang, 2011) and removes the impurities present on the surface of carbon nanotubes. Incorporation of new functional groups on the carbon nano-tubes' surface increases the capacity of adsorption from wastewater (Gupta et al., 2015). The addition of grafting functional molecules or groups, oxygen-containing group was done by plasma technique, chemical modification, and microwave on the surface of carbon nanotubes to improve their surface characteristics (Zhang et al., 2012; Shao et al., 2010 a, b; Chen et al., 2012). The plasma technique requires low energy and it is environmentally friendly. Carbon nanotubes were modified with metal or metal oxide like MnO_2 (Liang et al., 2004), Al_2O_3 (Gupta et al., 2015), and iron oxide (Tang et al., 2012; Zhang et al., 2012) shows impressive results for the removal of heavy metals from wastewater.

7.3.3 Graphene

Graphene is an allotrope of carbon which possesses lots of environmental applications. By the oxidation of graphite layer via the chemical method, graphene oxide is produced. In graphene oxide, carbon is a nanomaterial having a two-dimensional structure. Graphene oxide can be synthesized by the Hummers' method (Lingamdinne et al., 2016). The hydrophilic groups were induced in graphene oxide which required a special oxidation process (Gopalkrishnan et al., 2015). Induced hydroxyl and carboxyl groups are inserted as a functional group which increases the adsorption of graphene oxide for the removal of heavy metals (Lingamdinne et al., 2016; Gopalkrishnan et al., 2015; Li et al., 2009; Taherian et al., 2013; Santhosh et al., 2016; Zhao et al., 2011). Graphene oxides have mechanical strength, a light weight, high surface area, flexibility, and chemical stability. Graphene oxide has two-dimensional basal planes, due to which it shows for the maximum adsorption of heavy metals (Santhosh et al., 2016). Like carbon nanotubes, graphene oxide contains a hydrophilic functional group due to which it did not require any acid treatment to enhance its adsorption capacity (Zhao et al., 2011). Researchers suggest graphene-based nanomaterials are more useful for the adsorption of heavy metals from wastewater (Azamat et al., 2015; Dong et al., 2015; Vu et al., 2017; Zare-Dorabei et al., 2016). Graphene oxide-enabled sand filter in a column reactor is used for the removal of heavy metals from wastewater (Ding et al., 2014). Modification of the graphene oxide occurs when doped with TiO_2, and it is used for the adsorption of Pb^{2+}, Cd^{2+}, and Zn^{2+} ions from the water (Lee et al., 2012). Graphene and its composites have very high efficiency for the removal of heavy metals.

7.3.4 Nanomaterials as Photocatalysts

Interaction of light energy with metallic nanoparticles got attention due to their photocatalytic activities for various pollutants (Akhavan, 2009). These photocatalysts are made of semiconductor metals. It can degrade organic pollutants in wastewater like detergents, dyes, pesticides, and volatile organic compound (Lin et al., 2014). Semiconductor nanocatalysts are effective for the degradation of halogenated and non-halogenated organic compounds for heavy metals (Adeleye et al., 2016). The mechanism of photocatalysis is based on the photoexcitation of electron in the catalyst. The irradiation with light generates holes and exited electrons. In an aqueous medium, water molecules trapped the holes and hydroxyl radicals are generated (Anjum et al., 2016). The radicals act as a powerful oxidization agent. These hydroxyl radicals oxidize the organic pollutants and generate water and gaseous degradation products (Akhavan, 2009). Due to high reactivity under ultraviolet light (k< 390nm) and chemical stability, TiO_2 is most applicable in photocatalysis (Akhavan, 2009). ZnO has also been extensively studied by researchers (Lin et al., 2014). Efficiency of photocatalyst depends on the factors like particle size, band gap energy, dose, pollutant concentration, and pH. CdS nanoparticles as a photocatalyst also received attention for the treatment of industrial dyes in wastewater (Tristao et al., 2006; Zhu et al., 2009).

Photocatalytic treatment of wastewater in the presence of visible light is now an area of focus in research. By modification in the nano-material/semiconductor band, gap energy is decreased from the UV to the visible region (Anjum, 2016). These modifications can occur by dye-doping metal impurities, sensitization, hybrid nanoparticles or composites using narrow band-gap semiconductors or anions (Ni et al., 2007). The addition of new metals and anions creates a narrow band gap, as impurity energy levels in the presence of visible light conducts an electron into a semiconductor for the initiation of a catalytic reaction (Qu et al., 2013). ZnO and TiO_2 nanomaterials absorb a small portion of the UV region and can allow for the modification of loading metals on its surface. Some conductive metals are not effective for doping e.g. Pt and Ru. Au, Ag, and Pd show good photocatalytic activities (Barakat et al., 2013 a. b; Sathishkumar et al., 2011). CdS:Eu, ZnO:Co, ZnSe:Mn, ZnS:Mn, ZnS:Cu, CdS:Mn, ZnS:Pb etc are doped nanocatalysts which are already developed (Chandrakar et al., 2015) Cr, Si, Co, Mg, Mn, Fe, Fe, Al, In, and Ga are used as dopant which enhance the surface area of metal oxide nano-structure (Jamal et al., 2012).

Photocatalysis is a potential technique for the remediation of various kinds of wastewater (Yu et al., 2001). Photocatalysts also have the ability to render inactive pathogenic organisms like bacteria in the wastewater (Yu et al, 2003). TiO_2 has a high antimicrobial power with some drawbacks. So for photocatalysts with efficient antimicrobial activity incorporation of the nanoparticles, it is necessary to increase the surface area (Liu et al., 2008). As earlier reported (Akhavan, 2009), TiO_2 films covered with Ag nanoparticles achieved seven times higher antimicrobial activity fort *E. coli* bacteria than normal TiO_2 film under visible light. The mechanism behind this was the incorporation of other metal increases of the surface area which generates more active sites at mesoporous catalysts to enhance the degradation of microorganisms (Akhavan, 2009; Liu et al., 2008). The presence of an extracellular polymeric substance (EPS) may decrease the antimicrobial action of the catalyst. Thus, it is important to remove the extracellular polymeric substance to achieve high efficiency of photocatalysis for wastewater disinfection (Chaturvedi et al., 2012).

7.3.5 Nanomaterials as Electrocatalysts

The process of electrocatalysis is an emerging technology for wastewater remediation and direct electricity generation from microbial fuel cell. In microbial fuel cells, electrocatalysts play an important role (Chen et al., 2015). The use of nanosize electrocatalysts can improve the activity of fuel cells due to the larger surface area and uniform distribution of catalysts in the reaction media (Liu et al., 2005). Lots of research has been done to develop the carbon supported nanoelectrocatalysts for application in fuel cells (Chaturvedi et al., 2012; Tang et al. 2012). In glucose oxidation electro-catalysis reaction, carbon black XC72 supported Pt nanocatalyst shows a potential up to 6.2 mA cm^{-2} of current density (Chen et al., 2015). One major drawback in the use of Pt as a catalyst is its high cost and low availability. These problems can be removed by the use of Pd nanoparticles. Single-wall carbon nanotubes functionalized with Pd nanoparticles showed high electro-catalytic activity. Pd can also form composites with nanomaterials like pristine multi-walled carbon nanotubes (pMWCNT), amine-modified multi-walled carbon nanotubes, carboxylated multi-walled carbon nanotubes, hydroxyl-modified pristine multi-walled carbon nanotubes, carboxylated graphene, and XC72 carbon black (Chen et al., 2015). These hybrid electrocatalysts with reduced particle size have better dispersion, and show high catalytic reduction for pollutant 4-nitrophenol because of electrocatalysis of dioxygen reduction and 4-nitrophenol (Qiu et al., 2012).

7.3.6 Nanosize Fenton Catalyst

Fenton's reaction is also used in wastewater treatment by the oxidation of organic pollutants (Neyens and Baeyens, 2003). The main disadvantage in this reaction is the regular loss of catalyst material and acidic conditions (pH=3) required for optimum function (Kurain and Nair, 2015; Ferroudi et al., 2013). This problem can be removed by the use of nano-material. The nano-ferrites have crystalline size, synthesized by sol-gel and auto combustion method (Kurian et al., 2014). The spinel ferrites capped with Ni, Zn, Co, and Cu have a catalytic property because of special magnetic and electronic properties. Due to the high chemical and thermal stability, MFe_2O_4 is extensively used as catalyst (Kurain and Nair 2015). Nanoparticles of iron oxide which are magnetically separable can be used as Fenton catalysts (Shahwan et al., 2011; Sun and Lemeley, 2011). Ferrite with Ba, Co, and Mn, carbonaceous materials, and maghemite, etc. has a high capability to recover from water by use of a magnetic field. Magnetic separation is a more efficient fast method, as compared to the other reported methods (Ambashta and Sillanpaa, 2010).

7.3.7 Carbon Nanotube Membranes

Carbon nanotubes are used to synthesize polymer composite membrane. These polymer composites show an enhanced performance because of features like extremely high strength, low mass density, tensile modulus, high flexibility, and large aspect ratio (Liu et al., 2016). They can be classified as single-walled carbon nanotubes consisting of a singular-walled tube or multi-walled tube carbon nanotubes consisting of multi-walled tubes (Popov, 2004; Rajabi et al., 2013). Synthesis of modified nanotube membranes is already reported. Carboxyl multi-walled carbon nanotube/calcium alginate composite membrane had a high strength and good anti-fouling property which is synthesized by the hydrogel nano-filtration membrane method by using polyethyleneglycol 400 as a pore-forming agent (Jie et al., 2015). The composite membrane enhances the hydrophilic and tensile mechanical property, which helps in the removal of the selected pollutant. It was observed that there was 32.95 L/m2h and 98.20% of flux and rejection rate, respectively, of the composite membrane for the dye brilliant blue (Kurian and Nair, 2015). Chitosan-/silica-coated carbon nanotube-composite membranes are prepared by adopting a simple sol-gel method using the chitosan and silica-coated carbon nanotubes. The chitosan-/silica-coated carbon nanotubes' composite membrane showed enhanced mechanical properties, oxidative and thermal stability, and proton conductivity.

7.3.8 Electrospun Nano-Fiber Membranes

Electrospun nanofiber membranes (ENMs) are used for treating wastewater (Matsuura et al., 2010; Botes and Eugene et al., 2010; Qu et al., 2012). This technique includes low energy, low cost, and a lighter process. Higher porosity and surface area-to-volume ratio are advantages of this technique (Balamurugan et al., 2011; Tabe, 2014). The fiber diameter affects the surface area-to-volume ratio and affects membrane porosity. By electrospinning, fiber diameter can be changed by changing solution concentration, surface tension, applied voltage, and spinning distance (Theron et al., 2004; Tabe, 2014). Several natural and synthetic polymers have been electrospun into nanofibers like poly (vinyl chloride), polyvinylidene fluoride, polybenzimidazole, polyurethanes, Nylon-6, polystyrene, poly(vinyl phenol), Kevlar (poly (p-phenylene terephthalamide), poly(vinyl alcohol), polycarbonates, poly(e-capro-lactone), polysulfones, poly(ethylene terephthalate), and many others (Souhaimi and Matsuura, 2011; Feng et al., 2013). These electrospun nanofibers can be used in proton-exchange membrane-fuel cell catalysts. The nanofiber polymer supports catalysts and helps to access the reactant, proton transfer, and electronic continuity for fuel cells (Wang et al., 2014). Graphene is used as a catalyst with electrospun nanofibers

because of its two-dimensional single layer structure, unique electronic properties, and high surface area. Graphene-doped polyacrylonitrile/polyvinylidene fluoride electrospun nanofiber has an improved porosity and electrical conductivity (Wei et al., 2016).

Nanofiber membranes have wide application in the removal of heavy metals in wastewater and particulate microbes. Electrospun polysulfone fiber membrane helps remove the particles from biologically treated wastewater to reduce COD, ammonia, and suspended solids (Xu et al., 2008b). Electrospun nanofibrous membrane doped with polyvinylidene fluoride was used for the separation of approximately 90% of particles from wastewater. These membranes have potential application in reverse osmosis or the ultra-filtration step in a wastewater treatment plant (Gopal et al., 2006).

The electrospun membranes can remove metals like nickel, cadmium, copper, and chromium among others (Nasreen et al., 2013). Amine cellulose acetate/silica nanofiber membranes are used for the removal of chromium (VI) from wastewater (Taha et al., 2012; Lin et al., 2011), lead and copper are removed by using chitosan nanofiber memebarne (Teng et al., 2011). Nanofiber membranes are also used for the removal of salts from water in the desalination process; this requires lower operational pressure, improved flux, and low energy (Nasreen et al., 2013). Composite nanofibrous membranes of polyvinylidene fluoride-co-hexafluoropropene achieved almost 100% salt rejection (Shih, 2011). 99.95% of salt removal was done by using clay nanoparticles polyvinylidene fluoride composite membranes in the distillation process (Prince et al., 2012). The electrospun nano-fibrous membranes also show antimicrobial activity for both bacteria and viruses. Ultra-fine cellulose fibers infused with PAN electrospun nanofiber membranes trapped the virus, having a negative surface charge due to the presence of electrostatic positive change on the membrane (Sato et al., 2011). Similarly, the membrane also showed 100% removal of Escherichia coli bacteria from water. Electrospun membrane composed of Ag nanoparticle doped PAN nanofiber shows its antimicrobial action against gram-negative Escherichia coli and gram-positive Bacillus cereus (Shi et al., 2011).

7.3.9 Hybrid Nano-Membranes

Hybrid membranes have additional functionalities like adsorptions, photocatalysis, or antimicrobial activities. This can be formed by varying the hydrophilicity of membranes, their porosity, pore size, mechanical stability, and charge density. Zeolite-impregnated polysulfone (PSf) nanoparticle membrane removes lead and nickel from wastewater by filtration and adsorption process (Yurekli, 2016). Recently, Wen et al. (2016); applied the hybrid mechanism of adsorption for treatment of radiations tainted water and oil uptake, using sodium titanate nanobelt membrane (Na-TNB) for Sr^{2+} removal. titanate nanobelt membrane adsorption mechanism is based on the formation of radioactive cation stable solid and that was trapped inside the membrane. In addition, this multifunctional membrane has capability to adsorb oils. The removal of oil from wastewater has also been done by using a nanostructured polymer-based (styrene, divinyl benzene, potassium persulfate, sorbitan monooleate) membrane/sorbent (El Naggar et al., 2015). This polymer material membrane shows the efficiency up to 99.75% of oil removal from wastewater in just 75 min. Membrane based adsorbent i.e. carbonized PAN ENMs and multi-walled carbon nanotubes embedded membranes for application in removal of target contaminants (Singh et al., 2013).

7.4 Combination of Biological-Nanotechnology Processes

Nowadays nanocatalysts, nanoabsorbent, nanotubes, nanostructured catalytic membranes, nanopowder and micromolecules are used to improve water quality (Gupta, et al., 2006; Diallo and Savage, 2005). Experiment result already reported the biological wastewater treatment process with advance nanotechnology resulted in efficient water purification system (Gupta, et al., 2006; Diallo and Savage, 2005; Yin et al., 2013).

7.4.1 Nanoparticle-Doped Algal Membrane Bioreactor (A-MBR)

Water purification by algae cultivation is one of the potential techniques. Wastewater cantain micronutrients, trace metals, vitamins and salts of NO^{-3}, $PO_4{}^{3-}$ with Ca, Na, K, and NH^{4+}. These are essential for growth several types the algae (Abou-Shanab et al., 2013; Chong et al., 2000). Solutions are a mixture of these chemical salts and water. These solutions in the presence of light and carbon dioxide give suitable conditions for algal growth (Grima et al., 2003). Harvesting of algal biomass can be done by sedimentation, air flotation, and centrifugation in support with chemical flocculation. These techniques are costly (Brennan and Owende, 2010). Membrane technology is an advanced approach for algae cultivation, in which cultivation is simply done through a membrane bioreactor (Hu et al., 2015). Membrane technology does not require any chemicals so water can be reused after filtration and simplify the algal biomass separation (Ríos et al., 2012a). Algal biomass was recovered by this method without cell damage and low energy is required (Hu et al., 2015). The membranes are made of polysulfone, polyvinylidene fluoride, and polyethersulfone. They have chemical and physical stability. But the only problem is the membrane fouling due to the hydrophobic mechanism between membrane materials and microbial cells (Ríos et al., 2012b). Enhancement of hydrophilicity and overcoming membrane fouling can be done by plasma treatment (Maximous et al., 2009), surface coating (Kim et al., 2011), and by the incorporation of nanomaterials (Yin et al., 2013). Research studies revealed that nanoparticles enhance the hydrophilicity and reduce membrane fouling. For example, the blending of carbon nanotubes and TiO_2 nanoparticles with PSF hollow fiber membranes (HFMs), results in a modification in surface (hydrophilicity) and antifouling (Yin et al., 2013). Several studies reported (Kim et al., 2011) modification of the polyvinyl with TiO_2 nanoparticles on the top layer of the reverse osmosis, thus minimizing fouling by use of the self-cleaning mechanism under UV radiation. TiO_2 particles as photocatalysis have also been used in pollution control due to their high surface area and hydrophilic properties (Madaeni and Ghaemi, 2007). These characteristics are supports to reduce the hydrophobicity and fouling by incorporation of these nanoparticles with membranes (Madaeni and Ghaemi, 2007). The fabricated PVDF/TiO_2 nanocomposite membranes were used in algal membrane bioreactor (A–MBR) for wastewater treatment. Analysis results showed enhanced nutrient removal of up to 75% of phosphorus and nitrogen by A–MBRs, increased hydrophilic characteristic, and reduced membrane fouling of the PVDF (Moghimifar et al., 2014).

7.5 Conclusions and Future Perspectives

In a present scenario, there is a need for advanced water technologies to purify water, eliminate chemical and biological pollutants, and intensify the industrial production processes of wastewater. This can be achieved by the use of nanotechnology for advanced wastewater treatment processes. Several nanomaterials have been synthesized and tested successfully for wastewater treatment. Like nano-adsorbents (based on oxides, Fe, MnO, ZnO,MgO, CNT), photocatalysts (ZnO, TiO2, CdS, ZnS:Cu, CdS:Eu, CdS:Mn), electrocatalysts (Pt, Pd), and nano-membranes (multiwalled CNTs, electrospun PVDF, PVC, Na-TNB). Furthermore, these nanoparticles can be integrated with biological processes (algal membrane, anaerobic digestion, microbial fuel cell) to improve in water purification. Each technology has its own merits and specific pollutant removal efficiency. The nano-adsorbents have efficient potential to remove heavy metals such as Cr, As, Hg, Zn, Cu, Ni, Pb, and Vd from wastewater. Nanoparticle photocatalysts can be used for treatment of both toxic pollutants and heavy metals, where the modification in catalyst material can provide the capability of using visible region of solar light instead of high cost artificial ultraviolet radiation.

There is no doubt of efficiency of utilization nanomaterials in wastewater treatment; however, this technology has some serious downsides. In order to reduce the health risk there is need a future research to prepare such catalysts having least toxicity to the environment. More work is required to re-evaluate the ecotoxicity potential for each new modification in catalyst and for existing materials. Moreover, further work is required on developing cost-effective methods of synthesizing nanomaterials and testing the efficiency at large scale for successful field application Meanwhile, the impacts and risks of the environmental nanomaterials should be further focused on, and their synthesis via green chemistry to diminish their environmental impact should also be pursued in parallel. Protocols and guidelines should be established to regulate the use of nanomaterials to minimize their impact to human health and aquatic environment. These protocols, standards and guidelines are expected to be built based on our deep understanding on the nano impact.

REFERENCES

Abou-Shanab, R.A.I., Ji, M.K., Kim, H.C., Paeng, K.J., Jeon, B.H. (2013). Microalgal species growing on piggery wastewater as a valuable candidate for nutrient removal and biodiesel production. *J. Environ. Manage.*, 115, 257–264.

Adeleye, A.S., Conway, J.R., Garner, K., Huang, Y., Su, Y., Keller, A.A. (2016). Engineered nanomaterials for water treatment and remediation: Costs, benefits, and applicability. *Chem. Eng. J.*, 286, 640–662.

Akhavan, O. (2009). Lasting antibacterial activities of Ag–TiO2/Ag/a-TiO2 nanocomposite thin film photocatalysts under solar light irradiation. _J. Colloid Interface Sci._, _336_, 117–124.

Ali, I. (2012). New generation adsorbents for water treatment. _Chem. Rev._ _112_(10), 5073–5091.

Ambashta, R.D., Sillanpaa, M. (2010). Water purification using magnetic assistance: A review. _J. Hazard. Mater._, _180_, 38–49.

Amin, M.T., Alazba, A.A., Manzoor, U. (2014). A review of removal of pollutants from water/wastewater using different types of nanomaterials. _Adv. Mater. Sci. Eng._, 825910.

Anjum, M., Al-Makishah, N.H., Barakat, M.A. (2016). Wastewater sludge stabilization using pre-treatment methods. _Process Saf. Environ. Prot._, _102_, 615–632.

Azamat, J., Sattary, B.S., Khataee, A., Joo, S.W. (2015). Removal of a hazardous heavy metal from aqueous solution using functionalized graphene and boron nitride nanosheets: Insights from simulations. _J. Mol. Graphics Modell._, _61_, 13–20.

Babursah, S., Çakmakci, M., Kinaci, C. (2006). Analysis and monitoring: Costing textile effluent recovery and reuse. _Filtr. Sep._, _43_(5), 26–30.

Balamurugan, R., Sundarrajan, S., Ramakrishna, S. (2011). Recent trends in nanofibrous membranes and their suitability for air and water filtrations. _Membranes_, _1_(3), 232–248.

Barakat, M.A., Al-Hutailah, R.I., Hashim, M.H., Qayyum, E.,Kuhn, J.N. (2013a). Titania supported silver-based bimetallic nanoparticles as photocatalysts. _Environ. Sci. Pollut. Res._, _20_(6), 3751–3759.

Barakat, M.A., Ramadan, M.H., Alghamdi, M.A., Al-Garny, S.S., Woodcock, H.L., Kuhn, J.N. (2013b). Remediation of Cu (II), Ni (II), and Cr (III) ions from simulated wastewater by dendrimer/titania composites. _J. Environ. Manage._, _117_, 50–57.

Botes, M., Eugene Cloete, T. (2010). The potential of nanofibers and nanobiocides in water purification. _Crit. Rev. Microbiol._, _36_(1), 68–81.

Brennan, L., Owende, P. (2010). Biofuels from microalgae: A review of technologies for production, processing, and extractions of biofuels and co-products. _Renewable Sustainable. Energy Rev._, _14_, 557–577.

Brumfiel, G. (2003). Nanotechnology: A little knowledge. _Nature_, _424_, 246–248.

Burkhard, R., Deletic, A., Craig, A. (2000). Techniques for water and wastewater management: A review of techniques and their integration in planning. _Urban Water_, _2_(3), 197–221.

Chandrakar, R.K., Baghel, R.N., Chandra, V.K., Chandra, B.P. (2015). Synthesis, characterization and photoluminescence studies of Mn doped ZnS nanoparticles. _Superlattices Microstruct._, _86_, 256–269.

Chaturvedi, S., Dave, P.N., Shah, N.K. (2012). Applications of nanocatalyst in new era. _J. Saudi Chem. Soc._, _16_, 307–325.

Chen, C.C., Lin, C.L., Chen, L.C. (2015). Functionalized carbon-nanomaterial supported palladium nano-catalysts for electrocatalyticglucose oxidation reaction. _Electrochim. Acta_, _152_, 408–416.

Chen, H., Li, J., Shao, D., Ren, X., Wang, X. (2012). Poly (acrylic acid) grafted multiwall carbon nanotubes by plasma techniques for Co(II) removal from aqueous solution. _Chem. Eng. J._, _210_, 475–481.

Chong, A.M.Y., Wong, Y.S., Tam, N.F.Y. (2000). Performance of different microalgal species in removing nickel and zinc from industrial wastewater. _Chemosphere_, _41_, 251–257.

Crini, G., Badot, P.M. (2007). _Traitement et épuration des eaux industrielles polluées._ Besançon, France: Presses Universitaires de Franche-Comté.

Das, S., Sen, B., Debnath, N. (2015). Recent trends in nanomaterials applications in environmental monitoring and remediation. _Environ. Sci. Pollut. Res._, _22_(23), 18333–18344.

Diallo, M.S., Savage, N. (2005). Nanoparticles and water quality. _J. Nano. Res._, _7_(4), 325–330.

Ding, Z., Hu, X., Morales, V.L., Gao, B. (2014). Filtration and transport of heavy metals in graphene oxide enabled sand columns. _Chem. Eng. J._, _257_, 248–252.

Dong, Z., Zhang, F., Wang, D., Liu, X., Jin, J. (2015). Polydopamine mediated surface-functionalization of graphene oxide for heavy metal ions removal. _J. Solid State Chem._, _224_, 88–93.

Dutta, A.K., Maji, S.K., Adhikary, B. (2014). C-Fe2O3 nanoparticles: An easily recoverable effective photo-catalyst for the degradation of rose bengal and methylene blue dyes in the waste-water treatment plant. _Mater. Res. Bull._, _49_, 28–34.

Edelstein, A.S., Cammaratra, R.C. (1998). Nanomaterials: Synthesis, properties and applications. Boca Raton, FL: CRC Press.

El Naggar, A.M., Noor El-Din, M.R., Mishrif, M.R., Nassar, I.M. (2015). Highly efficient nano-structured polymer-based membrane/ sorbent for oil adsorption from O/W emulsion conducted of petroleum wastewater. _J. Dispersion Sci. Technol._, _36_, 118–128.

El Saliby, I.J., Shon, H., Kandasamy, J., Vigneswaran, S. (2008). Nanotechnology for wastewater treatment: In brief. _Encyclopedia of Life Support Syst. (EOLSS)._

Engates, K.E., Shipley, H.J. (2011). Adsorption of Pb, Cd, Cu, Zn and Ni to titanium dioxide nanoparticles: Effect of particle size, solidconcentration and exhaustion. _Environ. Sci. Pollut. Res._, _18_, 386–395.

Feng, C., Khulbe, K.C., Matsuura, T., Tabe, S., Ismail, A.F. (2013). Preparation and characterization of electro-spun nanofiber membranes and their possible applications in water treatment. _Sep. Purif. Technol._, _102_, 118–135.

Ferroudj, N., Nzimoto, J., Davidson, A., Talbot, D., Briot, E., Dupuis, V., Abramson, S. (2013). Maghemite nanoparticles and maghemite/silica nanocomposite microspheres as magnetic Fenton catalysts for the removal of water pollutants. Appl. Catal. B: Environ., _136_, 9–18.

Fu, F., Wang, Q. (2011). Removal of heavy metal ions from wastewaters: A review. _J. Environ. Manage._, _92_, 407–418.

Gao, C., Zhang, W., Li, H., Lang, L., Xu, Z., (2008). Controllable fabrication of mesoporous MgO with various morphologies andtheir absorption performance for toxic pollutants in water. _Cryst. Growth Des._, _8_, 3785–3790.

Ge, F., Li, M.M., Ye, H., Zhao, B.X. (2012). Effective removal of heavy metal ions Cd2+, Zn2+, Pb2+, Cu2+ from aqueous solution by polymer-modified magnetic nanoparticles. _J. Hazard. Mater._, _211_, 366–372.

Gopal, R., Kaur, S., Ma, Z., Chan, C., Ramakrishna, S., Matsuura, T. (2006). Electrospun nanofibrous filtration membrane. _J. Membr. Sci._, _281_, 581–586.

Gopalakrishnan, A., Krishnan, R., Thangavel, S., Venugopal, G., Kim, S.J. (2015). Removal of heavy metal ions from pharmaeffluents using graphene-oxide nanosorbents and study of their adsorption kinetics. _J. Ind. Eng. Chem._, _30_, 14–19.

Grey, D., Garrick, D., Blackmore, D., Kelman, J., Muller, M., Sadoff, C. (2013). Water security in one blue planet: Twenty-first century policy challenges for science. *Philos. Trans. R. Soc. London A*, *371*, 20120406.

Grima, E.M., Belarbi, E.H., Fernandez, F.A., Medina, A.R., Chisti, Y. (2003). Recovery of microalgal biomass and metabolites: Process options and economics. *Biotechnol. Adv.*, *20*(7), 491–515.

Gupta, S.K., Behari, J., Kesari, K.K. (2006). Low frequencies ultrasonic treatment of sludge. *Asian J. Water Environ. Pollut.*, *3*, 101–105.

Gupta, V.K., Agarwal, S., Saleh, T.A. (2011). Synthesis and characterization of alumina-coated carbon nanotubes and their application for lead removal. *J. Hazard. Mater.*, *185*, 17–23.

Gupta, V.K., Tyagi, I., Sadegh, H., Shahryari-Ghoshekand, R., Makhlouf, A.S.H., Maazinejad, B. (2015). Nanoparticles as adsorbent; a positive approach for removal of noxious metal ions: A review. *Sci. Technol. Dev.*, *34*: 195.

Hebeish, A.A., Abdelhady, M.M., Youssef, A.M. (2013). TiO2 nanowire and TiO2 nanowire doped Ag-PVP nanocomposite for antimicrobial and self-cleaning cotton textile. *Carbohydr. Polym.*, *91*(2), 549–559.

Higazy, A., Hashem, M., ElShafei, A., Shaker, N., Hady, M.A. (2010). Development of antimicrobial jute packaging using chitosan and chitosan-metal complex. *Carbohydr. Polym.*, *79*(4), 867–874.

Hilal, N., Al-Zoubi, H., Darwish, N.A., Mohamma, A.W., Arabi, M.A. (2004). A comprehensive review of nanofiltration membranes: Treatment, pretreatment, modelling, and atomic force microscopy. *Desalination*, *170*(3), 281–308.

Hu, W., Yin, J., Deng, B., Hu, Z. (2015). Application of nano TiO2 modified hollow fiber membranes in algal membrane bioreactors for high-density algae cultivation and wastewater polishing. *Bioresour. Technol.*, *193*, 135–141.

Ihsanullah, Al-Khaldi, F.A., Abusharkh, B., Khaled, M., Atieh, M.A., Nasser, M.S., Saleh, T.A., Agarwal, S., Tyagi, I., Gupta, V.K. (2015). Adsorptive removal of cadmium (II) ions from liquid phase using acid modified carbon-based adsorbents. *J. Mol. Liq.* 204, 255–263.

Jamal, A., Rahman, M.M., Khan, S.B., Faisal, M., Akhtar, K., Rub, M.A., Asiri, A.M., Al-Youbi, A.O. (2012). Cobalt doped antimony oxide nano-particles based chemical sensor and photocatalyst for environmental pollutants. *Appl. Surface Sci.*, *261*: 52–58.

Jie, G., Kongyin, Z., Xinxin, Z., Zhijiang, C., Min, C., Tian, C., Junfu, W. (2015). Preparation and characterization of carboxyl multi-walled carbon nanotubes/calcium alginate composite hydrogelnano-filtration membrane. *Mater. Lett.*, *157*, 112–115.

Khajeh, M., Laurent, S., Dastafkan, K. (2013). Nanoadsorbents: Classification, preparation, and applications (with emphasis on aqueous media). *Chem. Rev.*, *113*(10), 7728–7768.

Kim, E.S., Yu, Q., Deng, B. (2011). Plasma surface modification of nanofiltration (NF) thin-film composite (TFC) membranes to improve anti organic fouling. *Appl. Surf. Sci.*, *257*, 9863–9871.

Klug, K.L., Dravid, V.P. (2002). Observation of two-and three dimensional magnesium oxide nanostructures formed by thermal treatment of magnesium diboride powder. *Appl. Phys. Lett.*, *81*, 1687–1689.

Kumar, K.Y., Muralidhara, H.B., Nayaka, Y.A., Balasubramanyam, J., Hanumanthappa, H. (2013). Hierarchically assembled mesoporous ZnO nanorods for the removal of lead and cadmium by using differential pulse anodic stripping voltammetric method. *Powder Technol.*, *239*, 208–216.

Kurian, M., Nair, D.S. (2015). Heterogeneous Fenton behavior of nano nickel zinc ferrite catalysts in the degradation of 4-chlorophenol from water under neutral conditions. *J. Water Process Eng.*, *8*, 37–49.

Kurian, M., Nair, D.S. (2016). Effect of preparation conditions on nickel zinc ferrite nanoparticles: A comparison between sol–gel auto combustion and co-precipitation methods. *J. Saudi Chem. Soc.*, *20*(1), 517–522.

Kurian, M., Nair, D.S., Rahnamol, A.M. (2014). Influence of the synthesis conditions on the catalytic efficiency of NiFe2O4 and ZnFe2O4 nanoparticles towards the wet peroxide oxidation of 4- chlorophenol. *React. Kinet. Mech. Catal.*, *111*, 591–604.

Kyzas, G.Z., Matis, K.A. (2015). Nanoadsorbents for pollutants removal: A review. *J. Mol. Liq.*, *203*, 159–168.

Lee, Y.C., Yang, J.W. (2012). Self-assembled flower-like TiO2 on exfoliated graphite oxide for heavy metal removal. *J. Ind. Eng. Chem.*, *18*(3), 1178–1185.

Li, J., Guo, S., Zhai, Y., Wang, E. (2009). Nafion–grapheme nanocomposite film as enhanced sensing platform for ultrasensitive determination of cadmium. *Electrochem. Commun.*, *11*(5), 1085–1088.

Li, Q.L., Mahendra, S., Lyon, D.Y., Brunet, L., Liga, M.V., Li, D., Alvarez, P.J.J. (2008b). Antimicrobial nanomaterials for water disinfection and microbial control: Potential applications and implications. *Water Res.*, *42*(18), 4591–4602.

Li, Y.H., Ding, J., Luan, Z., Di, Z., Zhu, Y., Xu, C., Wu, D., Wei, B. (2003). Competitive adsorption of Pb2+, Cu2+ and Cd2+ ions from aqueous solutions by multiwalled carbon nanotubes. *Carbon*, *41*, 2787–2792.

Li, Y.H., Wang, S., Wei, J., Zhang, X., Xu, C., Luan, Z., Wu, D., Wei, B. (2002). Lead adsorption on carbon nanotubes. *Chem. Phys. Lett.*, *357*, 263–266.

Liang, P., Shi, T., Li, J. (2004). Nanometer-size titanium dioxideseparation/preconcentration and FAAS determination of trace Zn and Cd in water sample. Int. J. Environ. Anal. Chem., *84*(4), 315–321.

Lin, S.T., Thirumavalavan, M., Jiang, T.Y., Lee, J.F. (2014). Synthesis of ZnO/Zn nano photocatalyst using modified polysaccharides for photodegradation of dyes. *Carbohydr. Polym.*, *105*, 1–9.

Lin, Y., Cai, W., Tian, X., Liu, X., Wang, G., Liang, C. (2011). Polyacrylonitrile/ferrous chloride composite porous nanofibers and their strong Cr-removal performance. *J. Mater. Chem.*, *21*(4), 991–997.

Lingamdinne, L.P., Koduru, J.R., Roh, H., Choi, Y.L., Chang, Y.Y., Yang, J.K. (2016). Adsorption removal of Co(II) from wastewater using graphene oxide. *Hydrometallurgy*, *165*, 90–96.

Liu, H., Gong, C., Wang, J., Liu, X., Liu, H., Cheng, F., Wang, G.,Zheng, G., Qin, C., Wen, S. (2016). Chitosan-/silica-coated carbon nanotubes composite proton exchange membranes for fuel cellapplications. *Carbohydr. Polym.*, *136*, 1379–1385.

Liu, Y., Chen, X., Li, J., Burda, C. (2005). Photocatalytic degradationof azo dyes by nitrogen-doped TiO2 nanocatalysts. *Chemosphere*, *61*(1), 11–18.

Liu, Y., Wang, X., Yang, F., Yang, X. (2008). Excellent antimicrobial properties of mesoporous anatase TiO2 and Ag/TiO2 composite films. *Microporous. Mesoporous. Mater.*, *114*(1), 431–439.

Lubick, N., Betts, K. (2008). Silver socks have cloudy lining| Court bans widely used flame retardant. *Environ. Sci. Technol.*, *42*(11), 3910–3910.

Luo, T., Cui, J., Hu, S., Huang, Y., Jing, C. (2010). Arsenic removal and recovery from copper smelting wastewater using TiO2. *Environ. Sci. Technol.*, *44*(23), 9094–9098.

Madaeni, S.S., Ghaemi, N. (2007). Characterization of self-cleaning RO membranes coated with TiO2 particles under UV irradiation. *J. Membr. Sci.*, *303*(1), 221–233.

Madrakian, T., Afkhami, A., Ahmadi, M., Bagheri, H. (2011). Removal of some cationic dyes from aqueous solutions using magnetic-modified multi-walled carbon nanotubes. *J. Hazard. Mater.*, *196*, 109–114.

Martynkova, G.S., Valaskova, M. (2014). Antimicrobial nanocomposites based on natural modified materials: A review of carbons and clays. *J. Nanosci. Nanotechnol.*, *14*(1), 673–693.

Matsuura, T., Feng, C., Khulbe, K.C., Rana, D., Singh, G., Gopal, R., Kaur, S., Barhate, R.S., Ramakrishna, S., Tabe, S. (2010). Development of novel membranes based on electrospun nanofibers and their application in liquid filtration, membrane distillation, and membrane adsorption. *Maku (Japanese J. Membr.)*, *35*, 119–127.

Maximous, N., Nakhla, G., Wan, W. (2009). Comparative assessment of hydrophobic and hydrophilic membrane fouling in wastewater applications. *J. Membr. Sci.*, *339*(1), 93–99.

Mo, M., Yu, J.C., Zhang, L., Li, S.K. (2005). Self-assembly of ZnO nanorods and nanosheets into hollow microhemispheres and microspheres. *Adv. Mater.*, *17*(6), 756–760.

Moghimifar, V., Raisi, A., Aroujalian, A. (2014). Surface modification of polyethersulfone ultrafiltration membranes by corona plasma assisted coating TiO2 nanoparticles. J. Membr. Sci., *461*, 69–80.

Nasreen, S.A.A.N., Sundarrajan, S., Nizar, S.A.S., Balamurugan, R.,Ramakrishna, S. (2013). Advancement in electrospun nanofibrous membranes modification and their application in water treatment. *Membranes*, *3*(4), 266–284.

Neyens, E., Baeyens, J. (2003). A review of classic Fenton's peroxidation as an advanced oxidation technique. *J. Hazard. Mater.*, *98*(1), 33–50.

Ni, M., Leung, M.K., Leung, D.Y., Sumathy, K. (2007). A review and recent developments in photocatalytic water-splitting using TiO2 for hydrogen production. *Renewable Sustainable Energy Rev.*, *11*(3), 401–425.

Ozmen, M., Can, K., Arslan, G., Tor, A., Cengeloglu, Y., Ersoz, M. (2010). Adsorption of Cu (II) from aqueous solution by using modified Fe3O4 magnetic nanoparticles. *Desalination*, *254*(1), 162–169.

Palimi, M.J., Rostami, M., Mahdavian, M., Ramezanzadeh, B. (2014). Surface modification of Fe2O3 nanoparticles with 3-aminopropyltrimethoxysilane (APTMS): An attempt to investigate surface treatment on surface chemistry and mechanical properties of polyurethane/Fe2O3 nanocomposites. *Appl. Surf. Sci.*, *320*, 60–72.

Parida, K.M., Kanungo, S.B., Sant, B.R. (1981). Studies on MnO2-I. Chemical composition, microstructure and other characteristics of some synthetic MnO2 of various crystalline modifications. *Electrochim. Acta*, *26*, 435–443.

Parsons, S., Jefferson, B. (2006). *Introduction to Potable Water Treatment Processes*. Hoboken, NJ: Wiley-Blackwell.

Pendergast, M.M., Hoek, E.M.V. (2011). A review of water treatment membrane nanotechnologies. *Energy Environ. Sci.*, *4*(6), 1946–1971.

Petrinic, I., Andersen, N.P.R., Sostar-Turk, S., Le Marechal, A.M. (2007). The removal of reactive dye printing compounds using nanofiltration. *Dyes Pigm.*, *74*, 512–518.

Petronella, F., Truppi, A., Ingrosso, C., Placido, T., Striccoli, M., Curri, M.L., Agostiano, A., Comparelli, R. (2016). Nanocomposite materials for photocatalytic degradation of pollutants. *Catal. Today.* http://dx.doi.org/10.1016/j.cattod.2016.05.048.

Popov, V.N. (2004). Carbon nanotubes: Properties and application. *Mater. Sci. Eng. R*,. *43*, 61–102.

Pradeep, T., Anshup. (2009). Noble metal nanoparticles for water purification: A critical review. *Thin Solid Films*, *517*(24), 6441–6478.

Prince, J.A., Singh, G., Rana, D., Matsuura, T., Anbharasi, V., Shanmugasundaram, T.S. (2012). Preparation and characterization of highly hydrophobic poly (vinylidene fluoride)–clay nanocomposite nanofiber membranes (PVDF–clay NNMs) for desalination using direct contact membrane distillation. *J. Membr. Sci.*, *397*, 80–86.

Qiu, L., Peng, Y., Liu, B., Lin, B., Peng, Y., Malik, M.J., Yan, F. (2012). Polypyrrole nanotube-supported gold nanoparticles: An efficient electrocatalyst for oxygen reduction and catalytic reduction of 4-nitrophenol. *Appl. Catal. A*, *413*, 230–237.

Qu, X., Alvarez, P.J., Li, Q. (2013). Applications of nanotechnology in water and wastewater treatment. *Water Res.*, *47*(12), 3931–3946.

Qu, X., Brame, J., Li, Q., Alvarez, P.J. (2012). Nanotechnology for a safe and sustainable water supply: Enabling integrated water treatment and reuse. *Acc. Chem. Res.*, *46*(3), 834–843.

Qu, Z., Yan, L., Li, L., Xu, J., Liu, M., Li, Z., Yan, N. (2014). Ultraeffective ZnS nanocrystals sorbent for mercury (II) removal based on size-dependent cation exchange. *ACS Appl. Mater. Interfaces*, *6*(20), 18026–18032.

Rai, M., Yadav, A., Gade, A. (2009). Silver nanoparticles as a new generation of antimicrobials. *Biotechnol. Adv.*, *27*(1), 76–83.

Rajabi, Z., Moghadassi, A.R., Hosseini, S.M., Mohammadi, M. (2013). Preparation and characterization of polyvinylchloride based mixed matrix membrane filled with multi walled carbon nanotubes for carbon dioxide separation. *J. Indus. Eng. Chem.*, *19*(1), 347–352.

Rashidi, H.R., Sulaiman, N.M.N., Hashim, N.A., Hassan, C.R.C., Ramli, M.R. (2015). Synthetic reactive dye wastewater treatment by using nanomembrane filtration. *Desalin. Water Treat.*, *55*(1), 86–95.

Reddy, P.A.K., Reddy, P.V.L., Kwon, E., Kim, K.H., Akter, T., Kalagara, S. (2016). Recent advances in photocatalytic treatment of pollutants in aqueous media. *Environ. Int.*, *91*, 94–103.

Ren, X., Chena, C., Nagatsu, M., Wang, X. (2011). Carbon nanotubes as adsorbents in environmental pollution management: A review. *J. Chem. Eng.*, *170*, 395–410.

Ríos, S.D., Salvadó, J., Farriol, X., Torras, C. (2012). Antifouling microfiltration strategies to harvest microalgae for biofuel. *Bioresour. Technol.*, *119*, 406–418.

Sadegh, H., Shahryari-Ghoshekandi, R., Kazemi, M. (2014). Study in synthesis and characterization of carbon nanotubes decorated by magnetic iron oxide nanoparticles. *Int. Nano Lett.*, *4*, 129–135.

Santhosh, C., Velmurugan, V., Jacob, G., Jeong, S.K., Grace, A.N., Bhatnagar, A. (2016). Role of nanomaterials in water treatment applications: A review. *Chem. Eng. J.*, *306*, 1116–1137.

Sathishkumar, P., Sweena, R., Wu, J.J., Anandan, S. (2011). Synthesis of CuO-ZnO nanophotocatalyst for visible light assisted degradation of a textile dye in aqueous solution. *Chem. Eng. J.*, *171*, 136–140.

Sato, A., Wang, R., Ma, H., Hsiao, B.S., Chu, B. (2011). Novel nanofibrous scaffolds for water filtration with bacteria and virus removal capability. *J. Electron Microsc.*, *60*, 201–209.

Schwarzenbach, R.P., Escher, B.I., Fenner, K., Hofstetter, T.B., Johnson, C.A., Von Gunten, U., Wehrli, B. (2006). The challenge of micropollutants in aquatic systems. *Science*, *313*, 1072–1077.

Shahwan, T., Sirriah, S.A., Nairat, M., Boyacı, E., Eroğ lu, A.E., Scott, T.B., Hallam, K.R. (2011). Green synthesis of iron nanoparticles and their application as a Fenton-like catalyst for the degradation of aqueous cationic and anionic dyes. *Chem. Eng. J.*, *172*, 258–266.

Shamsizadeh, A.A., Ghaedi, M., Ansari, A., Azizian, S., Purkait, M.K. (2014). Tin oxide nanoparticle loaded on activated carbon as new adsorbent for efficient removal of malachite green-oxalate: Nonlinear kinetics and isotherm study. *J. Mol. Liq.*, *195*, 212–218.

Shao, D., Hu, J., Wang, X. (2010a). Plasma induced grafting multiwalled carbon nanotube with chitosan and its application for removal of UO2 2+, Cu2+, and Pb2+ from aqueous solutions. *Plasma Processes Polym.*, *7*, 977–985.

Shao, D., Jiang, Z., Wang, X. (2010b). SDBS modified XC-72 carbon for the removal of Pb (II) from aqueous solutions. *Plasma Processes Polym.*, *7*, 552–560.

Shi, Q., Vitchuli, N., Nowak, J., Caldwell, J.M., Breidt, F., Bourham, M., Zhang, X., McCord, M. (2011). Durable antibacterial Ag/ polyacrylonitrile (Ag/PAN) hybrid nanofibers prepared by atmosphericplasma treatment and electrospinning. *Eur. Polym. J.*, *47*, 1402–1409.

Shih, J.H. (2011). A Study of Composite Nanofiber Membrane Applied in Seawater Desalination by Membrane Distillation (Doctoral dissertation, Master's thesis). National Taiwan University of Science and Technology.

Singh, S., Barick, K.C., Bahadur, D. (2013). Fe3O4 embedded ZnO nanocomposites for the removal of toxic metal ions, organic dyes and bacterial pathogens. *J. Mater. Chem. A.*, *1*, 3325–3333.

Song, S., Liu, Z.W., He, Z.Q., Zhang, A.L., Chen, J.M. (2010). Impacts of morphology and crystallite phases of titanium oxide on the catalytic ozonation of phenol. *Environ. Sci. Technol.*, *44*(10), 3913–3918.

Souhaimi, M.K., Matsuura, T. (2011). *Membrane Distillation: Principles and Applications*. Great Britain: Elsevier.

Srivastava, S., Jadon, N., Jain, R. (2016). Next-generation polymer nanocomposite-based electrochemical sensors and biosensors: A review. *TrAC Trends Anal. Chem.*, *82*, 55–67.

Sun, S.P., Lemley, A.T. (2011). P-nitrophenol degradation by aheterogeneous Fenton-like reaction on nano-magnetite: Process optimization, kinetics, and degradation pathways. *J. Mol. Catal. A*, *349*, 71–79.

Tabe, S. (2014). Electrospun nanofiber membranes and their applications in water and wastewater treatment. In *Nanotechnology for Water Treatment and Purification* (pp. 111–143). Springer.

Taha, A.A., Wu, Y.N., Wang, H., Li, F. (2012). Preparation and application of functionalized cellulose acetate/silica composite nanofibrous membrane via electrospinning for Cr(VI) ion removal from aqueous solution. *J. Environ. Manage.*, *112*, 10–16.

Taherian, F., Marcon, V., van der Vegt, N.F., Leroy, F. (2013). What is the contact angle of water on graphene? *Langmuir*, *29*, 1457–1465.

Tang, W.W., Zeng, G.M., Gong, J.L., Liu, Y., Wang, X.Y., Liu, Y. Y., Liu, Z.F., Chen, L., Zhang, X.R., Tu, D.Z. (2012). Simultaneous adsorption of atrazine and Cu(II) from wastewater by magnetic multi-walled carbon nanotube. *Chem. Eng. J.*, *211*, 470–478.27.

Tang, X., Zhang, Q., Liu, Z., Pan, K., Dong, Y., Li, Y. (2014). Removal of Cu (II) by loofah fibers as a natural and low-cost adsorbent from aqueous solutions. *J. Mol. Liq.*, *199*, 401–407.

Tang, Z., Geng, D., Lu, G. (2005). Size-controlled synthesis of colloidal platinum nanoparticles and their activity for the electrocatalytic oxidation of carbon monoxide. *J. Colloid Interface Sci.*, *287*, 159–166.

Tarigh, G.D., Shemirani, F. (2013). Magnetic multi-wall carbon nanotube nanocomposite as an adsorbent for preconcentration and determination of lead (II) and manganese (II) in various matrices. *Talanta*, *115*, 744–750.

Tawabini, B.S., Khaldi, S.F.A., Khaled, M.M., Atieh, M.A. (2011). Removal of arsenic from water by iron oxide nanoparticlesimpregnated on carbon nanotubes. *J. Environ. Sci. Health*, *46*, 215–223.

Teng, M., Wang, H., Li, F., Zhang, B. (2011). Thioether-functionalized mesoporous fiber membranes: Sol-gel combined electrospun fabrication and their applications for Hg2+ removal. *J. Colloid Interface Sci.*, *355*, 23–28.

Theron, J., Walker, J.A., Cloete, T.E. (2008). Nanotechnology and water treatment: Applications and emerging opportunities. *Crit. Rev. Microbiol.*, 34: 43–69.

Theron, S.A., Zussman, E., Yarin, A.L. (2004). Experimental investigation of the governing parameters in the electrospinning of polymer solutions. *Polymer*, *45*(6), 2017–2030.

Tristao, J.C., Magalhaes, F., Corio, P., Terezinha, M. (2006). Electronic characterization and photocatalytic properties of CdS/ TiO2 semiconductor composite. *J. Photochem. Photobiol. A*, *181*, 152–157.

Turki, A., Guillard, C., Dappozze, F., Ksibi, Z., Berhault, G., Kochkar, H. (2015). Phenol photocatalytic degradation over anisotropic TiO2 nanomaterials: Kinetic study, adsorption isotherms and formal mechanisms. *Appl. Catal. B*, *163*, 404–414.

Tuzen, M., Soylak, M. (2007). Multiwalled carbon nanotubes for speciation of chromium in environmental samples. *J. Hazard. Mater.*, *147*, 219–225.

Vu, H.C., Dwivedi, A.D., Le, T.T., Seo, S.H., Kim, E.J., Chang, Y.S. (2017). Magnetite graphene oxide encapsulated in alginate beads for enhanced adsorption of Cr (VI) and As (V) from aqueous solutions: role of crosslinking metal cations in pH control. *Chem. Eng. J.*, *307*, 220–229.

Wang, H., Yuan, X., Wu, Y., Zeng, G., Chen, X., Leng, L., Li, H. (2015). Facile synthesis of amino-functionalized titanium metalorganic frameworks and their superior visible-light photocatalytic activity for Cr (VI) reduction. *J. Hazard. Mater.*, *286*, 187–194.

Wang, H.Q., Yang, G.F., Li, Q.Y., Zhong, X.X., Wang, F.P., Li, Z. S., Li, Y.H. (2011). Porous nano-MnO2 large scale synthesis via a facile quick-redox procedure and application in a supercapacitor. *New J. Chem.*, *35*, 469–475.

Wang, X., Cai, W., Lin, Y., Wang, G., Liang, C. (2010). Mass production of micro/nanostructured porous ZnO plates and their strong structurally enhanced and selective adsorption performance for environmental remediation. *J. Mater. Chem.*, *20*, 8582–8590.

Wang, X., Richey, F.W., Wujcik, K.H., Ventura, R., Mattson, K., Elabd, Y.A. (2014). Effect of polytetrafluoroethylene on ultralow platinum loaded electrospun/electrosprayed electrodes in proton exchange membrane fuel cells. *Electrochim. Acta*, *139*, 217–224.

Wei, M., Jiang, M., Liu, X., Wang, M., Mu, S. (2016). Graphene doped electrospun nanofiber membrane electrodes and proton exchange membrane fuel cell performance. *J. Power Sources*, *327*, 384–393.

Wen, T., Zhao, Z., Shen, C., Li, J., Tan, X., Zeb, A., Xu, A.W. (2016). Multifunctional flexible free-standing titanate nanobelt membranes as efficient sorbents for the removal of radioactive 90Sr2+ and 137Cs+ ions and oils. Sci. Rep., 6.

WHO (World Health Organization). (2015). Drinking-water: Fact sheet No. 391. http://www.who.int/mediacentre/factsheets/fs391/en/.

Xu, D., Tan, X., Chen, C., Wang, X. (2008a). Removal of Pb (II) from aqueous solution by oxidized multiwalled carbon nanotubes. *J. Hazard. Mater.*, *154*, 407–416.

Xu, Z., Gu, Q., Hu, H., Li, F. (2008b). A novel electrospun polysulfone fiber membrane: Application to advanced treatment of secondary bio-treatment sewage. *Environ. Technol.*, *29*, 13–21.

Yin, J., Deng, B.L. (2015). Polymer-matrix nanocomposite membranes for water treatment. *J. Membr. Sci.*, *479*, 256–275.

Yin, J., Zhu, G., Deng, B. (2013). Multi-walled carbon nanotubes (MWNTs)/polysulfone (PSU) mixed matrix hollow fiber membranes for enhanced water treatment. *J. Membr. Sci.*, *437*, 237–248.

Yin, Y., Zhang, G., Xia, Y. (2002). Synthesis and characterization of MgO nanowires through a vapor-phase precursor method. *Adv. Funct. Mater.*, *12*, 293–298.

Yu, J.C., Ho, W., Lin, J., Yip, H., Wong, P.K. (2003). Photocatalyticactivity, antibacterial effect, and photoinduced hydrophilicity of TiO2 films coated on a stainless steel substrate. *Environ. Sci. Technol.*, *37*, 2296–2301.

Yu, J.C., Yu, J., Ho, W., Zhang, L. (2001). Preparation of highly photocatalytic active nano-sized TiO2 particles via ultrasonic irradiation. *Chem. Commun.*, *19*, 1942–1943.

Yurekli, Y. (2016). Removal of heavy metals in wastewater by using zeolite nano-particles impregnated polysulfone membranes. *J. Hazard. Mater.*, *309*, 53–64.

Zare, K., Najafi, F., Sadegh, H. (2013). Studies of ab initio and Monte Carlo simulation on interaction of fluorouracil anticancer drug with carbon nanotube. *J. Nanostruct. Chem.*, *3*, 1–8.

Zare-Dorabei, R., Ferdowsi, S.M., Barzin, A., Tadjarodi, A. (2016). Highly efficient simultaneous ultrasonic-assisted adsorption of Pb (II), Cd (II), Ni (II) and Cu (II) ions from aqueous solutions by graphene oxide modified with 2, 20-dipyridylamine: central composite design optimization. *Ultrason. Sonochem.*, *32*, 265–276.

Zelmanov, G., Semiat, R. (2008). Phenol oxidation kinetics in water solution using iron (3)-oxide-based nano-catalysts. *Water Res.*, *42*, 3848–3856.

Zhang, C., Sui, J., Li, J., Tang, Y., Cai, W. (2012). Efficient removal of heavy metal ions by thiol-functionalized superparamagnetic carbon nanotubes. *Chem. Eng. J.*, *210*, 45–52.

Zhang, Q., Xu, R., Xu, P., Chen, R., He, Q., Zhong, J., Gu, X. (2014a). Performance study of ZrO2 ceramic micro-filtration membranes used in pretreatment of DMF wastewater. *Desalination*, *346*, 1–8.

Zhang, Y., Yan, L., Xu, W., Guo, X., Cui, L., Gao, L., Wei, Q., Du, B. (2014b). Adsorption of Pb (II) and Hg (II) from aqueous solution using magnetic CoFe2O4 reduced graphene oxide. *J. Mol. Liq.*, *191*, 177–182.

Zhao, G., Li, J., Ren, X., Chen, C., Wang, X. (2011). Few-layered graphene oxide nanosheets as superior sorbents for heavy metal ion pollution management. *Environ. Sci. Technol.*, *45*, 10454–10462.

Zhu, H., Jiang, R., Xiao, L., Chang, Y., Guan, Y., Li, X., Zeng, G. (2009). Photocatalytic decolorization and degradation of Congo Red on innovative crosslinked chitosan/nano-CdS composite catalyst under visible light irradiation. *J. Hazard. Mater.*, *169*, 933–940.

Zhu, Y.Q., Hsu, W.K., Zhou, W.Z., Terrones, M., Kroto, H.W., Walton, D.R.M. (2001). Selective Co-catalysed growth of novel MgO fishbone fractal nanostructures. *Chem. Phys. Lett.*, *347*, 337–343.

8

Functionalized Nanomaterials for Remediation and Environmental Applications

Ayushi Jain, Shweta Wadhawan, Vineet Kumar, and Surinder Kumar Mehta

CONTENTS

8.1 Introduction

The word 'nano' has been taken from the Greek word 'nanos' which means 'small' which is used for substances in the dimensions of a billionth of a meter (10^{-9}) in size. The particles whose one dimension are in the range of 1–100nm are known as nanoparticles. Nanoparticles have different physicochemical properties as compared to bulk materials. In the field of nanotechnology, nanostructures are gaining great recognition from researchers due to their unique properties in various fields viz. optical, physiochemical, and biological. Many researchers around the globe are working on the synthesis and functionalization of nanoparticles by various physical, chemical, and biological methods. The word functionalization means modification of synthesized nanoparticles for tuning their properties for various applications. For a specific application, a particular functionalization process is required. For example, for use in the field of biomedicine, NPs must outperform the traditional agents' context to their minimum toxicity for in vitro applications. On the other hand, for in vivo applications, NPs must avoid non-specific interactions with plasma proteins, so as to reach their projected target efficiently. In addition to this, colloidal stability of NPs must also be maintained under various physiological conditions and a wide range of pH.

In this chapter, we have described the use of various functionalized nanoparticles in the remediation process. The term 'remediation' refers to the removal of harmful and toxic materials from the environment.

8.2 Use of Functionalized NPs in Remediation

Functionalized nanomaterials can be used for the remediation of wastewater and ground water.

8.2.1 Wastewater Remediation

Wastewater remediation can further be categorized into organic, heavy metal and, uranium contaminant remediation.

8.2.1.1 Organic Compounds

Organic water remediation involves dyes and pesticides as main contaminants.

8.2.1.1.1 Dye Remediation

Various kinds of organic effluents viz. dyes, pesticides, insecticides are discharged from industries into water. These contaminants cause water pollution, which imposes harmful impacts on living organisms and aquatic life. Therefore, the remediation of contaminated water is crucial to make it fit for human consumption and for aquatic organisms. Dyes are colored pollutants mainly discharged from textile and paper industries. Dye-contaminated wastewater is hazardous for the health of aquatic animals, as well as other living organisms who use this water for life purposes like drinking, bathing, washing etc [1]. Basic dyes cause skin irritation, allergic dermatitis, shock, increased heartbeat, vomiting, cyanosis, jaundice, tissue necrosis [2]. They may also cause mutations and sometimes even cancer in humans [3]. Aside from this, dyes cause reduced sunlight absorption into water resources which inhibit bacterial growth and are responsible for the inefficient biological degradation of pollutants [4]. Therefore, these toxic substances should be removed prior to their discharge into receiving water bodies [5]. For the past few years, nanoparticles have emerged as attractive materials for the removal of colored pollutants from wastewater, due to their very high specific surface area and high adsorption capacity. Nanoparticles remove dyes from waste water by adsorption of dyes on their surfaces.

Sometimes this adsorption is followed by the degradation of dyes into simple degraded by-products. Nanoparticles are more convenient for the removal process, as compared to other conventional adsorbents like charcoal in terms of their reusability and recyclability. But the problem with bare nanoparticles is their colloidal dispensability and difficult recovery from the solution after adsorption [6, 7]. To overcome this problem, NPs are functionalized with a polymer, surfactants, resin, or other macromolecules [8, 9]. The functionalization process not only enhances the surface area of the NPs and available adsorption sites, but also improves the mechanical, thermal, and chemical stability, which increases the practical applicability and reusability of the adsorbent. To make the nanoparticles more water dispersible and stable they are functionalized with other hydrophilic molecules or polymers like poly(acrylic acid), poly(amido acid), poly(amido amine), and hyperbranched polyglycerol (HPG) [10–15]. He et al. reported a method for the synthesis of multi-hydroxy hyperbranched polyglycerol (HPG) capped Fe_3O_4 (Fe_3O_4/HPG) nanoparticles for the removal of methylene blue, rhodamine B, and Congo red [16]. In addition to the above-mentioned synthetic polymer, some natural polymers or resin-like agar, chitosan cellulose, and cellulose derivatives are also used to functionalize the bare NPs for dye removal. The advantages of using natural polymers are that they are plentiful, inexpensive, biodegradable, and eco-friendly [17]. Also, they have a high density of hydroxyl groups that can easily interact and bind with other functional groups of nanoparticles as well as pollutants [17]. For example, Wang et al. [18] reported hydroxyethyl cellulose (HEC) and hydroxyl propyl methyl cellulose (HPMC) as a stabilizer for the preparation of magnetic Nps (MNPs), which is used for dye discoloration. To prevent the agglomeration of MNPs in solution, naked MNPs are also functionalized with different surfactants. In a unique approach, functionalization of nanoparticles with magnetic nanomaterials is performed to enhance the recovery of nanoparticles after adsorption. For example, magnetic carbon nanomaterials have been prepared by magnetic functionalisation of CNT with Fe_3O_4 for the elimination of methylene blue dye from water [19]. Furthermore, the functionalization and deposition of nanadsorbents onto conventional adsorbents like activated carbon also results in a drastic increase in reactive centers and adsorption capacity. In one of the studies, CuS nanoparticles were loaded onto the activated carbon for the effective adsorption of methylene blue and bromophenol blue [20].

Another approach to remediate dyes from wastewater is degradation of dyes using nanoparticles. NPs have gained a lot of interest due to their catalytic role in the degradation of organic dyes. NPs catalyse the degradation process by absorbing a photon of adequate energy which is equal to their band-gap energy [21]. It results in excitation of an electron from the valence band to the conduction band of the photocatalyst, leading to creation of a hole in valence band. A photocatalyzed reaction is favored by preventing the excited electron and the hole recombination. The excited electron interacts with an oxidant to give a reduced product, and the hole interacts with a reductant to give an oxidized product. The photogenerated electrons reduce the organic pollutant or dissolved O_2 into a superoxide radical anion $O^{2-\bullet}$ [22]. On the other hand, the photogenerated holes oxidize the organic pollutant to carbocation i.e. R^+, or OH^-, and H_2O into $OH\bullet$ radicals.

The $\bullet OH$ radical formed is a very strong oxidizing agent (standard redox potential +2.8 V) which can oxidize most azo dyes to the mineral end-products. Therefore for nanostructure to be a good photocatalyst it should have a small band gap and large surface area. Furthermore, to enhance their degradation capacity various functionalization techniques are available for the complete degradation of dyes to water and carbon dioxide. To improve the properties of a nanophotocatalyst, they are functionalized with different materials like graphene. It provides a high specific surface area, mechanical stability, and high mobility of electrons, which enhances the photocatalytic activity of nanoparticles. For example, CdO nanoparticles were functionalized with graphene for the degradation of methylene blue [23]. The photocatalytic activity was enhanced due to the increased amount of O^{2-} and $\bullet OH$ radicals in the solution containing dye. During the photo catalytic process, NPs can be used in suspension form as well as in immobilized form. For practical application, photocatalyst immobilized on a support is preferred, since it does not require separation and recycling of the photocatalyst from the solution. So, many researchers have focused on the immobilization of photocatalyst on various support materials. For example, the widely used nano photocatalyst TiO_2 has been immobilised on different inorganic supports such as MCM-41 [18], SBA-15 [24], Na-HZSM-5 [25], NaA and CaA zeolite [26], rice husk silica nanocomposite [27]. But there is a limitation with the use of the above-mentioned inorganic supports that photocatalysts i.e TiO_2 may not have even distribution on the support surface because of high aggregation tendency resulting, in agglomeration of the particles. So, to eliminate this problem, functionalization of these inorganic supports like silicates can be done with organic polymers. The various functional groups from organic molecules coordinate with photocatalysts leading to even distribution of nanophotocatalysts on the support surface [28]. In this context, water-soluble tannins with the polyhydroxy groups have been proven to be appropriate, since they can coordinate with many types of metal ions by their many phenolic hydroxyls. One of the scientific studies has shown that tannin can form bonds with $-NH_2$ groups of aminated mesoporous silica through the crosslinking of aldehyde. In this study, TiO_2 supported on oak gall tannin-immobilized hexagonal mesoporous silicate (TiO_2–OGTHMS) were used for the catalytic degradation of DY86. They showed that immobilization of tannin on the surface of HMS results in uniform distribution of TiO_2 nanoparticles on the surface of OGT-HMS without agglomeration, which aggregated on the surface of tannin-free HMS. The photocatalytic performance was also enhanced with TiO_2–OGTx-HMS photocatalysts, as compared to tannin-free TiO_2–HMS [29].

After the photo degradation experiment, photocatalysts are difficult to separate from the solution due to their nano powder form. The separation process consumes a lot of money and time, as they require a very long time to settle, or sometimes centrifugation is employed for their separation. In order to overcome this problem, the magnetic property can be introduced into the photocatalyst by making their nanocomposites with magnetic materials. Once the catalyst gets magnetic behavior, it could be easily separated using a magnet or the magnetic particles settled at the bottom by joining together. Further, the incorporation of MNPs also enhances the photocatalytic activity due to better suppression of the photo-generated electron–hole recombination and the increased light absorption due to the surface plasmon effect of nanoparticles. In a study, Ni NPs were functionalized with SiO_2/TiO_2 for the enhanced degradation of AB1 dye [30]. Similarly, the introduction of Ag nanoparticles helped to suppress the electron–hole pair recombination into SiO_2/TiO_2 [31].

The photocatalytic efficiency of NPs can also be improved by other different approaches such as the modification of NPs by metal and non-metal doping and the use of coupled semiconductors. As the pollutant molecules should be adsorbed by the photocatalyst surface prior to photocatalytic reaction, the specific area of the surface and crystal defects play an important role in the degradation process. Doping or co-alloying of metal oxide NPs with metals (transition) and non-metals leads to an increase in the crystal defects and shifts the band gap energies towards a visible region, thus enhancing the photocatalytic activity. In addition to the band gap shift of the material, doping also leads to change in the structure and oxidation state of the nanomaterials. Dopants also create suitable trap states to capture these photo-generated electrons and holes, which prevent the recombination thus increasing the lifetime of the electron and hole, which enhances the degradation efficiency. Bhattacharya et al. demonstrated the enhanced photocatalytic activity of TiO_2 nanoparticles towards rhodamine B by doping TiO_2 with a different concentration of Mo dopants [32]. Similarly, Mn-doped ZnO NPs show better degradation performance due to the creation of defect states by Mn which act as intermediate states for the excitation of electrons from the valence band to the conduction band [33, 34]. The introduction of some isoelectronic transition metal cations such as 4d (Nb^{5+} or Mo^{6+}) and 5d (Ta^{5+} or W^{6+}) on the surface of TiO_2 leads to mixing of their orbital with the 3d (Ti^{2+}) orbital in conduction band. The introduction of anions such as N^{3-} or P^{3-} leads to mixing of 2p orbital of O^{2-} in the valence band. This mixing lowers the band-gap of the materials providing suitable band edges for many chemical reactions, like dye degradation or water splitting. Some TiO_2 coalloyed systems including Nb^5 and N^{3-} are found to have seven times better degradation efficiency towards methylene blue as compared to anatase TiO_2 and almost twice as compared to commercial P-25 [35]. In addition, Hoang et al. [36] demonstrated the co-incorporation of Ta and N into TiO_2 rutile nanowires for photoelectrochemical water oxidation. [36]

To decrease the band gap and widen absorption range into a visible light region, two or more semiconductor materials can be coupled with each other. The coupled semiconductor materials possess different energy-level systems that increase the charge separation resulting in the suppression of the electron–hole pair recombination under irradiation. This provides enhanced photocatalytic activity [37–39]. These systems also exhibit higher degradation of organic pollutants. For example, Saravanan et al. [40] prepared coupled ZnO/CdO nanocomposites for the efficient photodegradation of methylene blue in visible light.

Furthermore, for the remediation of dyes and other organic pollutants containing waster another new alternative technology i.e. advanced oxidation process (AOP) has been applied. This technique provides better discoloration and degradation of organic pollutants due to their efficiency, low cost, small waste and sludge production, the lack of toxic reagents, and the simplicity of the technology. The electro-Fenton (E-Fenton) process

is an efficient method among AOPs for the oxidation of dyes present in waste water [41–43]. It involves the formation of a strong oxidizing hydroxyl radical ($\bullet OH$) in aqueous solution by the reaction between hydrogen peroxide (H_2O_2) and iron ions as catalyst. In traditional Fenton processes, dissolved iron (Fe^{2+}) is used, which leads to the production of large amounts of sludge at pH above 4 and the formation of a large amount of anions in the treated wastewater. To overcome this limitation, Fe-containing NPs are used at circum neutral pH without sludge formation. In this process, Fe^{2+} ions are immobilized on the nanocatalyst surface – due to this they are not involved in complexation reactions, even at high pH [44–46]. In particular, magnetite (Fe_3O_4) has been employed for the oxidation of various organic compounds in the Fenton process. This activity is attributed to the fact that both Fe^{2+} and Fe^{3+} ions are present in the octahedral sites of the Fe_3O_4 crystal structure, which enhances the decomposition of the H_2O_2 molecule, resulting in the formation of $\bullet OH$ radicals [47, 48]. Furthermore, its magnetism makes it easily separable at the end of the reaction. In addition to this, the efficiency of the oxidation of organic compounds is enhanced with the incorporation of other transition metals in the spinel systems, like $Fe_{3-x}Co_xO_4$, $Fe_{3-x}Cr_xO_4$, $Fe_{3-x}Mn_xO_4$ [49, 50], and $Fe_{3-x}V_xO_4$ [51] during the heterogeneous Fenton process. This formation of substitutional solid solution results in an increase in the surface area of the NPs which increases the adsorptive removal of dye on NPs, as well as the conversion of the H_2O_2 to $\bullet OH$, which is also enhanced due to a larger number of exposed active sites [52, 53]. For example, Barros et al. [54] described the use of substituted magnetic nanoparticles i.e. $Fe_{3-x}Cu_xO_4$ ($0 \leq x \leq 0.25$) for the degradative removal of amaranth food dye.

8.2.1.1.2 Pesticide Remediation

Pesticides are the chemicals which are used to control or remove animal or plant pests and prevent disease caused by pests or insects. These can be classified either on the basis of their purpose or on the basis of different chemical compounds present in them. According to their mode of function, they can be classified as insecticides, herbicides, fungicides, etc. On the basis of the chemical compound, these involve arsenic, pyrethroid carbamates, organophosphates, organochlorides, coumarins, and nitrophenol derivatives. Pesticides can be classified on the basis of use and chemical structure (Figure 8.1). Pesticides pose hazardous effects on the central nervous system (CNS) in humans. Due to the high toxicity of these pesticides, environment protection agencies have set the maximum limits for their contamination level in drinking water. Pesticides mainly used for agricultural purposes either remain in soil (soil pollution) or they enter the water system, depending upon their solubility. Furthermore, degradation products of pesticide in animals, vegetables, and water sources undergo biomagnification at each level of the food chain. Once pesticides enter the environment through direct agricultural use or from industrial waste, they can undergo various changes into more toxic degradation products. Thus a persistent need for the development of the methods for pesticide remediation has attracted the attention of researchers (Figure 8.2).

Nanomaterials have emerged as the most promising tools for the removal and degradation of pesticides to remediate the environmental pollution, due to their unique physicochemical

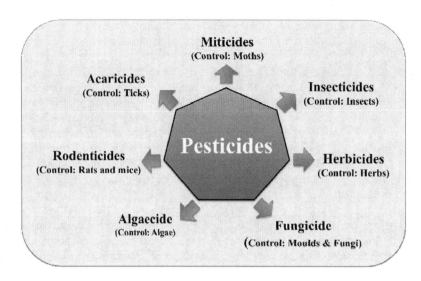

FIGURE 8.1 Types of pesticides on the basis of their use.

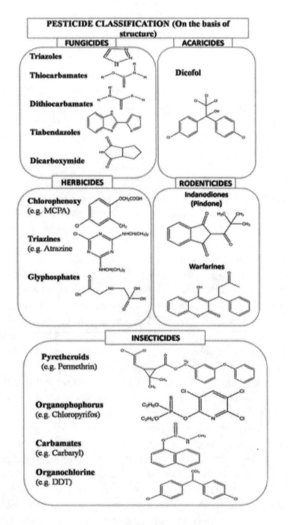

FIGURE 8.2 Classification of pesticides on the basis of their chemical structure.

properties. Pesticides can be degraded or removed as such from wastewater via magnetic separation or nanofiltration.

Pesticides can be degraded by photo catalytic degradation or advanced oxidation technologies. The use of nanoparticles in the above-mentioned techniques is to enhance the degradation capacity, due to their increased surface area. Furthermore, nanoparticles are functionalized to improve their catalytic, optical, magnetic, mechanical, thermal, and other properties. Depending upon the type of functionalizing material, these can be categorized into three classes i.e. ceramic nanocomposites (Al_2O_3/ SiO_3), metal (Fe/MgO), and polymer nanocomposites (polyester/TiO_2) [55, 56]. Many magnetic Fe_2O_3 nanoparticles functionalised with polystyrene for the efficient removal of different organochlorine pesticides [57] have been reported. In addition to polymers, zeolites have also been used to functionalise the metal nanoparticles (ZnO, TiO_2) for the removal of different organophosphate pesticides [58, 59].

The functionalization of nanoparticles with biopolymers is emerging as a more favorable approach for the removal of organic and inorganic pollutants from the environment. Due to the hydrophilic nature of biopolymers, their applicability and separation in aqueous medium is stepped up. Ag and ZnO nanoparticles functionalized with biopolymer chitosan have been used for the removal of pyrethroid and triazines respectively, through adsorption [60, 61].

8.2.1.2 Heavy Metal Ion Remediation

In the previous section, we have gone through the remediation of organic pollutants, i.e dyes using functionalized nanoparticles via adsorption and degradation. On similar lines, we are continuing the discussion on the remediation of heavy metal ions from wastewater using nanoparticles. Heavy metal ion contamination of wastewater poses a major threat to the environment and public health worldwide. Since they are

TABLE 8.1

Hazardous Effects of Heavy Metal Ions and Their Permissible Limits

Sr. No.	Metal Ion	Permissible Limit (mg/L)	Toxicity	Reference
1.	Pb(II)	0.01	Toxic, non-biodegradable, causes anemia, kidney disorder, cardiovascular disease, nervous system damage, cancer, and death	[7, 95]
2.	Cu(II)	1.2	Causes nerve disorders, arthritis, memory problems	[95]
3.	Hg(II)	0.0002	Affect Central nervous system, kidneys, causes paralysis, insomnia	[92]
4.	Co	0.1	Defects in bones, diarrhoea, pulmonary problems paralysis, low blood pressure, and cell mutations	[96]
5.	Cr	Surface water sources: 0.1; portable water sources: 0.05	Epidermal, kidney, and gastric damage	[94]
6.	Cd	0.01	Carcinogenic, bronchiolitis, chronic obstructive pulmonary disease (COPD), emphysema, fibrosis, and skeletal damage.	[7]
7.	Ni	0.020	Carcinogenic, non biodegradable	[97]

non-biodegradable and cannot be metabolised, they are even more hazardous than organic pollutants. The major problem with heavy metals is their tendency for accumulation in the environment, resulting in heavy metal toxicity. These heavy metal ions are introduced to the ecosystem through effluents from various industries such as fertilizer, paints, pigments, batteries, welding, alloy manufacturing, metal fabrication, electroplating, and mining, etc.

From here, these enter food chains through a number of pathways and get accumulated in living organisms by the process of bio magnification over their life span. Being non-biodegradable, they not only affect the aquatic life but also threaten the health of other living organisms, including humans. The hazardous effects of excess of these heavy metal ions, along with their permissible limits, are given in Table 8.1.

Therefore, the remediation of these noxious heavy metal ions from the environment has been an important issue in recent years. Furthermore, the increasing requirement of clean water with a very low concentration of heavy metal ions and environmental regulations on the discharge of these metal ions makes it highly important to build various potential and reliable methods to remediate these metal ions. In the literature, different traditional methods such as ion exchange, precipitation, reverse osmosis, electro dialysis, and adsorption have been used [62–67] for the removal of heavy metal ions from contaminated effluents. Although these methods are efficient and can attain the discharge standards, some of them generate a large quantity of secondary waste. Adsorption is a most feasible and widespread method for the removal of metal ions, due to its cost effectiveness and simple operational conditions. Many conventional adsorbents like metal oxides, clay, minerals, and activated carbon have been used widely for the adsorption of heavy metal ions from wastewater [68–70]. But these sorbents suffer from low adsorption efficiency, recyclability, and reusability. Nanomaterials have emerged as the most capable materials for a novel environmental remediation technique, due to producing little flocculants, recyclability, high surface area, enhanced active sites, large number of functional groups, and having the capability of treating a large amount of wastewater within a short period of time. Although bare nanoparticles are better adsorbents than conventional materials, the separation of nanoparticles from wastewater is difficult.

To overcome these problems, the functionalization of nanoparticles can be done. Further, adsorption capacity and the efficiency of nonmaterial is also enhanced by various functionalization processes. In the context of this, we are going to discuss various functionalized nanoparticles for the metal ion removal from wastewater.

In the field of water treatment, separation, and recycling of adsorbents, there are important aspects. But some of the nanoporous adsorbents suffer the problem of separation and recyclization due to dispensability issues. This problem can be overcome by immobilizing the nanoadsorbent on a polymer solid support. These polymeric supports not only improve the mechanothermal properties of the adsorbent but also provide functional groups to bind with the heavy metal ions. In a study, poly vinyl alcohol (PVA) was reacted with other organic [71, 72] and inorganic materials [73, 74], which supplied hydroxyl groups on the surface of the nanofibers. These hydroxyl groups can also help to adsorb heavy metal ions. For example, Hallaji et al. demonstrated the use of PVA/ZnO nano fibrous adsorbent for the enhanced removal of U (VI), Cu (II), and Ni (II) from aqueous solution [75].

In the last decade, surface modification of magnetic nanoparticles has attracted much attention, due to their easy magnetic separation from the solution. But bare magnetic nanoparticles face the problem of particle-particles aggregation. Therefore in recent years, much focus is on the surface-functionalized MNPs like MNPs-polymers with a core shell nanostructure. One of the best advantages of this functionalization is that the polymer shell prevents the core part from particle aggregation, and enhances the stability of nanostructures towards dispersion. For example, Fe_3O_4 MNPs modified with organosilane and copolymers of acrylic acid (AA) and crotonic acid (CA) have been reported for the removal of Cd^{2+}, Zn^{2+}, Pb^{2+}, Cu^{2+} ion from water [76].

Toxic metal like Cr (VI) existing in anionic form like $HCrO_4^-$, $Cr_2O_7^{2-}$ in a water system brings a greater challenge in separation. To overcome the technical limitation, an effective approach is the development of adsorbents of improved selectivity toward heavy metal ions by incorporation of the metal into complexation with various functional groups like hydroxyl, amino, carboxyl, phosphate, and amide. These functional groups have been chemically grafted by polymerisation on the surface of nanoadsorbents to improve their selectivity towards toxic metal ions.

For example, NH$_2$-functionalized magnetic polymer nanoadsorbents were used for the selective removal of Cr (VI) from aqueous solution [77].

Sometimes to enhance the chemical stability of adsorbent and to make them resistant to chemicals, functionalization is required. In this context, chitosan (CS) has been widely used as a low-cost adsorbent for the removal of heavy metal ions. Chitosan is soluble in diluted organic acid. Therefore, modification processes have been developed to increase its chemical stability in an acidic medium and to make it resistant to chemicals and organic acids. For example, chitosan functionalized with epichlorohydrin has been used for the removal of copper(II), lead(II), and zinc(II) ion from water. Thus, prepared adsorbent was chitosan cross-linked via epichlorohydrin.

Another functionalization technique, where biocompatible nanoadsorbent has been prepared by combining chitosan with another bio molecule like hydroxyapatite (HAp), for the efficient removal of Pb and Cd [78], has been developed. Hydroxyapatite is an important part of hard tissue and exhibit properties of bioactivity and biocompatibility. The combination of hydroxyapatite with chitosan containing amino functional groups enhances the adsorption capacity.

In recent years, adsorption using nano zero-valent iron (NZVI) has gained significant interest for use in the environmental remediation of heavy-metal ions. It is due to the insolubility of NZVI in water, high surface area, and reduction capability towards heavy metal ions. NZVI has been successfully used for the removal of As (III) from water [79].

Furthermore, NZVI gets oxidized in the presence of water and oxygen, followed by the production of ferrous ions which gets converted into magnetite under particular redox conditions and pH. The magnetite being magnetic can be rapidly separated from the solution. Due to its small particle size, the direct implication of NZVI in wastewater treatment systems might cause iron pollution. Further, high agglomeration, mobility, lack of stability leads to the need of some surface functionalization with stabilizers such as chitosan and cyclodextrins (CD). Functional groups like amino in CS and hydroxyl groups in CD have the ability to form chelate complexes with many heavy metal ions. The use of CS as a stabilizing agent can assist detoxification of metal ions in solution. Some studies report the removal of Cr (VI) using Fe-Chitosan complex nanostructures [80, 81]. In some methods, NZVI was entrapped by polymeric CS which was linked with CMβ-CD to produce CS-NZVI-CMβ-CD beads, which were used for the removal of Cr(VI) and Cu(II) [82].

Although nanoparticles play an important role in the removal of heavy metal ions from wastewater as mentioned above, the separation of the adsorbent load from the large volume of water is a critical issue. To overcome the bottleneck of this issue, polymer nanofiber membranes are being used as the base materials of the adsorbents due to their high gas permeability, high porosity, small inter-fibrous pore size and very high specific surface area, which leads to high adsorption capacity. Furthermore, these membranes are easily modified with functional groups. For example, the polyacrylonitrile (PAN) nanofibers modified by diethylene triamine were used as the adsorbent of Cu^{2+} [83]. In one study, Saeed et al. functionalized PAN nanofibers with chelating adsorbent aldoxime for the selective removal of Cu^{2+} and Pb^{2+} [84].

However, for application in heavy metal ion remediation, there is a need to develop a purification method which does not lead to the production of a huge amount of secondary waste and provide rapid recycling of adsorbents at a commercial scale. Magnetic separation has emerged as very promising method for nanoadsorbent separation. But to increase the number of binding sites, these have to be functionalized with ion chelating organic compounds. For example, magnetic nanoadsorbent was developed for the adsorption of heavy metal ions by the surface modification of Fe$_3$O$_4$-NPs with glutaraldehyde, hydroxyapatite [85, 86]. Furthermore, surface functionalization of magnetic NPs with ionic liquids has also been proven useful for the enhanced adsorption of metal ions, especially Cr (VI). Here ionic liquids are physically or chemically immobilized on the surface of NPs. Ionic liquids are first polymerised via free radical polymerisation before immobilisation on the surface of Fe$_3$O$_4$-NPs [87].

Nowadays, carbon-based functional nanomaterials have been emerging as a new type of adsorbent for the removal of pollutants from wastewater, due to ease of operation of the separation and high adsorption capacity. Carbon coating with silicate nanomaterial has attracted a great attention in the field of heavy metal ion remediation due to low cost, improved adsorption capacity, with reference to the conventional carbonaceous counterparts. Zhu et al. synthesized montmorillonite-functionalized carbon nanoparticles to remove Pb^{2+} ions from water [88]. Carbonaceous nanomaterial i.e graphene oxide (GO) is also a good choice for water remediation. Furthermore, its functionalization with some chelating ligands like EDTA enhances its capacity to bind with metal ion through chelation. In this process, functionalization of GO with EDTA is done by reacting N-(trimethoxysilylpropyl) ethylenediamine triacetic acid (EDTA-silane) with GO in a silylation process [89].

The above-mentioned functionalized nanoparticles involved the functionalization of nanoparticles with polymers, inorganic materials, carbon, and biomolecules. Proper surface functionalization can prevent the agglomeration of nanoparticles, thus enhancing the stability and adsorption capacity. In this respect, new functional materials are being explored for the surface functionalization. In some reports, MNPS have been functionalized with mineral oxides which are ubiquitous in soils. Nanosized manganese oxides (MnO$_2$) are among these minerals that have a high surface area, oxidizing capacity, adsorptive abilities, and better stability, even under acidic conditions [90]. Therefore, MnO$_2$-coated Fe$_3$O$_4$ nanoparticles have been prepared and used for the removal of Cu(II), Pb(II), Zn(II), and Cd(II)) from aqueous systems [91].

In addition to the above-mentioned functionalized nanoparticles, metal organic frameworks (MOF) are becoming popular for heavy metal remediation. These have an exceptionally high surface area, porosity, and controllable pore shape and size. Moreover, these can be easily functionalized with various functional groups like thiol, amino, hydroxyl, etc. which makes them versatile adsorbents. For example, thiol-functionalized Cu-MOF, amino-functionalized Cr-MOF and amino-functionalized Zr-MOF have been successfully used for the removal of Hg, Pb, and Cd ions, respectively [92–94]. In the synthesis process, metal salt is heated in a microwave with any organic compound containing functional groups. The product obtained is separated by centrifugation and used as such after washing.

Adsorption capacity and stability of adsorbent can be also be enhanced by capping them with surfactant molecules [95]. In some studies, enhanced adsorption capacity upon capping with surfactant has been demonstrated. Maximum adsorption capacity (q_m) of anatase mesoporous titanium oxide nanofiber was enhanced from 6.9 mg/g to 12.8 mg/g toward Cu(II) ions when capped with surfactant. [7].

Among various types of nanoparticles, cerium oxide is a good adsorbent [96] and poly vinyl pyrrolidone is an effective support for the removal of heavy metal ions from aqueous systems [7]. Additionally, the adsorption capacity for heavy metal ions increases remarkably after modification of surface adsorbent with -SH, -NH$_2$ and -S functional groups [97].

One of the most suitable nanomaterials used for the removal of inorganic and organic pollutants are hollow nanomaterials. Hollow nanomaterials have an internal cavity and different nanostructures. To improve their adsorption performance and recyclability, hollow nanomaterials can be synthesized with complex multi-shelled structures. Both organic and inorganic materials can be used to synthesize hollow nanomaterials. These can be classified as 0D(nanoparticles), 1D(nanowires), 2D(nanotubes), or 3-D hollow structures. Hollow nanostructures show better adsorption capacity than other normal nanomaterials [98]. For example, hollow sub-microspheres of poly (o-phenylenediamine) (PoPD) showed enhanced adsorptivity toward Pb^{2+} as compared to solid PoPD sub-microspheres. Similar trends were shown by hollow nanomaterials for the removal of other heavy metal ions like Hg^{2+}, Cd^{2+}, and Cu^{2+} [99].

8.2.1.3 Uranium Remediation

Growing demand for non-fossil energy has resulted in an increasing number of nuclear power resources. The main fuel for nuclear power plants is radioactive elements. Uranium is the most commonly occurring radioactive contaminant on the earth. The major source of uranium is sea water, which contains 3.3 ppb of uranium and possesses a total uranium reserve of approximately 4.5 billion tonnes [100–102]. Uranium accumulates in the environment by discharges from nuclear power plants, mining, and hydrometallurgical processing sites, hospitals, and research organisations. Improper nuclear waste management is also one of the reasons for the accumulation of uranium in the environment. Therefore, the separation and removal of trace level uranium from the environment is a great technological challenge. Uranium exists in two forms: U (IV) and U (VI). The U (VI) form is more toxic and more mobile than U (IV). Conventional methods for the removal of U are biotic and abiotic chemical reduction [103, 104] solvent extraction [105], ion exchange [106], precipitation [107] and adsorption, etc. But these methods face some limitations in their application, i.e. re-oxidation of reduced U compounds and not applicable at very low pH values. Therefore, there is a strong urge to build more effective technologies for the remediation of U-contaminated acidic and basic media. However, adsorption has been found to be effective and environmentally friendly in U remediation. Many adsorbents like activated carbon, zeolites, bentonite, montmorillonite, and polymeric materials have been developed for the removal of uranium [108–111]. But these conventional sorbents face the problem of sludge formation and recyclization. To overcome these issues, nanoparticles are emerging as a more efficient tool for

the removal of U. Due to their small size, they possess a high surface area and high sorption capacity. In addition to this, nano sorbents can be easily reused and recycled. Further, adsorption capacity can be enhanced by some functionalization processes. For example, adsorption capacity of silica nanoparticles can be improved by functionalizing mesoporous silica nanomaterials for the removal of organic and inorganic pollutants such as toxic metal anions [112], lanthanides [113], actinides [114], and U [115] In some cases, mesoporous silica has been functionalized with hard Lewis base ligands like salicylic acid, glycine, and phosphine for the effective removal of hard acid lanthanides 113 and actinides [116]. Along with actinides, U has also been selectively removed by incorporating hydroxypyridinones onto the surface of mesoporous silica. All these studies report the removal of U at neutral or near neutral pH (5–8) values where U tends to precipitate strongly. To overcome this pH problem and separation issues, researchers have combined the effect of magnetic separation with mesoporous silica and zeolites. Grafting new functional groups on the surface of mesoporous silica greatly enhances adsorption sites density and the selectivity for target metals. Furthermore, the mechanism of adsorption is also changed along with a wide pH range for efficient adsorption [117].

For example, in one of the studies, mesoporous silica was functionalized with iron oxide to make it magnetic mesoporous silica, which was further incorporated with functional group moieties like benzoylthiourea to remove U under acidic and alkaline conditions. Similarly, ethylenediamine-modified magnetic chitosan nanoparticles have been developed for the efficient removal of U from aqueous solution [118]. Using magnetic nanoparticles for the removal of U, one thing has to be kept in the mind: that the magnetic material (Fe$_2$O$_3$) should be capped to avoid its oxidation with air and leaching in acidic conditions.

8.3 Soil and Groundwater Remediation

Soil pollution is closely linked to groundwater contamination, therefore the pollutants or the methods used for soil remediation indirectly affect the quality of groundwater. Soil and groundwater are mainly contaminated with organic substances, metals and pesticides, or halogenated compounds. Landfill seepage, agriculture, and chemical accidents are also the sources of these pollutants. Therefore, maintaining and restoring the quality of water and soil has become one of the greatest challenges in current situation. In most of the cases, conventional remediation and treatment technologies have been proven to be effective up to a certain limit in reducing the levels of pollutants in soil. These involve an ex situ pump and treat method and soil washing. Other in situ conventional remediation methods are thermal treatment and chemical oxidation. These methods are expensive and take a long time for the remediation. The main problem with these methods is that the soil contamination is mainly due to polycyclic aromatic hydrocarbons (PAHs). Because of their hydrophobic nature, they strongly sorb to the soil particles and once they got sorbed they become resistant to degradation by microorganisms. Removal of PAHs from the subsurface of soil by the pump and treat method is not so effective because of the slow kinetics of desorption reactions and very slow dissolution of PAHs from non-aqueous phase mixtures. Somehow surfactants increase the solubility of PAHs and enhance the removal process in the pump and treat method. But the efficiency of surfactant-enhanced removal is limited due to the concentration and stability of surfactant micelles. So the concentration of surfactants must be greater than the critical micelle concentration to perform the remediation process effectively. To overcome the above limitations, there is always a need for more reliable, cheap and fast remediation methods. Nanotechnology here also has emerged as a revolutionary approach in soil remediation. Due to the small size of nanoparticles, they possess a very high surface area which makes the remediation process easy. Remediation process is classified into two categories i.e. in situ and ex situ technologies. The in situ method involves the remediation of contaminated soil and groundwater directly within the subsurface by injecting nanoparticles, and ex situ technologies involves the removal of soil and groundwater from the site followed by treatment on-site or off-site [119]. Both in situ and ex situ remediation of soil and ground water using nanoparticles can be a further one of two types: adsorptive and reactive remediation. In situ adsorptive remediation involves the sequestration of pollutants i.e. chemical removal of contaminants by making its complexes with a binding agent like metal ions or other nanoparticles (e.g. iron oxide) and ex situ adsorptive remediation involves the treatment of contaminated soil or solution with nanoadsorbent, followed by nano filtration. Another remediation method i.e. reactive remediation where contaminants are degraded to less toxic materials either by introducing NZVI directly into the soil (in situ), or by extraction of contaminated soil or solution followed by photo oxidation of target substances in soil (ex situ). As the nanoparticles are more reactive than larger particles – due to their high surface area – they are more effective in the remediation process. Nanoparticles such as NZVI donate their electrons, get easily oxidised and reduce the contaminant species to fewer toxic substances. In situ remediation is more common due its efficiency and cost effectiveness. In addition to NZVI, some carbon-based nanomaterials have also been explored for the removal of organics, phosphates, chlorates, and other metal ions from the soil through adsorption [120].

But there is a restriction with bare nanoparticles: they are under high van der Waals forces and MNPs possess magnetic properties which cause agglomeration of nanoparticles that results in reduction in the surface area. Due to this it becomes difficult for the nanoparticles to travel the required distances in soils and groundwater which leads to the need for the installation of a large number of injection points around the contaminated area. Therefore the remediation process becomes expensive. Traveling over large distance is required to achieve a good distribution of reactive particles in soil matrix. Therefore, to enhance the remediation process, nanoparticles are being functionalized and there is a need to develop a strategy to modify the surface properties of nanoparticles to enhance the stability and dispersion in the subsurface. Although this surface functionalization may reduce the reactivity of nanoparticles up to a certain extent, it promotes their dispersion and delivery to larger distances on contaminated sites. Therefore, researchers are functionalising NPs with stabilizers, i.e. surfactants or polymers to stop them from agglomeration. Commonly used functionalization processes are coating with a stabilizing agent or coupling with trace metals like Pd, Pt, and Ag, etc. Stabilization with negatively charged stabilizer is preferable due to its repulsive interaction with negatively charged

soil particles which reduces the aggregation of particles [121]. The stabilizing agent must provide a degradability feature also with low toxicity. In other words, it must be environmentally friendly in order to avoid the situation that its use is a problem rather than a solution. Coating of nanoparticles, especially iron oxide with stabilisers such as carboxymethyl cellulose (CMC), prevents the agglomeration of nanoparticles which enables them to travel through long distances. For example, phosphate immobilization capacity of nanoscale bare magnetite nanoparticles in the soil by adsorption has been compared with nanoscale particles coated with CMC [122]. Non-stabilized nano magnetite could not permeate through soil column under gravity due to its quick agglomeration into micro particles but the carboxymethyl cellulose-coated NPs were easily transported because of their small size and high negative charge due to the presence of CMC. In some approaches, NZVI are also combined with carbon materials, or embedding of NZVI in oil droplets has also been carried out to facilitate delivery of particles to the target area [123]. In addition to the above-mentioned methods, coupling of nano zerovalent iron with other metals increases their reactivity towards target degradation. For example, in the United States, the NZVI is commonly combined with palladium, to increase the reactivity. But in Europe, these bimetallic particles are not used due to their possible toxicity and the limited additional benefit.

We can conclude that nanoparticles can be effectively used for the remediation of soil and groundwater. The smaller the size of nanoparticles, the higher the reactivity and sorption capacity. But one thing should be kept in mind: decreasing the particle size and increasing reactivity may render a substance more toxic. So before the application of nanoparticles to the soil and groundwater, their toxicity towards the environment should be checked.

8.4 Conclusions

This chapter mainly focuses on the importance of functionalized nanoparticles in environment remediation. Salient features of functionalised NPs of different types such as polymer-coated metal and metal oxide nanoparticles, metal nanoparticles coated with organic functional groups, NZVI and MNPs functionalized with biopolymers, metal and non-metal coalloyed nanoparticles are illustrated. These NPs are superior when compared to bare nanoparticles, because of their enhanced stability and selectivity towards a wide variety of contaminants. In addition to this, another major advantage of functionalized nanoparticles is better removal efficiency. These functionalized nanomaterials are being extensively utilized in wastewater and soil remediation for the removal of dyes, pesticides, heavy metal ions, uranium, etc. Therefore, based on the above discussion, it can be said that functionalized nanoparticles have immense application in the environmental remediation process. Due to their powerful potential, it is expected that their application will increase significantly in the near future.

REFERENCES

1. Sharma, M.K. and Sobti, R.C. (2000). Rec effect of certain textile dyes in Bacillus subtilis mutant research genetics. Toxicology environment. *Mutagen*, *46*, 527–38.

2. Vadivelan, V. and Kumar, K.V. (2005). Equilibrium, kinetics, mechanism, and process design for the sorption of methylene blue onto rice husk. *Journal of Colloid and Interface Science, 286*, 90–100.

3. Eren, E. (2009). Investigation of a basic dye removal from aqueous solution onto chemically modified Unye bentonite. *Journal of Hazardous Materials, 166*, 88–93.

4. Mckay, G. and Duri, B. (1987). Simplified model for the equilibrium adsorption of dyes from mixtures using activated carbon. *Chemical Engineering Processes 22*, 145–56.

5. Iram, M., Guo, C., Guan, Y., Ishfaq, A. and Liu, H. (2010). Adsorption and magnetic removal of neutral red dye from aqueous solution using Fe_3O_4 hollow nanospheres. *Journal of Hazardous Materials, 181*, 1039–50.

6. Abbasizadeh, S., Keshtkar, A.R. and Mousavian, M.A. (2013). Preparation of a novel electrospun poly vinyl alcohol/titanium oxide nanofiber adsorbent modified with mercapto groups for uranium (VI) and thorium (IV) removal from aqueous solution. *Journal of Chemical Engineering, 220*, 161–71.

7. Vu, D., Li. Z., Zhang, H., Wang, W., Wang, Z., Xu, X., Dong, B. and Wang, C. (2012). Adsorption of Cu (II) from aqueous solution by anatase mesoporous TiO_2 nano fibers prepared via electrospinning. *Journal of Colloid and Interface Science, 367*, 429–35.

8. Abbasizadeh, S., Keshtkar, A.R. and Mousavian, M.A. (2014). Sorption of heavy metal ions from aqueous solution by a novel cast PVA/TiO_2 nanohybrid adsorbent functionalized with amine groups. Journal of Industrial and Engineering Chemistry, 1656–64.

9. Mittal, H., Maity, A. and Ray, S.S. (2015). Development of gum karaya and nanosilica based hydrogel nanocomposites for the decolourization of cationic dyes. *Chemical Engineering Journal, 279*, 166–79.

10. Ge, J., Hu, Y., Biasini, M., Dong, C., Guo, J. and Beyermann, W. (2007). One-step synthesis of highly water-soluble magnetite colloidal nanocrystals. *Chemical Engineering Journal, 13*, 7153–61.

11. Juang, T.Y., Kan, S.J., Chen, Y.Y., Tsai, Y.L., Lin, M.G. and Lin, L.L. (2014). Surface-functionalized hyperbranched poly (amido acid) magnetic nanocarriers for covalent immobilization of a bacterial γ-glutamyl transpeptidase. *Molecules, 19*, 4997–5012.

12. Kurtan, U., Baykal, A. and Sozeri, H.J. (2014). Synthesis and characterization of sulfamic-acid functionalized magnetic Fe_3O_4 nanoparticles coated by poly (amidoamine) dendrimer. *Inorganic and Organometallic Polymers, 24*, 948–53.

13. Zhao, L., Chano, T., Morikawa, S., Saito, Y., Shiino, A. and Shimizu, S. (2012). Hyper branched polyglycerol-grafted superparamagnetic iron oxide nanoparticles: Synthesis, characterization, functionalization, size separation, magnetic properties, and biological applications. *Advanced Functional Materials, 22*, 5107–17.

14. Wang, L., Neoh, K.G., Kang, E.T., Shuter, B. and Wang, S.C. (2009). Super paramagnetic hyperbranched polyglycerol-grafted Fe_3O_4 nanoparticles as a novel magnetic resonance imaging contrast agent: An in-vitro assessment. *Advanced Functional Materials, 19*, 2615–22.

15. Wang, S., Zhou, Y., Yang, S. and Ding, B. (2008). Growing hyperbranched polyglycerols on magnetic nanoparticles to resist nonspecific adsorption of proteins. *Colloids and Surfaces B: Interfaces, 67*(1), 122–26.

16. Yangang, He., Zehong, C., Yuting, Q., Bin, X., Linggui, N. and Li, Z. (2015). Facile synthesis and functionalization of hyperbranched polyglycerol capped magnetic Fe_3O_4 nanoparticles for efficient dye removal. *Materials Letters, 151*, 100–3.

17. Carpenter, A.W., Lannoy, C.F. de and Wiesner, M.R. (2015). Cellulose nanomaterials in water treatment technologies. *Environmental Science and Technology, 49*(9), 5277–88.

18. Wang, X.P., Ma, J., Liu, H., and Ning, P. (2015). Synthesis, characterization, and reactivity of cellulose modified nano zero-valent iron for dye discoloration. *Applied Surface Science, 345*, 57–66.

19. Bahgat, M., Farghali, A.A., Rouby, W. El, Khedr, M. and Mohassab Ahmed, M.Y. (2013). Adsorption of methyl green dye onto multi-walled carbon nanotubes decorated with Ni nanoferrite. *Applied Nanoscience, 3*, 251–61.

20. Mazaheri, H., Ghaedi, M., Asfaram, A. and Hajati, S. (2016). Performance of CuS nanoparticle loaded on activated carbon in the adsorption of methylene blue and bromophenol blue dyes in binary aqueous solutions: Using ultrasound power and optimization by central composite design. *Journal of Molecular Liquids, 219*, 667–76.

21. Akir, S., Barras, A., Coffinier, Y., Bououdina, M., Boukherroub, R., and Omrani, A.D. (2016). Eco-friendly synthesis of ZnO nanoparticles with different morphologies and their visible light photocatalytic performance for the degradation of Rhodamine B. *Ceramics International, 42*, 10259–65.

22. Khan, R. and Fulekar, M.H. (2016). Biosynthesis of titanium dioxide nanoparticles using *Bacillus amyloliquefaciens* culture and enhancement of its photocatalytic activity for the degradation of a sulfonated textile dye Reactive Red 31. *Journal of Colloid and Interface Science, 475*, 184–91.

23. Kumar, S., Ojha, A.K. and Walkenfort, B. (2016.) Cadmium oxide nanoparticles grown in situ on reduced graphene oxide for enhanced photocatalytic degradation of methylene blue dye under ultraviolet irradiation. *Journal of Photochemistry & Photobiology B: Biology, 159*, 111–9.

24. Lachheb, H., Ahmed, O., Houas, A. and Nogier, J.P. (2011). Photocatalytic activity of TiO2-SBA-15 under UV and visible light. *Journal of Photochemistry and Photobiology A: Chemical, 226*, 1–8.

25. Zhang, W., Wang, K., Yu, Y. and He, H. (2010). TiO_2/HZSM-5 nano-composite photocatalyst: HCl treatment of NaZSM-5 promotes photocatalytic degradation of methyl orange. *Chemical Engineering Journal, 163*, 62–7.

26. Petkowicz, D.I., Pergher, S.B.C., da Silva, C.D.S., da Rocha, Z.N. and dos Santos, J.H.Z. (2010). Catalytic photodegradation of dyes by in situ zeolite-supported titania. *Chemical Engineering Journal, 158*, 505–12.

27. Huang, X., Li, L., Liao, X. and Shi, B. (2010). Preparation of platinum nanoparticles supported on bayberry tannin grafted silica bead and its catalytic properties in hydrogenation. *Journal of Molecular Catalysis A: Chemical, 320*, 40–6.

28. Huang, X., Liao, X. and Shi, B. (2010). Tannin-immobilized mesoporous silica bead ($BT–SiO_2$) as an effective adsorbent of Cr (III) in aqueous solutions. *Journal of Hazardous Materials, 173*, 33–9.

29. Binaeian, E., Seghatoleslami, N., Chaichi, M.J. and Tayebi, H.A. (2016). Preparation of titanium dioxide nanoparticles supported on hexagonal mesoporous silicate (HMS) modified by oak gall tannin and its photocatalytic performance in degradation of azo dye. *Advanced Powder Technology, 27*, 1047–55.

30. Mahesh, K.P.O. and Kuo, D.H. (2015). Synthesis of Ni nanoparticles decorated SiO$_2$/TiO$_2$ magnetic spheres for enhanced photocatalytic activity towards the degradation of azo dye. *Applied Surface Science, 357*, 433–8.

31. Mahesh, K.P.O., Kuo, D.H. and Huang, B.R. (2015). Facile synthesis of hetero structured Ag-deposited SiO$_2$@TiO$_2$composite spheres with enhanced catalytic activity towards the photo degradation of AB 1 dye. *Journal of Molecular Catalysis A: Chemical, 396*, 290–6.

32. Bhattacharyya, K., Majeed, J., Dey, K.K., Ayyub, P., Tyagi, A.K. and Bharadwaj, S.R. (2014). Effect of Mo-incorporation in the TiO$_2$ lattice: A mechanistic basis for photocatalytic dye degradation. *Journal of Physical Chemistry C, 118*, 15946–62.

33. Mahmood, M.A., Baruah, S. and Dutta, J. (2011). Enhanced visible light photo catalysis by manganese doping or rapid crystallization with ZnO nanoparticles. *Materials Chemistry and Physics, 130*, 531–35.

34. Wang, X.L., Luan, C.Y., Shao, Q., Pruna, A., Leung, C.W., Lortz, R., Zapien, J.A. and Ruotolo, A. (2013). Effect of the magnetic order on the room-temperature band-gap of Mn doped ZnO thin films. *Applied Physical Letters, 102*, 102–12.

35. Breault, T.M. and Bartlett, B.M. (2012). Lowering the band gap of anatase-structured TiO$_2$ by coalloying with Nb and N: Electronic structure and photocatalytic degradation of methylene blue dye. *Journal of Physical Chemistry C, 116*, 5986–94.

36. Hoang, S., Guo, S., and Mullins, C.B. (2012). Co incorporation of N and Ta into TiO$_2$ nanowires for visible light driven photo electrochemical water oxidation. *Journal of Physical Chemistry C, 116*, 23283–90.

37. Xu, C., Cao, L., Su, G., Liu, W., Yu, Y. and Qu, X. (2010). Preparation of ZnO/Cu2O compound photocatalyst and application in treating organic dyes. *Journal of Hazardous Materials, 176*(1–3), 807–13.

38. Liu, Z.L., Deng, J.C., Deng, J.J. and Li, F.F. (2008). Fabrication and photocatalysis of CuO/ZnO nano-composites via a new method. *Material Science and Engineering B, 150*, 99–104.

39. Gopidas, K.R., Bohorquez, M. and Kamat P.V. (1990). Photophysical and photochemical aspects of coupled semiconductors: Charge-transfer processes in colloidal cadmium sulfide-titania and cadmium sulphide silver(I) iodide systems *Journal of Physical Chemistry, 94*, 6435–40.

40. Saravanan, R., Shankar, H., Prakash, T., Narayanan, V. and Stephen, A. (2011). ZnO/CdO composite nanorods for photocatalytic degradation of methylene blue under visible light. *Materials Chemistry and Physics, 125*, 277–80.

41. Garcia-Segura, S., Centellas, F. Arias, C. Garrido, J.A., Rodríguez, R.M., Cabot, P.L. and Brillas, E. (2011). Comparative decolorization of monoazo, diazo and triazo dyes by electro-Fenton process. *Electrochimica Acta, 58*, 303–11.

42. Barros, W.R.P., Franco, P.C., Steter, J.R., Rocha, R.S. and Lanza, M.R.V. (2014). Electro-Fenton degradation of the food dye amaranth using a gas diffusion electrode modified with cobalt (II) phthalocyanine. *Journal of Electro Analytical Chemistry, 722–723*, 46–53.

43. Barros, W.R.P., Alves, S.A., Franco, P.C., Steter, J.R., Rocha, R.S. and Lanza, M.R.V. (2014). Electrochemical degradation of tartrazine dye in aqueous solution using a modified gas electrode. *Journal of Electrochemical Society, 161*, H438–42.

44. Lin, S.S. and Gurol, M.D. (1996). Heterogeneous catalytic oxidation of organic compounds by hydrogen peroxide. *Water Science and Technol, 34*, 57–64.

45. Lin, S.S. and Gurol, M.D. (1998). Decomposition of hydrogen peroxide on iron oxide kinetics, mechanism, and implications. *Environmental Science and Technology, 32*, 1417–23.

46. Kwan, W.P. and Voelker, B.M.M. (2003). Rates of hydroxyl radical generation and organic compound oxidation in mineral-catalyzed fenton-like systems. *Environmental Science and Technology, 37*, 1150–8.

47. Costa, R.C.C. and Moura, F.C.C., Ardisson, J.D., Fabris, J.D. and Lago, R.M. (2008). Highly active heterogeneous Fenton-like systems based on Fe 0/Fe$_3$O$_4$ composites prepared by controlled reduction of iron oxides. *Applied Catalysis B: Environmental, 83*, 131–9.

48. Moura, F.C.C., Araújo, M.H., Costa, R.C.C., Fabris, J.D., Ardisson, J.D., Macedo, W.A.A. and Lago, R.M. (2005). Efficient use of Fe metal as an electron transfer agent in a heterogeneous Fenton system based on Fe0/Fe3O4 composites. *Chemosphere, 60*, 1118–23.

49. Silva, C.N. and Lago, R.M. (2006). Novel active heterogeneous Fenton system based on Fe$_{3-x}$M$_x$O$_4$ (Fe, Co, Mn, Ni): the role of M^{2+} species on the reactivity towards H$_2$O$_2$ reactions. *Journal of Hazardous Materials, 129*, 171–78.

50. Magalhães, F., Pereira, M.C., Botrel, S.E.C., Fabris, J.D., Macedo, W.A., Mendonc, R., Lago, R.M. and Oliveira, L.C.A. (2007). Cr-containing magnetites Fe$_{3-x}$Cr$_x$O$_4$: The role of Cr^{3+} and Fe^{2+} on the stability and reactivity towards H$_2$O$_2$ reactions. *Applied Catalysis A, 332*, 115–23.

51. Liang, X., Zhong, Y., Zhu, S., Ma, L., Yuan, P., Zhu, J., He, H. and Jiang, Z. (2012). The contribution of vanadium and titanium on improving methylene blue decolorization through heterogeneous UV-Fenton reaction catalyzed by their co-doped magnetite. *Journal of Hazardous Materials, 199–200*, 247–54.

52. Yang, S., He, H., Wu, D., Chen, D., Liang, X., Qin, Z., Fan, M., Zhu, J. and Yuan, P. (2009). Decolorization of methylene blue by heterogeneous Fenton reaction using Fe3–xTixO4 (0 ≤ x ≤ 0.78) at neutral pH values. *Applied Catalysis B, 89*, 527–35.

53. Zhong, X., Liang, Y., Zhong, J., Zhu, S., Zhu, P., He, H. and Zhang, J. (2012). Heterogeneous UV/Fenton degradation of TBBPA catalyzed by titanomagnetite: Catalyst characterization, performance and degradation products. *Water Research, 46*, 4633–44.

54. Barros, W.R.P., Steter, J.R. and Lanza, M.R.V. and Tavares, A.C. (2016). Catalytic activity of Fe$_{3-x}$Cu$_x$O$_4$ (0 ≤ x ≤ 0.25) nanoparticles for the degradation of Amaranth food dye by heterogeneous electro-Fenton process. *Applied Catalysis B: Environmental, 180*, 434–41.

55. Aragay, G., Pino, F. and Merkoçi, A. (2012). Nanomaterials for sensing and destroying pesticides. *Chemical Review, 112*, 5317–38.

56. Sahithya, K. and Nilanjana, D. (2015). Remediation of pesticides using nanomaterials: An overview. *International Journal of Chemical Technology Resources, 8*(8), 86–91.

57. Jing, L., Yang, C., and Zongshan, Z. (2013). Effective organochlorine pesticides removal from aqueous systems by magnetic nanospheres coated with polystyrene. *Journal of Wuhan University Technology, 29*, 168–73.

58. Anandan, S., Vinu, A., Venkatachalam, N., Arabindoo, B. and Murugesan V. (2006). Photocatalytic activity of ZnO impregnated Hβ and mechanical mix of ZnO/Hβ in the degradation

of monocrotophos in aqueous solution. *Journal of Molecular Catalysis*, 256, 312–20.

59. Gomez, S., Marchena, C.L., Renzini, M.S., Pizzio, L. and Pierella, L. (2015). In situ generated TiO₂ over zeolitie supports as reusable photocatalysts for the degradation of dichlorvos. *Applied Catalysis B: Environment*, 162, 167–73.

60. Saifuddin, N., Nian, C.Y., Zhan, L.W. and Ning, K.X. (2011). Chitosan silver nanoparticles composite as point of use drinking water filtration system for household to remove pesticides in water. *Asian Journal of Biochemistry*, 6, 142–59.

61. Dehaghi, S.M., Rahmanifar, B., Moradi, A.M. and Azar, P.A. (2014). Removal of permethrin pesticide from water by chitosan-zinc oxide nanoparticles composite as an adsorbent. *Journal of Saudi Chemical Society*, 18, 348–55.

62. Ning, R.Y. (2002). Arsenic removed by reverse osmosis. *Desalination*, 143, 237–41.

63. Von, G.U. (2003). Part 1 Oxidation kinetics and product formation. *Water Research*, 37, 1443–67.

64. Chen, X., Chen, G. and Yue, P.L. (2002). Novel electrode system for electroflotation of wastewater. *Environmental Science and Technology*, 36, 778–83.

65. Hu, Z., Lei, L., Li, Y. and Ni, Y. (2003). Chromium adsorption on high-performance activated carbons from aqueous solution. *Separation and Purification Technology*, 31, 13–8.

66. Reddad, Z., Gerente, C., Andres, Y., Thibault, J.F. and Le Cloirec, P. (2003). Cadmium and lead adsorption by natural polysaccharide in MF membrane reactor: Experimental analysis and modelling. *Water Research*, 37, 3983–91.

67. Ciesielski, W., Lii, C.Y., Yen, M.T. and Tomasik, P. (2003). Internation of starch with salt of metals from the transition groups. *Carbohydrate Polymer*, 51, 47–56.

68. Fonseca, B., Figueiredo, H., Rodrigues, J., Queiroz, A., Tavares, T. (2011). Mobility of Cr, Pb, Cd, Cu and Zn in a loamy sand soil: A comparative study. *Geodermal*, 164, 232–7.

69. Tan, X.L., Fang, M., Chen, C.L., Yu, S.M. and Wang, X.K. (2008). Counterion effects of Ni²⁺ and sodium dodecylbenzene sulfonate adsorption to multiwalled carbon nanotubes in aqueous solution. *Carbon*, 46, 1741–50.

70. Tan, X.L., Fan, Q.H., Wang, X.K. and Grambow, B. (2009). Eu(III) sorption to TiO₂ (anatase and rutile): Batch, XPS, and EXAFS study. *Environmental Science and Technology*, 43, 3115–21.

71. Demir, M.O., Zen, B.O. and Zcelik, S. (2009). Formation of pseudoisocyanine J-aggregates in poly (vinyl alcohol) fibers by electrospinning. *Journal of Physical Chemistry B*, 113, 11568–73.

72. Stoiljkovic, A., Venkatesh, R. Klimov, E., Raman, V., Wendorff, J.H. and Greiner, A. (2009). Poly (styrene-co-n-butyl acrylate) nanofibers with excellent stability against water by electrospinning from aqueous colloidal dispersions. *Macromolecules*, 42, 6147–51.

73. Wu, H., Zhang, R., Liu, X., Lin, D. and Pan, W. (2007). Electrospinning of Fe, Co, and Ni nanofibers: Synthesis, assembly, and magnetic properties. *Chemical Materials*, 19, 3506–11.

74. Pauporte, T. (2007). Highly transparent ZnO/polyvinyl alcohol hybrid films with controlled crystallographic orientation growth. *Crystal Growth and Design*, 7, 2310–5.

75. Hallaji, H., Keshtkar, A.R. and Moosavian, M.A. (2015). A novel electrospun PVA/ZnO nanofiber adsorbent for U(VI), Cu(II) and Ni(II) removal from aqueous solution. *Journal of the Taiwan Institute of Chemical Engineers*, 46, 109–18.

76. Ge, F., Li, M.M., Ye, H. and Zhao, B.X. (2012). Effective removal of heavy metal ions Cd²⁺, Zn²⁺, Pb²⁺, Cu²⁺ from aqueous solution by polymer-modified magnetic nanoparticles. *Journal of Hazardous Materials*, 211–212, 366–72.

77. Shen, H., Pan, S., Zhang, Y., Huang, X., Gong, H. (2012). A new insight on the adsorption mechanism of amino-functionalized nano-Fe₃O₄ magnetic polymers in Cu (II), Cr (VI). *Chemical Engineering Journal*, 183, 180–91.

78. Park, S., Flores, A.G., Chung, Y.S. and Kim, H. (2015). Removal of cadmium and lead from aqueous solution by hydroxyapatite/chitosan hybrid fibrous sorbent: Kinetics and equilibrium studies. *Journal of Chemistry*, 2015, 396290 (12 pages).

79. Rajkanel, S., Che, J. Markgrene, Chulchoi, H. (2006). Arsenic (V) Removal from groundwater using nano scale zero-valent iron as a colloidal reactive barrier material. *Environmental Science and Technology*, 40, 2045–50.

80. Shen, C., Chen, H., Wu, S., Wen, Y., Li, L., Jiang, Z., Li, M. and Liu, W. (2013). Highly efficient detoxification of Cr (VI) by chitosan Fe(III) complex: Process and mechanism studies. *Journal of Hazardous Materials*, 244–245, 689–97.

81. Zimmermann, A.C., Mecabo, Fagundes, A.T. and Rodrigues, C.A. (2010). Adsorption of Cr (VI) using Fe-crosslinked chitosan complex (Ch–Fe). *Journal of Hazardous Materials*, 179, 192–96.

82. Sikder, M.T., Miharac, Y., Islama, M.S., Saitod, T., Tanaka, S. and Kurasaki, M. (2014). Preparation and characterization of chitosan-caboxymethyl-β-cyclodextrin entrapped nano zero-valent iron composite for Cu (II) and Cr (IV) removal from wastewater. *Chemical Engineering Journal*, 236, 378–87.

83. Neghlani, P.K., Rafizadeh, M. and Taromi, F. A. (2011). Preparation of aminated-polyacrylonitrile nanofiber membranes for the adsorption of metal ions: Comparison with microfibers. *Journal of Hazardous Materials*, 186(1), 182–9.

84. Saeed, K., Haider, Oh, S. and Park, T.J. (2008). Preparation of amidoxime-modified polyacrylonitrile (PAN-oxime) nanofibers and their applications to metal ions adsorption. *Journal of Membrane Science*, 322, 400–5.

85. Ozmen, M., Can, K., Arslan, G., Tor, A., Cengeloglu, Y. and. Ersoz, M. (2010). Adsorption of Cu (II) from aqueous solution by using modified Fe₃O₄ magnetic nanoparticles. *Desalination*, 254, 162–9.

86. Feng, Y., Gong, J.L., Zeng, G.M., Niu, Q.Y. Zhang, H.Y., Niu, C. G., Deng, J.H. and Yan, M. (2010). Adsorption of Cd (II) and Zn (II) from aqueous solutions using magnetic hydroxyapatite nanoparticles as adsorbents. *Chemical Engineering Journal*, 162, 487–94.

87. Ferreira, T.A., Rodriguez, J.A., Hernandez, M.E.P., Lara, A.G., Barrado, E. and Hernandez, P. (2017). Chromium (VI) removal from aqueous solution by magnetite coated by a polymeric ionic liquid-based adsorbent. *Materials*, 10, 502.

88. Zhu, K., Jia, H., Wang, F., Zhu, Y., Wang, C. and Ma, C. (2017). Efficient removal of Pb (II) from aqueous solution by modified montmorillonite/carbon composite: Equilibrium, kinetics, and thermodynamics. *Journal of Chemical Engineering Data*, 62, 333–40.

89. Carpio, Isis E.M., Mangadlao, Joey D., Nguyen, Hang N., Advincula, Rigoberto C. and Rodrigues, Debora F. (2014). Graphene oxide functionalized with ethylenediamine triacetic acid for heavy metal adsorption and anti-microbial applications. *Carbon*, 77, 289–301.

90. Post, J.E. (1999). Manganese oxide minerals: Crystal structures and economic and environmental significance. *Proceedings of the National Academy of Sciences of the United States of America*, 96, 3447–54.

91. Kim, E.J., Lee, C.S. Chang, Y.Y., and Chang, Y.S. (2013). Hierarchically structured manganese oxide-coated magnetic nano composites for the efficient removal of heavy metal ions from aqueous systems. *ACS Applied Material Interfaces*, 5, 9628–34.

92. Ke, F., Qiu, L.G., Yuan, Y.P., Peng, F.M., Jiang, X., Xie, A.J., Shen, Y.H. and Zhu, J.F. (2011). Thiol–functionalization of metal–organic framework by a facile coordination based post synthetic strategy and enhanced removal of Hg^{2+} from water. *Journal of Hazardous Materials*, 196, 36–43.

93. Luo, X., Ding, L. and Luo, J. (2015). Adsorptive removal of Pb(II) ions from aqueous samples with amino-functionalization of metal–Organic frameworks MIL–101 (Cr). *Journal of Chemical and Engineering Data*, 60, 1732–43.

94. Wang, K., Gu, J. and Yin, N. (2017). Efficient removal of Pb (II) and Cd(II) using NH_2functionalized Zr-MOFs via rapid microwave-promoted synthesis. *Industrial Engineering and Chemical Research*, 56, 1880–7.

95. Yari, S., Abbasizadeh, S., Mousavi, S.E., Moghaddam, M.S. and Moghaddam, A.Z. (2015). Adsorption of Pb(II) and Cu(II) ions from aqueoussolution by an electrospun CeO_2nanofiberadsorbent functionalized with mercapto groups. *Process Safety and Environmental Protection*, 94, 159–71.

96. Hua, M., Zhang, Sh., Pan, B., Zhang, W., Lv, L., Zhang, Q., (2012). Heavy metal removal from water/wastewater by nano-sized metal oxides: A review. *Journal of Hazardous Materials*, 211–212, 317–31.

97. Irani, M., Keshtkar, A.R., Mousavian, M.A. (2011). Removal of Cd(II)and Ni(II) from aqueous solution by PVA/TEOS/TMPTMShybrid membrane. *Chemical Engineering Journal*, 175, 251–9.

98. Zhang, Y., He, Z., Wang, H., Qi, L., Liu, G. and Zhang, X. (2015). Applications of hollow nanomaterials in environmental remediation and monitoring: A review. *Frontiers of Environmental Science and Engineering*, 9(5), 770–83.

99. Han, J., Dai, J. and Guo, R. (2011). Highly efficient adsorbents of poly (ophenylenediamine) solid and hollow sub-microspheres towards lead ions: A comparative study. *Journal of Colloid and Interface Science* 356(2), 749–56.

100. Davies, R.V., Kennedy, J., Mcilroy, R.W., Spence, R., Hill, K.M. (1964). Extraction of uranium from sea water. *Nature*, 203, 1110–5.

101. Sather, A.C., Berryman, O.B., Rebek, J.J. (2010). Selective recognition and extraction of the uranyl ion. *Journal of American Chemical Society*, 132, 13572–4.

102. Manolis, J.M. and Mercouri, G.K. (2012). Layered metal sulfides capture uranium from seawater. *Journal of American Chemical Society*, 134, 16441–6.

103. Gu, B., Liang, L., Dickey, M.J., Yin, X. and Dai, S. (1998). Reductive precipitation of uranium (VI) by zero-valent iron. *Environmental Science and Technology*, 32, 3366–73.

104. Wu, W.M., Carley, J., Luo, J., Ginder-Vogel, M.A., Cardenas, E., Leigh, M.B., Hwang, C.C., Kelly, S.D., Ruan, C.M., Wu, L.Y., Van Nostrand, J., Gentry, T., Lowe, K., Mehlhorn, T., Carroll, S., Luo, W.S., Fields, M.W., Gu, B.H., Watson, D., Kemner, K.M., Marsh, T., Tiedje, J., Zhou, J.Z., Fendorf, S.,

Kitanidis, P.K., Jardine, P.M. and Criddle, C.S. (2007). In situ bioreduction of uranium (VI) to submicromolar levels and reoxidation by dissolved oxygen. *Environmental Science and Technology*, 41, 5716–23.

105. Lapka, J.L., Paulenova, A., Alyapyshev, M.Y., Babain, V.A., Herbst, R.S., Law, J.D. (2009). Extraction of uranium (VI) with diamides of dipicolinic acid from nitric acid solutions. *Radiochimia Acta*, 97, 291–6.

106. Krestou, A., Xenidis, A. and Panias, D. (2004). Mechanism of aqueous uranium (VI) uptake by hydroxyapatite. *Minerals Engineering*, 17, 373–81.

107. Baeza, A., Fernandez, M., Herranz, M., Legarda, F., Miro, C. and Salas, A. (2006). Removing uranium and radium from a natural water. *Water Air and Soil Pollution*, 173, 57–69.

108. Kilincarslan, A. and Akyil, S. (2005). Uranium adsorption characteristic and thermodynamic behaviour of clinoptilolite zeolite. *Journal of Radioanalytical and Nuclear Chemistry*, 264, 541–8.

109. Psareva, T.S., Zakutevskyy, O.I., Chubar, N.I., Strelko, V.V., Shaposhnikova, T.O., Carvalho, J.R., Correia, M.J.N. (2005). Uranium sorption on cork biomass. *Colloids and Surfaces A: Physicochemical and Engineering Aspects*, 252, 231–6.

110. Mellah, A., Chegrouche, S. and Barkat, M. (2006). The removal of uranium (VI) from aqueous solutions onto activated carbon: Kinetic and thermodynamic investigations. *Journal of Colloid and Interface Science*, 296, 434–41.

111. Camtakan, Z., Erenturk, S.A. and. Yusan, S. (2012). Magnesium oxide nanoparticles: Preparation, characterization, and uranium sorption properties. *Environmental Progress & Sustainable Energy*, 31(4), 536–43.

112. Fryxell, G.E., Liu, J., Hauser T.A., Nie, Z.M., Ferris, K.F., Mattigod, S., Gong, M.L. and Hallen, R.T. (1999). Design and synthesis of selective mesoporous anion traps. *Chemical Materials*, 11, 2148–54.

113. Fryxell, G.E., Wu, H., Lin, Y.H., Shaw, W.J., Birnbaum, J.C., Linehan, J.C., Nie, Z.M., Kemner, K. and Kelly, S. (2004). Lanthanide selective sorbents: Self assembled monolayers on meso porous supports (SAMMS). *Journal of Material Chemistry*, 14, 3356–63.

114. Lin, Y.H., Fiskum, S.K., Yantasee, W., Wu, H., Mattigod, S.V., Vorpagel, E., Fryxell, G.E., Raymond, K.N. and Xu, J.D. (2005). Incorporation of hydroxyl pyridinone ligands into self-assembled monolayers on mesoporous supports for selective actinide sequestration. *Environmental Science and Technology*, 39, 1332–7.

115. Li, D., Egodawatte, S., Kaplan, D.I., Larsen, S.C., Serkiz, S.M., Seaman, J.C. (2016). Functionalized magnetic mesoporous silica nanoparticles for U removal from low and high pH of ground water. *Journal of Hazardous Materials* 317, 494–502.

116. Fryxell, G.E., Lin, Y.H., Fiskum, S., Birnbaum, J.C., Wu, H., Kemner, K. and Kelly, S. (2005). Actinide sequestration using self-assembled monolayers on mesoporous supports. *Environmental Science and Technology* 39, 1324–31.

117. Zhou, L., Zou, H., Wang, Y., Liu, Z., Huang, Z., Luo, T., Adesina, A.A. (2016). Adsorption of uranium (VI) from aqueous solution using phosphonic acid-functionalized silica magnetic microspheres. *Journal of Radioanalytical and Nuclear Chemistry*, 310, 1155–63.

118. Wang, J., Peng, R., Yang, J., Liu, Y. and Hu, X. (2011). Preparation of ethylenediamine-modified magnetic chitosan complex for adsorption of uranyl ions. *Carbohydrate Polymers*, 84, 1169–75.

119. Sharma, H.D. and Reddy, K.R. (2004). *Geo Environmental Engineering: Site Remediation, Waste Containment, and Emerging Waste Management Technologies* (p. 961). Hoboken, NJ: Wiley.

120. Mueller, N.C. and Nowack, B. (2009). Nanotechnology developments for the environment sector report of the observatory NANO EU FP7 project. Available at www.observatorynano.eu/project/document/2790.

121. Petosa, R., Jaisi, D.P., Quevedo, I.R., Elimelech, M. and Tufenkji, N. (2010). Aggregation and deposition of engineered nanomaterials in aquatic environments: Role of physicochemical interactions. *Environmental Science and Technology*, *44*(17), 6532–49.

122. Mueller, N.C. and Nowack, B. (2010). Nanoparticles for remediation: Solving big problems with little particles. *Elements*, *6*, 395–400.

123. Mueller, N.C. and Nowack, B. (2010). Nano zero valent iron: The solution for water and soil remediation? Report of the Observatory NANO. Available at www.observatorynano.eu/project/catalogue/2EV.FO.

9

Quantum Dots: Fabrication, Functionalization, and Applications

Kulvinder Singh and Shikha Sharma

CONTENTS

9.1 Introduction: General Overview and History

In recent years, there is more emphasis on the miniaturization of devices due to advancements in the field of material science, especially nanotechnology (Henglein 1989). After the first synthesis of nanoparticles, and continuing now, various research groups have been actively engaged in this field to develop more concise and developed solid devices that make life more comfortable, the first developed computer was quite huge in size as compared to today's laptop or notebook (Khan et al. 2017). Today, you can store hundreds of GB of data in a very fine chip called a memory card, which is only possible

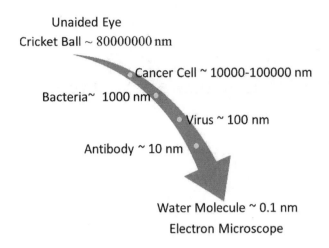

Unaided Eye
Cricket Ball ~ 80000000 nm

Cancer Cell ~ 10000-100000 nm

Bacteria~ 1000 nm

Virus ~ 100 nm

Antibody ~ 10 nm

Water Molecule ~ 0.1 nm
Electron Microscope

FIGURE 9.1   Schematic representation of the size variation.

due to developments in solid state devices. With the advanced development of nanotechnology, there is not only a revolution in electronic devices but it also has a major contribution in the biological, chemical, and physical sciences. Richard Feynman in his historical talk on nanotechnology in 1959 quoted that 'There's Plenty of Room at the Bottom' explained his desire to explore the lower dimensions of the materials (Junk and Riess 2006). Figure 9.1 demonstrates the understanding of nano dimensions.

Today, nanotechnology provides a major contribution towards the improvement of various industries (agricultural, food, biomedical), transport, energy, etc., and makes these fields prosperous day by day (Kuchibhatla et al. 2007; Trindade et al. 2001; Bera et al.2004). Tailoring the size of materials to a nano regime opens up new windows of great possibility for diverse applications, that results in the sustained development of mankind and various sectors of industry. In recent years, more research has been devoted towards controlling the size of nanomaterials at an atomic level that leads to a special class of nanomaterials called 'quantum dots'. Mark Reed named these special classes of nanoparticles as quantum dots (Owen and Brus 2017). These dots were first explored in a glass matrix by Alexey Ekimov in 1981 (Ekimov and Onushchenko 1981). After three years of research, a revolution came in quantum dots when Louis E. Brus in 1984 successfully fabricated the colloidal solution of quantum dots that led to the enhancement in the application field of quantum dots (Brus 1984). When the nanoparticles are constrained in the zero-dimensional, i.e. quantum dots, there is a definite effect on the energy states, as well as photoluminescence emission of nanocrystals that can be understood through the density of the energy states as given in Figure 9.2.

The band gap or the distance between the highest occupied molecular orbital and lowest unoccupied molecular orbital is dependent on dimension (diameter), thus it can be tuned to a different diameter by synthetic conditions promising good optical properties with continuous band gap tune ability. Due to intense working on the modification or functionalization of the surface of quantum dots in the last decades, there is an increase in the applicability of the quantum dots in diverse fields, with improvements in specification.

9.2 Classification of Quantum Dots

On the basis of constituents, broadly quantum dots can be classified as metal-based (cadmium telluride, cadmium sulfide, cadmium selenide, zinc sulfide, zinc telluride, zinc selenide, etc.); metalloids (silicon quantum dots); and carbonaceous quantum dots (carbon dots, graphitic carbon nitride quantum dots, graphene quantum dots). Metallic-based quantum dots are generally fabricated of groups III–V, II–VI, or IV–VI elements of the periodic table and are highly explored nanomaterials in these fields. These traditional quantum dots have three major kinds of structures: homogenous (cadmium sulfide, cadmium telluride); core shell (cadmium selenide @ zinc sulfide and cadmium sulfide @ zinc sulfide); and ternary structures (cadmium tellurium selenide), depending on the synthetic conditions. However, due to their cytotoxic issues and the leakage of heavy metal ions, it hampers their usage in biomedical applications and various other applications. Since then, in order to overcome this issue, other types of quantum dots have been developed having low cytotoxicity and more biocompatibility, i.e. silicon quantum dots. Various methods have been developed to fabricate silicon quantum dots which are elaborated on in the next sections of this chapter. In addition to this, a new emerging class of quantum dots has come into existence, i.e. carbonaceous quantum dots which are more biocompatible, and provide excellent water stability, as compared to any other class of quantum dots. They are also widely used in various biomedical and other applications.

9.3 Properties of Quantum Dots

In a bulk nanomaterial, an electron-hole (exciton) pair is characteristically bound within a specific dimension called the Bohr exciton radius. If the Bohr exciton radius of the semiconductor material is further constrained to a certain limit, major properties (optical, magnetic, and conductive) of semiconductor materials change (Brus 1983). Quantum dots are the semiconducting nanoparticles with a size restriction of 1–10 nm. This name is referred to due to the reason that these quantum dots show a quantum size effect or quantum confinement effect (Chand et al. 2017). With the decrease in size, there is a blue shift in the photoluminescence emission of quantum dots which reveals its quantum confinement effect, as displayed in Figure 9.3 (Chand et al. 2017).

9.3.1 Quantum Confinement

One of the most exciting properties of quantum dots is the quantum confinement effect that tailors the energy states of the nanomaterials. Figure 9.2 shows the change in energy states of the nanomaterials when the size of the nanostructures decreases to a quantum regime (Tan et al. 2014). When the nanomaterials approach the quantum size, their atomic levels are discrete in comparison to the bulk materials, where the atomic levels are continuous (Jayanthi et al. 2007). This effect is predominant when the size of material is sufficiently equal to up to a few nm (>10 nm) and their band gap surpasses kT (k is Boltzmann constant and T is temperature), resulting in restriction of the electron and hole mobility in the crystals. Due to the dependence of the band gap on particle size,

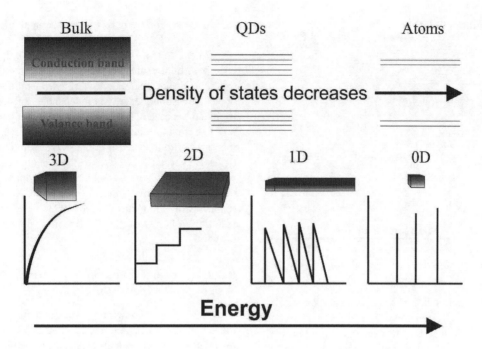

FIGURE 9.2 Schematic presentation of change in density states with energy.

there is a blue shift (decrease in wavelength) in the UV-visible spectrum of the quantum dots, with a decrease in particle size – the peak position depending on the type and composition of the semiconductor material (Ali et al. 2018). This blue shift in the spectrum is optical proof of the quantum confinement effect and is helpful in tuning the band gap of the quantum dots (Li et al. 2018). These quantum confinement effects have consequences in the broadening of the band gap with a reduction in the size of particles. In addition to this, size-dependent band gap property,

i.e. quantum confinement effect, leads to the electronic levels of each energy level demonstrating atomic-like wave character as the solution of the Schrödinger wave equation is quite similar to the electron bound to the nucleus (Ciftja 2019). That is why they are known as 'artificial atoms' with atom-like emission peaks that are very sharp (Ciftja 2019). The band gap of the materials is the energy essential to generate an electron and a hole at ground state, i.e. with no kinetic energy, at a sufficient distance apart that their Coulombic attraction is minimum. When one of the carriers, i.e. electron or hole, approaches the other, this results in the formation of an electron-hole bound pair, which is known as an exciton, with the energy of a few meV, inferior than the band gap of the materials. These excitons act as a hydrogen atom with one exception, in the case of exciton, a hole forms the nucleus. Comparing the masses of both hole and proton, proton is heavier than the hole that affects the results of the Schrödinger wave (Wiegand et al. 2018). The equation expressing exciton Bohr radius is given as

$$r_b = \frac{\hbar^2 \mathcal{E}}{e^2} \left(\frac{1}{m_e} + \frac{1}{m_h} \right) \tag{9.1}$$

Where r_b is the exciton Bohr radius, m_e and m_h are the masses of electron and hole respectively, and the constants $\mathcal{E}$, e, and $\hbar$ optical dielectric constants, charge on an electron, and reduced plank constants respectively. When the dimensions of the nanomaterials approach quantum dots and the radius of the quantum dots meets the Bohr radius (R_b), the motion of the holes and electrons are confined to quantum dots that causes an upsurge of the excitonic energy; this results in a decrease in the wavelength of the UV-vis spectrum (blue shift) of quantum dots and their photoluminescence (Wiegand et al. 2018). In the case of quantum dots, the excitonic energy and their interaction with other excitons is much higher as compared to the bulk counterpart (Klimov 2006). In other words, these particles behave similarly to particles in a one-/two-/three-dimensional box. The energy levels of quantum dots are well-defined, similar to particles in a box (Figure 9.4).

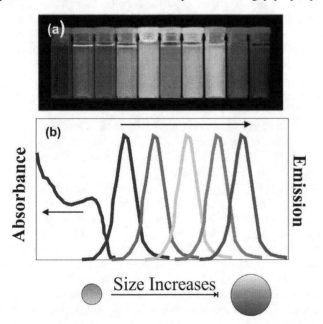

FIGURE 9.3 (a) Ten distinguishable emission colors of ZnS-capped CdSe QDs by a near-UV lamp. From left to right (blue to red), the emission maxima are located at 443, 473, 481, 500, 518, 543, 565, 587, 610, and 655 nm. (b) Size-tunable emission spectra of QDs (Han, M.; Gao, X.; Su, J.Z.; Nie, S. Quantum-dot-tagged microbeads for multiplexed optical coding of biomolecules. *Nat. Biotechnol.* 2001, 19, 631–635. With Copyrights).

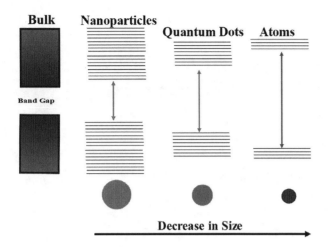

FIGURE 9.4 Effect of size on the band gap and emission of the nanoparticles.

When the nanoparticle's size acquires a dimension similar to the wavelength of the electron wave function, this leads to the quantum confinement effect. Due to their unmatched properties like high surface area, excellent optical properties, and electronic properties, these quantum dots have proven to be a better candidate as compared to the nanoparticles, as well as the bulk counterpart (Bajorowicz et al. 2018). Due to this effect, quantum dots of the same materials with a different size distribution emit a different color of light, e.g. zinc oxide quantum dots of a different size emits a different color (blue to orange) (Asok et al. 2012). There are two main theories for the exciton interactions, i.e. effective mass theory and linear combination of atomic/molecular orbital theory.

9.3.1.1 Effective Mass Theory

The present approach is also called the effective mass theory model, which is based on the theory of the particle in a box and supports the quantum confinement effect of the quantum dots. This theory was developed by Efros and Efros in 1982 and is further advanced by Brus (Brus 1983; Hayrapetyan et al. 2015). The basic assumption of the theory is that the particles when present in the particular boundaries have a finite potential energy and outside that boundary the energy of the particle is assumed to be zero. The relation between energy and the wave vector (k) is given by the relation:

$$E = \frac{\hbar^2 k^2}{2m} \qquad (9.2)$$

Equation (9.2) holds good for an exciton of the semiconductor, resulting in the energy band which is parabolic in nature. The shift of the exciton energy due to the decrease in size of the quantum dots can be evaluated using the following equation:

$$\Delta E_g = \frac{\hbar \pi^2}{2\mu R^2} - \frac{1.8 e^2}{\varepsilon R}$$

$$= \frac{\hbar \varepsilon}{e^2} \left(\frac{1}{m_e} + \frac{1}{m_h} \right) - \frac{1.78 e^2}{\varepsilon R} - 0.248 E_{Ry} \qquad (9.3)$$

Here μ is the reduced mass of the exciton and E_{Ry} is the Rydberg energy.

In Equation 9.3, the first term represents the confinement of energy with the radius of quantum dots. The second term reveals the dependence interaction energy with radius of the quantum dots. The Rydberg energy is independent of the size and is negligible in the case of the semiconductor with low dielectric properties of quantum dots. Equation 9.3 also reveals that band gap increases with a decrease in the size of semiconductor nanoparticles. However, the effective mass model has certain limitations, when the size of the semiconductors nanoparticles approaches that of the quantum regime, the energy and wave vector no longer remain parabolic (Murray et al. 1993; Wang and Herron 1991), resulting in failure of the effective mass model theory for quantum dots.

9.3.1.2 Linear Combination of Atomic/ Molecular Orbital Theory

This theory is also known as the linear combination of the atomic/molecular orbital model, and is more successful in developing the relationship between the electronic structures of semiconductors clusters between atoms, quantum dots, nanoparticles, and bulk structures. In addition, this theory is also much effective in explaining the size and exciton energy relationship of the crystals. Wang and Herron reveal the dependence of the electronic structures on the band gap as well as the size of the crystals (Wang and Herron 1991). In the case of diatomic molecules, the atomic orbitals of the individual atoms combine to form bonding as well as antibonding molecular orbitals. According to this approach, the quantum dots are considered to be a combination of a large number of atoms. As the number of the atoms increases in the crystals, the number of the bonding as well as the antibonding atomic orbitals increases and also the size of quantum dots increases. With an increase in atomic orbitals, both bonding and well as antibonding orbital increase and becomes denser, resulting in the conversion of discrete energy levels to continuous energy bands. The orbitals with higher energy will remain vacant and are called the lowest unoccupied molecular orbital and the orbitals that contain bonding electrons and are of lower energy are called the highest occupied molecular orbital. The energy difference between the upper edge of the highest occupied molecular orbital and the lowest edge of the lowest unoccupied molecular orbital is called the "band gap" or "energy gap" of the materials. The bands are reduced to discrete values as the number of the atomic orbitals is a lot less – that means the smaller nanoparticles or quantum dots have quantized energy levels of excitons or band gap, and their electronic structures and band structures are in between an individual atom and bulk materials. This linear combination of an atomic/molecular orbital model provides better approximation to calculate electronic distribution and the exciton energy or band gap of much smaller nanoparticles, i.e. quantum dots. At the same time, this methodology, due to mathematical complexity and other limitations, is unable to predict the energy levels of those crystals bigger than quantum dots. However, the determination of quantum confinement degree is calculated by the ratio of the radius of quantum dots to the radius of the bulk exciton. When the crystal size is bigger, then the Bohr diameter of semiconductor crystals display a translational motion of coupled exciton that arises due to Coulombic interactions between hole and electron and give rise to single particle behavior. When the radius lies in between the Bohr radius and bulk radius, the adsorption and emission energy of the photo excited electron is evaluated by the magnitude of kinetic energy of the confinement and the exciton interaction. The band gap of the quantum dots can be estimated by using electrochemistry with the aid of cyclic voltammetry using three electrode systems (Shuichiro Ogawa et al. 1997). The technique helps to determine the reduction potential of oxidation of the films of the quantum dots resulting in the calculation of band gap. The quantum dots are first coated on the working electrode surface with platinum electrode that acts as a counter electrode. By using cyclic voltammetry, the band gap of various quantum dots has been calculated successfully like cadmium sulfide, cadmium selenide, cadmium telluride, etc. (Santosh et al. 2001; Kucur et al. 2003; Bera et al. 2004; Sergey et al. 2005; Inamdar et al. 2008).

9.3.2 Photoluminescence Property

Photoluminescence is also one of the major features of various properties of quantum dots. When the quantum dots are excited with the external source of energy (photon of light of different energy), the electron in the valence bond or highest occupied molecular orbital gets excited to the conduction band or lowest unoccupied molecular orbital, resulting in the electron-hole pair i.e. exciton. Depending on the type of energy used for the excitation of the different type, this results in different types of photoluminescence, e.g. the photon is used to get photoluminescence, electric field is used for electroluminescence, electron to get cathodoluminescence, etc. (Lide and Mo 1995). These optical absorption energies are evaluated with the electronic structures of the quantum dots. Once the electron is excited to a higher energy level, it will emit its excessive energy via radiative or non-radiative decay, i.e. photon emission or Auger electron emission respectively, resulting in a recombination of the electron-hole pair to reach ground state energy level.

9.3.2.1 Radiative Relaxation

The bright and highly intense photoluminescence of quantum dots originates due to the radiative relaxation of the quantum dots. These radiative emissions may arise due to band edge emission or formation of some defective states in the quantum dots, and are major factors or additional states in quantum dots that give birth to these radiative emissions. This near band emission is basically known to be the exciton energy or electron hole recombination energy that results in the band gap energy of the quantum dots. In the case of defective emission, the photoluminescence is always at a higher wavelength as compared to near band edge emission, due to the intermediate state formed in between the valence and conduction, giving an excited electron an additional pathway for the radiative emission. Details of both processes are explained in the next section.

9.3.2.1.1 Band Edge Emission

Band edge emission and near band edge emission are common processes for radiative emission in the case of semiconductors and insulators. The process of recombination of the excited electron and the hole in the ground state is called a band edge emission. The exciton is bound by a few meV which reveals that the recombination of the electron-hole pair leads near band edge emission which is slightly less than the band gap of the quantum dots. The ground state of the quantum dots is denoted as $1s_e$–$1s_h$ (exciton state). At room temperature, the full width at half-maxima of near band edge emission of quantum dots lies in between 15 to 30 nm, depending on the diameter of the quantum dots. The photoluminescence can be tailored by the size of zinc selenide quantum dots over the spectrum range 390–440 nm with full width at half maxima as 12.7–16.9 nm (Reiss et al. 2004; Georgios et al. 2004). In the case of bulk semiconductor materials, the photoluminescence emission is easily predictable and quite simply explained by the effective mass model, but at the same time photoluminescence emission of quantum dots is a little bit complicated, e.g. the lifetime of the excited electron in the case of quantum dots is in ns, but in the case of the bulk semiconductors the lifetime is in μs which can be due to the presence of surface states in quantum dots (Bawendi et al. 1990; Efros et al. 1996). The band gap can be either evaluated via absorption spectrum or emission spectrum. In the case of cadmium selenide quantum dots, the photoluminescence shows two emission peaks that originated due to different electron transition centers i.e. $1s_e$-$1s_h$ and $1s_e$-$2s_h$ when observed at a very low temperature (Bawendi et al. 1990). The stoke shift present in the emission is also size-dependent, observed experimentally, e.g. the stoke emission of 5.6 nm cadmium selenide quantum dots and 1.7 nm cadmium selenide quantum dots appears at 2 and 20 meV, respectively. These results can be interpreted via experiments or theoretically that with lowering of the size of quantum dots there is elevation in the distance among the optically prohibited ground exciton state and optical energetic states (Efros and Rosen 2000). In addition to near band edge emission and stoke emission, there are also some irregular emission centers in the quantum dots, i.e. blinking. This term refers to the emission of the quantum dots for some time followed by the dark emission (Nirmal et al. 1996). The reason for the blinking is basically due to the photo-induced ionization process, and further charged quantum dots resulting in the partition between holes and electrons (Efros and Rosen 1997). This theory proposes that the quantum dots show black emission during their ionization state. When cadmium telluride quantum dots were studied as the base model, it was observed the Auger process dominates over the dampening of the photoluminescence emission of ionized quantum dots (Efros and Rosen 1997). The blinking process is not fully reliable with the experimental findings, e.g. experimentally the Auger process displays a linear relationship, while theoretically this process should be linked quadratically to the time of blinking. It is further explored via the Monte Carlo method, when applied theoretically to zinc cadmium selenide quantum dots the blinking process follows the following law, as given in Equation 9.4 (Stefani et al. 2005):

$$P(t) = A.t^{-m} \qquad (9.4)$$

Various other mechanisms have been developed to elaborate the blinking proess but still the precise theory for this model is unknown and there is a lack of proper explanation in the literature (Kuno et al. 2000; van Sark et al. 2002; Stefani et al. 2005; Issac, von Borczyskowski, and Cichos 2005). Carbon dots show UV-visible emission from 290 to 320 nm. This blue shift in the UV-visible absorption appears due to the decrease in the size of quantum dots. Depending on the absorbance of the different sized quantum dots, the emission spectra also changes, e.g. the absorbance at 295, 300, and 325 nm gives an emission at 470, 450, and 383 nm, respectively. Also, the emission spectrum is red-shifted, with an increase in excitation wavelength which is favored by the two mechanisms, i.e. multi-phonon emission and anti-stoke photoluminescence (Sun et al. 2013). The photoluminescence emission of mostly studied carbonaceous quantum dots is endorsed for quantum confinement and/or the surface defects developed in the carbonaceous quantum dots. Still, the origin of photoluminescence emission of carbonaceous quantum dots remains unclear (Xu et al. 2014). One report on the photoluminescence emission of carbon dots reveals that the bright emission originates due to the quantum confinement of the graphitic structure, and not due to the defective emission (Li et al. 2010).

9.3.2.1.2 Defect Emission

Radiative emission other than near band edge emission often originates due to various reasons; impurities or defective states in the crystal lattices is one of the major contributors to photoluminescence emission (Issac et al. 2005). These defects may be either donor or acceptor, depending on the type of impurities, i.e. excessive electron and electron-deficient. These defective states exert Coulombic attraction to the exciton pair, i.e. the hole is attracted towards the donor type defect, and the electron is attracted towards the acceptor type defect. When these types of interactions are modeled on the hydrogen atom, this results in the decrease of binding energy and approaches to the dielectric constant of the materials (Gfroerer 2006). These defective states are of two types, i.e. shallow states that lie in the conduction band of the nanocrystal while the other, i.e. the deep state, lies in the valence band. The shallow defect displays radiative emission and effective generally at a low temperature because at room or elevated temperature the exciton gains energy and leaves the defective states. In the case of deep level defective states, the excitons are long-lived (high life time) and display typically non-radiative emission. Emissions from these two defective states are dependent on the concentration and are helpful in examining their energy. The photoluminescence emission is the combined contribution of both band gap and defective emission, as on tailoring the excitation energy both the distribution and intensity of the emission spectra changes. In addition, the excitation wavelength is also helpful in evaluating the photoexcited state of quantum dots but the lifetime of these is very short (Gfroerer 2006). Due to the large volume-to-surface ratio, these defective states are also expected to be on the surface of the quantum dots and are generally called surface defects. Various methodologies have been developed to improve the surface and make it defect-free, but still there is always a lack of one hundred percent defect free surface of the crystals (Cheng et al. 2006). These defective states arise due to synthesis processes and/or passivation of the surfaces. These surface defects damper the luminescence as well as the electrical property of the quantum dots via non-radiative emission (Djurišić et al. 2004). Yet there are some quantum dots where the surface defect leads to highly intense radiative emission, e.g. zinc oxide. In fact, zinc oxide nanocrystals are a perfect example to understand the surface defects, donor, and

acceptor defects. In addition, the photoluminescence emission, as well as the size control of zinc oxide nanoparticles is very much dependent on the solvent system. When water is taken as a solvent, various defective emissions originate in the crystals which are elaborated below. Mishra et al. have synthesized zinc oxide nanocrystals and decoded the origin of peaks in the photoluminescence spectrum (Mishra et al. 2010). The excited electron dissipates via five different pathways, as shown in the photoluminescence of zinc oxide. Emission appears at 396 nm and originates due to the near band edge emission, i.e. exciton peak (Vanheusden et al. 1998). In addition to peak at 396 nm, i.e. 416, 445, 481, 524 nm, it contributes to the emission spectrum, and originates due to the presence of various defective states in the crystals, such as oxygen vacancy (V_o), zinc vacancy (V_{Zn}), interstitial oxygen (O_i), interstitial zinc (Zn_i), and oxygen antisites (O_{Zn}). A peak centered at 416 nm appears due to the electronic shift from the donor level of Zn_i to the valence band (Fan et al. 2005). Luminescence appears at 481 nm, and is due to the contribution of radiative transition of the shallow donor level to Zn_i to a higher acceptor level of neutral Zn (Tatsumi et al. 2004). Different theories have been developed by various research groups to describe the origin of the final peak, i.e. that at 524 nm. Dingle, in his experiments, reveals that the origin of this peak depends on the presence of a trace amount of copper ions (as impurity) in the lattice structure of the zinc oxide (Dingle 1969). Similarly, another theory explained the presence of oxygen deficient environment that gives rise to the vacancies of oxygen and hence leads transition between V_o-acceptor V_{Zn}, which is responsible for the emission of the peak at 524 nm (Heo et al.2005). In addition to these two theories, another is also present in literature for the origin of the peak at 524 nm and is highly cited – that explains the origin of the green emission due to the presence of interstitial oxygen in the crystal lattice (Wen et al. 2005). The detailed scheme of different photoluminescence emission centers of zinc oxide nanocrystal is illustrated in the scheme given in Figure 9.5.

Liqiang and his group have explained the effect of diverse excitations on the luminescence spectra of zinc oxide (Liqiang Jing et al. 2005). The PL emission ranges from 400 to 500 nm have two intense peaks at 420 and 480 nm, characteristically due to the band edge free and binding excitons, respectively (Zhang et al. 2003; Lide and Mo 1995). Interestingly, when the reaction temperature increases, these defective states diminish and the defective emissions merge into a single peak except near band edge emission (Tam et al. 2006). By increasing the annealing temperature, the defective peak intensity increases and the shape of the emission peak also changes. When the temperature is elevated from 400 to 600 °C in air, both orange and green emission of zinc oxide appears. In addition to the reaction temperature, the reaction atmosphere also plays an important role. At 200 °C, in the presence of forming gas (N_2 and H_2), there is no signal of defective emission of zinc oxide (Figure 9.6). When the temperature is elevated from 400 and 600 °C in an inert atmosphere, green emission with a strong intensity appears.

From above, it is clear that in the aqueous system there is a distinct photoluminescence emission of zinc oxide but the intensity of these emissions is quite low – in fact the main disadvantage is the hydrolysis of Zn^{2+} ions to $Zn(OH)_2$ and $Zn(OH)_4^{2-}$ is very fast and basic. Therefore, controlling the photoluminescence as

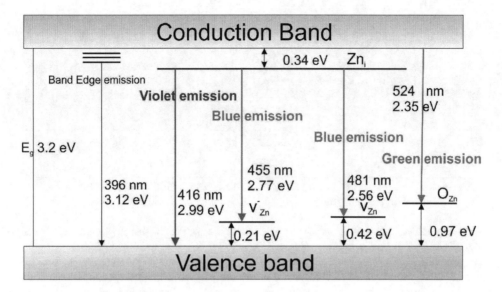

FIGURE 9.5 Origin of emission peaks in ZnO nanoparticles.

well as the growth of nanocrystals in an aqueous medium is a very tedious process. Moving further, Hu et al. have studied the effect of various non-aqueous solvents on the PL emission of zinc oxide nanoparticles (Hu et al. 2010). Figure 9.7 depicts the photoluminescence emission of zinc oxide nanocrystals in different solvents showing intense and broad green emission. Near band

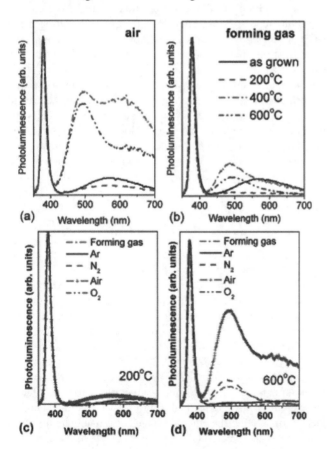

FIGURE 9.6 PL spectra of ZnO nanorods at different annealing temperatures for samples annealed in: (a) air, (b) forming gas, different annealing atmospheres for annealing at (c) 200 °C, and (d) 600 C. (Source: Tam et al. 2006)

edge emission disappeared in the case of methanol and acetone while other solvents show a distinctive near band edge emission at 390 nm. In addition, the photoluminescence emission intensity increases with increases. The intensity and position of the photoluminescence emission of zinc oxide nanoparticles increased with an increase in the size of nanoparticles and by changing the solvent respectively. Actually, the excessive hydroxy groups present are due to the basic solution on the surface of zinc oxide nanoparticles which give rise to the surface defects during the growth process. The larger surface area and more surface defects give rise to lower ultra violet or near band edge emission of zinc oxide nanostructures.

In the case of carbonaceous quantum dots, the origin of photoluminescence is still a mystery as explained above. When the reason for photoluminescence emission is explored, there is other published literature that reveals excitons of carbon, aromatic systems, quantum confinement, trapped states, defects dues to edges, oxygen-containing groups, and zigzag sites contribute to photoluminescence emission. Bao et al. observed electrochemically that irrespective of the particle size, the increase or decrease in wavelength happens due to the surface of the carbon dots (Bao et al. 2011).

9.3.2.1.3 Activator Emission

Photoluminescence emission of quantum dots also arises due to the incorporation of impurities, which is known to be extrinsic photoluminescence. The mechanism that favors extrinsic photoluminescence is exciton transition which can occur between the donor to valence state or the donor to acceptor level or the conduction to acceptor state. In certain cases, this emission center is localized on the activator atom center. In the majority of cases, the rule for the emission is hassle-free in ligand or crystal field, due to orbital's intermixing, such as d-p where the orbitals are fragmented into hyperfine splitting. In addition, d-d transition is allowed in the case of transition metals (Yang et al. 2004). When manganese (II) is doped in cadmium sulfide and the surface is passivated with zinc sulfide, the increase in the lifetime of the excited electron is of ms which arises due to forbidden

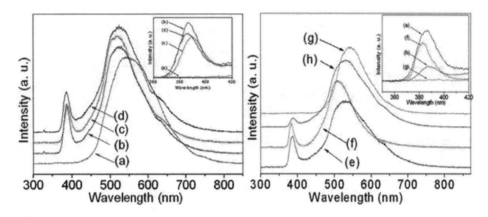

FIGURE 9.7 PL spectra of the ZnO nanostructures in (a) methanol, (b) ethanol, (c) 1-propanol, (d) 1-butanol, (e) 1-pentanol, (f) 1-hexanol, (g) acetone, and (h) isopropanol solutions of NaOH, respectively. (Source: Adapted from Hu et al. 2010)

d-d transition of manganese (II) impurity (Yang et al. 2004). Similarly, when f block ionic impurities are added to the quantum dots structures, there is often f-f transition which is examined in the photoluminescence emission (Lee 1996). The f level impurities are isolated by the crystal field of the host crystal lattice via shielding of exterior p and s orbitals, resulting in the emission similar to atomic spectra. A lot of work has been done on the extrinsic optical emission of doped zinc oxide quantum dots that include transition as well as rare earth elements (Xiu et al. 2006; Ranjani Viswanatha et al. 2006; Wang et al. 2006). When zinc oxide quantum dots have been doped with terbium ions, two types of emission have been observed in the photoluminescence emission: one corresponds to terbium ions, and the other corresponds to the defective state (Liu et al. 2001). When dopant concentration increases, the emission of terbium increases, while at the same time the emission peak of defects decrease. Nevertheless, when zinc oxide is doped with dysprosium ions, it exhibits a relatively strong ultra violet emission with a very weak emission from dysprosium (Wu et al. 2006). The emission of manganese-doped zinc oxide quantum dots depends strongly on the reaction condition (Zhang et al. 2003). The manganese-doped zinc oxide quantum dots leads to dampening of the green emission, as well as other defective peaks in the crystals; at the same time some results also report the blue shift, as well as an increase in the intensity of the near band edge emission. Zinc sulfide based-doped quantum dots are also a very important class of semiconductors. Manganese-doped zinc sulfide quantum dots are a highly explored combination for phosphor. When zinc sulfide is doped, the manganese results in the enhancement of the quantum yield, as well as an emission intensity of the zinc sulfide quantum dots with ns of the lifetime of the excited electron (Bol and Meijerink 1998; Su et al. 2003). The reason for the increase in intensity is accredited by the effectual energy transfer from the zinc sulfide quantum dots to the manganese (II) ions, enabled via the mixing of the orbitals. The intermixing of the orbitals of zinc sulfide and manganese also attributes to the short lifetime of the excited electron. The effect of nitrogen doping on the photoluminescence emission of carbon dots has also been explored by Gong et al. to explore the spectral and luminescence properties of carbon dots (Gong et al. 2015). The emission peak bare carbon dots remain constant when excited from 280–340 nm, however when the excitation wavelength increases from 360–480 nm,

there is a red shift in the emission spectrum of carbon dots. A similar type of emission results was obtained for nitrogen-doped carbon dots. The independent nature of the emission spectra on excitation is due to the core π→π* transitions of graphitic nature, while the excitation-dependent nature of the quantum dots is due to the surface defective state which arises due to the functionalities present on the surface of the quantum dots (n→π*). In addition, the photoluminescence emission intensity of nitrogen-doped carbon dots is much higher as compared to the bare carbon dots, revealing the higher density of surface defects on the surface of carbon dots due to various functionalities of nitrogen. In another report, when carbon dots are doped with nitrogen and sulfur, nitrogen-doped carbon dots show a higher quantum yield as well as the lifetime of the excited electron as compared to sulfur-doped, as well as the bare carbon dots. This is due to the reason that the nitrogen-doped carbon dots fabricate a new kind of surface defective state, giving new a pathway for the excited electron to radiative decay (Wenjing Lu et al. 2015). The photoluminescence emission of nitrogen-doped carbon dots can be tailored from blue, green, and yellow via the reduction of the concentration of doping of nitrogen, and is similar to the result obtained from graphene quantum dots (Wu et al. 2014; Tetsuka et al. 2012).

9.3.2.2 Quantum Yield of Quantum Dots

The measurement of the quantum yield is vital for quantum dots and is also an important property of the quantum dots. It has been observed that the quantum yield values differ for particular quantum dots among different reports. There may be several reasons that include: a different approach to measure quantum yield; a difference in the concentration of quantum dots or sample or both; alteration of the slit width for measurement; instrumental error may be there (van Sark et al. 2002). The inorganic semiconductors may not perform as dye, but in the case of carbonaceous quantum dots, they behave somewhat similarly to dye in terms of quantum yield. The methodology to determine the quantum yield is by relating the emission intensity of the quantum yield with that of the standard taken, followed by the measurement of the absorbance of the quantum dots as well as the standard, with one condition that the absorbance value of both the quantum dots and standard are kept below 0.08. The following relation was used to determine the quantum yield of the quantum dots (Bera et al. 2008a; Qian et al. 2008):

$$QY = QY_{std} \frac{1-10^{-A_{std}}}{1-10^{-A}} \times \frac{\eta^2}{\eta^{2}_{std}} \times \frac{I}{I_{std}} \qquad (9.5)$$

In the above equation, std refers to standard, η refers to refractive index, I stands for emission intensity. When measuring the intensity for standard and sample, the excitation wavelength should be the same.

Dai et al. (2018) have synthesized cadmium selenide quantum dots that have been shown to be the quantum yield of about 64% using rhodamine 101 in ethanol which is shown to be stable for up to 120 days (Dai et al. 2018). When core shell quantum dots have been prepared, i.e. cadmium selenide/zinc sulfide, the quantum yield is hampered to the value of 50%, with the aid of rhodamine 560 in ethanol solution (excitation 560 nm, emission 480–850 nm). In another example, when the zinc sulfide is replaced by cadmium sulfide, i.e. cadmium selenide/cadmium sulfide, there is improvement in the quantum yield (84%) (Xiaogang Peng et al. 1997). A 50% quantum yield has been observed for zinc selenide emitting blue light using stilbene as the standard and methanol as solvent (Margaret and Guyot-Sionnest 1998). When the zinc selenide is doped with manganese, there is dampening of the quantum yield i.e. 22% (Norris et al. 2000). It has been reported in the literature that the composition as well as the stoichiometric ratio of the precursor are significant areas to improve the quantum yield of the quantum dots (Qu and Peng 2002). When the particle size of the zinc oxide quantum dots is tailored from 0.7 to 1 nm, there is a decrease in the quantum yield of quantum dots (van Dijken et al. 2001). Quantum dots that are appropriately surface-passivated have an elevated quantum yield in the visible region (400–700 nm), e.g. cadmium selenide (65-85%), cadmium sulfide (60%), and indium phosphide (10–40%). The quantum yield of carbonaceous quantum dots are carbon dots and nitrogen-doped carbon dots are 5.6 and 15.8% (Wu et al. 2014). Sun et al. have synthesized nitrogen-doped graphene quantum dots with excellent water stability that show an enormously high quantum yield of about 74% with the aid of rhodamine B (Sun et al. 2015).

9.4 Synthesis of Quantum Dots

A lot of synthetic methods have been developed for the synthesis of traditional quantum dots, i.e. metal-based quantum dots which come under two classes, i.e. top-down approach and bottom-up approach. In top-down methodology, the bulk material is finely grounded to particles with the size range in nanometers that include sonication, physical vapor deposition, ball milling, etc. In the case of bottom-up methodology, the quantum dots are prepared using a chemical reaction of various metal ions using certain types of techniques that include hydrothermal synthesis, solvothermal techniques, microemulsion, thermal decomposition, etc.

9.4.1 Top-Down Approach

In the case of the top-down approach, bulk material is finely grounded to form very small and fine materials that are in the nano range of dimensions, especially quantum dots. Lithography, particularly electron beam lithography, is commonly used in top-down methods to fabricate quantum dots. The lithographic techniques used on semiconductor materials are generally helpful for constructing integrated circuits. In this technique, the semiconductor wafer is treated with photosensitive material which is further protected with a stencil and treated with UV light. The part which remains untreated with the UV light is then washed away. The wafer is further treated with hydrogen fluoride that reacts with the semiconductor with a photosensitive coating. This technique carries various advantages that include control over shape and size with particular packing geometries that are essential for exploring the quantum confinement effect. In addition, there are also some other types of lithographic techniques available for the effective fabrication of quantum dots, i.e. focused laser or ion beam lithography. Schnauber et al. have fabricated indium arsenide quantum dots using electron beam lithography for the development of a multi-node quantum optical circuit (Schnauber et al. 2018). Fatimy et al. have fabricated epitaxial graphene quantum dots using lithographic techniques for bolometric sensors (Fatimy et al. 2016). Instead of carrying so many advantages, there are certain major disadvantages with these techniques such as structural deficiencies by patterning, integration of impurities.

Etching also plays a vital role in the fabrication of quantum dots. In the case of dry etching, the reactive gas molecules are introduced in the etching compartment, followed by applying the radio frequency of desired voltage to generate plasma which disintegrates the molecules (gas) to more responsive fragments. These highly energetic species collide on the surface and fabricate a reactive product to etch the patterned sample. If these energetic fragments are ions, the etching methodology is called reactive ion etching. Gallium arsenide/aluminum gallium arsenide quantum dots are reported to be synthesized by using reactive ion etching with the aid of boron trichloride and argon (Scherer et al. 1987). Zinc telluride quantum dots have been synthesized by the same reaction protocol using methane and hydrogen gas (Tsutsui et al. 1993).

In the case of lithography for high lateral precision of synthesized quantum dots, focused ion beam techniques are used. In this methodology, the semiconductor substrate surface is sputtered by the extremely focused beam form melted metal sources, such as palladium/arsenic/boron, gallium, gold/silicon/barium, or gold/silicon. The size of the ion beam in lithography tailors the size, shape, and interparticle distances in quantum dots (Peng et al. 1997), and is also helpful to selective deposition of material from a reacting gas. To monitor the established patterning lithography as well as other parameters, scanning ion beam images can be established by ion beam nanolithography but due to certain constrains like slow processing, expensive equipment, surface damage etc., limit its use. In this context, one more methodology has been developed to attain patterns with quantum dots, i.e. exposure of electron beam lithography followed by lift off, and is called the 'etching process'. This methodology gives a low degree of stiffness in the design of nanostructures i.e. wire, rings spheres etc. and is effectively applied for the synthesis of II-VI and III-V types of quantum dots.

There are various methodologies for the preparation of carbonaceous quantum dots under the top-down approach that include laser ablation, arc-discharge, electrochemical oxidation, etc. It was first prepared in 2004 accidentally by Xu et al. while preparing a caron nanotube using the arc-discharge method (Xu et al. 2004). Later on, Sun et al. explored the laser ablation method for the synthesis of fluorescent carbon dots (Sun et al. 2006). Zhou and his coworkers fabricated fluorescent carbon dots with the aid of the electrochemical method in which multi-wall carbon nanotube is used as a working electrode (Zhou et al. 2007). The obtained quantum dots are of 2.8 nm size with the quantum yield of about 6.4%. Also, efforts have been made to fabricate fluorescent quantum dots by replacing multi-wall carbon nanotube with graphite as a working electrode (Zhao et al. 2008; Zheng et al. 2009). To further improve the efficacy of the electrochemical method for the synthesis, ionic liquids have been used as electrolytes and graphite as working electrodes. It has been observed that when ionic liquids have been used as electrolytes, the reaction rate improves to a greater extent. Also, when different mole fractions of ionic liquids and water have been taken as electrolytes, the emission wavelength of the carbon dots can be tailored (Lu et al. 2009). Graphene quantum dots have been prepared by using graphene oxide by using a rapid continuous hydrothermal technique and the particles' size has been controlled with the aid of a surface directing agent (Kellici et al. 2017).

Similarly, in another report, the graphene quantum dots were synthesized by using laser treatment with a 1064 nm pulsed laser beam with the pulse rate of 10 ns before the hydrothermal treatment of graphene oxide solution (Qin et al. 2015). The detailed methodologies, as well as transmission electron microscope images, are displayed in Figure 9.8. This top-down route from pulsed laser ablation in liquid is used for the fabrication of the graphene quantum dots. In the typical reaction procedure, firstly graphene nanosheets have been prepared and then treated with laser fragmentation for a certain time. Figure 9.8 shows the prepared nanosheets of graphene. When the nanosheets have been exploited using a laser pulse, the sheets are fragmented to a smaller sized graphene, i.e. graphene quantum dots. A large amount of heat has been liberated by the laser pulse resulting in the cutting of long graphene sheets to a smaller size.

In the case of graphitic carbon nitride quantum dots, the top-down approach to size-controlled synthesis has been done by the hydrothermal treatment of an ethanolic solution of bulk carbon nitride in the presence of potassium hydroxide as an exfoliator (Zhan et al. 2017). With the aid of ethanol and potassium hydroxide, the interlayer spacing of bulk carbon nitride material is greatly extended via weakening of the interlayer interaction that helps in the ease of exfoliation of bulk carbon nitride to graphitic carbon nitride quantum dots when both the temperature and pressure is raised. The inter-lamellar spacing of bulk material is lower than potassium and hydroxide ions, ensuring the intercalation of the bulk carbon nitride materials; also, the potassium ion is a π electron acceptor, while bulk graphitic carbon nitride is a π electron donor, due to which it is efficient in intercalation. In addition to the intercalating agent, it oxidizes the surface of the bulk carbon nitride, resulting in increases of functional groups containing oxygen on the surface of the carbon nitride. Tian et al. (2013) have used an exfoliation process to fabricate graphitic carbon nitride quantum dots using an ultrasonication method. In the typical reaction, 1 mg/ml solution of bulk graphitic carbon nitride quantum dots has been prepared and is treated in ultrasound for continuous 10 h. followed by centrifugation to remove

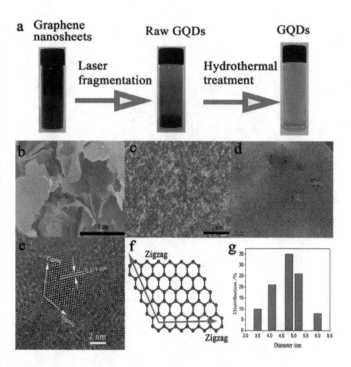

FIGURE 9.8   The photos of the graphene nanosheet solution target (left), after laser fragmentation (middle) and the raw GQDs after hydrothermal treatment (right). (b) SEM image of the graphene nanosheet. (c) SEM image of the raw GQDs. (d) TEM image of the GQDs. (e) High-resolution TEM image of the individual GQDs. (f) Orientation of the hexagonal GQDs network and the relative zigzag directions. (g) Histogram of diameter distribution. (Source: Qin et al. 2015)

bulk material. The prepared graphitic carbon nitride quantum dots show the quantum yield of 14.5% (Tian et al. 2013).

9.4.2 Bottom-Up Approach

Various methodologies have been developed under this category. Broadly speaking, there are two main categories that fall under the bottom-up approach, i.e. wet chemical route and vapor phase route. Wet chemical methodology is further classified into various categories, i.e. hydrothermal route, microemulsion, sol-gel chemistry, hot solution decomposition, electrochemical, etc., while the vapor phase route mainly includes sputtering, liquid metal ions, molecular beam epitaxy, etc.

9.4.2.1 Wet Chemical Route

This is a widely used methodology for synthesizing quantum dots, i.e. conventional precipitation methods with watchful control over the parameters of the reaction solution (Brian et al. 2004). The precipitation methods involve both nucleation of seeds and growth of the nanostructures. The nucleation process is either a homogenous or heterogeneous process (Burda et al. 2005). In the case of the homogenous process, the solute molecules/atoms, ions or molecules react and reach an optimum size without the support of a surface directing agent. In this process, the size, shape, and composition of the desired quantum dots can be optimized by tailoring various factors that include temperature, time, surface directing agent, concentration, etc. Some of the wet chemical routes are discussed below.

9.4.2.1.1 Sol-Gel Synthesis

Sol-gel methodology has been explored very much for the synthesis of different nanostructures, especially quantum dots (Bang et al. 2006; Bera et al. 2008b; Spanhel and Anderson 1991; Bera et al. 2010). In this reaction method, a sol, i.e. nanoparticles stabilized in the solvent by Brownian motion, is synthesized by taking metal salt in either acidic or basic solution. This methodology involves three steps, i.e. hydrolysis, condensation, and growth. Firstly, the metal precursor is hydrolyzed in the medium (either acidic or basic) and condensed to form sol via polymerization and finally networked in gel. This process is successfully applied to fabricate I-VI & IV-VI quantum dots, e.g. cadmium sulfide @ zinc sulfide (Selvan et al. 2001), cadmium telluride (Chunliang and Murase 2003). In the case of zinc oxide quantum dot synthesis, a basic solution of zinc ion salt has been added to basic ethanolic solution which is further kept for controlled aging in the air (Bang et al. 2006). For better control over size some polymers have been added to the solution (Ye 2018). The theoretical understanding and mechanism behind the synthesis of zinc oxide quantum dots is given as follows:

The chemical reactions that affect the nucleation and growth of zinc oxide quantum dots are given as (Yi Zhang and Wang 2012):

$$Zn^{2+} + OH^- \rightarrow \left[Zn(OH) \right]^+ \tag{9.6}$$

$$2n \left[Zn(OH) \right]^+ \rightarrow (ZnO)_n + nZn^{2+} + nH_2O \tag{9.7}$$

$$(ZnO)_n + 2k \left[Zn(OH) \right]^+ \rightarrow (ZnO)_{n+k} + kZn^{2+} + kH_2O \tag{9.8}$$

The reactions given in Equations 9.6 and 9.7 favor the nucleation step, while the step shown in Equation 9.8 favors the growth of the zinc oxide nanostructures. These reaction steps reveal that first zinc oxide crystallite is formed in large density, on the surface of that crystallite growth takes place and zinc oxide quantum dots grows in size. One other reaction pathway that favors the formation of zinc oxide quantum dots is given by the equations below and is considered acceptable by certain research groups (Joo et al. 2005):

$$Zn^{2+} + 2OH^- \rightarrow Zn(OH)_2 \tag{9.9}$$

$$Zn(OH)_2 \leftrightarrow ZnO + H_2O \tag{9.10}$$

Here the reaction given by Equation 9.9 shows the intermediate product formation, while Equation 9.10 reveals both nucleation and growth. Those supportive of this reaction predict instantaneous nucleation of quantum dots. However, the fabricated nuclei do not have undeviating size usually due to Ostwald ripening during the development of quantum dots, i.e. the dissipation of small nuclei to form the larger ones. Further, Equation 9.10 is reversible which shows that if the concentration of water is greater, a reversible reaction is favorable, leaving the lesser concentration of zinc oxide in solution. This is an important parameter to understand the poor stability of unmodified zinc oxide in water systems during synthesis. There are certain other reactions that have been proposed for the synthesis of zinc oxide quantum dots (Vergés et al. 1990):

$$Zn^{2+} + 4OH^- \rightarrow Zn(OH)_4^{2-} \tag{9.11}$$

$$Zn(OH)_4^{2-} + Zn(OH)_4^{2-} \rightarrow Zn_2O(OH)_6^{4-} + H_2O \tag{9.12}$$

$$Zn_xO_y(OH)_z^{(x+2y-2x)-} + Zn(OH)_4^{2-}$$
$$\rightarrow Zn_{x+1}O_{y+1}(OH)_{z+2}^{(x+2y-2x+2)-} + H_2O \tag{9.13}$$

In the above three reactions, Equations 9.11 and 9.12 favor the nucleation, while that of Equation 9.13 favors the growth of zinc oxide. The followers of the above reaction have faith in the excessive number of hydroxide ions dangling bonds existing on the surfaces of zinc oxide surfaces. The growth and photoluminescence properties of zinc oxide quantum dots are explained when experiments are conducted on the above mechanism. The sol-gel chemistry is cost-effective, simple and easily scaled up, but at the same time the size distribution in sol-gel synthesis is very broad with high defective states in the nanocrystals, hence limiting its use (Sashchiuk et al. 2002).

9.4.2.1.2 Microemulsion Methodology

To fabricate quantum dots at room temperature, the microemulsion route is the most popular one. The assemblies are classified as oil in a water system, i.e. microemulsion and water in an oil system, i.e. reverse microemulsions. It is a three-component system, i.e. oil, water, and an emulsifying agent. Instead of water, some other polar solvents can also be used for the fabrication of these assemblies. Out of the two, reverse microemulsion is widespread for fabricating quantum dots. The water is dispersed in oil with the aid surface active agents, e.g. triton-X, tween 80, sodium dodecyl sulphate, cetyltrimethyl ammonium bromide, aerosol OT, etc. These surface-active agents carry both hydrophobic as well as hydrophilic groups in its chain that form small micelles in oil. These assemblies are physiochemically stable and are used as a reactor for the nanocrystals' growth.

When these reverse microemulsions are forcefully stirred in the presence of a precursor of quantum dots, there is an incessant pass on of reactants due to active collision. The growth of quantum dots is restricted by the dimensions of the cavity of microemulsion that is dependent on the ratio of surfactant and water which is denoted by W. The relationship between W and radius of a cavity is given as (Hoener et al. 1992):

$$\left(\frac{r+15}{r} \right)^3 - 1 = \frac{27.5}{W} \tag{9.14}$$

This cavity of reverse micelle has been very effective for the synthesis of IV-VI core, II-VI core, and core/shell quantum dots, such as cadmium selenide (Liu and Park 2014), silver iodide (Liu et al. 2005), silica @ zinc selenide (Liu et al. 2013), cadmium sulfide (Prasad 2013), zinc sulfide/cadmium selenide (Kortan et al. 1990), cadmium selenide/zinc selenide (Hoener et al. 1992), zinc selenide (Lee et al. 2012; Georgios et al. 2004; Shuichiro Ogawa et al. 1997). By controlling the ratio of surfactant, volume-added concentration of surfactant, various morphologies of cadmium selenide have been obtained (Liberato Manna et al. 2000). In addition, silicon quantum dots have also been synthesized in a reverse microemulsion reactor, using a reducing agent. The photoluminescence emission as well as the size of silicon quantum dots can be controlled by using different reducing agents (hydrides) with a strong reducing agent; the silicon quantum dots formed are narrower in size as compared to other ones (Shiohara et al. 2011). By controlling the ratio of water to surfactant, one can easily control the size of quantum dots but due to a low yield, as well as the addition of certain impurities, it carries some disadvantages.

9.4.2.1.3 Thermal Decomposition Methodology

Decomposition of organometallic compounds at higher temperatures is a well-known route for the fabrication of quantum dots which was first explored in 1993 (Murray et al. 1993). Metal ion precursors such as halides, carbonates oxides, and alkyl of alkaline earth metals or bis(trimethyl-silyl) and phosphene precursor of group VI are typically explored for the fabrication of quantum dots using a hot solution decomposition process (Figure 9.9). Let us try to understand the reaction protocol of the hot solution decomposition process. The reaction proceeds as heating of two solutions A and B, where solution A is formed by dissolving chalcogenides X (sulfur, selenium, tellurium) in a particularly organic solvent. Solution B is fabricated by dissolving metal precursor in an organic solvent, i.e. trioctylphosphine and the surfactant, i.e. trioctylphosphineoxide. The prepared solution B is then thermally treated at elevated temperature (200–300 °C) in a three-neck round bottom flask. Wait till the solution B attains equilibrium with the temperature. When the solution attains the desired temperature, inject solution A to solution B quickly. The whole process is carried out in an inert atmosphere. The reaction process is observable with the change in color of the reaction mixture. The reaction progress can be monitored by taking a small volume of sample at regular intervals and observed under a UV-visible or photoluminescence spectrophotometer. Here the surfactant, i.e. trioctylphosphineoxide helps in controlling the size of the quantum dots over time and the reaction time controls the growth of the quantum dots (Pu and Hsu 2014; Talapin et al. 2010; Lianhua Qu et al.2001). Also, there is a change in the photoluminescence emission with the change in the size of the quantum dots. This change in the photoluminescence emission of quantum dots with size variation is due to the quantum

confinement effect and due to the surfactant molecules that are dynamically adsorbed and desorbed onto the surface of nanocrystals via their polar head.

In another example, lead selenide quantum dots have also been synthesized by the same methodology (Čapek et al. 2015). The synthetic route of lead selenide quantum dots has been divided into three steps that include the preparation of a trioctylphosphineselenide solution, preparation of lead oleate solution, and finally the two solutions are mixed and allowed to treat at high temperature. The trioctylphosphine selenide solution is prepared by mixing selenium powder in trioctylphosphine under an inert atmosphere and stirred for 16 h. resulting in a change of color from yellow to colorless. The preparation of the lead solution is done by taking lead oxide and oleic acid in hexadecane and then heating to get a colorless solution. The two solutions are then added at an elevated temperature to a three-neck flask and then the desired amount of trioctylphosphine and diphenylphosphine are added. After a certain interval of time, the reaction temperature is decreased for the growth of quantum dots. To observe the quantum confinement effect, aliquots are taken from the reaction vessel after certain intervals of time. The precipitates are obtained by treating the sample with toluene, ethanol, and acetonitrile mixture. In the synthesis, trioctylphosphine and oleic acid lower the reaction rate, and diphenylphosphine increases the rate of formation of diphenylphosphineselenide. The size of quantum dots has been controlled with a decrease in the concentration of trioctylphosphine. A similar approach has been implemented for the fabrication of quantum dots of cadmium selenide (Kim et al.2014). In this synthetic route, the solution of selenium-triphenylphosphine is added to the solution of cadmium oleate, octadecene, and ocetylamine which is preheated at 250 °C that results in the formation of cadmium selenide quantum dots. In addition, numerous quantum dots have been prepared by the same approach that includes selenides of lead, cadmium, zinc, telluride of cadmium, and so on (Pu and Hsu 2014; Čapek et al. 2015; Moreels et al. 2011; Hou et al. 2014; Zhang et al. 2010; Gao et al. 2009; Bansal et al. 2016). Generally, the development of quantum dots is classified in two elementary stages, i.e. nucleation and growth of crystallite. When the reaction proceeds with the lowering of the free energy, homogenous nucleation takes place instinctively. Figure 9.10 displays the relation between free energy relationships with the nucleation of the crystallite.

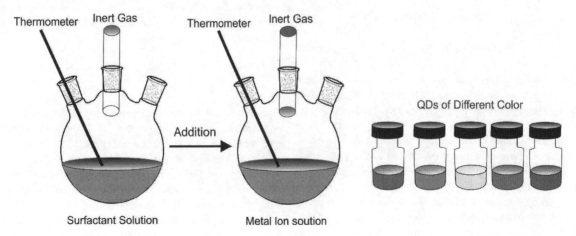

FIGURE 9.9   Presentation of thermal decomposition reaction for the synthesis of quantum dots.

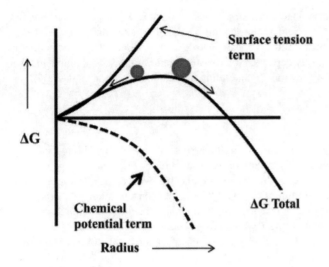

FIGURE 9.10 Variation in free energy with radius for a system due to the formation of QDs. (Source: Chand et al. 2017)

The surface of the quantum dots and the solution that surrounds the quantum dots experience a tension, i.e. surface tension. The reactive species within the solution have their chemical potential that generates the overall chemical potential of the solution. The surface tension is positive before the reaction proceeds as the growth of quantum dots starts the surface tension increase. With the proceeding of the reaction, the value of chemical potential is less than zero and starts increasing in a negative direction with the addition of reactive species within the reaction mixture. Both these parameters, i.e. surface tension and chemical potential of the reactive species cumulatively give the overall change in free energy in the synthesis of quantum dots (Burda et al. 2005). When the two species are injected in the hot solution monomers of the desired quantum dots will form and direct the nucleation of the quantum dots within the reaction mixture. The classical and thermodynamical behavior of the quantum dots elucidates the basics of the growth of crystallites of the quantum dots. In classical theory, crystallites develop uninterruptedly as they overcome the particular size dimensions known as Ostwald ripening. The nanocrystals' size starts increasing when they acquire a particular surface energy. With the processing of the reaction, the monomer concentration starts decreasing which is compromised by the growth, and this growth is dominated by the Ostwald ripening. The size of quantum dots is inversely proportional to the addition of monomers (Li et al. 2015). But this theory fails to afford the adequate explanation of the hot solution approach but it provides an effective explanation for the formation of small-sized monodisperse nuclei. The certain limitations that restrict this theory, the surface tension factor involvement depends upon the arrangements and interaction with the ligands present on the surface of quantum dots but in the case of bulkier surfactants like trioctylphosphine oxide, there is a steric hindrance between them that contributes positively towards the surface tension factor. Another factor that limits the theory is the hot solution process which is based on controlling the nucleation by quenching the temperature. It is believed that the nucleation size can be controlled by quenching of temperature but there is also a possibility that the size may surpass the critical size of nucleation (de Mello Donegá et al. 2005). In this context, a new

theory has been developed for tailoring the size of quantum dots smaller than the size reached after focusing on the hot solution process (Abe et al. 2012). The free energy change is related to critical size by the relation given as

$$\Delta G = \frac{4\pi}{3} \gamma r_c^2 \tag{9.15}$$

It has been explored experimentally that with an increase in the size of the crystallite, the change in free energy is not consistent as in Equation 9.15. This happens due to the decrease in participation of surface energy with an upsurge in crystallite diameter. In addition, continued nucleation also specifies that a hot solution process with abrupt nucleation is not essential for the low poly dispersity; also, there is no relevant effect on the post-focus size on the sample addition and temperature. It is rather complicated to envisage both nucleation rate as well as time span for the number of crystallites fabricated on the basis of temperature (Abe et al. 2012; Brauser et al. 2016). The initial concentration of the reactive species for the formation of quantum dots is the major factor for the uniform and desirable size distribution of quantum dots and by controlling the initial concentration of the precursor highly mono dispersible quantum dots can be attained which is only possible by considering two stages, i.e. nucleation and growth (Talapin et al. 2001). Overall, this route is advantageous in providing enough thermal energy to anneal the defective state with a highly monodisperse particle size. In addition, there are some limitations to the process, i.e. cost of precursors and surfactant, high temperature, toxicity, and the major one is the aqueous dispersibility that limits its application in a water system.

9.4.2.1.4 Ion Exchange Methodology

The ion exchange process is a direct reaction of the predesigned ligands with the appropriate metal ion species (Deng et al. 2009). This methodology has many disadvantages, like predesigning of the ligand, prolonged time, and instability of ligands in solution, etc. To overcome these limitations, a substitute approach has been adopted for the synthesis of water unstable quantum dots, i.e. ion exchange process (Cheng et al. 2015). This approach is further classified into two categories, i.e. anion exchange and cation exchange. In both these cases, the anion and cation of parent quantum dots can be exchanged by the anion and cation present in the solution respectively. In addition to exchange to cation with the cation of solution, the anion lattice structure of the parent structure remains unaffected in both cases. The exchange of cation is very fast as compared to anion because in the case of anion exchange their difference is that the anion radius is very prominent, anion exchange is favored at an elevated temperature, and anion exchange requires a longer time as compared to cation exchange. Also, due to internal strain as well as being refractive in their mismatch in the anion exchange can allow the regression of the two materials (Nag et al. 2014). In the case of cation exchange, the reaction depends upon the solubility product of both reactive species as well as final products. The driving force for cation exchange is decreased in the solubility product. Lattice parameters of reactants and products decide the volume change during cation exchange (Jaiswal et al. 2012). An example of such a type of synthesis is the fabrication of silver sulfide quantum dots (Gui et al. 2014; Tan et al. 2014). In the typical reaction synthesis, cadmium sulfide quantum dots have been prepared by a co-precipitation method, followed by the cation exchange that can be achieved by dispersing cadmium sulfide quantum dots in silver nitrate solution. The appearance of a dark brown color from orange indicates the progress of the cation exchange. Using cation exchange methodology, 99.5% of yield has been obtained that displays that the efficacy of the present process is excellent. In addition, the cation exchange protocol introduces some defective states in the quantum dots that lead to a decrease in quantum yield as compared to quantum dots prepared with conventional techniques. Also, if the crystal structures of initial as well as final quantum dots are different, it leads to the creation of defective states in quantum dots. This methodology, irrespective of its easiness, also carries some drawbacks, e.g. generation of lattice defects, and reaction completion is hard to find out, as some initial ionic species are also present in the final product that leads to dampening of the photoluminescence emission, etc.

9.4.2.1.5 Co-Precipitation Methodology

This process is the simplest synthetic route for the preparation of quantum dots due to its ease of handling, low temperature, no specific instrument is required under this process. The co-precipitation technique involves the precipitation of the desired quantum dots using a particular ionic species. The reaction route follows the dropwise addition of metal ions to another solution that contains anions under stirring. If the product formed is sensitive to air, then the reaction is preceded in an inert atmosphere. In some cases, if required, the pH of the solution has to increase to form a basic solution by adding different bases, e.g. sodium hydroxide, potassium hydroxide, ammonium hydroxide, etc. during the progress of the reaction, followed by stirring. Once the reaction is completed, the precipitates are washed several times with water and kept for drying at 50–60°C. Various quantum dots, i.e. sulfides, selenides, tellurides of cadmium, and zinc have been reported to be synthesized by the same process (Dasari Ayodhya et al. 2013; Peternele et al. 2014; Li et al. 2009; Sharma et al. 2015; Song et al. 2015; Mishra et al. 2011). A reactant solution is first prepared by dissolving the two reactants, i.e., cadmium oxide as well as sulfide/selenide/telluride in the desired concentration so that the reactants disperse completely in the solvent medium. The final solution is stirred for a few minutes to make to solution homogenous. Finally, the precipitating agent i.e., sodium hydroxide, potassium hydroxide, and ammonium hydroxide are added to get the desired pH and to start the precipitation of the quantum dots. The prepared quantum dots were then washed with water to remove any other impurities. The final pH of the solution is the deciding factor for photoluminescence emission of the quantum dots (Wang et al. 2014; Chang et al. 2015). The reaction time for stirring finalizes the size of nanocrystals (Chang et al. 2015; Wang et al. 2014; Tan et al. 2014). In addition to the stirring time, the introduction of a doping agent and a capping agent to the solution also controls the particle size as well as the morphology of the crystals (Rahdar et al. 2012; Dasari Ayodhya et al. 2013). Also, the precursor concentration of anion controls the size as well as density of the quantum dots prepared (Gao et al. 2013). The theoretical understanding of the co-precipitation process can be explained by using the three processes, i.e. nucleation, growth, and Ostwald ripening. In the initial phase of reaction, i.e. precipitation, there is a generation of a huge concentration of crystallites known as the nucleation process. This nucleation process promotes the creation of new nuclei spontaneously within the solution. This nucleation is mostly dependent on the super saturation of the solution and can be explained by the following relation (Cushing et al. 2004):

$$S = \frac{A_a B_b}{K_{sp}}$$ (9.16)

where A_a and B_b are the activities of solute a and b respectively, and K_{sp} is the solubility product of the reaction.

Certain research groups also believe that the difference in concentration $\Delta C = C - C_{eq}$, (C is solute concentration; C_{eq} is the equilibrium concentration) is the major factor for precipitation in the solution (Mark Roelands et al. 2006). The difference in concentration is related to the super saturation by the following relation:

$$S = \frac{C}{C_{eq}}$$ (9.17)

If the radius of the newly formed crystallite is less or equal to the critical radius, then it will dissipate in the solution. If it is higher, then the crystallite size will increase within the reaction mixture. The equation that follows nucleation rate to be homogenous is given as (R_n) (Cushing et al. 2004; Furedi-Milhofer 1981):

$$R_n = -\frac{16\pi V^2 \sigma_{sl}}{3K^3 T^3 \ln^2 S}$$ (9.18)

where K is the Boltzmann constant, V is the atomic volume (solute), σ_{sl} is the surface tension at the liquid-solid surface. Equation 9.18 displays that the nucleation process is insignificant before the super saturation attained. The dissipation of smaller particles on top of larger particles during growth is termed Ostwald ripening. For the restriction/stabilization of particle size, generally two approaches have been used:

(1) Steric repulsion caused by surface directing agents such as surfactants, polymers, or any other surface-active agent, between the particles.
(2) Electrostatic repulsion between charged species present at the surface of nanocrystals.

For stabilizing quantum dots, capping agents are mainly used to control the size. During synthesis of quantum dots, the surface energy of the particles is very high due to the high ratio of surface and volume – this increased energy is neutralized by the addition of capping agent. The detained co-precipitation method has been successfully employed for the fabrication of different types of quantum dots (Li et al. 2009; Mishra et al. 2011; Chang et al. 2015; Tan et al. 2014; B. Gao et al. 2013). Still, this method is not appropriate for the synthesis of a suitable and accurate stoichiometric phase, and controlling the crystallinity is also a bit difficult.

9.4.2.1.6 Hydrothermal Methodology

The synthetic route which is appropriate to create an extremely crystalline phase which is unstable at the melting point under a different route, is known as hydrothermal synthesis. Materials that have high vapor pressure close to their melting point can also be synthesized using the hydrothermal route. This methodology is also helpful for the synthesizing of high-density crystalline nanomaterials though preserving control over their composition. The crystallization in this process takes place at a higher temperature (>100 °C) and pressure (10–80 MPa). Under that elevated temperature and pressure materials show an increase in solubility which is generally insoluble under normal reaction conditions. Initial pH, temperature, reaction time, and the pressure are the key factors that play an important role in nucleation as well as the growth of the nanocrystals (Jagtap et al. 2015; Zhao et al. 2014; Yang et al. 2012; Han et al. 2010; Ma et al. 2015; Liu et al. 2013). The synthesis is operated in autoclaves made up of a Teflon vessel that is a completely sealed steel cylinder so that there is no leakage of the solution from the cylinder. These hydrothermal vessels can work up to 300 °C and 250 bars of pressure depending on the quality of Teflon used in the assembly. A typical example of quantum dot synthesis using a hydrothermal vessel has been explored by Zao et al. who successfully synthesized zinc selenide/zinc sulfide core shell quantum dots (Zhao et al. 2014). For this synthesis, solution zinc chloride is solubilized in the presence of N-acetyl-L-cysteine and stirred. In the second step, selenium is reduced by treating it with sodium borohydride to form sodium hydroselenide, followed by the addition of prepared sodium hydroselenide solution to the first solution and mixed under vigorous stirring to get a homogenous solution.

The prepared homogenous solution is then transferred to the Teflon vessel sealed in steel cylinder and heated at 170–210 °C for 40 to 80 min. The solution conditions in hydrothermal vessels efficiently lower the energy barrier for the synthesis and accelerate the synthesis rate that otherwise requires an elevated temperature. Finally, after the reaction is over, the vessel is cooled to room temperature and finally the prepared quantum dots are centrifuged and washed with water. Various researchers have used the hydrothermal approach for the synthesis of different quantum dots (Jagtap et al. 2015; Zhao et al. 2014; Liu et al. 2013; Dunne et al. 2014). The approach of the synthetic route is also very helpful in fabricating carbonaceous quantum dots. Wang et al. have fabricated fluorescent carbon dots using glutathione and glucose, keeping the molar ratio 3.5 solution, and then transferring to the hydrothermal vessel (Wang et al. 2015). The hydrothermal vessel is then treated at 180 °C for 20 h. to get highly fluorescent carbon dots. Green synthesis of carbon dots has also been carried out by using this assembly. In one particular reaction protocol, rose-heart radish is hydrothermally treated at 180 °C for 3h to get blue light-emitting carbon dots (Liu et al. 2017) and the graphical route for synthesis is given in Figure 9.11.

Other green precursors like peach juice (Atchudan et al. 2018), *H. undatus* fruits (Arul et al. 2017), *Prunuspersica* (Atchudan et al. 2017a), bamboo leaves (Liu et al. 2014), Jinhua bergamot (Yu et al. 2015), pig skin (Wen et al. 2016), *Ocimum sanctum* (Kumar et al. 2017), *Chionanthusretusus* (Atchudan et al. 2017b), *Saccharumofficinarum* (Mehta et al. 2014), sweet potato (Wenbo Lu et al. 2013), corn stalk (Shi et al. 2017), etc. The detailed mechanism for the synthesis of carbon dots using peach juice is displayed in Figure 9.12.

The use of an aqueous solvent is the main advantage of this synthetic route. Also, water is polar in nature in normal conditions but when the temperature and pressure of the water system increase, there is a rise in the density of water that leads to the dissolution of various inorganic compounds to form a uniform and highly crystalline growth of the nanocrystals. Control of size as well as the shape of nanomaterials is also possible in this methodology via controlling the precursor concentration, as well as the addition of a surface directing agent. Shortcomings of this method comprise

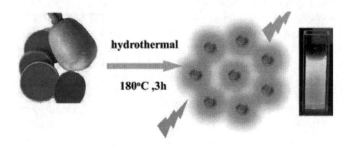

FIGURE 9.11 Diagram for the synthesis of N-CDs from rose-heart radish, along with photograph of the corresponding sample under 365 Nm UV lamp excitation. (Source: Liu et al. 2017)

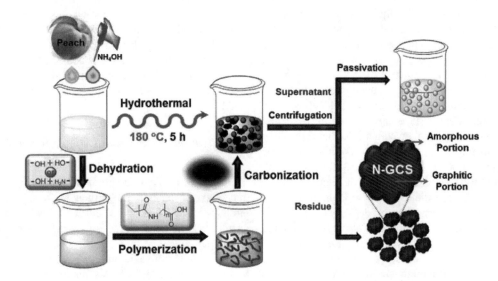

FIGURE 9.12 Plausible formation mechanism of N-GCSs from the extract of peach by hydrothermal process. (Source: Atchudan et al. 2017)

the prior information on solubility of precursors, corrosive nature of slurries to the Teflon vessel, and explosion of the Teflon vessel may also happen. In addition, the expensive cost of the hydrothermal assembly and the non-observable crystal growth are the main drawbacks of the present technique.

9.4.3 Other Synthetic Methodologies

When the precursor solution (in aqueous or non-aqueous medium) is treated with micro and sonic waves, this also leads to the formation of quantum dots (Qian et al. 2005). These energetic waves offer a sufficient amount of energy to the precursor solution for the growth of quantum dots. Ultrasonic waves are reported to be helpful in synthesizing quantum dots (1–10 nm) by providing implosive bursts of bubbles in a liquid (Junjie Zhu et al. 1999). This acoustic cavitation develops a localized hotspot through adiabatic compression with the gas inside the collapsing bubble that progresses the reaction to form quantum dots (Junjie et al. 1999). One approach for fabricating quantum dots using ultrasonic radiation is given as follows: acetate precursor of metal ion is added to the seleno-urea solution and then sonicated for 1 h. under an inert atmosphere. During synthesis the solution temperature rises to 80°C (Issac et al. 2005). Similarly, silver indium sulfide quantum dots have been prepared using sonochemistry (Panda et al. 2017). Various other quantum dots have been synthesized using sonochemical synthesis like zinc sulfide (Goharshadi et al. 2012), cadmium selenide@ zinc sulfide (Murcia et al. 2006).

9.5 Functionalization of Quantum Dots

The surface engineering of quantum dots has recently gained a lot of attention in various fields of science. The development of this field is basically focused on the functionalization of quantum dots. This functionalization can be tuned from one to another on the basis of the requirement of the application of nanoparticles/ quantum dots (Mehta et al. 2012; Brad et al. 2008). Although quantum dots show good photostability, bright luminescence, and tunable emission, a longer lifetime of an excited electron, still there are certain issues like water dispersibility, specificity in sensing, biocompatibility, etc. that limits its use as bare quantum dots, so the surface engineering of quantum dots is essential to make it more pronounced in specific applications (Cinteza 2010). In addition, due to smaller size, quantum dots have a high surface energy that can generate surface defects, leading to the dampening of the PL emission of bare quantum dots (Jamieson et al. 2007; Liberato Manna et al. 2002; Hess et al. 2001). In the case of bare quantum dots, surface also leaches out by oxidation when it is exposed to a prolonged ionic medium, resulting in the release of metal ions in the solution which can be toxic in biomedical application (Katari et al. 1994; Kloepfer et al. 2005; Singh et al. 2012). Stable dispersion of quantum dots in an aqueous medium is also a pronounced shortcoming of the bare quantum dots. Above all, the poor target delivery or specificity in the quantum dots is the major limitation of the bare quantum dots. All these issues can be resolved by surface engineering, in fact by proper functionalization of the quantum dots. So, the functionalization of the quantum dots is very important to gain maximum output from the quantum dots.

The surface defects of the quantum dots can be minimized via capping the quantum dots with a stable compound that can also help in reducing the photobleaching process and act as a sacrificial layer over the quantum dots. In this context, zinc sulfide is frequently used as a surface protector that increases the stability and performance of the quantum dots with enhancement in quantum yield but still solubility issues remain as such because zinc sulfide capping does not improve the aqueous stability of the quantum dots (Clapp et al. 2006). Those quantum dots which are synthesized by hot solution decomposition, as discussed in the thermal decomposition section, are synthesized in an organic solvent like hexane, octane, toluene, etc., and have hydrophobic surface functionalities like phosphines, amines, etc. to prevent these quantum dots from agglomeration. In order to enhance their aqueous dispersibility, these quantum dots have been functionalized with the hydrophilic surface-active agent (Xiaohu Gao et al. 2004; Christina Graf et al. 2006; Brad et al. 2008; Thanh and Green 2010). To cover the dispersibility issues, three main strategies have been adopted: (1) amphiphilic functionalization; (2) ligand exchange; (3) surface silanization.

9.5.1 Amphiphilic Functionalization

In the amphiphilic approach, the hydrophobic molecules present on the surface are preserved by introducing another hydrophobic surface-active agent on top of the hydrophobic ligands, generally block co-polymers have been used. These block co-copolymers form an addition layer on the quantum dots and enhance their dissolution in an aqueous system, and has been reported by several research groups (Thanh and Green 2010). Multifunctional capping agents show supplementary benefits, i.e. improved stability in aqueous, more binding sites for modification, which is supported by a number of groups (Kimihiro Susumu et al. 2007; Sperling and Parak 2010). Due to a number of contact points on the surface of quantum dots, as well as thermal fluctuations, using amphiphilic surface-active agents are proven to be beneficial. The amphiphilic molecules bind (hydrophobic tail) with the hydrophobic ligands present on the surface of quantum dots via hydrophobic interactions and these interactions are independent of nature as well as the composition of the quantum dots. These hydrophobic interactions are known as van der Waals forces of attraction and occur between the hydrophobic tail of amphiphilic molecules and the hydrophobic ligand present on the surface of quantum dots. Quantum dots coated with hydrophobic ligand and then with amphiphilic molecules display similar chemical as well as physical properties to that of core quantum dots (Sperling and Parak 2010).

9.5.1.1 Poly(1-Carboxyethylene)-Based Polymer

Poly(1carboxyetheylene) is also known as poly acrylic acid; a long chain carboxylic acid containing a highly charged polymer and can be easily modified to form amide with amine groups (Wang et al. 1988; Serdar Celebi et al. 2007). Long chain aliphatic amine (octylamine) that otherwise is hydrophobic in nature can be tuned to hydrophilic ligand by modifying it with poly acrylic acid and makes quantum dots water stable (Xingyong Wu et al. 2003). This phase transfer reaction is a simple methodology that increases the applicability of quantum dots in an aqueous system. To further improve the stability of quantum dots, quantum dots capped with octylamine and then with poly acrylic acid have been further reacted with lysine by carbodiimide chemistry (1-ethyl-3-(3-dimethylaminopropyl) carbodiimide) (Luccardini et al. 2006). In addition, poly acrylic acid polymer can also be altered with the mixture of isopropylamine and octylamine for effectual and water stable functionalization on quantum dots. Kairdolf et al. have synthesized cadmium telluride/cadmium selenide quantum dots using poly acrylic acid functionalized with dodecylamine, resulting inamphiphilic quantum dots stable in both aqueous and organic phase (Kairdolf et al. 2008).

9.5.1.2 Poly(maleic Anhydride) Polymers

Poly(maleic anhydride) is an amphiphilic polymer, synthesized by copolymerization olefin and maleic anhydride (Sperling and Parak 2010). In the case of poly(maleic anhydride) polymer, the densities of carboxylic groups are higher and are aligned in an alternating layer as compared to poly acrylic acid. During the phase change, the anhydride ring reacts with water that leads to the ring opening and the generation of two acid functional groups (Pellegrino et al. 2003). Several derivatives of poly(maleic anhydride) are commercially available and successfully been used for the modification of quantum dots yielding stable aqueous dispersion, such as poly(maleic anhydride alt-1-octadecene), poly(maleic anhydride alt-1-tetradecene), and so on (Di Corato et al. 2008). Before these poly(maleic anhydride) derivates are used for surface modification of quantum dots, the reactivity of maleic anhydride with other functional groups such as amine as well as alcohol has also been explored (Lin et al. 2008). Some other modifications have also been published where hydrophobic ligands (dodecylamine) on quantum dots are functionalized covalently by poly ethylene glycols, sugars, dyes, biotin, etc. (Lin et al. 2008).

9.5.1.3 Other Amphiphilic Polymers

Another class of amphiphilic polymers such as block-copolymers have also been used for the surface engineering of quantum dots that involves a hydrophobic as well as hydrophilic groups (Stevenson et al. 2001; Möller et al. 1996). These polymers willingly form micellar-like assemblies with either of the group held inside or outside of the assemblies, depending on the polarity of the solvent. These assemblies have been quite useful for phase transfer reactions, nanoparticle synthesis and capping (Nann 2005; Nann et al. 2005; Dong et al. 2014). Various block co-polymers have been reported to be cross-linked over the hydrophobic surface of quantum dots (Yoojin Kim et al. 2005; Byeong-Su Kim et al. 2005; Cheng et al. 2008). The thickness over the quantum dots can tuned by selecting a suitable polymer with appropriate block length (Kang and Taton 2005). In the recent years, quantum dots were encapsulated in the assemblies to form pluronic-quantum dot assemblies, thus fabricating an innovative micro reactor (Jia et al. 2012).

9.5.2 Ligand Substitution Functionalization

The ligand substitution reaction can basically be known as the substitution of an already present ligand by a flexible functionality that offers quantum dots with extra targeting, mobility, stability, and dispersibility (Hedi Mattoussi et al. 2000; Parak et al. 2002; Medintz et al. 2005; Tetsuo Uyeda et al. 2005). As already discussed in the section of the hot-solution decomposition process, the quantum dots prepared using methodology have various organic ligands present on the surface that are hydrophobic in nature. These ligands prevent aggregation of the quantum dots and passivate the surface defects to preserve quantum yield (Hammer et al.2007; Bera et al. 2010; Talapin et al. 2010). These hydrophobic ligands provide excellent stability in organic solvents but at the same time have very low stability in an aqueous medium. To address this issue, ligand exchange is implemented in which organic ligand is exchanged with amphiphilic ligands through mass action (Thanh and Green 2010; Mei et al. 2008). Ligands that are used for the modification of quantum dots and its inherent solvent are always in dynamic equilibrium with each other, resulting in the substitution of the ligand present on the surface and the free one present in the solution. Due to this dynamic equilibrium, ligand exchange methods take place and occupy the vacant site available on the surface of quantum dots, but there are certain conditions that exchange ligands must possess to undergo this exchange mechanism: the concentration should be double, and there should be extra affinity towards the surface of quantum dots (Hedi Mattoussi et al. 2000; Medintz et al. 2005; Clapp et al. 2006). Thus, for substituting a current ligand with an anticipated ligand, the most significant property is the metal–ligand affinity (Alivisatos 1996; Thanh and Green 2010). If somehow the affinity of the exchange ligand is lesser towards the surface, still with increase in the concentration of the exchange ligand, this methodology can be carried out (Yanjie Zhang and Clapp 2011). Carbodithiolates, dithiothreitol, dihydrolipoic acid, mercaptopropanoic acid, and mercaptoacetic acid are the thiols that are in high demand (biomedical applications) for quantum dot functionalization and can be easily obtained via ligand exchange methodology. When these thiols are used as capping agents, thiol groups specifically bind with the quantum dot surface and the carboxylic group usually carries negative charge at neutral pH, and thus provides the additional stability to the quantum dots by ionic repulsion. Computational modeling studies support the modification of thiols on quantum dots and confirm that the thiol firstly forms a bond with the metal on the surface; in addition, it has a weaker S-S bond (Craig et al. 1998; Show-Jen Chiou et al. 2000; Hammes and Carrano 2001; Pejchal and Ludwig 2004). The driving force for this linkage is the affinity of a sulfur atom towards the metal that helps in the ligand exchange process and can be further improved by the sulfur-sulfur bond. This ligand exchange methodology for improving the dispersibility of the quantum dots is the simplest approach. Also, the size of quantum dots remains unchanged during the whole methodology as compared to other techniques. These advantages prove to be helpful in exploring various application aspects of quantum dots. Instead of these benefits, the ligand exchange process also carries various shortcomings:

i. It takes several hours for thiols to undergo ligand exchange reaction due to their weak interaction with surface of the metal.

ii. This affinity is dependent on the nature of quantum dots.

iii. The ionic stability of quantum dots is imperfect in highly concentrated salt medium and leads to aggregation (few hundred mM).

iv. Over a long period of time, thiols' molecules form S-S bonds that lead to the detachment from the surface and aggregation takes place.

v. QY is also affected (dampening) by the ligand exchange process.

9.5.3 Silanization

The methodology of fabricating an amorphous silica layer on top of nanocrystals is called silanization. The electrostatic response of silica allows it as an appropriate modification for enhancing the dispersibility in of quantum dots in water system without affecting the PL emission of the materials. Silica displays an inconsistent response in an aqueous medium, as it does not merge overhead its isoelectric point like other oxides and reveal its excellent stability at pH~7.0 or in the presence of a high concentration of salt. The enhanced stability over the isoelectric point is developed due to the highly energetic short-range interactions. The water molecules and the surface of silica experiences H bonding which leads to the strong repulsive interactions with the free water molecules, resulting in the minimization of the aggregation and enhancement of the dispersion. Thus, the presence of silica at the surface of nanocrystals is more beneficial as compared to other capping agents. In addition, although the particle size increases by silica coating, it prevents agglomeration, making the quantum dots less interactive with the oxygen in the solution (this prevents photo bleaching), also it does not alter the optical emission as well as absorption of the quantum dots. Also, the surface of quantum dots has been preserved from the oxidation that leads to enhancement in the photostability of nanoparticles. Silica-modified quantum dots display exceptionally a hundred times extra stability with photobleaching as compared to its bare counterpart. Figure 9.13 displays the comparison of citrated coated cadmium sulfide and silica encapsulated cadmium sulfide (Correa-Duarte et al. 1998).

Comparing silica and citrate modified cadmium sulfide, the latter system degrades within 24 h. when exposed to visible light, while the previous one is highly stable to visible light. The degradation of CdS happens by the oxygen at the surface in which sulfide is converted to sulfate ions. Silica modification avoids the contact of oxygen to the surface of quantum dots so reducing the degradation, as shown in Figure 9.13.

The present methodology is a little bit difficult and can be achieved through many steps as shown in Figure 9.14. Firstly, the activation of the surface has been done by the first layer of silane molecules. This step involves the exchange of organic ligand with silane than can help the dispersion in water or ethanol. Maximum research is dedicated to the surface activation process, e.g. a citrate group on quantum dots has been exchanged with (3-sulfanylprosulfanylpropyl), then by the solvent (Correa-Duarte et al. 2001). The silica layer on the surface of quantum dots is then grown with the aid of tetraethyl orthosilicate as per Stober's process. Similarly, (3-sulfanylprosulfanylpropyl) trimethoxysilane has been used for the phase transfer of quantum dots that are capped with trioctylphosphine oxide and hexadecylamine in tetrahydrofuran (Nann 2005). These (3-sulfanylprosulfanylpropyl) trimethoxysilane coated quantum dots are stable in polar solvent (methanol, ethanol). The silica thickness over the quantum dots has been controlled by varying the tetra ethyl orthosilicate concentration, reaction time, water content, base (ammonia) concentration, etc.

The proper ratio of water and ammonia is required to control the thickness of the silica on the surface. Water catalysis and the hydrolysis of tetra ethyl ortho silicate trigger the nucleation of the silica and provide the stability to the silica matrix (Arriagada and Osseo-Asare 1999; Serrano et al. 2011). This methodology provides a wide variety of advantages as given above, and in addition, there are a few drawbacks, i.e. the size of quantum dots increases with an increase in silanization as there are some specific applications in the biomedical field that require smaller size quantum dots. Secondly, silanization provides the insulating layer over the quantum dots which limits quantum dot use in electrochemistry.

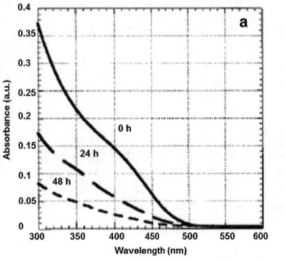
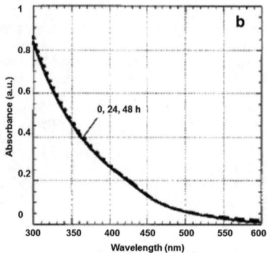

FIGURE 9.13 UV-visible spectra at different times after preparation of citrate stabilized particles (a) and silica-coated particles (b). Solid line, 5 min; dashed line, 24h; dotted line, 48h. (Source:Correa-Duarte et al. 1998)

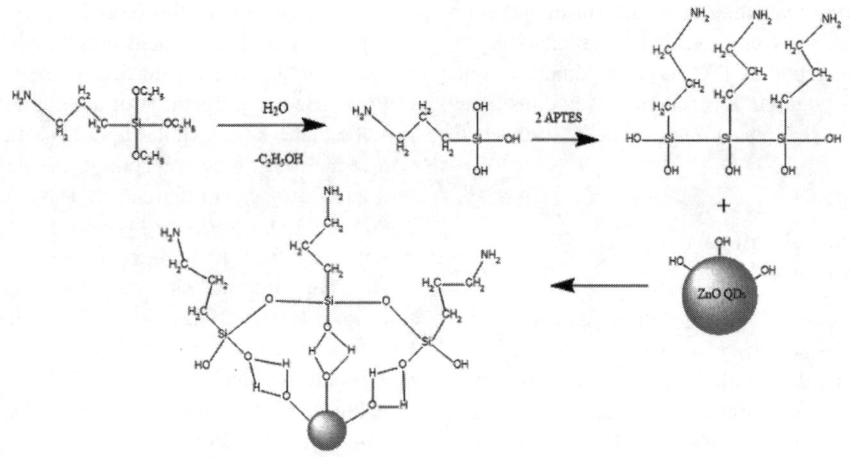

FIGURE 9.14   Demonstration of surface silanization of ZnO Quantum Dots.

9.5.4 Coating Strategies

9.5.4.1 1,2-Bis(2-iodoethoxy)ethane

1,2-Bis(2-iodoethoxy)ethane is a bifunctional cross-linker that delivers controllable hydrophilicity which can be tuned by the pH of solution in micelles. 1,2-Bis(2-iodoethoxy)ethane provides a micellar shell over a pH receptive core (Shizhong Luo et al. 2005). Limited reports are present on 1,2-Bis(2-iodoethoxy)ethane modification on the surface of quantum dots. Luo et al. fabricate gold nanoparticles stabilized with poly(2-dimethylamino) ethyl methacrylate)-b-poly(ethylene oxide) with thiol functionality at the terminals (Shizhong Luo et al. 2005). The thiolated poly(2-dimethylamino) ethyl methacrylate)-b-poly(ethylene oxide) has been switched with citrate on surface gold nanoparticles and 1,2-Bis(2-iodoethoxy)ethane has been employed for selective cross-link residues. Fascinatingly, it was detected that there is a change in displaying alterable pH response in both cross-linked and without cross-linked gold nanoparticles. Likewise, a dendritic-linear block copolymer functionalized magnetite nanoparticle was fabricated by Wu et al. (Xiaomeng Wu et al. 2011).

9.5.4.2 Glutaraldehyde

Glutaraldehyde is one of the chemicals that reacts with biomolecules that carries the functionalities such as imidazole, phenol, thiol, and amine where the side chain amino acid are the most sensitive ones and act as nucleophiles (Habeeb and Hiramoto 1968). These nucleophiles form imine by attacking the carbonyl group of glutaraldehyde. This route has been successful implemented by Santos et al., who synthesized cadmium sulfide/cadium hydroxide quantum dots modified with concanavalin-A (Con-A) lectin with the help of glutaraldehyde functionalized above quantum dots (Santos et al. 2006). The same group in another attempt synthesized glutaraldehyde functionalized cadmium sulfide/cadium hydroxide quantum dots with effective fluorescent centers for bioimaging of red blood cells (Farias et al. 2008).

9.5.4.3 Disulfide Bridge

The functionalization of antibodies or the aptamers on the surface of quantum dots by cross-linking of amine functionality

with the thiol functional group has been a very efficient approach. Poly amidoamine and polyisoprene modified quantum dots have revealed excessive ability for target drug delivery as well as bioimaging (Geraldo et al. 2011; Pöselt et al. 2012; Akin et al. 2012). Although this approach is highly beneficial in biomedical applications, this approach often alters the physical as well as the chemical properties of the quantum dots that results in dampening of the quantum yield. Jin et al. have shown that surface functionalization of cadmium selenide/telluride cadium sulfide quantum dots with glutathione shows only 22% quantum yield (Jin et al. 2008). Further, there is no clear understanding of the stability of the system under in vivo conditions. Oxidation due to photobleaching also beaks the disulfide bridge between the surface of quantum dots and the thiol functional group.

9.5.4.4 Micellar Phospholipid Functionalization

Quantum dots capped with liposomes and phospholipid micelles are proven to be highly beneficial for biomedical applications as they carry several advantages like no change in the surface of quantum dots after encapsulation, optical properties remain as they are, due to denser functionalization of liposome and phospholipids which restricts the unavoidable adsorption on the surface (Williams et al. 2018; Al-Alwani et al. 2019). Due to these positive behaviors of functionalization, it is highly used for in vivo drug delivery systems. Dubertret et al. 2002 explained that phospholipid modified cadmium selenide/zinc sulfide quantum dots can be used for in vitro and in vivo imaging (Dubertret et al. 2002). In addition, when the same system is conjugated with deoxyribonucleic acid, it reveals the in vitro fluorescent probe and interacts with the specific sequence. This methodology provides a better alternative as compared to others for stability and reduces photodegradation in various biological systems.

9.5.4.5 Polyacrylate Functionalization

Polymeric modification provides an effective and efficient steric stabilization via electrostatic repulsion and gives a reaction center to further functionalize with biomolecules. Moreover, an individual polymer chain delivers a large number of adsorption sites with numerous carboxyl groups free for modification. Celebi et al. (2007) explored the fabrication of poly(acrylic acid) functionalized cadmium sulfide quantum dots in an aqueous medium with a quantum yield of 17%, and remains stable for eight months. Wei et al. reported TAT-modified quantum dots using polyacrylate TAT-quantum dots (polyacrylate) considered for localization of TAT-QD(polyacrylate) in cells (Wei et al. 2009).

9.5.5 Surface Functionalization with Biological Molecules

Improvement of bioavailability, sensing, drug delivery, water stability, etc. of quantum dots have also been done by the functionalization of quantum dots with numerous bioactive molecules that include: proteins, peptides, enzymes, antibodies, aptamers, nucleic acids, sugars, etc. (Xing and Rao 2008). When the quantum dots have been functionalized with these biological molecules, enough binding sites have been available for the drugs, analytes, toxics, that enhance the utility of quantum dots (Rosenthal et al. 2011). Figure 9.15 shows various routes for bio as well as chemical functionalization of quantum dots (Algar et al. 2010). In particular, peptides and antibodies include several carboxyl and amine moieties that can bind to the surface with the help of amide bonds (Sperling and Parak 2010). Also, deoxyribonucleic acid with amine functionality can be introduced over the quantum dots by amine conjugation; deoxyribonucleic acid itself carries several binding sites like hydroxyl, amines, phosphates, etc.; those are helpful in binding it with the quantum dots

(Murcia et al. 2008). These binding are entropically favorable. These deoxyribonucleic acid functionalized quantum dots have been explored in the field of gene detection (Yi Zhang and Wang 2012). Surface modification of quantum dots with poly ethylene glycol enhances the cellular uptake of nanoparticles in the body (Oh et al. 2011). However, poly ethylene glycol molecules contain an ethoxy group that shows lesser interaction (nonspecific binding) with the surface of quantum dots and cannot be further modified, so for activation of poly ethylene glycol, it is required to generate different functional groups such as amines, carboxyl, thiols, etc. (Harris 1985). Similarly, dextran, when used as a surface modifier, offers essential stability and biocompatibility to quantum dots (Thanh and Green 2010). Dextran-modified quantum dots provide a wide pH working range quantum dots without undergoing any change in the emission properties of quantum dots (Wilson et al. 2010). Thus, the selection of functionalization, for the modification of quantum dots, is dependent on the required application of the quantum dots.

9.5.5.1 Adsorption

Adsorption is also one of the common techniques to modify the surface of quantum dots with biologically active molecules (Sperling and Parak 2010; Delong et al. 2010). Due to the huge surface-to-volume ratio or high surface energy biomaterials such as enzymes, proteins, antibodies easily adsorb on the surface of nanoparticles that are non-specific in nature and independent of

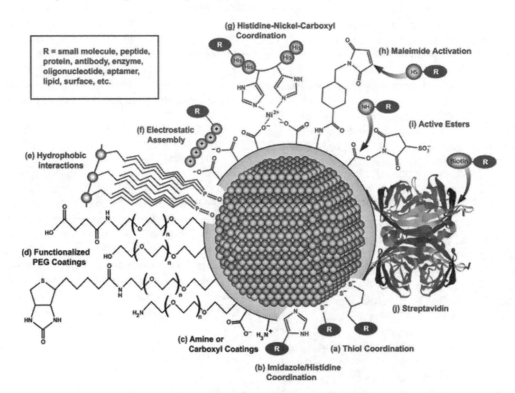

FIGURE 9.15 An illustration of some selected surface chemistries and conjugation strategies that are applied to QDs. The grey periphery around the QD represents a general coating. This coating can be associated with the surface of the QD via (e) hydrophobic interactions, or ligand coordination. Examples of the latter include: (a) mono dentate or bidentate thiols; (b) imidazole, polyimidazole (e.g. polyhistidine), or dithiocarbamate (not shown) groups. The exterior of the coating mediates aqueous solubility by the display of (c) amine or carboxyl groups; or (d) functionalized PEG. Common strategies for bioconjugation include: (a) thiol modifications; or (b) polyhistidine or metallothionein (not shown) tags that penetrate the coating and interact with the surface of the QD; (f) electrostatic association with the coating; (g) nickel mediated assembly of polyhistidine to carboxyl coatings; (h) maleimide activation and coupling; (i) active ester formation and coupling; (j) biotin-labeling and streptavidin–QD conjugates. Figure not to scale. (Source: Algar et al. 2010)

the functional groups. The driving forces for these types of interactions are ionic interactions between the charges on the bio molecules and the ligand on top of the quantum dots. Appropriate proteins can be tailored to develop a positive charge on their surface – that is helpful in self-assembling on the negative surfaced (carboxyl group) quantum dots via electrostatic interactions (Clapp et al. 2006; Hitchcock et al. 2012). Another scheme engages chemical interaction such as hydrogen bonding, the ligand on the surface, and the biomolecules and is dependent of the type of biomolecules used for the functionalization, such as the adsorption of oligonucleotides on mecapto acetic acid coated cadmium selenide/zinc sulfide quantum dots and the adsorption profile is dependent on the pH, ionic strength, as well nature of the solvent (Sungjee and Bawendi 2003; Algar and Krull 2010). pH plays a critical role in adsorption mechanism; at lower pH, adsorption is higher due to protonation of the carboxyl groups, while with an increase in pH there is a decrease in adsorption that reveals the involvement of hydrogen bonding in the adsorption process.

9.5.5.2 Covalent Coupling of Biomolecules

The use of a carboxyl end functional group on the surface of quantum dots is quite common as it can be easily coupled with amine to form amides which are the common functional groups in proteins, antibodies, aptamers, deoxyribonucleic acid, etc. (Pereira and Lai 2008). The benefit of making amide bonds is the simple aqueous process, without using any spacer to conserve the hydrodynamic radius that typically occurs in basic conditions which conserve properties as well as the structure of proteins (Sperling and Parak 2010; Kairdolf et al. 2013). This coupling of amine and carboxyl group is called carbodiimide conjugation and is often used for modification quantum dots with amine end group biomolecules. 1-Ethyl-3-(3-dimethylaminopropyl) carbodiimide is generally used conjugating agent for the development of amide bonds as shown in Figure 9.16 (Donegan and Rakovich 2013). Typically, amide coupling proceeds with high effectiveness and the yield can be amplified by stabilizing the intermediate. On the other hand, 1-Ethyl-3-(3-dimethylaminopropyl) carbodiimide single-handedly is not very competent as a cross-linking agent as its rate of reaction is quite slow. This slow rate permits the O-acyl

isourea as a reactive intermediate to undertake hydrolysis that proceeds with the revival of carboxylic acid.

The effectiveness of the process can be enhanced by stabilizing or trapping intermediate that helps in the progress of reaction (Khorana 1953; DeTar et al. 1966). The intermediate (O-acyl isourea) can be trapped by the help of N-hydroxysuccinamide ester to form an amine-activated intermediate that enhances the formation of the amide bond (Anderson et al. 1964). Another alternative of N-hydroxysuccinamide is sulfo-N-hydroxysuccinamide which can also be used for the formation of amide bond (Staros 1982). The sulfo- N-hydroxysuccinamide ester has been shown to retain 50–80% of the activity of enzyme (Ferguson and Duncan 2009).

9.6 Applications of Quantum Dots

The unexceptional and unique properties of quantum dots make them special and smart for various practical utilities in various fields of science that include photonics, biomedical, optoelectronics, sensing, dye sensitized solar cells, fuel cells, etc. A few of them are discussed below.

9.6.1 Solar Cells

The development of the quantum dots for solar cell applications is one of the most promising and needed applications of quantum dots. This fact is supported by the increase in publications per year on solar cell application of quantum dots (Kamat 2007; Bera et al. 2010; Talapin et al. 2010; Emin et al. 2011; Tang et al. 2011; Prabhakaran et al. 2012; Nikolenko and Razumov 2013). There is a very rapid progress in this application of quantum dots. This progress can be easily evaluated by the maximum efficiency of solar cells in the past few years: in 2013, the maximum efficacy was 6% which increased in 2014 to 8.55% and was achieved with lead sulfide quantum dots (Tang et al. 2011; Chuang et al. 2014). In 2015, the efficiency increased to 9% (Ren et al. 2015). In 2016, the efficiency increased to 11.1% using zinc-copper-indium-selenium quantum dots (Du et al. 2016). This data reveals that solar cell application is a growing field within quantum dot applications.

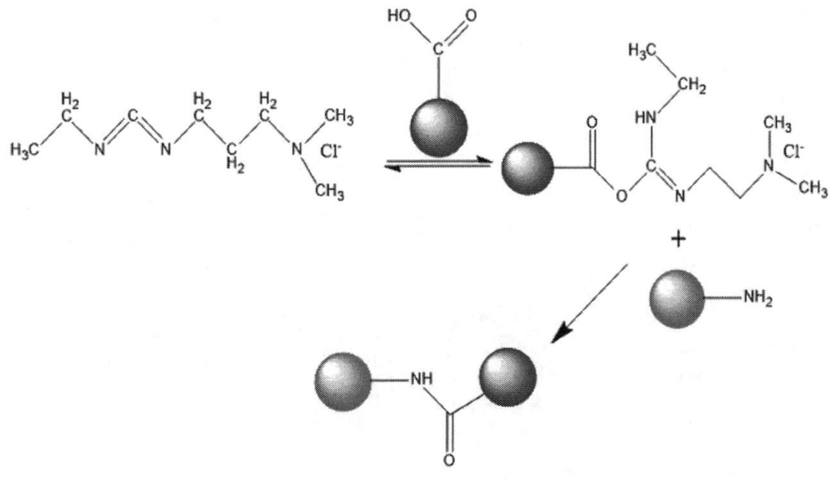

FIGURE 9.16 Demonstration of amide coupling (in reactant sp hybridized C is taken as bent due to sake of simplicity)

9.6.2 Light-Emitting Diodes

Quantum dots have also found application in developing light-emitting diodes due to their light-emitting, unique color tuning, as well as increasing the wavelength properties of quantum dots (Luo et al. 2015). Liquid phase synthesis of cadmium selenide @ zinc selenide shows light-emitting diodes and has a luminance of 28.76 cd/cm^2 with a current density of 4.9 cd/A, the obtained values are higher than the reference material (Kang et al. 2015). Further, these results have been improved by incorporating the gold nanoparticles into the electron transfer layer of the quantum dot system due to the resonant electron hole pair interaction with the surface Plasmon of gold nanoparticles (Kim et al. 2015). In addition, various white light-emitting quantum dots have found interest in light-emitting diodes (Torriss et al. 2009; Chien and Tien 2012; Ruan et al. 2016).

9.6.3 Biomedical Applications

Enhanced water stability, biocompatibility, and the high luminescence properties of quantum dots help in finding a place in various biomedical applications (Chin et al. 2010; Rosenthal et al. 2011; Regulacio et al. 2013; Lihong Jing et al. 2014). In a particular example, when quantum dots have been modified with luminescent drug molecules, no photoluminescence emission has been observed in the system due to Forster energy transfer (Vaishali Bagalkot et al. 2007). The aptamers used for the modification of quantum dots help in the detection of specific malignant cells and deliver the quantum dots to the specific site. When the drug is released from the system, quantum dots regain their photoluminescence emission due to the absence of Forster energy transfer. The regain of photoluminescence reveals the successful delivery of the drug. This study shows a modern development in the direction of considerable obstacles in biomedical application, such as for detection as well as drug delivery (Kelkar and Reineke 2011). Intensive efforts have been made in the field of biomedical application, particularly in vivo, where detailed investigation of toxicity is required (Jin et al. 2011). While studying quantum dots, particularly cadmium metal-based quantum dots, due to their toxicity, it is necessary to provide cadmium into the core of core shell quantum dots. Yet, cadmium ions gradually penetrate in the shell of the system and enhance its toxicity. So, in the case of biomedical application of in vivo studies, cadmium-free quantum dots are preferred; in this field, indium phosphide quantum dots are the potential semiconductor to replace the cadmium-based quantum dots (Regulacio et al. 2013; Anc et al. 2013; Brichkin 2015). On the basis of toxicity, it has been shown that indium phosphide @ zinc sulfide provides a better alternative as compared to cadmium selenide@ zinc sulfide quantum dots in terms of in vivo applications (Brunetti et al. 2013). In this context, various studies have been successfully done by different groups (Zhou et al. 2015; Frecker et al. 2016). The internalization of the therapeutic molecules into the cells is most important for the treatment and diagnosis. Therefore, in this context, the small size gold nanoparticles were functionalized with cell penetrating peptides, and this novel biofunctional material is internalized into Gram-positive and Gram-negative bacterial strains (Kumar et al. 2018b). Also, the biocompatible molecules (amino acid, food preservatives) functionalized on the surface of nanoparticles have great potential in vitro and in vivo biomedical applications (Kumar et al. 2018a; Chhibber et al. 2017, 2019). However, for the treatment of cancer diseases, a facile method has been developed in the intracellular delivery of TAT functionalized small gold nanoparticles into the cancer cells (Bansal et al. 2018).

9.6.4 Sensors

Quantum dots, due to their composition, inherent bright luminescence, and other electronic properties, have also been widely explored in the field of sensing applications for various environmental toxic chemicals, clinical sensing, food quality monitoring (Shtykov and Rusanova 2008; Yi Zhang and Wang 2012; Lou et al. 2014; Singh et al. 2014; Shorie et al. 2019; Sangar et al. 2019). In this context, free chlorine has been successfully detected using highly water stable and fluorescent zinc oxide quantum dots at room temperature. Due to the high oxidizing power of free chlorine, it degrades the defective states of zinc oxide, which leads to dampening of the photoluminescence emission of zinc oxide quantum dots (Singh and Mehta 2016). Lead sulfide quantum dots have been successfully employed for the sensing of deoxyribonucleic acid, nitrogen dioxide, and hydrogen sulfide gas with high sensitivity and selectivity (Lesiak et al. 2019; Xin et al. 2019; Liu et al. 2015; Liu et al. 2014; Wang et al. 2015). Manganese-doped zinc sulfide quantum dots stabilized with chitosan have been developed for the sensing of harmful bacteria that causes food poisoning and other waterborne diseases (Mazumder et al. 2010). A simple, selective, and sensitive approach has been developed for the detection of prostate-specific antigen based on the Förster Resonance Energy Transfer mechanism in which energy is transferred from cadmium telluride quantum dots modified with antibody- to gold-functionalized with aptamers (Kavosi et al. 2018). Carbon dots synthesized using lemon extract and L-arginine have been used for the efficient sensing of cupric ions in an aqueous medium (Das et al. 2017). Various carbonaceous quantum dots have also been developed for the sensing of various toxic metal ions/chemicals (Fan et al. 2017; Xiaohui Gao et al. 2016; Chen et al. 2016). A simple and sensitive method has been designed for the detection of malathionpestiside, using cadmium telluride@ cadmium sulfide quantum dots functionalized with melathion specific aptamer and guanidinium containing homopolymer. The limit of detection of malathionpestiside was observed to be 4 pM using this novel nanoprobe (Bala et al. 2018).

9.7 Conclusions

In this chapter, the particular characteristic of structures, properties, application, and performance of quantum dots have been explained. Among the various structures of nanoparticles, zero-dimensional structures have covered the way for the abundant developments in both functional and applied applications. This is due to the fact that zero-dimensional nanoparticles, i.e. quantum dots, display diverse optical as well as electronic properties that leads to numerous applications in various field of sciences. Significant advancement has been made in both experimental as well as theoretical studies to understand the nucleation and growth of quantum dots. Various methodologies from gas, solution, and solids have been developed for the synthesis of quantum dots. Most commonly, the bottom-up approach has been utilized for the synthesis of quantum dots due to its wide advantages over the top-down approach. In the case of the bottom-up approach (sol-gel, microemulsion, co-precipitation, etc.), a different route leads to a change in the properties of quantum dots. Also, there are certain limitations for each methodology that lead to a change in the physicochemical properties of quantum dots. This change in physicochemical properties affects the efficacy of the quantum dots in particular applications. The photoluminescence emission of quantum dots gives a slight idea about the arrangement as well as the internal structure of bands. Further, to enhance the application and the water stability of the quantum dots, different methodologies have been adopted as discussed in this chapter. In addition, quantum dot-based biomedical applications have proven to be a major research area. To make these quantum dots favorable to biomedical application, there are certain hurdles that need to be overcome to ensure a close collaboration of biotechnology with nanotechnology. Surface engineering of quantum dots is an essential step for the betterment of nanobiotechnology. In addition to making the quantum dots more specific to certain analytes or target delivery, surface engineering is an essential step that not only enhances its water dispersion but also increases its bioavailability as well as its specificity. Endless opportunities for biotechnical applications of quantum dots exist – the foremost necessity will be the design of functional biomolecules and their modification approaches to make them nontoxic and multifunctional.

REFERENCES

Abe, Sofie, Richard Karel Čapek, Bram De Geyter, and Zeger Hens. (2012). "Tuning the Postfocused Size of Colloidal Nanocrystals by the Reaction Rate: From Theory to Application." *ACS Nano*, 6(1), 42–53. doi:10.1021/nn204008q.

Akin, Mehriban, Rebecca Bongartz, Johanna G. Walter, Dilek Odaci Demirkol, Frank Stahl, Suna Timur, and Thomas Scheper. (2012). "PAMAM-Functionalized Water Soluble Quantum Dots for Cancer Cell Targeting." *Journal of Materials Chemistry*, 22(23), 11529–11536. doi:10.1039/c2jm31030a.

Al-Alwani, Ammar J., O.A. Shinkarenko, A.S. Chumakov, M.V. Pozharov, N.N. Begletsova, Anna S. Kolesnikova, V.P. Sevostyanova, and E.G. Glukhovskoy. (2019). "Influence of Capping Ligands on the Assembly of Quantum Dots and Their Properties." *Materials Science and Technology*, 35, 1053–1060. doi:10.1080/02670836.2019.1612141.

Algar, W. Russ, and Ulrich J. Krull. (2010). "Developing Mixed Films of Immobilized Oligonucleotides and Quantum Dots for the Multiplexed Detection of Nucleic Acid Hybridization Using a Combination of Fluorescence Resonance Energy Transfer and Direct Excitation of Fluorescence." *Langmuir*, 26(8), 6041–6047. doi:10.1021/la903751m.

Algar, W. Russ, Anthony J. Tavares, and Ulrich J. Krull. (2010). "Beyond Labels: A Review of the Application of Quantum Dots as Integrated Components of Assays, Bioprobes, and Biosensors Utilizing Optical Transduction." *Analytica Chimica Acta*, 673(1), 1–25. doi:10.1016/J.ACA.2010.05.026.

Ali, Rai Nauman, Hina Naz, Jing Li, Xingqun Zhu, Ping Liu, and Bin Xiang. (2018). "Band Gap Engineering of Transition Metal (Ni/Co) Codoped in Zinc Oxide (ZnO) Nanoparticles." *Journal of Alloys and Compounds*, 744, 90–95. doi:10.1016/J.JALLCOM.2018.02.072.

Alivisatos, A.P. (1996). "Semiconductor Clusters, Nanocrystals, and Quantum Dots." *Science*, 271(5251), 933–937. doi:10.1126/science.271.5251.933.

Anc, M.J., N.L. Pickett, N.C. Gresty, J.A. Harris, and K.C. Mishra. (2013). "Progress in Non-Cd Quantum Dot Development for Lighting Applications." *ECS Journal of Solid State Science and Technology*, 2(2), R3071–R3082. doi:10.1149/2.016302jss.

Anderson, George W., Joan E. Zimmerman, and Francis M. Callahan. (1964). "The Use of Esters of N-Hydroxysuccinimide in Peptide Synthesis." *Journal of the American Chemical Society*, 86(9), 1839–1842. doi:10.1021/ja01063a037.

Arriagada, F.J., and K. Osseo-Asare. (1999). "Synthesis of Nanosize Silica in a Nonionic Water-in-Oil Microemulsion: Effects of the Water/Surfactant Molar Ratio and Ammonia Concentration." *Journal of Colloid and Interface Science*, 211(2), 210–220. doi:10.1006/JCIS.1998.5985.

Arul, Velusamy, Thomas Nesakumar Jebakumar Immanuel Edison, Yong Rok Lee, and Mathur Gopalakrishnan Sethuraman. (2017). "Biological and Catalytic Applications of Green Synthesized Fluorescent N-Doped Carbon Dots Using Hylocereus Undatus." *Journal of Photochemistry and Photobiology B*, 168(March), 142–148. doi:10.1016/J.JPHOTOBIOL.2017.02.007.

Asok, Adersh, Mayuri N. Gandhi, and A.R. Kulkarni. (2012). "Enhanced Visible Photoluminescence in ZnO Quantum Dots by Promotion of Oxygen Vacancy Formation." *Nanoscale*, 4(16), 4943. doi:10.1039/c2nr31044a.

Atchudan, Raji, Thomas Nesakumar Jebakumar Immanuel Edison, Dasagrandhi Chakradhar, Suguna Perumal, Jae-Jin Shim, and Yong Rok Lee. (2017a). "Facile Green Synthesis of Nitrogen-Doped Carbon Dots Using Chionanthus Retusus Fruit Extract and Investigation of Their Suitability for Metal Ion Sensing and Biological Applications." *Sensors and Actuators B*, 246(July), 497–509. doi:10.1016/J.SNB.2017.02.119.

Atchudan, Raji, Thomas Nesakumar Jebakumar Immanuel Edison, Suguna Perumal, Namachivayam Karthik, Dhanapalan Karthikeyan, Mani Shanmugam, and Yong Rok Lee. (2018). "Concurrent Synthesis of Nitrogen-Doped Carbon Dots for Cell Imaging and ZnO@nitrogen-Doped Carbon Sheets for Photocatalytic Degradation of Methylene Blue." *Journal of Photochemistry and Photobiology A*, 350(January), 75–85. doi:10.1016/J.JPHOTOCHEM.2017.09.038.

Atchudan, Raji, Thomas Nesakumar Jebakumar Immanuel Edison, Suguna Perumal, and Yong Rok Lee. (2017b). "Green Synthesis of Nitrogen-Doped Graphitic Carbon Sheets with Use of Prunus Persica for Supercapacitor Applications." *Applied Surface Science*, 393(January), 276–286. doi:10.1016/J.APSUSC.2016.10.030.

Bajorowicz, Beata, Marek P. Kobylański, Anna Gołąbiewska, Joanna Nadolna, Adriana Zaleska-Medynska, and Anna Malankowska. (2018). "Quantum Dot-Decorated Semiconductor Micro- and Nanoparticles: A Review of Their Synthesis, Characterization and Application in Photocatalysis." *Advances in Colloid and Interface Science*, 256(June), 352–372. doi:10.1016/J.CIS.2018.02.003.

Bala, Rajni, Anuradha Swami, Ilja Tabujew, Kalina Peneva, Nishima Wangoo, and Rohit K. Sharma. (2018). "Ultra-Sensitive Detection of Malathion Using Quantum Dots-Polymer Based Fluorescence Aptasensor." *Biosensors and Bioelectronics*, *104*(May), 45–49. doi:10.1016/J.BIOS.2017.12.034.

Bang, Jungsik, Heesun Yang, and Paul H Holloway. (2006). "Enhanced and Stable Green Emission of ZnO Nanoparticles by Surface Segregation of Mg." *Nanotechnology*, *17*(4), 973–978. doi:10.1088/0957-4484/17/4/022.

Bansal, A.K., F. Antolini, S. Zhang, L. Stroea, L. Ortolani, M. Lanzi, E. Serra, S. Allard, U. Scherf, and I.D.W. Samuel. (2016). "Highly Luminescent Colloidal CdS Quantum Dots with Efficient Near-Infrared Electroluminescence in Light-Emitting Diodes." *The Journal of Physical Chemistry C*, *120*(3), 1871–1880. doi:10.1021/acs.jpcc.5b09109.

Bansal, Kavita, Mohammad Aqdas, Munish Kumar, Rajni Bala, Sanpreet Singh, Javed N. Agrewala, O.P. Katare, Rohit K. Sharma, and Nishima Wangoo. (2018). "A Facile Approach for Synthesis and Intracellular Delivery of Size Tunable Cationic Peptide Functionalized Gold Nanohybrids in Cancer Cells." *Bioconjugate Chemistry*, *29*(4), 1102–1110. doi:10.1021/acs.bioconjchem.7b00772.

Bao, Lei, Zhi-Ling Zhang, Zhi-Quan Tian, Li Zhang, Cui Liu, Yi Lin, Baoping Qi, and Dai-Wen Pang. (2011). "Electrochemical Tuning of Luminescent Carbon Nanodots: From Preparation to Luminescence Mechanism." *Advanced Materials*, *23*(48), 5801–5806. doi:10.1002/adma.201102866.

Bawendi, M.G., W.L. Wilson, L. Rothberg, P.J. Carroll, T.M. Jedju, M.L. Steigerwald, and L.E. Brus. (1990). "Electronic Structure and Photoexcited-Carrier Dynamics in Nanometer-Size CdSe Clusters." *Physical Review Letters*, *65*(13), 1623–1626. doi:10.1103/PhysRevLett.65.1623.

Bera, Debasis, Suresh C. Kuiry, and Sudipta Seal. (2004). "Synthesis of Nanostructured Materials Using Template-Assisted Electrodeposition." *JOM*, *56*(1), 49–53. doi:10.1007/s11837-004-0273-5.

Bera, Debasis, Lei Qian, and Paul H Holloway. (2008a). "Time-Evolution of Photoluminescence Properties of ZnO/MgO Core/Shell Quantum Dots." *Journal of Physics D: Applied Physics*, *41*(18), 182002. doi:10.1088/0022-3727/41/18/182002.

Bera, Debasis, Lei Qian, Subir Sabui, Swadeshmukul Santra, and Paul H. Holloway. (2008b). "Photoluminescence of ZnO Quantum Dots Produced by a Sol–Gel Process." *Optical Materials*, *30*(8), 1233–1239. doi:10.1016/J.OPTMAT.2007.06.001.

Bera, Debasis, Lei Qian, Teng-Kuan Tseng, and Paul H. Holloway. (2010). "Quantum Dots and Their Multimodal Applications: A Review." *Materials*, *3*(4), 2260–2345. doi:10.3390/ma3042260.

Bol, A.A., and A. Meijerink. (1998). "Long-Lived Mn^{2+} Emission in Nanocrystalline ZnS:Mn^{2+}." *Physical Review B*, *58*(24), R15997–R16000. doi:10.1103/PhysRevB.58.R15997.

Brauser, Eric M., Trevor D. Hull, John D. McLennan, Jacqueline T. Siy, and Michael H. Bartl. (2016). "Experimental Evaluation of Kinetic and Thermodynamic Reaction Parameters of Colloidal Nanocrystals." *Chemistry of Materials*, *28*(11), 3831–3838. doi:10.1021/acs.chemmater.6b00878.

Brichkin, S.B. (2015). "Synthesis and Properties of Colloidal Indium Phosphide Quantum Dots." *Colloid Journal*, *77*(4), 393–403. doi:10.1134/S1061933X15040043.

Brunetti, Virgilio, Hicham Chibli, Roberto Fiammengo, Antonio Galeone, Maria Ada Malvindi, Giuseppe Vecchio, Roberto Cingolani, Jay L. Nadeau, and Pier Paolo Pompa. (2013). "InP/ZnS as a Safer Alternative to CdSe/ZnS Core/Shell Quantum Dots: In Vitro and in Vivo Toxicity Assessment." *Nanoscale*, *5*(1), 307–317. doi:10.1039/C2NR33024E.

Brus, L.E. (1983). "A Simple Model for the Ionization Potential, Electron Affinity, and Aqueous Redox Potentials of Small Semiconductor Crystallites." *Journal of Chemical Physics*, *79*(11), 5566–5571. doi:10.1063/1.445676.

Brus, L.E. (1984). "Electron–Electron and Electron-hole Interactions in Small Semiconductor Crystallites: The Size Dependence of the Lowest Excited Electronic State." *Journal of Chemical Physics*, *80*(9), 4403–4409. doi:10.1063/1.447218.

Burda, Clemens, Xiaobo Chen, Radha Narayanan, and Mostafa A. El-Sayed. (2005). "Chemistry and Properties of Nanocrystals of Different Shapes." *Chemical Reviews*, *105*(4), 1025–1102. doi:10.1021/cr030063a.

Čapek, Richard Karel, Dianna Yanover, and Efrat Lifshitz. (2015). "Size Control by Rate Control in Colloidal PbSe Quantum Dot Synthesis." *Nanoscale*, *7*(12), 5299–5310. doi:10.1039/C5NR00028A.

Chand, Subhash, Nagesh Thakur, S.C. Katyal, P.B. Barman, Vineet Sharma, and Pankaj Sharma. (2017). "Recent Developments on the Synthesis, Structural and Optical Properties of Chalcogenide Quantum Dots." *Solar Energy Materials and Solar Cells*, *168*(August), 183–200. doi:10.1016/J.SOLMAT.2017.04.033.

Chang, Yajing, Xudong Yao, Longfei Mi, Guopeng Li, Shengda Wang, Hui Wang, Zhongping Zhang, and Yang Jiang. (2015). "A Water–Ethanol Phase Assisted Co-Precipitation Approach toward High Quality Quantum Dot–Inorganic Salt Composites and Their Application for WLEDs." *Green Chemistry*. *17*(8), 4439–4445. doi:10.1039/C5GC01109D.

Chen, Jian, Ya Li, Kun Lv, Weibang Zhong, Hong Wang, Zhan Wu, Pinggui Yi, and Jianhui Jiang. (2016). "Cyclam-Functionalized Carbon Dots Sensor for Sensitive and Selective Detection of Copper(II) Ion and Sulfide Anion in Aqueous Media and Its Imaging in Live Cells." *Sensors and Actuators B*, *224*(March), 298–306. doi:10.1016/J.SNB.2015.10.046.

Cheng, Cheng, Hua Wei, Bao-Xian Shi, Han Cheng, Cao Li, Zhong-Wei Gu, Si-Xue Cheng, Xian-Zheng Zhang, and Ren-Xi Zhuo. (2008). "Biotinylated Thermoresponsive Micelle Self-Assembled from Double-Hydrophilic Block Copolymer for Drug Delivery and Tumor Target." *Biomaterials*, *29*(4), 497–505. doi:10.1016/J.BIOMATERIALS.2007.10.004.

Cheng, Hsin-Ming, Kuo-Feng Lin, Hsu-Cheng Hsu, and Wen-Feng Hsieh. (2006). "Size Dependence of Photoluminescence and Resonant Raman Scattering from ZnO Quantum Dots." *Applied Physics Letters*, *88*(26), 261909. doi:10.1063/1.2217925.

Cheng, Jinghui, Xianggge Zhou, and Haifeng Xiang. (2015). "Fluorescent Metal Ion Chemosensors via Cation Exchange Reactions of Complexes, Quantum Dots, and Metal–Organic Frameworks." *Analyst*, *140*(21), 7082–7115. doi:10.1039/C5AN01398D.

Chhibber, Sanjay, Vijay S. Gondil, Samrita Sharma, Munish Kumar, Nishima Wangoo, and Rohit K. Sharma. (2017). "A Novel Approach for Combating Klebsiella Pneumoniae Biofilm Using Histidine Functionalized Silver Nanoparticles." *Frontiers in Microbiology*, *8*(June), 1104. doi:10.3389/fmicb.2017.01104.

Chhibber, Sanjay, Vijay Singh Gondil, Love Singla, Munish Kumar, Tanya Chhibber, Gajanand Sharma, Rohit Kumar Sharma, Nishima Wangoo, and Om Prakash Katare. (2019). "Effective Topical Delivery of H-AgNPs for Eradication of Klebsiella Pneumoniae–Induced Burn Wound Infection." *AAPS PharmSciTech*, 20(5), 169. doi:10.1208/s12249-019-1350-y.

Chien, Ming-Chin, and Chung-Hao Tien. (2012). "Multispectral Mixing Scheme for LED Clusters with Extended Operational Temperature Window." *Optics Express*, 20(S2), A245. doi:10.1364/OE.20.00A245.

Chin, Patrick T.K., Tessa Buckle, Arantxa Aguirre de Miguel, Stefan C.J. Meskers, René A.J. Janssen, and Fijs W.B. van Leeuwen. (2010). "Dual-Emissive Quantum Dots for Multispectral Intraoperative Fluorescence Imaging." *Biomaterials*, 31(26), 6823–6832. doi:10.1016/J.BIOMATERIALS.2010.05.030.

Chuang, Chia-Hao M., Patrick R. Brown, Vladimir Bulović, and Moungi G. Bawendi. (2014). "Improved Performance and Stability in Quantum Dot Solar Cells through Band Alignment Engineering." *Nature Materials*, 13(8), 796–801. doi:10.1038/nmat3984.

Ciftja, Orion. (2019). "Properties of Quantum Dots and Their Biological Applications." In *Nano-Sized Multifunctional Materials* (pp. 21–45). Elsevier. doi:10.1016/B978-0-12-813934-9.00002-5.

Cinteza, Ludmila Otilia. (2010). "Quantum Dots in Biomedical Applications: Advances and Challenges." *Journal of Nanophotonics*, 4(1), 042503. doi:10.1117/1.3500388.

Clapp, Aaron R., Igor L. Medintz, and Hedi Mattoussi. (2006). "Förster Resonance Energy Transfer Investigations Using Quantum-Dot Fluorophores." *ChemPhysChem*, 7(1), 47–57. doi:10.1002/cphc.200500217.

Correa-Duarte, Miguel A., Michael Giersig, and Luis M. Liz-Marzán. (1998). "Stabilization of CdS Semiconductor Nanoparticles against Photodegradation by a Silica Coating Procedure." *Chemical Physics Letters*, 286(5–6), 497–501. doi:10.1016/S0009-2614(98)00012-8.

Correa-Duarte, Miguel A., Yoshio Kobayashi, Rachel A. Caruso, and Luis M. Liz-Marzán. (2001). "Photodegradation of SiO₂-Coated CdS Nanoparticles within Silica Gels." *Journal of Nanoscience and Nanotechnology*, 1(1), 95–99. doi:10.1166/jnn.2001.010.

Cushing, Brian L., Vladimir L. Kolesnichenko, and Charles J. O'Connor. (2004). "Recent Advances in the Liquid-Phase Syntheses of Inorganic Nanoparticles." *Chemical Reviews*, 104(9), 3893–3946. doi:10.1021/CR030027B.

Dai, Sheng, Yu-Sheng Su, Shu-Ru Chung, Kuan-Wen Wang, and Xiaoqing Pan. (2018). "Controlling the Magic Size of White Light-Emitting CdSe Quantum Dots." *Nanoscale*, 10(21), 10256–10261. doi:10.1039/C8NR01455H.

Das, Poushali, Sayan Ganguly, Madhuparna Bose, Subhadip Mondal, Amit Kumar Das, Susanta Banerjee, and Narayan Chandra Das. (2017). "A Simplistic Approach to Green Future with Eco-Friendly Luminescent Carbon Dots and Their Application to Fluorescent Nano-Sensor 'turn-off' Probe for Selective Sensing of Copper Ions." *Materials Science & Engineering C*, 75(June), 1456–1464. doi:10.1016/j.msec.2017.03.045.

DasariAyodhya, MaragoniVenkatesham, Amrutham Santoshi Kumari, Kotu Girija Mangatayaru, and Guttena Veerabhadram. (2013). "Synthesis , Characterization of ZnS Nanoparticles by Coprecipitation Method Using Various Capping Agents-Photocatalytic Activity and Kinetic Study." https://www.semanticscholar.org/paper/Synthesis-%2C-Characterization-of-ZnS-nanoparticles-DasariAyodhya-Kumari/eac4de28ae9618e3d75e9ac0ccda837090428b14.

de Mello Donegá, Celso, Peter Liljeroth, and Daniel Vanmaekelbergh. (2005). "Physicochemical Evaluation of the Hot-Injection Method, a Synthesis Route for Monodisperse Nanocrystals." *Small*, 1(12), 1152–1162. doi:10.1002/smll.200500239.

Delong, Robert K, Christopher M Reynolds, Yaneika Malcolm, Ashley Schaeffer, Tiffany Severs, and Adam Wanekaya. (2010). "Functionalized Gold Nanoparticles for the Binding, Stabilization, and Delivery of Therapeutic DNA, RNA, and Other Biological Macromolecules." *Nanotechnology, Science and Applications*, 3(September), 53–63. doi:10.2147/NSA.S8984.

Deng, Zhengtao, Fee Li Lie, Shengyi Shen, Indraneel Ghosh, Masud Mansuripur, and Anthony J. Muscat. (2009). "Water-Based Route to Ligand-Selective Synthesis of ZnSe and Cd-Doped ZnSe Quantum Dots with Tunable Ultraviolet A to Blue Photoluminescence." *Langmuir*, 25(1), 434–442. doi:10.1021/la802294e.

DeTar, DeLos F., Richard Silverstein, and Fulton F. Rogers. (1966). "Reactions of Carbodiimides. III. The Reactions of Carbodiimides with Peptide Acids." *Journal of the American Chemical Society*, 88(5), 1024–1030. doi:10.1021/ja00957a029.

Di Corato, Riccardo, Alessandra Quarta, Philomena Piacenza, Andrea Ragusa, Albert Figuerola, Raffaella Buonsanti, Roberto Cingolani, Liberato Manna, and Teresa Pellegrino. (2008). "Water Solubilization of Hydrophobic Nanocrystals by Means of Poly(Maleic Anhydride-Alt-1-Octadecene)." *Journal of Materials Chemistry*, 18(17), 1991. doi:10.1039/b717801h.

Dingle, R. (1969). "Luminescent Transitions Associated With Divalent Copper Impurities and the Green Emission from Semiconducting Zinc Oxide." *Physical Review Letters*, 23(11), 579–581. doi:10.1103/PhysRevLett.23.579.

Djurišić, Aleksandra B., Yu Hang Leung, Wallace C. H. Choy, Kok Wai Cheah, and Wai Kin Chan. (2004). "Visible Photoluminescence in ZnO Tetrapod and Multipod Structures." *Applied Physics Letters*, 84(14), 2635–2637. doi:10.1063/1.1695633.

Dmitri V. Talapin, Andrey L. Rogach, Andreas Kornowski, Markus Haase, and Horst Weller. (2001). "Highly Luminescent Monodisperse CdSe and CdSe/ZnS Nanocrystals Synthesized in a Hexadecylamine–Trioctylphosphine Oxide–Trioctylphospine Mixture." *Nano Letters*, 1(4), 207–211. doi:10.1021/NL0155126.

Donegan, John F., and Yury P. Rakovich. (2013). *Cadmium Telluride Quantum Dots : Advances and Applications*. CRC Press.

Dong, Jiaqi, Jiaying Li, and Jian Zhou. (2014). "Interfacial and Phase Transfer Behaviors of Polymer Brush Grafted Amphiphilic Nanoparticles: A Computer Simulation Study." *Langmuir*, 30(19), 5599–5608. doi:10.1021/la500592k.

Du, Jun, Zhonglin Du, Jin-Song Hu, Zhenxiao Pan, Qing Shen, Jiankun Sun, Donghui Long, et al. (2016). "Zn–Cu–In–Se Quantum Dot Solar Cells with a Certified Power Conversion Efficiency of 11.6%." *Journal of the American Chemical Society*, 138(12), 4201–4209. doi:10.1021/jacs.6b00615.

Dubertret, Benoit, Paris Skourides, David J Norris, Vincent Noireaux, Ali H Brivanlou, and Albert Libchaber. (2002). "In Vivo Imaging of Quantum Dots Encapsulated in Phospholipid Micelles." *Science*, *298*(5599), 1759–1762. doi:10.1126/science.1077194.

Dunne, Peter W., Chris L. Starkey, Miquel Gimeno-Fabra, and Edward H. Lester. (2014). "The Rapid Size- and Shape-Controlled Continuous Hydrothermal Synthesis of Metal Sulphide Nanomaterials." *Nanoscale*, *6*(4), 2406–2418. doi:10.1039/C3NR05749F.

Efros, Al.L., and M. Rosen. (1997). "Random Telegraph Signal in the Photoluminescence Intensity of a Single Quantum Dot." *Physical Review Letters*, *78*(6), 1110–1113. doi:10.1103/PhysRevLett.78.1110.

Efros, Al.L., and M. Rosen. (2000). "The Electronic Structure of Semiconductor Nanocrystals." *Annual Review of Materials Science*, *30*(1), 475–521. doi:10.1146/annurev.matsci.30.1.475.

Efros, Al.L., M. Rosen, M. Kuno, M. Nirmal, D.J. Norris, and M. Bawendi. (1996). "Band-Edge Exciton in Quantum Dots of Semiconductors with a Degenerate Valence Band: Dark and Bright Exciton States." *Physical Review B*, *54*(7), 4843–4856. http://www.ncbi.nlm.nih.gov/pubmed/9986445.

Ekimov, A.I., and A.A. Onushchenko. (1981). "Quantum Size Effect in Three-Dimensional Microscopic Semiconductor Crystals." *Journal of Experimental and Theoretical Physics Letters*, *34*, 345. http://adsabs.harvard.edu/abs/1981JETPL..34..345E.

El Fatimy, Abdel, Rachael L. Myers-Ward, Anthony K. Boyd, Kevin M. Daniels, D. Kurt Gaskill, and Paola Barbara. (2016). "Epitaxial Graphene Quantum Dots for High-Performance Terahertz Bolometers." *Nature Nanotechnology*, *11*(4), 335–338. doi:10.1038/nnano.2015.303.

Emin, Saim, Surya P Singh, Han Liyuan, Norifusa Satoh, and Ashraful Islam. (2011). "Colloidal Quantum Dot Solar Cells." *Solar Energy*, *85*, 1264–1282. doi:10.1016/j.solener.2011.02.005.

Fan, X.M., J.S. Lian, L. Zhao, and Y.H. Liu. (2005). "Single Violet Luminescence Emitted from ZnO Films Obtained by Oxidation of Zn Film on Quartz Glass." *Applied Surface Science*, *252*(2), 420–424. doi:10.1016/J.APSUSC.2005.01.018.

Fan, Yu Zhu, Ying Zhang, Na Li, Shi Gang Liu, Ting Liu, Nian Bing Li, and Hong Qun Luo. (2017). "A Facile Synthesis of Water-Soluble Carbon Dots as a Label-Free Fluorescent Probe for Rapid, Selective and Sensitive Detection of Picric Acid." *Sensors and Actuators B*, *240*(March), 949–955. doi:10.1016/J.SNB.2016.09.063.

Farias, Patrícia M.A., Adriana Fontes, André Galembeck, Regina C.B.Q. Figueiredo, and Beate S. Santos. (2008). "Fluorescent II-VI Semiconductor Quantum Dots: Potential Tools for Biolabeling and Diagnostic." *Journal of the Brazilian Chemical Society*, *19*(2), 352–356. doi:10.1590/S0103-50532008000200023.

Ferguson, Elaine L., and Ruth Duncan. (2009). "Dextrin–Phospholipase A$_2$: Synthesis and Evaluation as a Bioresponsive Anticancer Conjugate." *Biomacromolecules*, *10*(6), 1358–1364. doi:10.1021/bm8013022.

Frecker, Talitha, Danielle Bailey, Xochitl Arzeta-Ferrer, James McBride, and Sandra J. Rosenthal. (2016). "Review—Quantum Dots and Their Application in Lighting, Displays, and Biology." *ECS Journal of Solid State Science and Technology*, *5*(1), R3019–R3031. doi:10.1149/2.0031601jss.

Furedi-Milhofer, H. (1981). "Spontaneous Precipitation from Electrolytic Solutions." *Pure and Applied Chemistry*, *53*(11), 2041–2055. doi:10.1351/pac198153112041.

Gao, Bing, Chao Shen, Shuanglong Yuan, Yunxia Yang, and Guorong Chen. (2013). "Synthesis of Highly Emissive CdSe Quantum Dots by Aqueous Precipitation Method." *Journal of Nanomaterials*, *2013*(November), 1–7. doi:10.1155/2013/138526.

Gao, Xian-Feng, Hong-Bo Li, Wen-Tao Sun, Qing Chen, Fang-Qiong Tang, and Lian-Mao Peng. (2009). "CdTe Quantum Dots-Sensitized TiO$_2$ Nanotube Array Photoelectrodes." *Journal of Physical Chemistry C*, *113*(18), 7531–7535. doi:10.1021/jp810727n.

Gao, Xiaohu, Yuanyuan Cui, Richard M Levenson, Leland W K Chung, and Shuming Nie. (2004). "In Vivo Cancer Targeting and Imaging with Semiconductor Quantum Dots." *Nature Biotechnology*, *22*(8), 969–976. doi:10.1038/nbt994.

Gao, Xiaohui, Cheng Du, Zhihua Zhuang, and Wei Chen. (2016). "Carbon Quantum Dot-Based Nanoprobes for Metal Ion Detection." *Journal of Materials Chemistry C*, *4*(29), 6927–6945. doi:10.1039/C6TC02055K.

Geraldo, Daniela A, Esteban F Duran-Lara, Daniel Aguayo, Raul E Cachau, Jaime Tapia, Rodrigo Esparza, Miguel J Yacaman, Fernando Danilo Gonzalez-Nilo, and Leonardo S Santos. (2011). "Supramolecular Complexes of Quantum Dots and a Polyamidoamine (PAMAM)-Folate Derivative for Molecular Imaging of Cancer Cells." *Analytical and Bioanalytical Chemistry*, *400*(2), 483–492. doi:10.1007/s00216-011-4756-2.

Gfroerer, Timothy H. (2006). "Photoluminescence in Analysis of Surfaces and Interfaces." In *Encyclopedia of Analytical Chemistry*. Wiley. doi:10.1002/9780470027318.a2510.

Goharshadi, Elaheh K., Sayyed Hashem Sajjadi, Roya Mehrkah, and Paul Nancarrow. (2012). "Sonochemical Synthesis and Measurement of Optical Properties of Zinc Sulfide Quantum Dots." *Chemical Engineering Journal*, *209*(October), 113–117. doi:10.1016/J.CEJ.2012.07.131.

Gong, Xiaojuan, Wenjing Lu, Man Chin Paau, Qin Hu, Xin Wu, Shaomin Shuang, Chuan Dong, and Martin M.F. Choi. (2015). "Facile Synthesis of Nitrogen-Doped Carbon Dots for Fe3+ Sensing and Cellular Imaging." *Analytica Chimica Acta*, *861*(February), 74–84. doi:10.1016/J.ACA.2014.12.045.

Graf, Christina, Sofia Dembski, Andreas Hofmann, and Eckart Ruhl. (2006). "A General Method for the Controlled Embedding of Nanoparticles in Silica Colloids." *Langmuir*, *22*(13), 5604–5610. doi:10.1021/LA060136W.

Grapperhaus, Craig A., Thawatchai Tuntulani, Joseph H. Reibenspies, and Marcetta Y. Darensbourg. (1998). "Methylation of Tethered Thiolates in [(Bme-Daco)Zn]2 and [(Bme-Daco)Cd]2 as a Model of Zinc Sulfur-Methylation Proteins." *Inorganic Chemistry*, *37*(16), 4052–4058. doi:10.1021/IC971599F.

Gui, Rijun, Jie Sun, Dexiu Liu, Yanfeng Wang, and Hui Jin. (2014). "A Facile Cation Exchange-Based Aqueous Synthesis of Highly Stable and Biocompatible Ag$_2$S Quantum Dots Emitting in the Second near-Infrared Biological Window." *Dalton Transactions*, *43*(44), 16690–16697. doi:10.1039/c4dt00699b.

Habeeb, A.F.S.A., and R. Hiramoto. (1968). "Reaction of Proteins with Glutaraldehyde." *Archives of Biochemistry and Biophysics*, *126*(1), 16–26. doi:10.1016/0003-9861(68)90554-7.

Hammer, Nathan I., Todd Emrick, and Michael D. Barnes. (2007). "Quantum Dots Coordinated with Conjugated Organic Ligands: New Nanomaterials with Novel Photophysics." *Nanoscale Research Letters*, 2(6)., 282–290. doi:10.1007/s11671-007-9062-8.

Hammes, B.S., and C.J. Carrano. (2001). "Methylation of (2-Methylethanethiol-Bis-3,5-Dimethylpyrazolyl)Methane Zinc Complexes and Coordination of the Resulting Thioether: Relevance to Zinc-Containing Alkyl Transfer Enzymes." *Inorganic Chemistry*, 40(5), 919–927. http://www.ncbi.nlm.nih.gov/pubmed/11258999.

Han, Hyunjoo, Gianna Di Francesco, and Mathew M. Maye. (2010). "Size Control and Photophysical Properties of Quantum Dots Prepared via a Novel Tunable Hydrothermal Route." *The Journal of Physical Chemistry C*, 114(45), 19270–19277. doi:10.1021/jp107702b.

Harris, J. Milton. (1985). "Laboratory Synthesis of Polyethylene Glycol Derivatives." *Journal of Macromolecular Science Part C*, 25(3), 325–373. doi:10.1080/07366578508081960.

Hayrapetyan, D.B., A.V. Chalyan, E.M. Kazaryan, and H.A. Sarkisyan. (2015). "Direct Interband Light Absorption in Conical Quantum Dot." *Journal of Nanomaterials*, 2015(November), 1–6. doi:10.1155/2015/915742.

Henglein, Arnim. (1989). "Small-Particle Research: Physicochemical Properties of Extremely Small Colloidal Metal and Semiconductor Particles." *Chemical Reviews*, 89(8), 1861–1873. doi:10.1021/cr00098a010.

Heo, Y.W., D.P. Norton, and S.J. Pearton. (2005). "Origin of Green Luminescence in ZnO Thin Film Grown by Molecular-Beam Epitaxy." *Journal of Applied Physics*, 98(7), 073502. doi:10.1063/1.2064308.

Hess, B.C., I.G. Okhrimenko, R.C. Davis, B.C. Stevens, Q.A. Schulzke, K.C. Wright, C.D. Bass, C.D. Evans, and S.L. Summers. (2001). "Surface Transformation and Photoinduced Recovery in CdSe Nanocrystals." *Physical Review Letters*, 86(14), 3132–3135. doi:10.1103/PhysRevLett.86.3132.

Hitchcock, Adam P., Bonnie O. Leung, John L. Brash, Andreas Scholl, and Andrew Doran. (2012). "Soft X-Ray Spectromicroscopy of Protein Interactions with Phase-Segregated Polymer Surfaces." In *Proteins at Interfaces III State of the Art* (pp. 731–760). American Chemical Society. doi:10.1021/bk-2012-1120.ch034.

Hoener, Carolyn F., Kristi Ann Allan, Allen J. Bard, Alan Campion, Marye Anne Fox, Thomas E. Mallouk, Stephen E. Webber, and J. Michael White. (1992). "Demonstration of a Shell-Core Structure in Layered Cadmium Selenide-Zinc Selenide Small Particles by x-Ray Photoelectron and Auger Spectroscopies." *Journal of Physical Chemistry*, 96(9), 3812–3817. doi:10.1021/j100188a045.

Hou, Bo, David Benito-Alifonso, Richard Webster, David Cherns, M. Carmen Galan, and David J. Fermín. (2014). "Rapid Phosphine-Free Synthesis of CdSe Quantum Dots: Promoting the Generation of Se Precursors Using a Radical Initiator." *Journal of Materials Chemistry A*, 2(19), 6879–6886. doi:10.1039/C4TA00285G.

Hu, Q.R., S.L. Wang, P. Jiang, H. Xu, Y. Zhang, and W.H. Tang. (2010). "Synthesis of ZnO Nanostructures in Organic Solvents and Their Photoluminescence Properties." *Journal of Alloys and Compounds*, 496(1–2), 494–499. doi:10.1016/J.JALLCOM.2010.02.086.

Inamdar, Shaukatali N., Pravin P. Ingole, and Santosh K. Haram. (2008). "Determination of Band Structure Parameters and the Quasi-Particle Gap of CdSe Quantum Dots by Cyclic Voltammetry." *ChemPhysChem*, 9(17), 2574–2579. doi:10.1002/cphc.200800482.

Issac, Abey, Christian von Borczyskowski, and Frank Cichos. (2005). "Correlation between Photoluminescence Intermittency of CdSe Quantum Dots and Self-Trapped States in Dielectric Media." *Physical Review B*, 71(16), 161302. doi:10.1103/PhysRevB.71.161302.

Jagtap, Amardeep M., Jayakrishna Khatei, and K. S. R. Koteswara Rao. (2015). "Exciton–Phonon Scattering and Nonradiative Relaxation of Excited Carriers in Hydrothermally Synthesized CdTe Quantum Dots." *Physical Chemistry Chemical Physics*, 17(41), 27579–27587. doi:10.1039/C5CP04654H.

Jaiswal, Amit, Siddhartha Sankar Ghsoh, and Arun Chattopadhyay. (2012). "Quantum Dot Impregnated-Chitosan Film for Heavy Metal Ion Sensing and Removal." *Langmuir*, 28(44), 15687–15696. doi:10.1021/la3027573.

Jamieson, Timothy, Raheleh Bakhshi, Daniela Petrova, Rachael Pocock, Mo Imani, and Alexander M. Seifalian. (2007). "Biological Applications of Quantum Dots." *Biomaterials*, 28(31), 4717–4732. doi:10.1016/J.BIOMATERIALS.2007.07.014.

Jayanthi, K., S. Chawla, H. Chander, and D. Haranath. (2007). "Structural, Optical and Photoluminescence Properties of ZnS:Cu Nanoparticle Thin Films as a Function of Dopant Concentration and Quantum Confinement Effect." *Crystal Research and Technology*, 42(10), 976–982. doi:10.1002/crat.200710950.

Jia, Feng, Yanjie Zhang, Balaji Narasimhan, and Surya K. Mallapragada. (2012). "Block Copolymer-Quantum Dot Micelles for Multienzyme Colocalization." *Langmuir*, 28(50), 17389–17395. doi:10.1021/la303115t.

Jin, Shan, Yanxi Hu, Zhanjun Gu, Lei Liu, and Hai-Chen Wu. (2011). "Application of Quantum Dots in Biological Imaging." *Journal of Nanomaterials*, 2011(August), 1–13. doi:10.1155/2011/834139.

Jin, Takashi, Fumihiko Fujii, Yutaka Komai, Junji Seki, Akitoshi Seiyama, and Yoshichika Yoshioka. (2008). "Preparation and Characterization of Highly Fluorescent, Glutathione-Coated near Infrared Quantum Dots for in Vivo Fluorescence Imaging." *International Journal of Molecular Sciences*, 9(10), 2044–2061. doi:10.3390/ijms9102044.

Jing, Lihong, Ke Ding, Stephen V. Kershaw, Ivan M. Kempson, Andrey L. Rogach, and Mingyuan Gao. (2014). "Magnetically Engineered Semiconductor Quantum Dots as Multimodal Imaging Probes." *Advanced Materials*, 26(37), 6367–6386. doi:10.1002/adma.201402296.

Jing, Liqiang, Fulong Yuan, Haige Hou, Baifu Xin, Weimin Cai, and Honggang Fu. (2005). "Relationships of Surface Oxygen Vacancies with Photoluminescence and Photocatalytic Performance of ZnO Nanoparticles." *Science in China Series B*, 48(1), 25–30. doi:10.1007/BF02990909.

Joo, J., S.G. Kwon, J.H. Yu, and T. Hyeon. (2005). "Synthesis of ZnO Nanocrystals with Cone, Hexagonal Cone, and Rod Shapes via Non-Hydrolytic Ester Elimination Sol–Gel Reactions." *Advanced Materials*, 17(15), 1873–1877. doi:10.1002/adma.200402109.

Junk, Andreas, and Falk Riess. (2006). "From an Idea to a Vision: There's Plenty of Room at the Bottom." *American Journal of Physics*, *74*(9), 825–830. doi:10.1119/1.2213634.

Kairdolf, Brad A., Michael C. Mancini, Andrew M. Smith, and Shuming Nie. (2008). "Minimizing Nonspecific Cellular Binding of Quantum Dots with Hydroxyl-Derivatized Surface Coatings." *Analytical Chemistry*, *80*, 3029–3034. doi:10.1021/AC800068Q.

Kairdolf, Brad A., Andrew M. Smith, and Shuming Nie. (2008). "One-Pot Synthesis, Encapsulation, and Solubilization of Size-Tuned Quantum Dots with Amphiphilic Multidentate Ligands." *Journal of the American Chemical Society*, *130*(39), 12866–12867. doi:10.1021/ja804755q.

Kairdolf, Brad A., Andrew M. Smith, Todd H. Stokes, May D. Wang, Andrew N. Young, and Shuming Nie. (2013). "Semiconductor Quantum Dots for Bioimaging and Biodiagnostic Applications." *Annual Review of Analytical Chemistry*, *6*(1), 143–162. doi:10.1146/annurev-anchem-060908-155136.

Kamat, Prashant V. (2007). "Meeting the Clean Energy Demand: Nanostructure Architectures for Solar Energy Conversion." *Journal of Physical Chemistry C*, *111*(7), 2834–2860. doi:10.1021/JP066952U.

Kang, Byoung-Ho, Ju-Seong Kim, Jae-Sung Lee, Sang-Won Lee, Gopalan Sai-Anand, Hyun-Min Jeong, Seung-Ha Lee, Dae-Hyuk Kwon, and Shin-Won Kang. (2015). "Solution Processable CdSe/ZnS Quantum Dots Light-Emitting Diodes Using ZnO Nanocrystal as Electron Transport Layer." *Journal of Nanoscience and Nanotechnology*, *15*(9), 7416–7420. http://www.ncbi.nlm.nih.gov/pubmed/26716347.

Kang, Youngjong, and T. Andrew Taton. (2005). "Core/Shell Gold Nanoparticles by Self-Assembly and Crosslinking of Micellar, Block-Copolymer Shells." *Angewandte Chemie International Edition*, *44*(3), 409–412. doi:10.1002/anie.200461119.

Karanikolos, Georgios N,, Paschalis Alexandridis, Grigorios Itskos, Athos Petrou, and T. J. Mountziaris. (2004). "Synthesis and Size Control of Luminescent ZnSe Nanocrystals by a Microemulsion−Gas Contacting Technique." *Langmuir*, *20*(3), 550–553. doi:10.1021/LA035397+.

Katari, J.E. Bowen, V.L. Colvin, and A.P. Alivisatos. (1994). "X-Ray Photoelectron Spectroscopy of CdSe Nanocrystals with Applications to Studies of the Nanocrystal Surface." *Journal of Physical Chemistry*, *98*(15), 4109–4117. doi:10.1021/j100066a034.

Kavosi, Begard, Aso Navaee, and Abdollah Salimi. (2018). "Amplified Fluorescence Resonance Energy Transfer Sensing of Prostate Specific Antigen Based on Aggregation of CdTe QDs/Antibody and Aptamer Decorated of AuNPs-PAMAM Dendrimer." *Journal of Luminescence*, *204*(December), 368–374. doi:10.1016/J.JLUMIN.2018.08.012.

Kelkar, Sneha S., and Theresa M. Reineke. (2011). "Theranostics: Combining Imaging and Therapy." *Bioconjugate Chemistry*, *22*(10), 1879–1903. doi:10.1021/bc200151q.

Kellici, Suela, John Acord, Nicholas P. Power, David J. Morgan, Paolo Coppo, Tobias Heil, and Basudeb Saha. (2017). "Rapid Synthesis of Graphene Quantum Dots Using a Continuous Hydrothermal Flow Synthesis Approach." *RSC Advances*, *7*(24), 14716–14720. doi:10.1039/C7RA00127D.

Khan, Ibrahim, Khalid Saeed, and Idrees Khan. (2017). "Nanoparticles: Properties, Applications and Toxicities." *Arabian Journal of Chemistry*, *12*(7), 908–931. doi:10.1016/J.ARABJC.2017.05.011.

Khorana, H. G. (1953). "The Chemistry of Carbodiimides." *Chemical Reviews*, *53*(2), 145–166. doi:10.1021/cr60165a001.

Kim, Byeong-Su, Jiao-Ming Qiu, Jian-Ping Wang, and T. Andrew Taton. (2005). "Magnetomicelles: Composite Nanostructures from Magnetic Nanoparticles and Cross-Linked Amphiphilic Block Copolymers." *Nano Letters*, *5*(10), 1987–1991. doi:10.1021/NL0513939.

Kim, Na-Yeong, Sang-Hyun Hong, Jang-Won Kang, NoSoung Myoung, Sang-Youp Yim, Suhyun Jung, Kwanghee Lee, Charles W. Tu, and Seong-Ju Park. (2015). "Localized Surface Plasmon-Enhanced Green Quantum Dot Light-Emitting Diodes Using Gold Nanoparticles." *RSC Advances*, *5*(25), 19624–19629. doi:10.1039/C4RA15585H.

Kim, Taekeun, Yun Ku Jung, and Jin-Kyu Lee. (2014). "The Formation Mechanism of CdSe QDs through the Thermolysis of Cd(Oleate) $_2$ and TOPSe in the Presence of Alkylamine." *Journal of Materials Chemistry C*, *2*(28), 5593–5600. doi:10.1039/C4TC00254G.

Klimov, Victor I. (2006). "Mechanisms for Photogeneration and Recombination of Multiexcitons in Semiconductor Nanocrystals: Implications for Lasing and Solar Energy Conversion." *Journal of Physical Chemistry B*, *110*(34), 16827–16845. doi:10.1021/JP0615959.

Kloepfer, J.A., R.E. Mielke, and J.L. Nadeau. (2005). "Uptake of CdSe and CdSe/ZnS Quantum Dots into Bacteria via Purine-Dependent Mechanisms." *Applied and Environmental Microbiology*, *71*(5), 2548–2557. doi:10.1128/AEM.71.5.2548-2557.2005.

Kortan, A.R., R. Hull, R.L. Opila, M.G. Bawendi, M.L. Steigerwald, P.J. Carroll, and Louis E. Brus. (1990). "Nucleation and Growth of CdSe on ZnS Quantum Crystallite Seeds, and Vice Versa, in Inverse Micelle Media." *Journal of the American Chemical Society*, *112*(4), 1327–1332. doi:10.1021/ja00160a005.

Kuchibhatla, Satyanarayana V.N.T., A.S. Karakoti, Debasis Bera, and S. Seal. (2007). "One Dimensional Nanostructured Materials." *Progress in Materials Science*, *52*(5), 699–913. doi:10.1016/J.PMATSCI.2006.08.001.

Kucur, Erol, Jürgen Riegler, Gerald A. Urban, and Thomas Nann. (2003). "Determination of Quantum Confinement in CdSe Nanocrystals by Cyclic Voltammetry." *Journal of Chemical Physics*, *119*(4), 2333–2337. doi:10.1063/1.1582834.

Kumar, Amit, Angshuman Ray Chowdhuri, Dipranjan Laha, Triveni Kumar Mahto, Parimal Karmakar, and Sumanta Kumar Sahu. (2017). "Green Synthesis of Carbon Dots from Ocimum Sanctum for Effective Fluorescent Sensing of Pb2+ Ions and Live Cell Imaging." *Sensors and Actuators B*, *242*(April), 679–686. doi:10.1016/J.SNB.2016.11.109.

Kumar, Munish, Kavita Bansal, Vijay S. Gondil, Samrita Sharma, D.V.S. Jain, Sanjay Chhibber, Rohit K. Sharma, and Nishima Wangoo. (2018a). "Synthesis, Characterization, Mechanistic Studies and Antimicrobial Efficacy of Biomolecule Capped and PH Modulated Silver Nanoparticles." *Journal of Molecular Liquids*, *249*(January), 1145–1150. doi:10.1016/J.MOLLIQ.2017.11.143.

Kumar, Munish, Werner Tegge, Nishima Wangoo, Rahul Jain, and Rohit K. Sharma. (2018b). "Insights into Cell Penetrating Peptide Conjugated Gold Nanoparticles for Internalization into Bacterial Cells." *Biophysical Chemistry*, *237*(June), 38–46. doi:10.1016/J.BPC.2018.03.005.

Kuno, M., D.P. Fromm, H.F. Hamann, A. Gallagher, and D.J. Nesbitt. (2000). "Nonexponential 'Blinking' Kinetics of Single CdSe Quantum Dots: A Universal Power Law Behavior." *Journal of Chemical Physics*, *112*(7), 3117. doi:10.1063/1.480896.

Lee, Areum, Ting Wang, Ji Hyeon Kim, Hyon Hee Yoon, and Sang Joon Park. (2012). "Synthesis of Silica Encapsulated ZnSe Quantum Dots by Microemulsion Method." *Molecular Crystals and Liquid Crystals*, *564*(1), 10–17. doi:10.1080/15421406.2012.690634.

Lee, J.D. (1996). *Concise Inorganic Chemistry*. Chapman & Hall. https://www.wiley.com/en-us/Concise+Inorganic+Chemistry%2C+5th+Edition-p-9780632052936.

Lesiak, Anna, Kamila Drzozga, Joanna Cabaj, Mateusz Bański, Karol Malecha, Artur Podhorodecki, Anna Lesiak, et al. (2019). "Optical Sensors Based on II-VI Quantum Dots." *Nanomaterials*, *9*(2), 192. doi:10.3390/nano9020192.

Li, Chunliang, and Norio Murase. (2003). "Synthesis of Highly Luminescent Glasses Incorporating CdTe Nanocrystals through Sol–Gel Processing." *Langmuir*, *20*(1), 1–4. doi:10.1021/LA035546O.

Li, Haitao, Xiaodie He, Zhenhui Kang, Hui Huang, Yang Liu, Jinglin Liu, Suoyuan Lian, Chi Him A. Tsang, Xiaobao Yang, and Shuit-Tong Lee. (2010). "Water-Soluble Fluorescent Carbon Quantum Dots and Photocatalyst Design." *Angewandte Chemie International Edition*, *49*(26), 4430–4434. doi:10.1002/anie.200906154.

Li, Ruifeng, Zhenyu Ye, Weiguang Kong, Huizhen Wu, Xing Lin, and Wei Fang. (2015). "Controllable Synthesis and Growth Mechanism of Dual Size Distributed PbSe Quantum Dots." *RSC Advances*, *5*(3), 1961–1967. doi:10.1039/C4RA11012A.

Li, Shan Shan, Fu Tian Liu, Qun Wang, Xiu Xiu Chen, and Ping Yang. (2009). "Synthesis and Characterization of Water-Soluble Cu²⁺-Doped ZnSe Quantum Dots." *Advanced Materials Research*, *79–82*(August), 2043–2046. doi:10.4028/www.scientific.net/AMR.79-82.2043.

Li, Shihao, Jie Jiang, Yinan Yan, Ping Wang, Gang Huang, Nam hoon Kim, Joong Hee Lee, and Dannong He. (2018). "Red, Green, and Blue Fluorescent Folate-Receptor-Targeting Carbon Dots for Cervical Cancer Cellular and Tissue Imaging." *Materials Science and Engineering C*, *93*(December), 1054–1063. doi:10.1016/J.MSEC.2018.08.058.

Lianhua Qu, Z. Adam Peng, and Xiaogang Peng. (2001). "Alternative Routes toward High Quality CdSe Nanocrystals." *Nano Letters*, *1*(6), 333–337. doi:10.1021/NL0155532.

Lide, Zhang, and Chi-mei Mo. (1995). "Luminescence in Nanostructured Materials." *Nanostructured Materials*, *6*(5–8), 831–834. doi:10.1016/0965-9773(95)00188-3.

Lin, Cheng-An J., Ralph A. Sperling, Jimmy K. Li, Ting-Ya Yang, Pei-Yun Li, Marco Zanella, Walter H. Chang, and Wolfgang J. Parak. (2008). "Design of an Amphiphilic Polymer for Nanoparticle Coating and Functionalization." *Small*, *4*(3), 334–341. doi:10.1002/smll.200700654.

Lin, Wanjuan, Karolina Fritz, Gerald Guerin, Ghasem R. Bardajee, Sean Hinds, Vlad Sukhovatkin, Edward H. Sargent, Gregory D. Scholes, and Mitchell A. Winnik. (2008). "Highly Luminescent Lead Sulfide Nanocrystals in Organic Solvents and Water through Ligand Exchange with Poly(Acrylic Acid)." *Langmuir*, *24*(15), 8215–8219. doi:10.1021/la800568k.

Liu, Huan, Min Li, Gang Shao, Wenkai Zhang, Weiwei Wang, Huaibing Song, Hefeng Cao, Wanli Ma, and Jiang Tang. (2015). "Enhancement of Hydrogen Sulfide Gas Sensing of PbS Colloidal Quantum Dots by Remote Doping through Ligand Exchange." *Sensors and Actuators B*, *212*(June), 434–439. doi:10.1016/J.SNB.2015.02.047.

Liu, Huan, Min Li, Oleksandr Voznyy, Long Hu, Qiuyun Fu, Dongxiang Zhou, Zhe Xia, Edward H. Sargent, and Jiang Tang. (2014). "Physically Flexible, Rapid-Response Gas Sensor Based on Colloidal Quantum Dot Solids." *Advanced Materials*, *26*(17), 2718–2724. doi:10.1002/adma.201304366.

Liu, Juncheng, Poovathinthodiyil Raveendran, Zameer Shervani, Yutaka Ikushima, and Yukiya Hakuta. (2005). "Synthesis of Ag and AgI Quantum Dots in AOT-Stabilized Water-in-CO2 Microemulsions." *Chemistry–A European Journal*, *11*(6), 1854–1860. doi:10.1002/chem.200400508.

Liu, Kang, Ji Hyeon Kim, and Sang Joon Park. (2013). "Optical Properties of Silica-Encapsulated ZnSe Nanocrystals Prepared with Water-in-Oil Microemulsions." *Japanese Journal of Applied Physics*, *52*(1S), 01AN01. doi:10.7567/JJAP.52.01AN01.

Liu, Kang, and Sang Joon Park. (2014). "Preparation of Highly Luminescent CdSe Quantum Dots by Reverse Micelles." *Japanese Journal of Applied Physics*, *53*(8S2), 08ME03. doi:10.7567/JJAP.53.08ME03.

Liu, Shu-Man, Feng-Qi Liu, and Zhan-Guo Wang. (2001). "Relaxation of Carriers in Terbium-Doped ZnO Nanoparticles." *Chemical Physics Letters*, *343*(5–6), 489–492. doi:10.1016/S0009-2614(01)00740-0.

Liu, Wen, Haipeng Diao, Honghong Chang, Haojiang Wang, Tingting Li, and Wenlong Wei. (2017). "Green Synthesis of Carbon Dots from Rose-Heart Radish and Application for Fe3+ Detection and Cell Imaging." *Sensors and Actuators B*, *241*(March), 190–198. doi:10.1016/J.SNB.2016.10.068.

Liu, Xinlin, Changchang Ma, Yan Yan, Guanxin Yao, Yanfeng Tang, Pengwei Huo, Weidong Shi, and Yongsheng Yan. (2013). "Hydrothermal Synthesis of CdSe Quantum Dots and Their Photocatalytic Activity on Degradation of Cefalexin." *Industrial & Engineering Chemistry Research*, *52*(43), 15015–15023. doi:10.1021/ie4028395.

Liu, Yingshuai, Yanan Zhao, and Yuanyuan Zhang. (2014). "One-Step Green Synthesized Fluorescent Carbon Nanodots from Bamboo Leaves for Copper(II) Ion Detection." *Sensors and Actuators B*, *196*(June), 647–652. doi:10.1016/J.SNB.2014.02.053.

Lou, Yongbing, Yixin Zhao, Jinxi Chen, and Jun-Jie Zhu. (2014). "Metal Ions Optical Sensing by Semiconductor Quantum Dots." *Journal of Materials Chemistry C*, *2*(4), 595–613. doi:10.1039/C3TC31937G.

Lu, Jiong, Jia-xiang Yang, Junzhong Wang, Ailian Lim, Shuai Wang, and Kian Ping Loh. (2009). "One-Pot Synthesis of Fluorescent Carbon Nanoribbons, Nanoparticles, and Graphene by the Exfoliation of Graphite in Ionic Liquids." *ACS Nano*, *3*(8), 2367–2375. doi:10.1021/nn900546b.

Lu, Wenbo, Xiaoyun Qin, Abdullah M. Asiri, Abdulrahman O. Al-Youbi, and Xuping Sun. (2013). "Green Synthesis of Carbon Nanodots as an Effective Fluorescent Probe for Sensitive and Selective Detection of Mercury(II) Ions." *Journal of Nanoparticle Research*, *15*(1), 1344. doi:10.1007/s11051-012-1344-0.

Lu, Wenjing, Xiaojuan Gong, Ming Nan, Yang Liu, Shaomin Shuang, and Chuan Dong. (2015). "Comparative Study for N and S Doped Carbon Dots: Synthesis, Characterization and

Applications for Fe3+ Probe and Cellular Imaging." *Analytica Chimica Acta*, *898*(October), 116–127. doi:10.1016/J. ACA.2015.09.050.

Luccardini, C., C. Tribet, F. Vial, V. Marchi-Artzner, and M. Dahan. (2006). "Size, Charge, and Interactions with Giant Lipid Vesicles of Quantum Dots Coated with an Amphiphilic Macromolecule." *Langmuir*, *22*(5), 2304–2310. doi:10.1021/ LA052704Y.

Luo, Zhenyue, Haiwei Chen, Yifan Liu, Su Xu, and Shin-Tson Wu. (2015). "Color-Tunable Light Emitting Diodes Based on Quantum Dot Suspension." *Applied Optics*, *54*(10), 2845. doi:10.1364/AO.54.002845.

Ma, Changchang, Mingjun Zhou, Dan Wu, Mengyao Feng, Xinlin Liu, Pengwei Huo, Weidong Shi, Zhongfei Ma, and Yongsheng Yan. (2015). "One-Step Hydrothermal Synthesis of Cobalt and Potassium Codoped CdSe Quantum Dots with High Visible Light Photocatalytic Activity." *CrystEngComm*, *17*(7), 1701– 1709. doi:10.1039/C4CE02414A.

Manna, Liberato, Erik C. Scher, and A. Paul Alivisatos. (2000). "Synthesis of Soluble and Processable Rod-, Arrow-, Teardrop-, and Tetrapod-Shaped CdSe Nanocrystals." *Journal of the American Chemical Society*, *122*(51), 12700–12706. doi:10.1021/JA003055+.

Manna, Liberato, Erik C. Scher, Liang-Shi Li, and A. Paul Alivisatos. (2002). "Epitaxial Growth and Photochemical Annealing of Graded CdS/ZnS Shells on Colloidal CdSe Nanorods." *Journal of the American Chemical Society*, *124*(24), 7136– 7145. doi:10.1021/JA025946I.

Margaret A. Hines, and Philippe Guyot-Sionnest. (1998). "Bright UV-Blue Luminescent Colloidal ZnSe Nanocrystals." *Journal of Physical Chemistry B*, *102*(19), 3655–3657. doi:10.1021/ JP9810217.

Mattoussi, Hedi, J. Matthew Mauro, Ellen R. Goldman, George P. Anderson, Vikram C. Sundar, Frederic V. Mikulec, and Moungi G. Bawendi. (2000). "Self-Assembly of CdSe– ZnS Quantum Dot Bioconjugates Using an Engineered Recombinant Protein." *Journal of the American Chemical Society*, *122*(49), 12142–12150. doi:10.1021/JA002535Y.

Mazumder, Sonal, Jhimli Sarkar, Rajib Dey, M.K. Mitra, S. Mukherjee, and G.C. Das. (2010). "Biofunctionalised Quantum Dots for Sensing and Identification of Waterborne Bacterial Pathogens." *Journal of Experimental Nanoscience*, *5*(5), 438–446. doi:10.1080/17458081003588010.

Medintz, Igor L., H. Tetsuo Uyeda, Ellen R. Goldman, and Hedi Mattoussi. (2005). "Quantum Dot Bioconjugates for Imaging, Labelling and Sensing." *Nature Materials*, *4*(6), 435–446. doi:10.1038/nmat1390.

Mehta, S.K., Sakshi Gupta, Kulvinder Singh, and G.R. Chaudhary. (2012). "Multicomponent Gold Hybrid Structures : Synthesis and Applications." *Reviews in Advanced Sciences and Engineering, 1*(2), 103-118. doi:10.1166/rase.2012.1008.

Mehta, Vaibhavkumar N., Sanjay Jha, and Suresh Kumar Kailasa. (2014). "One-Pot Green Synthesis of Carbon Dots by Using Saccharum Officinarum Juice for Fluorescent Imaging of Bacteria (Escherichia Coli) and Yeast (Saccharomyces Cerevisiae) Cells." *Materials Science and Engineering C*, *38*(May), 20–27. doi:10.1016/J.MSEC.2014.01.038.

Mei, Bing C., Kimihiro Susumu, Igor L. Medintz, James B. Delehanty, T.J. Mountziaris, and Hedi Mattoussi. (2008). "Modular Poly(Ethylene Glycol) Ligands for Biocompatible

Semiconductor and Gold Nanocrystals with Extended PH and Ionic Stability." *Journal of Materials Chemistry. 18*(41), 4949. doi:10.1039/b810488c.

Mishra, S.K., R.K. Srivastava, S.G. Prakash, R.S. Yadav, and A.C. Panday. (2010). "Photoluminescence and Photoconductive Characteristics of Hydrothermally Synthesized ZnO Nanoparticles." *Opto-Electronics Review*, *18*(4), 467–473. doi:10.2478/s11772-010-0037-4.

Mishra, Sheo K., Rajneesh K. Srivastava, S.G. Prakash, Raghvendra S. Yadav, and A.C. Panday. (2011). "Structural, Photoconductivity and Photoluminescence Characterization of Cadmium Sulfide Quantum Dots Prepared by a Co-Precipitation Method." *Electronic Materials Letters*, *7*(1), 31–38. doi:10.1007/s13391-011-0305-6.

Möller, Martin, Joachim P. Spatz, and Arno Roescher. (1996). "Gold Nanoparticles in Micellar Poly(Styrene)-b-Poly(Ethylene Oxide) Films—Size and Interparticle Distance Control in Monoparticulate Films." *Advanced Materials*, *8*(4), 337–340. doi:10.1002/adma.19960080411.

Moreels, Iwan, Yolanda Justo, Bram De Geyter, Katrien Haustraete, José C. Martins, and Zeger Hens. (2011). "Size-Tunable, Bright, and Stable PbS Quantum Dots: A Surface Chemistry Study." *ACS Nano*, *5*(3), 2004–2012. doi:10.1021/nn103050w.

Murcia, Michael J., Daniel. E. Minner, Gina-Mirela Mustata, Kenneth Ritchie, and Christoph A. Naumann. (2008). "Design of Quantum Dot-Conjugated Lipids for Long-Term, High-Speed Tracking Experiments on Cell Surfaces." *Journal of the American Chemical Society*, *130*(45), 15054–15062. doi:10.1021/ja803325b.

Murcia, Michael J., David L. Shaw, Heather Woodruff, Christoph A. Naumann, Bruce A. Young, and Eric C. Long. (2006). "Facile Sonochemical Synthesis of Highly Luminescent ZnS–Shelled CdSe Quantum Dots." *Chemistry of Materials*, *18*(9), 2219– 2225. doi:10.1021/CM0505547.

Murray, C.B., D.J. Norris, and M.G. Bawendi. (1993). "Synthesis and Characterization of Nearly Monodisperse CdE (E = Sulfur, Selenium, Tellurium) Semiconductor Nanocrystallites." *Journal of the American Chemical Society*, *115*(19), 8706– 8715. doi:10.1021/ja00072a025.

Nag, Angshuman, Janardan Kundu, and Abhijit Hazarika. (2014). "Seeded-Growth, Nanocrystal-Fusion, Ion-Exchange and Inorganic-Ligand Mediated Formation of Semiconductor-Based Colloidal Heterostructured Nanocrystals." *CrystEngComm*, *16*(40), 9391–9407. doi:10.1039/ C4CE00462K.

Nann, Thomas. (2005). "Phase-Transfer of CdSe@ZnS Quantum Dots Using Amphiphilic Hyperbranched Polyethylenimine." *Chemical Communications*, (13), 1735. doi:10.1039/b414807j.

Nann, Thomas, Jurgen Riegler, Peter Nick, and Paul Mulvaney. (2005). "Quantum Dots with Silica Shells." *Proceedings of SPIE*, *5705*, 77. doi:10.1117/12.582871.

Nikolenko, L.M., and Vladimir F. Razumov. (2013). "Colloidal Quantum Dots in Solar Cells." *Russian Chemical Reviews*, *82*(5), 429–448. doi:10.1070/RC2013v082n05ABEH004337.

Nirmal, M., B.O. Dabbousi, M.G. Bawendi, J.J. Macklin, J.K. Trautman, T.D. Harris, and L.E. Brus. (1996). "Fluorescence Intermittency in Single Cadmium Selenide Nanocrystals." *Nature*, *383*(6603), 802–804. doi:10.1038/383802a0.

Norris, D.J., Nan Yao, F.T. Charnock, and T.A. Kennedy. (2000). "High-Quality Manganese-Doped ZnSe Nanocrystals." *Nano Letters 1*(1) 3-7 doi:10.1021/NL005503H.

Oh, Eunkeu, James B. Delehanty, Kim E. Sapsford, Kimihiro Susumu, Ramasis Goswami, Juan B. Blanco-Canosa, Philip E. Dawson, et al. (2011). "Cellular Uptake and Fate of PEGylated Gold Nanoparticles Is Dependent on Both Cell-Penetration Peptides and Particle Size." *ACS Nano*, 5(8), 6434–6448. doi:10.1021/nn201624c.

Owen, Jonathan, and Louis Brus. (2017). "Chemical Synthesis and Luminescence Applications of Colloidal Semiconductor Quantum Dots." *Journal of the American Chemical Society*, 139(32), 10939–10943. doi:10.1021/jacs.7b05267.

Panda, B.B., R.K. Rana, and B. Sharma. (2017). "Sonochemical Synthesis of AgInS2 Quantum Dots and Characterisation." *Journal of Nano- and Electronic Physics*, 9(2), 02002-1-02002–02004. doi:10.21272/jnep.9(2).02002.

Parak, W.J., R. Boudreau, M. Le Gros, D. Gerion, D. Zanchet, C.M. Micheel, S.C. Williams, A.P. Alivisatos, and C. Larabell. (2002). "Cell Motility and Metastatic Potential Studies Based on Quantum Dot Imaging of Phagokinetic Tracks." *Advanced Materials*, 14(12), 882. doi:10.1002/1521-4095(20020618)14:12<882::AID-ADMA882>3.0.CO;2-Y.

Pejchal, Robert, and Martha L Ludwig. (2004). "Cobalamin-Independent Methionine Synthase (MetE): A Face-to-Face Double Barrel That Evolved by Gene Duplication." *PLoS Biology*, 3(2), e31. doi:10.1371/journal.pbio.0030031.

Pellegrino, Teresa, Wolfgang J. Parak, Rosanne Boudreau, Mark A. Le gros, Daniele Gerion, A. Paul Alivisatos, and Carolyn A. Larabell. (2003). "Quantum Dot-Based Cell Motility Assay." *Differentiation*, 71(9–10), 542–548. doi:10.1111/j.1432-0436.2003.07109006.x.

Pereira, Mark, and Edward PC Lai. (2008). "Capillary Electrophoresis for the Characterization of Quantum Dots after Non-Selective or Selective Bioconjugation with Antibodies for Immunoassay." *Journal of Nanobiotechnology*, 6(1), 10. doi:10.1186/1477-3155-6-10.

Peternele, Wilson Sacchi, Victoria Monge Fuentes, Maria Luiza Fascineli, Jaqueline Rodrigues da Silva, Renata Carvalho Silva, Carolina Madeira Lucci, and Ricardo Bentes de Azevedo. (2014). "Experimental Investigation of the Coprecipitation Method: An Approach to Obtain Magnetite and Maghemite Nanoparticles with Improved Properties." *Journal of Nanomaterials*, 2014(May), 1–10. doi:10.1155/2014/682985.

Pöselt, Elmar, Christian Schmidtke, Steffen Fischer, Kersten Peldschus, Johannes Salamon, Hauke Kloust, Huong Tran, et al. (2012). "Tailor-Made Quantum Dot and Iron Oxide Based Contrast Agents for *in Vitro* and *in Vivo* Tumor Imaging." *ACS Nano*, 6(4), 3346–3355. doi:10.1021/nn300365m.

Prabhakaran, Prem, Won Jin Kim, Kwang-Sup Lee, and Paras N. Prasad. (2012). "Quantum Dots (QDs) for Photonic Applications." *Optical Materials Express*, 2(5), 578. doi:10.1364/OME.2.000578.

Prasad, Shri S. (2013). "Synthesis and Characterization of CdS Quantum Dots by Reverse Micelles Method." *Scholars Research Library Der Pharma Chemica*, 5(5), 1-4. www.der-pharmachemica.com.

Pu, Ying-Chih, and Yung-Jung Hsu. (2014). "Multicolored Cd1–xZnxSe Quantum Dots with Type-I Core/Shell Structure: Single-Step Synthesis and Their Use as Light Emitting Diodes." *Nanoscale*, 6(7), 3881. doi:10.1039/c3nr06158b.

Qian, Huifeng, Liang Li, and Jicun Ren. (2005). "One-Step and Rapid Synthesis of High Quality Alloyed Quantum Dots (CdSe–CdS) in Aqueous Phase by Microwave Irradiation with Controllable Temperature." *Materials Research Bulletin*, 40(10), 1726–1736. doi:10.1016/J.MATERRESBULL.2005.05.022.

Qian, Lei, Debasis Bera, and Paul H Holloway. (2008). "Temporal Evolution of White Light Emission from CdSe Quantum Dots." *Nanotechnology*, 19(28), 285702. doi:10.1088/0957-4484/19/28/285702.

Qin, Yuancheng, Yuanyuan Cheng, Longying Jiang, Xiao Jin, Mingjun Li, Xubiao Luo, Guoqing Liao, Taihuei Wei, and Qinghua Li. (2015). "Top-down Strategy toward Versatile Graphene Quantum Dots for Organic/Inorganic Hybrid Solar Cells." *ACS Sustainable Chemistry & Engineering*, 3(4), 637–644. doi:10.1021/sc500761n.

Qu, Lianhua, and Xiaogang Peng. (2002). "Control of Photoluminescence Properties of CdSe Nanocrystals in Growth." *Journal of the American Chemical Society*, 124(9), 2049–2055. http://www.ncbi.nlm.nih.gov/pubmed/11866620.

Rahdar, A., H. Asnaasahri Eivari, and R. Sarhaddi. (2012). "Study of Structural and Optical Properties of ZnS:Cr Nanoparticles Synthesized by Co-Precipitation Method." *Indian Journal of Science and Technology*, 5(1), 1855–1858. doi:10.17485/IJST/2012/V5I1/30945.

Ranjani Viswanatha, S. Chakraborty, S. Basu, and D.D. Sarma. (2006). "Blue-Emitting Copper-Doped Zinc Oxide Nanocrystals." *Journal of Physical Chemistry B*, 110(45), 22310-22312. doi:10.1021/JP065384F.

Regulacio, Michelle D., Khin Yin Win, Seong Loong Lo, Shuang-Yuan Zhang, Xinhai Zhang, Shu Wang, Ming-Yong Han, and Yuangang Zheng. (2013). "Aqueous Synthesis of Highly Luminescent AgInS2–ZnS Quantum Dots and Their Biological Applications." *Nanoscale*, 5(6), 2322. doi:10.1039/c3nr34159c.

Reiss, P., G. Quemard, S. Carayon, J. Bleuse, F. Chandezon, and A. Pron. (2004). "Luminescent ZnSe Nanocrystals of High Color Purity." *Materials Chemistry and Physics*, 84(1), 10–13. doi:10.1016/J.MATCHEMPHYS.2003.11.002.

Ren, Zhenwei, Jin Wang, Zhenxiao Pan, Ke Zhao, Hua Zhang, Yan Li, Yixin Zhao, Ivan Mora-Sero, Juan Bisquert, and Xinhua Zhong. (2015). "Amorphous TiO 2 Buffer Layer Boosts Efficiency of Quantum Dot Sensitized Solar Cells to over 9%." *Chemistry of Materials*, 27(24), 8398–8405. doi:10.1021/acs.chemmater.5b03864.

Roelands, C.P. Mark, Joop H. ter Horst, Herman J.M. Kramer, and Pieter J. Jansens. (2006). "Analysis of Nucleation Rate Measurements in Precipitation Processes." *Crystal Growth & Design*, 6(6), 1380–1392. doi:10.1021/CG050678W.

Rosenthal, Sandra J., Jerry C. Chang, Oleg Kovtun, James R. McBride, and Ian D. Tomlinson. (2011). "Biocompatible Quantum Dots for Biological Applications." *Chemistry & Biology*, 18(1), 10–24. doi:10.1016/j.chembiol.2010.11.013.

Ruan, Cheng, Yu Zhang, Min Lu, Changyin Ji, Chun Sun, Xiongbin Chen, Hongda Chen, et al. (2016). "White Light-Emitting Diodes Based on AgInS2/ZnS Quantum Dots with Improved Bandwidth in Visible Light Communication." *Nanomaterials*, 6(1), 13. doi:10.3390/nano6010013.

Sangar, Sugandha, Shikha Sharma, Virender Kumar Vats, S.K. Mehta, and Kulvinder Singh. (2019). "Biosynthesis of Silver Nanocrystals, Their Kinetic Profile from Nucleation to

Growth and Optical Sensing of Mercuric Ions." *Journal of Cleaner Production*, 228(August), 294–302. doi:10.1016/j. jclepro.2019.04.238.

Santos, Beate S., Patrícia M.A. de Farias, Frederico D. de Menezes, Ricardo de C. Ferreira, Severino A. Júnior, Regina C.B.Q. Figueiredo, Luiz B. de Carvalho, and Eduardo I.C. Beltrão. (2006). "CdS-Cd(OH)2 Core Shell Quantum Dots Functionalized with Concanavalin A Lectin for Recognition of Mammary Tumors." *Physica Status Solidi C*, 3(11), 4017–4022. doi:10.1002/pssc.200671568.

Santosh K. Haram, Bernadette M. Quinn, and Allen J. Bard. (2001). "Electrochemistry of CdS Nanoparticles: A Correlation between Optical and Electrochemical Band Gaps." *Journal of American Chemical Society, 123*(36), 8860-8861. doi:10.1021/JA0158206.

Sashchiuk, Aldona, Efrat Lifshitz, Renata Reisfeld, Tsiala Saraidarov, Marina Zelner, and Avi Willenz. (2002). "Optical and Conductivity Properties of PbS Nanocrystals in Amorphous Zirconia Sol-Gel Films." *Journal of Sol-Gel Science and Technology*, 24(1), 31–38. doi:10.1023/A:1015157431754.

Scherer, A., H.G. Craighead, and E.D. Beebe. (1987). "Gallium Arsenide and Aluminum Gallium Arsenide Reactive Ion Etching in Boron Trichloride/Argon Mixtures." *Journal of Vacuum Science & Technology B*, 5(6), 1599. doi:10.1116/1.583635.

Schnauber, Peter, Johannes Schall, Samir Bounouar, Theresa Höhne, Suk-In Park, Geun-Hwan Ryu, Tobias Heindel, et al. (2018). "Deterministic Integration of Quantum Dots into On-Chip Multimode Interference Beamsplitters Using in Situ Electron Beam Lithography." *Nano Letters*, 18(4), 2336–2342. doi:10.1021/acs.nanolett.7b05218.

Selvan, S.T., C. Bullen, M. Ashokkumar, and P. Mulvaney. (2001). "Synthesis of Tunable, Highly Luminescent QD-Glasses Through Sol-Gel Processing." *Advanced Materials*, 13(12–13), 985–988. doi:10.1002/1521-4095(200107)13:12/13<985::AID-ADMA985>3.0.CO;2-W.

Serdar Celebi, A. Koray Erdamar, Alphan Sennaroglu, Adnan Kurt, and Havva Yagci Acar. (2007). "Synthesis and Characterization of Poly(Acrylic Acid) Stabilized Cadmium Sulfide Quantum Dots." *Journal of Physical Chemistry B, 111*(44), 12668–12675. doi:10.1021/JP0739420.

Sergey K. Poznyak, Nikolai P. Osipovich, Alexey Shavel, Dmitri V. Talapin, Mingyuan Gao, Alexander Eychmüller, and Nikolai Gaponik. (2005). "Size-Dependent Electrochemical Behavior of Thiol-Capped CdTe Nanocrystals in Aqueous Solution." *Journal of Physical Chemistry B, 109*(3), 1094–1100. doi:10.1021/JP0460801.

Serrano, Iván Castelló, Qiang Ma, and Emilio Palomares. (2011). "QD-'Onion'-Multicode Silica Nanospheres with Remarkable Stability as PH Sensors." *Journal of Materials Chemistry*, 21(44), 17673. doi:10.1039/c1jm13125g.

Sharma, Navneet, Himanshu Ojha, Ambika Bharadwaj, Dharam Pal Pathak, and Rakesh Kumar Sharma. (2015). "Preparation and Catalytic Applications of Nanomaterials: A Review." *RSC Advances*, 5(66), 53381–53403. doi:10.1039/C5RA06778B.

Shi, Jing, Gang Ni, Jinchun Tu, Xiaoyong Jin, and Juan Peng. (2017). "Green Synthesis of Fluorescent Carbon Dots for Sensitive Detection of Fe2+ and Hydrogen Peroxide." *Journal of Nanoparticle Research*, 19(6), 209. doi:10.1007/s11051-017-3888-5.

Shiohara, Amane, Sujay Prabakar, Angelique Faramus, Chia-Yen Hsu, Ping-Shan Lai, Peter T. Northcote, and Richard D. Tilley. (2011). "Sized Controlled Synthesis, Purification, and Cell Studies with Silicon Quantum Dots." *Nanoscale*, 3(8), 3364. doi:10.1039/c1nr10458f.

Shizhong Luo, Jian Xu, Yanfeng Zhang, Shiyong Liu, and Chi Wu. (2005). "Double Hydrophilic Block Copolymer Monolayer Protected Hybrid Gold Nanoparticles and Their Shell Cross-Linking." *Journal of Physical Chemistry B*, 109(47), 22159–22166. doi:10.1021/JP0549935.

Shorie, Munish, Harmanjit Kaur, Gaganpreet Chadha, Kulvinder Singh, and Priyanka Sabherwal. (2019). "Graphitic Carbon Nitride QDs Impregnated Biocompatible Agarose Cartridge for Removal of Heavy Metals from Contaminated Water Samples." *Journal of Hazardous Materials*, 367(April), 629–638. doi:10.1016/J.JHAZMAT.2018.12.115.

Show-Jen Chiou, Julie Innocent, Charles G. Riordan, Kin-Chung Lam, Louise Liable-Sands, and Arnold L. Rheingold. (2000). "Synthetic Models for the Zinc Sites in the Methionine Synthases." *Inorganic Chemistry*, 39(19), 4347–4353. doi:10.1021/IC000505Q.

Shtykov, S.N., and T. Yu. Rusanova. (2008). "Nanomaterials and Nanotechnologies in Chemical and Biochemical Sensors: Capabilities and Applications." *Russian Journal of General Chemistry*, 78(12), 2521–2531. doi:10.1134/S1070363208120323.

Shuichiro Ogawa, Kai Hu, Fu-Ren F. Fan, and Allen J. Bard. (1997). "Photoelectrochemistry of Films of Quantum Size Lead Sulfide Particles Incorporated in Self-Assembled Monolayers on Gold." *Journal of Physical Chemistry B*, 101(29), 5707–5711.. doi:10.1021/JP970737J.

Singh, Kulvinder, G. R. Chaudhary, Sukhjinder Singh, and S. K. Mehta. (2014). "Synthesis of Highly Luminescent Water Stable ZnO Quantum Dots as Photoluminescent Sensor for Picric Acid." *Journal of Luminescence*, 154(October), 148-154. doi:10.1016/j.jlumin.2014.03.054.

Singh, Kulvinder, and S.K. Mehta. (2016). "Luminescent ZnO Quantum Dots as an Efficient Sensor for Free Chlorine Detection in Water." *Analyst, 141*(8), 2487-2492. doi:10.1039/C5AN02599K.

Singh, S., A. Sharma, and G. P. Robertson. 2012. "Realizing the Clinical Potential of Cancer Nanotechnology by Minimizing Toxicologic and Targeted Delivery Concerns." *Cancer Research* 72 (22): 5663–5668. doi:10.1158/0008-5472. CAN-12-1527.

Song, Liqing, Jingjing Shi, Jun Lu, and Chao Lu. (2015). "Structure Observation of Graphene Quantum Dots by Single-Layered Formation in Layered Confinement Space." *Chemical Science*, 6(8), 4846–4850. doi:10.1039/C5SC01416F.

Spanhel, Lubomir, and Marc A. Anderson. (1991). "Semiconductor Clusters in the Sol-Gel Process: Quantized Aggregation, Gelation, and Crystal Growth in Concentrated Zinc Oxide Colloids." *Journal of the American Chemical Society*, 113(8), 2826–2833. doi:10.1021/ja00008a004.

Sperling, R.A., and W.J. Parak. (2010). "Surface Modification, Functionalization and Bioconjugation of Colloidal Inorganic Nanoparticles." *Philosophical Transactions of the Royal Society A*, 368(1915), 1333–1383. doi:10.1098/rsta.2009.0273.

Staros, James V. (1982). "N-Hydroxysulfosuccinimide Active Esters: Bis(N-Hydroxysulfosuccinimide) Esters of Two Dicarboxylic Acids Are Hydrophilic, Membrane-Impermeant, Protein Cross-Linkers." *Biochemistry*, *21*(17), 3950–3955. doi:10.1021/bi00260a008.

Stefani, Fernando D, Xinhua Zhong, Wolfgang Knoll, Mingyong Han, and Maximilian Kreiter. (2005). "Memory in Quantum-Dot Photoluminescence Blinking." *New Journal of Physics*, *7*(1), 197–197. doi:10.1088/1367-2630/7/1/197.

Stevenson, J.P, M Rutnakornpituk, M Vadala, A.R Esker, S.W Charles, S Wells, J.P Dailey, and J.S Riffle. (2001). "Magnetic Cobalt Dispersions in Poly(Dimethylsiloxane) Fluids." *Journal of Magnetism and Magnetic Materials*, *225*(1–2), 47–58. doi:10.1016/S0304-8853(00)01227-0.

Su, F.H., Z.L. Fang, B.S. Ma, K. Ding, G.H. Li, and W. Chen. (2003). "Pressure Dependence of Mn2+ Luminescence in Differently Sized ZnS:Mn Nanoparticles." *Journal of Physical Chemistry B*, *107*(29), 6991–6996. doi:10.1021/JP0278566.

Sun, Dong, Rui Ban, Peng-Hui Zhang, Ge-Hui Wu, Jian-Rong Zhang, and Jun-Jie Zhu. (2013). "Hair Fiber as a Precursor for Synthesizing of Sulfur- and Nitrogen-Co-Doped Carbon Dots with Tunable Luminescence Properties." *Carbon*, *64*(November), 424–434. doi:10.1016/J.CARBON.2013.07.095.

Sun, Jing, Siwei Yang, Zhongyang Wang, Hao Shen, Tao Xu, Litao Sun, Hao Li, et al. (2015). "Ultra-High Quantum Yield of Graphene Quantum Dots: Aromatic-Nitrogen Doping and Photoluminescence Mechanism." *Particle & Particle Systems Characterization*, *32*(4), 434–440. doi:10.1002/ppsc.201400189.

Sungjee Kim, and Moungi G. Bawendi. (2003). "Oligomeric Ligands for Luminescent and Stable Nanocrystal Quantum Dots." *Journal of the American Chemical Society*, *125*(48), 14652–14653. doi:10.1021/JA0368094.

Susumu, Kimihiro, H. Tetsuo Uyeda, Igor L. Medintz, Thomas Pons, James B. Delehanty, and Hedi Mattoussi. (2007). "Enhancing the Stability and Biological Functionalities of Quantum Dots via Compact Multifunctional Ligands." *Journal of the American Chemical Society*, *129*(45), 13987–13996. doi:10.1021/JA0749744.

Talapin, Dmitri V., Jong-Soo Lee, Maksym V. Kovalenko, and Elena V. Shevchenko. (2010). "Prospects of Colloidal Nanocrystals for Electronic and Optoelectronic Applications." *Chemical Reviews*, *110*(1), 389–458. doi:10.1021/cr900137k.

Tam, K.H., C.K. Cheung, Y.H. Leung, A.B. Djurišić, C.C. Ling, C.D. Beling, S. Fung, et al. (2006). "Defects in ZnO Nanorods Prepared by a Hydrothermal Method." *Journal of Physical Chemistry B*, *110*(42), 20865–20871. doi:10.1021/ JP063239W.

Tan, Lianjiang, Ajun Wan, Tingting Zhao, Ran Huang, and Huili Li. (2014). "Aqueous Synthesis of Multidentate-Polymer-Capping Ag$_2$ Se Quantum Dots with Bright Photoluminescence Tunable in a Second Near-Infrared Biological Window." *ACS Applied Materials & Interfaces*, *6*(9), 6217–6222. doi:10.1021/am5015088.

Tang, Jiang, Kyle W. Kemp, Sjoerd Hoogland, Kwang S. Jeong, Huan Liu, Larissa Levina, Melissa Furukawa, et al. (2011). "Colloidal-Quantum-Dot Photovoltaics Using Atomic-Ligand Passivation." *Nature Materials*, *10*(10), 765–771. doi:10.1038/nmat3118.

Tatsumi, Tomohiko, Miki Fujita, Noriaki Kawamoto, Masanori Sasajima, and Yoshiji Horikoshi. (2004). "Intrinsic Defects in ZnO Films Grown by Molecular Beam Epitaxy." *Japanese Journal of Applied Physics*, *43*(5A), 2602–2606. doi:10.1143/JJAP.43.2602.

Tetsuka, Hiroyuki, Ryoji Asahi, Akihiro Nagoya, Kazuo Okamoto, Ichiro Tajima, Riichiro Ohta, and Atsuto Okamoto. (2012). "Optically Tunable Amino-Functionalized Graphene Quantum Dots." *Advanced Materials*, *24*(39), 5333–5338. doi:10.1002/adma.201201930.

Thanh, Nguyen T.K., and Luke A.W. Green. (2010). "Functionalisation of Nanoparticles for Biomedical Applications." *Nano Today*, *5*(3), 213–230. doi:10.1016/J.NANTOD.2010.05.003.

Tian, Jingqi, Qian Liu, Abdullah M. Asiri, Abdulrahman O. Al-Youbi, and Xuping Sun. (2013). "Ultrathin Graphitic Carbon Nitride Nanosheet: A Highly Efficient Fluorosensor for Rapid, Ultrasensitive Detection of Cu $^{2+}$." *Analytical Chemistry*, *85*(11), 5595–5599. doi:10.1021/ac400924j.

Torriss, Badr, Alain Haché, and Serge Gauvin. (2009). "White Light-Emitting Organic Device with Electroluminescent Quantum Dots and Organic Molecules." *Organic Electronics*, *10*(8), 1454–1458. doi:10.1016/J.ORGEL.2009.08.007.

Trindade, Tito, Paul O'Brien, and Nigel L. Pickett. (2001). "Nanocrystalline Semiconductors: Synthesis, Properties, and Perspectives." *Chemistry of Materials*, *13*(11), 3848-3858. doi:10.1021/CM000843P.

Tsutsui, Kazuo, Evelyn L. Hu, and Chris D. W. Wilkinson. (1993). "Reactive Ion Etched II-VI Quantum Dots: Dependence of Etched Profile on Pattern Geometry." *Japanese Journal of Applied Physics*, *32*(Part 1, No. 12B), 6233–6236. doi:10.1143/JJAP.32.6233.

Uyeda, H. Tetsuo, Igor L. Medintz, Jyoti K. Jaiswal, Sanford M. Simon, and Hedi Mattoussi. (2005). "Synthesis of Compact Multidentate Ligands to Prepare Stable Hydrophilic Quantum Dot Fluorophores." *Journal of the American Chemical Society*, *127*(11), 3870–3878. doi:10.1021/JA044031W.

Vaishali Bagalkot, Liangfang Zhang, Etgar Levy-Nissenbaum, Sangyong Jon, ‖ Philip W. Kantoff, Robert Langer, and Omid C. Farokhzad. (2007). "Quantum Dot–Aptamer Conjugates for Synchronous Cancer Imaging, Therapy, and Sensing of Drug Delivery Based on Bi-Fluorescence Resonance Energy Transfer." *Nano Letters*, *7*(10), 3065-3070. doi:10.1021/NL071546N.

van Dijken, A., J. Makkinje, and A. Meijerink. (2001). "The Influence of Particle Size on the Luminescence Quantum Efficiency of Nanocrystalline ZnO Particles." *Journal of Luminescence*, *92*(4), 323–328. doi:10.1016/S0022-2313(00)00262-3.

van Sark, W.G.J.H.M., P.L.T.M. Frederix, D.J. van den Heuvel, A.A. Bol, J.N.J. van Lingen, C. de Mello Donegá, H.C. Gerritsen, and A. Meijerink. (2002). "Time-Resolved Fluorescence Spectroscopy Study on the Photophysical Behavior of Quantum Dots." *Journal of Fluorescence*, *12*(1), 69–76. doi:10.1023/A:1015315304336.

Vanheusden, K., W.L. Warren, C.H. Seager, D.R. Tallant, J.A. Voigt, and B.E. Gnade. (1998). "Mechanisms behind Green Photoluminescence in ZnO Phosphor Powders." *Journal of Applied Physics*, *79*(10), 7983. doi:10.1063/1.362349.

Vergés, M. Andrés, A. Mifsud, and C.J. Serna. (1990). "Formation of Rod-like Zinc Oxide Microcrystals in Homogeneous Solutions." *Journal of the Chemical Society, Faraday Transactions*, *86*(6), 959–963. doi:10.1039/FT9908600959.

Wang, Chuanxi, Zhenzhu Xu, Hao Cheng, Huihui Lin, Mark G. Humphrey, and Chi Zhang. (2015). "A Hydrothermal Route to Water-Stable Luminescent Carbon Dots as Nanosensors for PH and Temperature." *Carbon*, *82*(February), 87–95. doi:10.1016/J.CARBON.2014.10.035.

Wang, Guang-Li, Kang-Li Liu, Jun-Xian Shu, Tian-Tian Gu, Xiu-Ming Wu, Yu-Ming Dong, and Zai-Jun Li. (2015). "A Novel Photoelectrochemical Sensor Based on Photocathode of PbS Quantum Dots Utilizing Catalase Mimetics of Bio-Bar-Coded Platinum Nanoparticles/G-Quadruplex/Hemin for Signal Amplification." *Biosensors and Bioelectronics*, *69*(July), 106–112. doi:10.1016/J.BIOS.2015.02.027.

Wang, K.T., I. Iliopoulos, and R. Audebert. (1988). "Viscometric Behaviour of Hydrophobically Modified Poly(Sodium Acrylate)." *Polymer Bulletin*, *20*(6), 577–582. doi:10.1007/BF00263675.

Wang, Meihua, Weifen Niu, Xin Wu, Lixia Li, Jun Yang, Shaomin Shuang, and Chuan Dong. (2014). "Fluorescence Enhancement Detection of Uric Acid Based on Water-Soluble 3-Mercaptopropionic Acid-Capped Core/Shell ZnS:Cu/ZnS." *RSC Advances*, *4*(48), 25183–25188. doi:10.1039/C4RA02819H.

Wang, Y.S., P. John Thomas, and P.O'Brien. (2006). "Optical Properties of ZnO Nanocrystals Doped with Cd, Mg, Mn, and Fe Ions." *Journal of Physical Chemistry B*, *110*(43), 21412–21415. doi:10.1021/JP0654415.

Wang, Ying, and N. Herron. (1991). "Nanometer-Sized Semiconductor Clusters: Materials Synthesis, Quantum Size Effects, and Photophysical Properties." *Journal of Physical Chemistry*, *95*(2), 525–532. doi:10.1021/j100155a009.

Wei, Yifeng, Nikhil R. Jana, Shawn J. Tan, and Jackie Y. Ying. (2009). "Surface Coating Directed Cellular Delivery of TAT-Functionalized Quantum Dots." *Bioconjugate Chemistry*, *20*(9), 1752–1758. doi:10.1021/bc8003777.

Wen, Fushan, Wenlian Li, Jong-Ha Moon, and Jin Hyeok Kim. (2005). "Hydrothermal Synthesis of ZnO:Zn with Green Emission at Low Temperature with Reduction Process." *Solid State Communications*, *135*(1–2), 34–37. doi:10.1016/J.SSC.2005.03.066.

Wen, Xiangping, Lihong Shi, Guangming Wen, Yanyan Li, Chuan Dong, Jun Yang, and Shaomin Shuang. (2016). "Green and Facile Synthesis of Nitrogen-Doped Carbon Nanodots for Multicolor Cellular Imaging and Co2+ Sensing in Living Cells." *Sensors and Actuators B*, *235*(November), 179–187. doi:10.1016/J.SNB.2016.05.066.

Wiegand, J., D.S. Smirnov, J. Osberghaus, L. Abaspour, J. Hübner, and M. Oestreich. (2018). "Hole-Capture Competition between a Single Quantum Dot and an Ionized Acceptor." *Physical Review B*, *98*(12), 125426. doi:10.1103/PhysRevB.98.125426.

Williams, Denise N., Sunipa Pramanik, Richard P. Brown, Bo Zhi, Eileen McIntire, Natalie V. Hudson-Smith, Christy L. Haynes, and Zeev Rosenzweig. (2018). "Adverse Interactions of Luminescent Semiconductor Quantum Dots with Liposomes and *Shewanella Oneidensis*." *ACS Applied Nano Materials*, *1*(9), 4788–4800. doi:10.1021/acsanm.8b01000.

Wilson, Robert, David G. Spiller, Alison Beckett, Ian A. Prior, and Violaine Sée. (2010). "Highly Stable Dextran-Coated Quantum Dots for Biomolecular Detection and Cellular Imaging." *Chemistry of Materials*, *22*(23), 6361–6369. doi:10.1021/cm1023635.

Wu, G.S., Y.L. Zhuang, Z.Q. Lin, X.Y. Yuan, T. Xie, and L.D. Zhang. (2006). "Synthesis and Photoluminescence of Dy-Doped ZnO Nanowires." *Physica E*, *31*(1), 5–8. doi:10.1016/J.PHYSE.2005.08.015.

Wu, Mingbo, Yue Wang, Wenting Wu, Chao Hu, Xiuna Wang, Jingtang Zheng, Zhongtao Li, Bo Jiang, and Jieshan Qiu. (2014). "Preparation of Functionalized Water-Soluble Photoluminescent Carbon Quantum Dots from Petroleum Coke." *Carbon*, *78*(November), 480-489 doi:10.1016/j.carbon.2014.07.029.

Wu, Xiaomeng, Xiaohua He, Liang Zhong, Shaoliang Lin, Dali Wang, Xinyuan Zhu, and Deyue Yan. (2011). "Water-Soluble Dendritic-Linear Triblock Copolymer-Modified Magnetic Nanoparticles: Preparation, Characterization and Drug Release Properties." *Journal of Materials Chemistry*, *21*(35), 13611. doi:10.1039/c1jm11613d.

Wu, Xingyong, Hongjian Liu, Jianquan Liu, Kari N. Haley, Joseph A. Treadway, J Peter Larson, Nianfeng Ge, Frank Peale, and Marcel P. Bruchez. (2003). "Erratum: Immunofluorescent Labeling of Cancer Marker Her2 and Other Cellular Targets with Semiconductor Quantum Dots." *Nature Biotechnology*, *21*(1), 41–46. doi:10.1038/nbt764.

Xiaogang Peng, Michael C. Schlamp, Andreas V. Kadavanich, A. P. Alivisatos. (1997). "Epitaxial Growth of Highly Luminescent CdSe/CdS Core/Shell Nanocrystals with Photostability and Electronic Accessibility." *Journal of the American Chemical Society*, *119*(30), 7019–7029. doi:10.1021/JA970754M.

Xin, Xin, Yong Zhang, Xiaoxiao Guan, Juexian Cao, Wenli Li, Xia Long, and Xin Tan. (2019). "Enhanced Performances of PbS Quantum-Dots-Modified MoS $_2$ Composite for NO $_2$ Detection at Room Temperature." *ACS Applied Materials & Interfaces*, *11*(9), 9438–9447. doi:10.1021/acsami.8b20984.

Xing, Yun, and Jianghong Rao. (2008). "Quantum Dot Bioconjugates for in Vitro Diagnostics & in Vivo Imaging." *Cancer Biomarkers*, *4*(6), 307–319. http://www.ncbi.nlm.nih.gov/pubmed/19126959.

Xiu, F.X., Z. Yang, L.J. Mandalapu, J.L. Liu, and W.P. Beyermann. (2006). "P-Type ZnO Films with Solid-Source Phosphorus Doping by Molecular-Beam Epitaxy." *Applied Physics Letters*, *88*(5), 052106. doi:10.1063/1.2170406.

Xu, Xiaoyou, Robert Ray, Yunling Gu, Harry J. Ploehn, Latha Gearheart, Kyle Raker, and Walter A. Scrivens (2004) "Electrophoretic analysis and purification of fluorescent single-walled carbon nanotube fragments." *Journal of American Chemical Society*, *126*(40), 12736–12737.

Xu, Zi-Qiang, Li-Yun Yang, Xiao-Yang Fan, Jian-Cheng Jin, Jie Mei, Wu Peng, Feng-Lei Jiang, Qi Xiao, and Yi Liu. (2014). "Low Temperature Synthesis of Highly Stable Phosphate Functionalized Two Color Carbon Nanodots and Their Application in Cell Imaging." *Carbon*, *66*(January), 351–360. doi:10.1016/J.CARBON.2013.09.010.

Ya-Ping Sun, Bing Zhou, Yi Lin, Wei Wang, K. A. Shiral Fernando, Pankaj Pathak, Mohammed Jaouad Meziani, et al. (2006). "Quantum-Sized Carbon Dots for Bright and Colorful Photoluminescence." *Journal of the American Chemical Society*, *128*(24), 7756–7757. doi:10.1021/JA062677D.

Yang, Heesun, Paul H. Holloway, Garry Cunningham, and Kirk S. Schanze. (2004). "CdS:Mn Nanocrystals Passivated by ZnS: Synthesis and Luminescent Properties." *Journal of Chemical Physics*, *121*(20), 10233–10240. doi:10.1063/1.1808418.

Yang, Lin, Jianguo Zhu, and Dingquan Xiao. (2012). "Microemulsion-Mediated Hydrothermal Synthesis of ZnSe and Fe-Doped ZnSe Quantum Dots with Different Luminescence Characteristics." *RSC Advances*, 2(21), 8179. doi:10.1039/c2ra21401f.

Ye, Yuanfeng. (2018). "Photoluminescence Property Adjustment of ZnO Quantum Dots Synthesized via Sol–Gel Method." *Journal of Materials Science: Materials in Electronics*, 29(6), 4967–4974. doi:10.1007/s10854-017-8457-2.

Yoojin Kim, Jeffrey Pyun, Jean M. J. Fréchet, Craig J. Hawker, and Curtis W. Frank. (2005). "The Dramatic Effect of Architecture on the Self-Assembly of Block Copolymers at Interfaces." *Langmuir*, 21(23), 10444-10458 . doi:10.1021/LA047122F.

Yu, Jing, Na Song, Ya-Kun Zhang, Shu-Xian Zhong, Ai-Jun Wang, and Jianrong Chen. (2015). "Green Preparation of Carbon Dots by Jinhua Bergamot for Sensitive and Selective Fluorescent Detection of Hg2+ and Fe3+." *Sensors and Actuators B*, 214(July), 29–35. doi:10.1016/J.SNB.2015.03.006.

Zhan, Yan, Zhiming Liu, Qingqing Liu, Di Huang, Yan Wei, Yinchun Hu, Xiaojie Lian, and Chaofan Hu. (2017). "A Facile and One-Pot Synthesis of Fluorescent Graphitic Carbon Nitride Quantum Dots for Bio-Imaging Applications." *New Journal of Chemistry*, 41(10), 3930–3938. doi:10.1039/C7NJ00058H.

Zhang, Lai-Jun, Xing-Can Shen, Hong Liang, and Jia-Ting Yao. (2010). "Multiple Families of Magic-Sized ZnSe Quantum Dots via Noninjection One-Pot and Hot-Injection Synthesis." *Journal of Physical Chemistry C*, 114(50), 21921–21927. doi:10.1021/jp1044282.

Zhang, X.T., Y.C. Liu, J.Y. Zhang, Y.M. Lu, D.Z. Shen, X.W. Fan, and X.G. Kong. (2003). "Structure and Photoluminescence of Mn-Passivated Nanocrystalline ZnO Thin Films." *Journal of Crystal Growth*, 254(1–2), 80–85. doi:10.1016/S0022-0248(03)01143-6.

Zhang, Yanjie, and Aaron Clapp. (2011). "Overview of Stabilizing Ligands for Biocompatible Quantum Dot Nanocrystals." *Sensors*, 11(12), 11036–11055. doi:10.3390/s111211036.

Zhang, Yi, and Tza-Huei Wang. (2012). "Quantum Dot Enabled Molecular Sensing and Diagnostics." *Theranostics*, 2(7), 631–654. doi:10.7150/thno.4308.

Zhao, Dan, Jiao-Tian Li, Fang Gao, Cui-ling Zhang, and Zhi-ke He. (2014). "Facile Synthesis and Characterization of Highly Luminescent UV-Blue-Emitting ZnSe/ZnS Quantum Dots via a One-Step Hydrothermal Method." *RSC Advances*, 4(87), 47005–47011. doi:10.1039/C4RA06077F.

Zhao, Qiao-Ling, Zhi-Ling Zhang, Bi-Hai Huang, Jun Peng, Min Zhang, and Dai-Wen Pang. (2008). "Facile Preparation of Low Cytotoxicity Fluorescent Carbon Nanocrystals by Electrooxidation of Graphite." *Chemical Communications*, (41), 5116. doi:10.1039/b812420e.

Zheng, Liyan, Yuwu Chi, Yongqing Dong, Jianpeng Lin, and Binbin Wang. (2009). "Electrochemiluminescence of Water-Soluble Carbon Nanocrystals Released Electrochemically from Graphite." *Journal of the American Chemical Society*, 131(13), 4564–4565. doi:10.1021/ja809073f.

Zhou, Jigang, Christina Booker, Ruying Li, Xingtai Zhou, Tsun-Kong Sham, Xueliang Sun, and Zhifeng Ding. (2007). "An Electrochemical Avenue to Blue Luminescent Nanocrystals from Multiwalled Carbon Nanotubes (MWCNTs)." *Journal of the American Chemical Society*, 129(4), 744–745. doi:10.1021/JA0669070.

Zhou, Juan, Yong Yang, and Chun-yang Zhang. (2015). "Toward Biocompatible Semiconductor Quantum Dots: From Biosynthesis and Bioconjugation to Biomedical Application." *Chemical Reviews*, 115(21), 11669–11717. doi:10.1021/acs.chemrev.5b00049.

Zhu, Junjie, Yuri Koltypin, and A. Gedanken. (1999). "General Sonochemical Method for the Preparation of Nanophasic Selenides: Synthesis of ZnSe Nanoparticles." *Chemistry of Materials*, 12(1), 73–78. doi:10.1021/CM990380R.

10

Silica Coating of Metal-Related Nanoparticles and Their Properties

Yoshio Kobayashi and Kohsuke Gonda

CONTENTS

10.1 Introduction

Colloidal particles with a size of a nanometer, i.e. nanoparticles, are especially interesting because they are intermediary between atoms or molecules and bulk material and therefore can be presumed to show unique characteristics that are not found in those of bulk material (Brus 1986; Ekimov et al. 1993).

Nanoparticles have three main problems. The first is based on the aggregation of nanoparticles. Nanoparticles tend to form their bulk through aggregation. Traditionally, the colloids of nanoparticles are stabilized by surface modification ways that use chemicals like surfactants and polymers to form a block between the nanoparticles and other nanoparticles approaching the nanoparticles. It is difficult to colloidally stabilize nanoparticles in the long term because most surfactants and polymers are organic chemicals, which are chemically and thermally unstable, and their molecular motion damages the block over time. The second is related to chemical stability. Nanoparticles are chemically unstable compared with bulk materials because the specific surface area and surface energy of nanoparticles are larger. The

third concerns toxicity. If the nanoparticles or the components released from the nanoparticles are toxic, the toxicity must be controlled for their practical use in our society.

The coating of nanoparticles with solid shells has three unique properties, as shown in Figure 10.1. One is the prevention of degradation. Degradation often occurs via the contact of the nanoparticles with oxygen. The coated nanoparticles cannot contact oxygen because of the physical barrier, which chemically stabilizes them. Another is the prevention of nanoparticle aggregation. The nanoparticles cannot come into contact with each other based on the block of the shells, so they do not aggregate. The third is the prevention of the dissolution of nanoparticles. The nanoparticles cannot contact the solvent because of the physical barrier, which controls the dissolution of the nanoparticles and then prevents their components from being released. Consequently, the toxicity derived from the nanoparticles or components released from the nanoparticles is reduced by the coating due to the block of the shells.

From these perspectives based on the three unique properties of solid shells, the coating of nanoparticles with a shell of solid

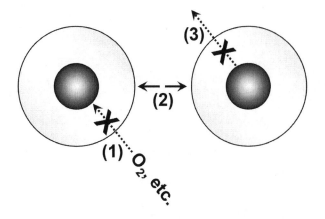

FIGURE 10.1 Effects of silica-coating of nanoparticles. (1) prevention of degradation, (2) prevention of aggregation, (3) prevention of dissolution.

material is a candidate for solving the above-mentioned problems of the nanoparticles. Among various solid materials, silica is suitable as the material of the solid shell because silica is inert in the living body, is chemically stable, and can be fabricated easily by chemical reactions (Stöber et al. 1968; Plumeré et al. 2012; Gholami et al. 2013; Joshi et al. 2014; Parpaite et al. 2014). In addition, silica particles are colloidally stable (Massé et al. 2013; Lipani et al. 2013; Qu et al. 2013; Chou et al. 2014), which expects that silica-coated nanoparticles should also be colloidally stable, because both silica particles and silica-coated nanoparticles have silica surfaces. The colloidal stability of silica-coated nanoparticles has an advantage in applications requiring colloidal stability such as medical diagnosis and the formation of devices consisting of functional nanoparticles.

The silica coating of nanoparticles that is regarded as a stabilizing technique has hitherto been addressed by several researchers (Liz-Marzán et al. 1996a, b; Correa-Duarte et al. 1998a; Ung et al. 1998). Extensive studies on the silica coating of metallic nanoparticles have been performed (Liz-Marzán et al. 1996b; Correa-Duarte et al. 1998a; Ung et al. 1998; Hall et al. 2000; Mulvaney et al. 2000). Many reported silica-coating methods are liquid-phase procedures, such as a sol-gel process, which consists of four steps. The first is the preparation of a colloidal solution of metallic nanoparticles by reducing metal ions with a reductant in an aqueous solution containing stabilizers such as surfactants and polymers. The second is the surface modification of the metal nanoparticles that uses silane coupling agents such as silicon alkoxides with an amino or thiol group as surface primers. Silica easily binds to the oxide surface via OH groups on its surface. However, it is difficult for silica to be deposited on pure metal particles by means of a slight amount of OH groups present on the metal surface. Therefore, the surface is required to be made 'vitreophilic' by using the silane coupling agents (Liz-Marzán et al. 1996a). The third is the slow deposition of silica produced from a sodium silicate in water. The fourth is the formation of the silica shells followed by extensive silica deposition via the sol-gel reaction of silicon alkoxide in alcohol/amine mixtures, which is based on the Stöber method (Stöber et al. 1968; Plumeré et al. 2012; Gholami et al. 2013; Joshi et al. 2014; Parpaite et al. 2014). We have also developed techniques for silica coating various nanoparticles in a liquid phase, as shown in Table 10.1. According to some of our studies on silica coating (Mine et al. 2003; Kobayashi et al. 2005a, 2012b), the second and third steps are not always necessary, if the parameters of silica coating, e.g., the concentrations of raw chemicals and reaction temperatures, are optimized in the third step.

TABLE 10.1

Authors' Work on Silica Coating of Nanoparticles

Core-Shell Particles	References
Au/SiO_2	Kobayashi et al. 2001, 2011, 2012b, 2013d, 2013f, 2014a, 2015b, 2016a, 2016d, Selvan et al. 2002, Mine et al. 2003, Park et al. 2006, Bahadur et al. 2011
$AuNR/SiO2$	Inose et al. 2017
$Au/SiO_2/polystyrene$	Gu et al. 2004
$Au/SiO_2/GdC$	Kobayashi et al. 2017
$Au/SiO_2/GdC/SiO_2$	Kobayashi et al. 2013b
Ag/SiO_2	Kobayashi et al. 2005a
AgI/SiO_2	Kobayashi et al. 2004b, 2005b, 2007c, 2008c, 2008d, 2010c, 2012c, 2012d, 2013c, 2014a, 2016c, Ayame et al. 2011, Sakurai et al. 2012
$Cu/SiO2$	Kobayashi and Sakuraba 2008
Co/SiO_2	Kobayashi and Liz-Marzàn 2001, Kobayashi et al. 2003
$CoPt/SiO_2$	Kobayashi et al. 2006, 2008b, 2009b
$PtRu/SiO_2$	Shimazaki et al. 2006
quantum dot/SiO_2	Correa-Duarte et al. 2001, Kobayashi et al. 2010a, 2010b, 2013a, 2014a, 2015a, Zhou et al. 2005
quantum dot/SiO_2/AU	Kobayashi et al. 2016b
quantum dot/GdC/SiO_2	Kobayashi et al. 2012a
GdC/SiO_2	Kobayashi et al. 2013e, Morimoto et al. 2011
$SiO_2/GdC/SiO_2$	Kobayashi et al. 2007b, 2014a
$SiO_2/Y:Eu/SiO_2$	Kobayashi et al. 2009a
Fe_3O_4/SiO_2	Kobayashi et al. 2007a, 2008a
nitrogen-doped TiO_2/SiO_2	Kobayashi et al. 2014b, Kobayashi and Iwasaki 2016b
fluorescent polystyrene/SiO_2	Kobayashi et al. 2004a, Cong et al. 2010

In this chapter, we introduce our research on the development of silica coating of various metal-related nanoparticles, including not only metallic nanoparticles but also metallic alloy nanoparticles and nanoparticles containing metallic nanoparticles (Kobayashi and Liz-Marzán 2001; Kobayashi et al. 2001, 2003, 2005a, 2006, 2008a, 2009b, 2011, 2012b, 2013b, c, d, 2014c, 2015b, 2016e, 2017; Selvan et al. 2002; Mine et al. 2003; Gu et al. 2004; Shimazaki et al. 2006; Park et al. 2006; Kobayashi and Sakuraba 2008; Bahadur et al. 2011; Inose et al. 2017), and on their properties such as optics (Mine et al. 2003; Kobayashi et al. 2005a, 2011, 2012b), magnetism (Kobayashi and Liz-Marzán 2001; Kobayashi et al. 2003, 2008a, 2009b), imaging (Kobayashi et al. 2011, 2012b, 2013b, d, 2014c, 2015b, 2016b, e, 2017; Bahadur et al. 2011), catalysis (Shimazaki et al. 2006), and photothermal conversion (Inose et al. 2017).

10.2 Effect of Silica Coating on the Chemical Stability of Metallic Nanoparticles

Nanoparticles of metals such as cobalt (Co) and copper (Cu) tend to be oxidized in water exposed to air and in air. The oxidation deteriorates their unique abilities to function as magnetic materials, catalysts, and fillers in metal-metal bonding. Silica coating will control the oxidation because the nanoparticles cannot contact oxygen through the physical barrier. In this section, we focus on these metals and introduce our previous research that mainly examined the effects of silica coating on the chemical stability of nanoparticles of these metals.

10.2.1 Metallic Cobalt Nanoparticles

Magnetic materials become superparamagnetic when their size decreases down to nanometer. It is well known that the magnetic dipole of nanoparticles that have superparamagnetic properties can be aligned under an outward magnetic field. This ability results in very interesting applications of magnetic nanoparticles. Chiefly, nanoparticles are likely to aggregate to become bulk. Hence, magnetic nanoparticles have to be separated in some way in order to preserve the individual properties and to make each nanoparticle behave as a single magnetic dipole. Pure metals, e.g., metallic iron (Fe), metallic Co, and metallic nickel (Ni), have good magnetic characteristics (Park et al. 2000; Racka et al. 2005; Osuna et al. 1996; Giersig and Hilgendorff 1999; Ely et al. 1999; Cordente et al. 2001; Sun et al. 1999). Mainly, it is difficult to use pure metals, especially their nanoparticles, due to their instability like oxidation in air. Hence, methods for bettering the chemical stability of nanoparticles are required to be developed. In the case of metallic Co nanoparticles, it is also difficult to overcome oxidation by oxygen dissolved in a liquid phase. The silica coating of nanoparticles is a path to solve the problem of chemical stabilization (Ohmori and Matijević, 1992, 1993; Philipse et al. 1994; Correa-Duarte et al. 1998b; Liu et al. 1998; Tago et al. 2002; Lu et al. 2002a). This section mentions the preparation of a colloidally stable aqueous colloid solution of silica-coated metallic Co (Co/SiO$_2$) nanoparticles with various sizes of metallic Co nanoparticles and various thicknesses of silica shells (Kobayashi et al. 2003).

The metallic Co nanoparticle colloidal solution was prepared by injecting an aqueous cobalt chloride (CoCl$_2$) solution to an aqueous solution of sodium borohydride (NaBH$_4$) and citric acid (citrate) under nitrogen bubbling. The mixture became a colloid solution with a color of gray, which implied particle formation. Silica coating was carried out by including an ethanolic solution containing (3-aminopropyl)trimethoxysilane (APMS) and tetraethyl orthosilicate (TEOS) in the mixture, so that the APMS would have been attached to the surface of metallic Co nanoparticles, the hydrolysis/condensation of both APMS and TEOS occurred, and finally the silica shells formed on the nanoparticles.

The optimal procedure and conditions for preparation for the synthesis of stable colloid solutions were found by performing the synthesis at various procedures and conditions for sysnthesis. Figure 10.2 shows the transmission electron microscope (TEM) images of Co/SiO$_2$ nanoparticles fabricated at various mole ratios of [citrate]/[Co]. A high contrast was observed between the Co core and the silica shell. Particles with an almost perfect core-shell structure were obtained with our method. These results indicated that our method was quite efficient despite its simplicity, compared with other silica-coating techniques proposed by several researchers (Liz-Marzán et al. 1996b; Correa-Duarte et al. 1998b). As shown in Figure 10.2(f), the core size decreased with increasing [citrate]/[Co] ratio. The dependence of the core size on the ratio was considered to be related to the adsorption of citrate ions on the metallic Co nanoparticle surface. The adsorption diminished double-layer repulsion between the metallic Co nanoparticles, which prevented further growth. The silica shell thickness also decreased with increasing citrate concentration, because the total surface area of the metallic Co nanoparticles rose with the decrease in core size.

For the crystallization of Co/SiO$_2$, the powder of Co/SiO$_2$ nanoparticles obtained by separating the nanoparticles from the dispersant by centrifugation and then drying them was annealed in atmosphere. Figure 10.3 gives TEM images of Co/SiO$_2$ nanoparticles (sample (e) in Figure 10.2) annealed in the atmosphere at 500–700°C. For the Co/SiO$_2$ nanoparticles annealed at 500°C, it was easy to distinguish between the core and the shell due to a high contrast. The contrast was lost with annealing at 600°C, and only the center of the particle was slightly darker. The annealing at this high temperature made the particles sintered, which deformed the surface of the particles more or less. At 700°C, no particles with a core-shell structure were obtained, and the obtained particles were no longer spherical.

The X-ray diffraction (XRD) patterns of Co/SiO$_2$ nanoparticles are given in Figure 10.4. As-prepared particles were poorly crystallized or were amorphous, because of the featureless pattern. The temperature dependences of the patterns were not the same. However, peaks attributed to metallic Co were detected in all the particles annealed at 400 or 500°C. The detection of metallic Co peaks is evidence of the protection of the core with the silica shell against its oxidation. In the [citrate]/[Co] ratios of 0.01 and 0.05, not only some peaks for metallic Co (cubic) but also several peaks assigned to cobalt boride like Co$_3$B or Co$_4$B (orthorhombic) were detected at 400°C. Such cobalt boride peaks vanished with raising the annealing temperature up to 500 and 600°C. In contrast, the metallic Co peaks became larger. This result pointed that metallic Co was produced via decomposition of cobalt boride. For 600°C, a peak assigned to cobalt (II)

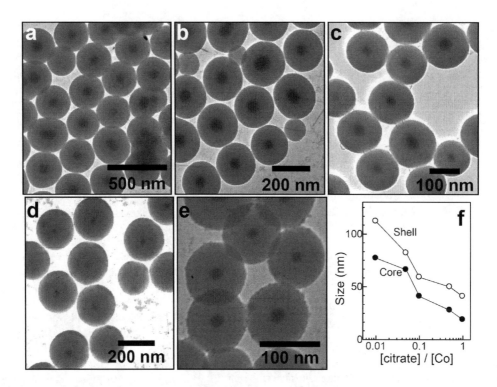

FIGURE 10.2   TEM images of Co/SiO$_2$ nanoparticles fabricated at [citrate]/[Co] ratios of (a) 0.01, (b) 0.05, (c) 0.1, (d) 0.5, and (e) 1. Figure (f) shows average core diameter and shell thickness as a function of [citrate]/[Co] ratio. (Source: Reprinted from *J. Phys. Chem. B*, **2003**, *107*, 7420, with kind permission from the American Chemical Society.)

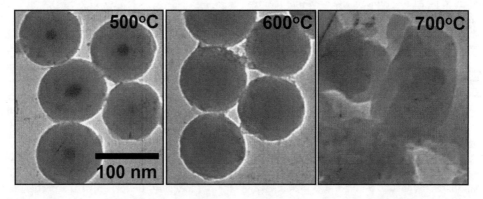

FIGURE 10.3   TEM images of Co/SiO$_2$ nanoparticles from sample (e) in Figure 10.2 after annealing at 500–700°C. (Source: Originated from *J. Phys. Chem. B*, **2003**, *107*, 7420, with kind permission from the American Chemical Society.)

oxide (CoO) (cubic) was also recorded, which showed that the Co cores were partially oxidized. The metallic Co peaks disappeared at 700°C. Conversely, the silica shells were transformed to cristobalite, i.e. were crystallized. Pure silica gel in the amorphous state is commonly transformed to the cristobalite phase by annealing above 1000°C (Iler 1979). The transformation temperature, or the crystallization temperature is lowered with the presence of sodium or borate ions because of the production of a network structure around such ions (Lin et al. 1998). In this work, a similar tendency was observed, since the particles probably contained such ions derived from starting chemicals. Cracks in the silica shell formed by the high-temperature annealing due to the crystallization, which resulted in the diffusion of oxygen through the cracks followed by oxidation of the core. According to the Scherrer equation applied for the peak at 75.9°C, the crystal sizes of metallic Co were between 10 and 20 nm, and they

were nearly the same in spite of the different core particle sizes and the different annealing temperatures. This result on the same crystal sizes indicated that the large cores were composed of aggregates of small metallic Co nanoparticles. In the synthesis at the small [citrate]/[Co] ratios, the large metallic Co nanoparticles were produced due to an amount of citrate ions insufficient for prevention of aggregation of metallic Co core. In the particles fabricated at [citrate]/[Co] ratios of 0.1, 0.5 and 1, no peaks due to cobalt borides were recorded at any annealing temperatures, so that the presence of citrate was found to give a tendency for borides not to be produced.

This section summarizes that our work proposed a simple method for synthesizing metallic Co nanoparticles coated with uniform silica shells, i.e. Co/SiO$_2$ core-shell nanoparticles. The metallic Co cores were quite chemically stable, even annealing in atmosphere.

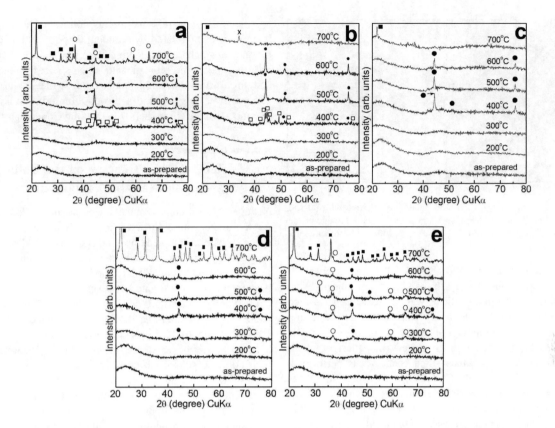

FIGURE 10.4 Evolution of XRD patterns of Co/SiO$_2$ nanoparticles upon annealing for 2 h in air at the temperatures indicated, for [citrate]/[Co] ratios as described in Figure 10.2. The symbols point to characteristic peaks of various compounds: ● Co (cubic), ○ Co$_3$O$_4$ (cubic), ■ SiO$_2$ (cristobalite), □ Co$_3$B or Co$_4$B (orthorombic), × CoO (cubic). (Source: Reprinted from *J. Phys. Chem. B*, **2003**, *107*, 7420, with kind permission from the American Chemical Society.)

10.2.2 Metallic Copper Nanoparticles

Metallic Cu nanoparticles can be applied to materials used in various fields, e.g. catalysis, electronics, and photoelectronics (de Oliveira et al. 2007; Huang et al. 1997; Liu and Bando 2003). It is difficult to use the metallic Cu nanoparticles because they are likely to be oxidized, or chemically unstable in atmosphere. Hence, methods for making them chemically stable are required to be developed. One is a method using organic liquid as a solvent or dispersant, because the organic liquid minimizes oxidation of metallic Cu surface (Lisiecki and Pileni 1993; Qi et al. 1997; Song et al. 2004; Park et al. 2007; Anžlovar et al. 2007). However, many kinds of organic compounds are harmful and hazardous. Another is a method using polymers or surfactant as stabilizers, because they protect metallic Cu nanoparticles to control oxidation of metallic Cu surface (Lisiecki et al. 1996; Huang et al. 1997; Wu and Chen 2004; Zhang et al. 2006; Khanna et al. 2007). However, the addition of stabilizer causes the metallic Cu nanoparticles to contain the stabilizer as an impurity, which may cause deterioration of their properties. As described in Section 10.2.1, the silica coating chemically stabilizes the metallic Co nanoparticles on the grounds that the blocks made of the silica shell prevent the nanoparticles from contacting oxygen molecules. In this section, we introduce our research that verified that silica coating also heightened the chemical stability of metallic Cu nanoparticles even in water and air (Kobayashi and Sakuraba 2008).

A colloidal solution of metallic Cu nanoparticles was prepared by reducing Cu ions with reductant in an aqueous solution. An aqueous solution containing hydrazine, cetyltrimethylammonium bromide (CTAB) and citric acid was added to an aqueous solution containing copper (II) chloride (CuCl$_2$), CTAB, and citric acid while stirring the former aqueous solution vigorously at room temperature. The color of the mixture gradually turned red. The red color was probably assigned to the surface plasmon resonance (SPR) of the metallic Cu nanoparticles (Lisiecki et al. 1996; Yeh et al. 1999; Khanna et al. 2007), which provided evidence of the generation of metallic Cu particles. Fabrication of silica-coated metallic Cu nanoparticles (Cu/SiO$_2$) was carried out by using a sodium silicate solution as the silica source. APMS dissolved in water was added to the Cu colloid. Then, the sodium silicate solution was added to the colloid solution of metallic Cu nanoparticles, and then the pH was adjusted to ca. 10 by adding the cation exchange resin to the colloid solution. The red color did not diminish even after the silica coating, which implied that the metallic Cu nanoparticles existed in the colloidal solution even after the procedure of Cu/SiO$_2$ fabrication.

Figure 10.5(a) gives a TEM image of the Cu/SiO$_2$ particles. The darker and lighter parts of the particles were attributed to Cu and silica, respectively, on the basis of electron density of Cu larger than that of silica. The metallic Cu nanoparticle size was 51.2±23.6 nm. The majority of metallic Cu nanoparticles were incorporated in a matrix with a gel-like structure. Some metallic Cu nanoparticles were coated with uniform silica shells, as shown in an inset of Figure 10.5(a). Figure 10.5(b) gives a TEM image of the Cu/SiO$_2$ particles fabricated in the absence of citrate. In contrast to the particles in the presence of citrate, no Cu/SiO$_2$ particles were produced. Our previous work on the silica coating

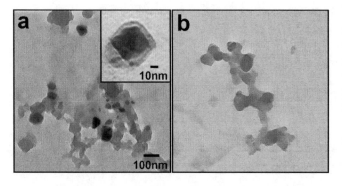

FIGURE 10.5 TEM images of Cu/SiO₂ particles prepared with (a) and without (b) citrate. Inset of the (a) shows high-magnification image of the particle. (Source: Reprinted from *Colloids Surf. A*, **2008**, *317*, 756, with kind permission from Elsevier.)

of metallic Au nanoparticles (Mine et al. 2003) and metallic silver (Ag) nanoparticles (Kobayashi et al. 2005a) indicated that silica coating of such nanoparticles was successfully performed by using citrates. Similarly to these results, the citrate was found to also contribute to promotion of the silica coating of metallic Cu nanoparticles. Figure 10.6(a) gives the XRD pattern of metallic Cu nanoparticles, of which silica coating was performed. The pattern based on metallic Cu (JCPDS card No. 4-0836) was recorded. Strong peaks assigned to cuprous oxide (Cu_2O) (JCPDS card No. 5-0667) were detected, too. This XRD result suggested that many of the Cu nanoparticles were oxidized in air during or after the preparation of nanoparticle powder for the XRD measurements by drying. Figure 10.6(b) gives the XRD pattern of Cu/SiO_2 nanoparticles recorded a few days after the preparation of the nanoparticle powder. Clear peaks attributed to metallic Cu and faint peaks due to Cu_2O were detected. Figure 10.6(c) shows the XRD pattern of Cu/SiO_2 nanoparticles recorded one month after the preparation of the nanoparticle powder. The pattern was almost the same as that recorded a few days after the preparation of the nanoparticle powder. Consequently, the silica coating was found to protect the metallic Cu nanoparticles and control the oxidation of metallic Cu nanoparticles, in contrast to the silica shell-free metallic Cu nanoparticles.

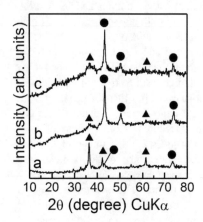

FIGURE 10.6 XRD patterns of metallic Cu nanoparticles (a) and Cu/SiO_2 nanoparticles (b). Symbol (c) stands for a pattern of Cu/SiO_2 nanoparticles, which was measured 1 month after preparation. ●: metallic Cu, ▲: Cu_2O. (Source: Reprinted from *Colloids Surf. A*, **2008**, *317*, 756, with kind permission from Elsevier.)

A summary of this section as follows. The metallic Cu nanoparticles fabricated by reducing Cu ions with hydrazine in the aqueous solution dissolving the citric acid and the CTAB in atmosphere were coated with silica shells by reacting with sodium silicate solution. The metallic Co in the obtained Cu/SiO_2 nanoparticles was not oxidized even after aging in atmosphere, in contrast to the silica shell-free metallic Cu nanoparticles.

10.3 Effect of Silica Coating on the Colloidal Stability of Metallic Nanoparticles

Nanoparticles of coinage metals, e.g. metallic gold (Au), metallic Ag, and metallic Cu have an SPR band in the wavelength range of visible light, and this band is quite dependent on particle morphology such as size and shape and to the characteristics of medium around the metallic nanoparticles. The dependence has been intensively investigated by several researchers (Liz-Marzán et al. 1996b; Kobayashi et al. 2001; Doremus 1964; Kreibig and Genzel 1985; Farbman et al. 1992; Underwood and Mulvaney 1994). Section 10.2.2 introduced the use of the sensitive SPR band to show the existence of metallic Cu nanoparticles in their colloidal solution after silica coating, or to show the successful silica coating of metallic Cu nanoparticles. It was difficult to investigate the SPR of metallic Cu nanoparticles both experimentally and theoretically because the size of metallic Cu nanoparticles, which is one of the parameters governing the SPR, was not uniform due to the low stability of metallic Cu nanoparticles. In contrast, it is easier to fabricate monodispersed metallic Au nanoparticles and metallic Ag nanoparticles because such nanoparticles are more chemically stable than metallic Cu nanoparticles. This section introduces our research on the development of methods for fabricating silica-coated metallic Au nanoparticles (Au/SiO_2) and silica-coated metallic Ag nanoparticles (Ag/SiO_2) and on the theoretical calculation of the SPR peak position, which explained the effect of silica coating on the colloidal stability of the nanoparticles.

10.3.1 Metallic Gold Nanoparticles

10.3.1.1 Direct Coating

Liz-Marzán et al. have performed the fabrication of Au/SiO_2 by a liquid-phase process consisting of several steps (Liz-Marzán et al. 1996b). Their method uses a coupling agent to activate the metallic Au nanoparticle surface in advance of pre-coating with silica by using sodium silicate. However, both the sodium silicate and the coupling agent can be incorporated as impurities into the final particles. This section describes a method for direct silica coating of citrate-stabilized metallic Au nanoparticles by a sol-gel process (Mine et al. 2003). The method is based on a method for synthesizing monodispersed silica particles that has been proposed by our group (Nagao et al. 2000; Mine and Konno et al. 2001), and requires no coupling agent.

Metallic Au nanoparticles having an average size of ca. 15 nm, which were synthesized by reducing tetrachloroauric (III) acid (HAuCl₄) with sodium citrate (Na-cit) in vigorously-stirred water at a constant temperature of 80°C (Enüstün and Turkevich 1963), were used as core particles. The mixture turned to a colloid solution with a color of wine red in a few minutes, which implied the production of metallic Au nanoparticles. An aqueous ammonia solution and TEOS/ethanol/H$_2$O were one by one put into the metallic Au nanoparticle colloidal solution.

Figure 10.7 gives TEM images of the Au/SiO$_2$ nanoparticles formed at various TEOS concentrations. Some core-free silica particles were obtained, and an increase in their amount occurred by decreasing the TEOS concentration. This phenomenon was possibly concerned with a decrease in the ionic strength of the solution leading to an increase in electrostatic repulsion between the metallic Au nanoparticles and silica nuclei generated from TEOS. At the high TEOS concentrations, most of the metallic Au cores were coated by silica shells. The shell thickness increased in a range of 29±5 – 88±10 nm with increasing initial TEOS concentration, which meant that the shell thickness could be varied within a certain range. Figure 10.8(A) gives the extinction spectra of colloidal solutions of metallic Au nanoparticles and of Au/SiO$_2$ nanoparticles with different silica shell thicknesses. SPR absorption bands red-shifted from 520 to 530 nm with an increase in silica shell thickness in a range of thickness smaller than 40 nm. Above 40 nm shell thickness, the band blue-shifted back with an increase in silica shell thickness (Figure 10.8(B)). The effects of silica shells with various thicknesses in various solvents have been explained by Mie theory (Liz-Marzán et al. 1996b). According to the literature, a local increase in the refractive index and scattering from large silica shells contribute to red-shift and blue-shift, respectively. On the assumption that those nanometer-sized particles behave as dipole oscillators, the extinction cross-section Q_{ext} can be expressed as a sum of the absorption cross-section, Q_{abs}, and the scattering cross-section, Q_{sca}, as follows:

$$Q_{ext} = Q_{abs} + Q_{sca}$$
$$= \frac{8\pi R \varepsilon_m^{1/2}}{\lambda} \text{Im} \left[\frac{(\varepsilon_s - \varepsilon_m)(\varepsilon_c + 2\varepsilon_s) + (1-g)(\varepsilon_c - \varepsilon_s)(\varepsilon_m + 2\varepsilon_s)}{(\varepsilon_s + 2\varepsilon_m)(\varepsilon_c + 2\varepsilon_s) + (1-g)(2\varepsilon_s - 2\varepsilon_m)(\varepsilon_c - \varepsilon_s)} \right]$$
$$+ \frac{128\pi^4 R^4 \varepsilon_m^2}{3\lambda^4} \left| \frac{(\varepsilon_s - \varepsilon_m)(\varepsilon_c + 2\varepsilon_s) + (1-g)(\varepsilon_c - \varepsilon_s)(\varepsilon_m + 2\varepsilon_s)}{(\varepsilon_s + 2\varepsilon_m)(\varepsilon_c + 2\varepsilon_s) + (1-g)(2\varepsilon_s - 2\varepsilon_m)(\varepsilon_c - \varepsilon_s)} \right|^2$$

$$(10.1)$$

where R, ε and λ represent the metallic nanoparticle radius, the dielectric function, and the wavelength, respectively. The subscripts c, s, and m stand for the core, the shell, and the medium material, respectively, and g is the volume fraction of the shell layer. In addition, ε_c is calculated from the Drude expression, as follows:

$$\varepsilon_c = \varepsilon_{cb} - \frac{\omega_p^2}{\omega^2 + i\omega/\tau} \qquad (10.2)$$

where ε_{cb} is the high-frequency dielectric constant due to interband and core transitions, and ω_p and τ are the plasma frequency and free electron relaxation time for the metal, respectively. Here, τ is dependent on the size of the metallic nanoparticles, as follows:

$$\frac{1}{\tau} = \frac{1}{\tau_b} + \frac{V_f}{R} \qquad (10.3)$$

where τ_b is the free electron relaxation time for the bulk metal, and V_f is the velocity of electrons at the Fermi energy. Each parameter was in ref. (Underwood and Mulvaney 1994). The dotted line in the same inset stands for the maximum position given by the above calculations using Mie theory and data taken from ref. (Liz-Marzán et al. 1996b), showing that the optical properties of the Au/SiO$_2$ nanoparticles almost follow Mie theory. Consequently, this agreement of the experimental data with the theoretical calculations confirmed that the metallic Au nanoparticles were successfully coated with silica shells with no aggregation of the metallic Au nanoparticles, which guaranteed the colloidal stability of the metallic Au nanoparticles.

10.3.1.2 Amine-Free Coating

Several methods for fabricating Au/SiO$_2$ nanoparticles have been reported (Mine et al. 2003; Lu et al. 2002b; Xu and Perry 2007; Ye et al. 2008). Most reported methods are on the basis of a Stöber method, which uses silicon alkoxide and amine as a silica source and a catalyst, respectively. Because of the harm of amines to living bodies, they should be taken away from the Au/SiO$_2$ nanoparticles. However, it is difficult to take them away from the final nanoparticles because the amines are everywhere in the silica shell. Accordingly, amine-free methods are required for producing Au/SiO$_2$ nanoparticles. In this section, we introduce our research on the development of an amine-free method for fabricating Au/SiO$_2$ nanoparticles (Kobayashi et al. 2011).

The Au/SiO$_2$ nanoparticles were synthesized by a sol-gel process using TEOS, APMS, and sodium hydroxide (NaOH) in the colloid solution of metallic Au nanoparticles having an average particle size of 16.9±1.2 nm that were obtained by performing a process similar to that in Section 10.3.1.1. The metallic Au nanoparticle colloidal solution was put into a H$_2$O/ethanol solution. Then, the APMS dissolved in ethanol, TEOS/ethanol solution, and NaOH aqueous solution were injected into the Au/H$_2$O/ethanol solution at a constant temperature of 35°C.

Figure 10.9 gives the TEM images of Au/SiO$_2$ nanoparticles synthesized at various TEOS concentrations. In the examined concentration range, core-shell particles were produced, and the core-shell structure was nearly perfect. The shell was thickened in the thickness range of 6.0–61.0 nm thickness with raising the TEOS concentration. Figure 10.10(A) presents extinction spectra of the colloidal solutions of metallic Au nanoparticles and Au/SiO$_2$ nanoparticles having various silica shell thicknesses. The wavelength of ca. 525 nm for the SPR band appeared to be related to the TEOS concentration or the shell thickness. The band wavelength vs. shell thickness is plotted in Figure 10.10(B). The SPR wavelength red-shifted as the shell thickened in the range 0 (the uncoated metallic Au nanoparticles) – 22.2 nm. Over 22.2 nm, the SPR wavelength tended to blue-shift back with increasing shell thickness. This tendency for the shifts was also observed in our work on the direct silica coating that was introduced in Section 10.3.1.1 (Mine et al. 2003). As well as the direct coating, the maximum wavelength as a function of shell thickness was obtained by performing the calculations using Equations (10.1)–(10.3), as also shown in Figure 10.10(B). The red-shift and blue-shift tendencies in the experimentally obtained wavelengths could be qualitatively followed by the calculations, which supported that silica coating was successfully performed because of

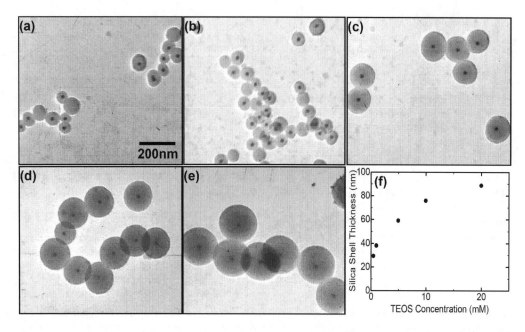

FIGURE 10.7 TEM images of Au/SiO$_2$ nanoparticles synthesized at TEOS concentrations of (a) 0.0005, (b) 0.001, (c) 0.005, (d) 0.01, and (e) 0.02 M. Figure (f) shows plot of silica shell thickness as a function of TEOS concentration. (Source: Originated from *J. Colloid Interface Sci.*, **2003**, *264*, 365, with kind permission from Elsevier.)

the ability to vary the silica shell thickness. Thus, the colloidal stability of the metallic Au nanoparticles was maintained by the silica coating.

10.3.2 Metallic Silver Nanoparticles

Metallic Ag nanoparticles show the SPR in the range of visible light, and can be more widely applied than Au because of the unique properties of the metallic Ag nanoparticles. The metallic Ag nanoparticles have a narrower SPR band. Its extinction coefficient is ca. five times larger than that for Au. The SPR wavelength for Ag is ca. 400 nm, which is well separated from the band-to-band transition energy. These are not seen for Au (Mulvaney 1996). All these properties are related to performance of the metallic Ag nanoparticles in fields such as surface-enhanced Raman scattering and nonlinear optical response. Though the metallic Ag nanoparticles coated with shells are

promising for various applications, they have a problem. The metallic Ag nanoparticles in the coated particles are chemically unstable for ammonia, because they are easily oxidized and then form water-soluble and colorless complex ions (Ung et al. 1998; Yin et al. 2002). One process to solve the problem has been to use nanoparticles of alloys of Ag and Au, which resulted in excessively grown shells at the Au content higher than 25% (Rodríguez-González et al. 2004). Hence, a good technique for fabricating silica-coated nanoparticles remains lacking. This section describes our research on the direct silica coating of metallic Ag nanoparticles, which is on the basis of the sol-gel process using alkoxysilanes (Kobayashi et al. 2005a) and prevents the Ag nanoparticles from dissolving.

A colloidal solution of metallic Ag nanoparticles having an average particle size of 10 nm, which was synthesized by reducing silver (I) perchlorate (AgClO$_4$) with NaBH$_4$, was used as the core nanoparticles in the following silica coating. A AgClO$_4$

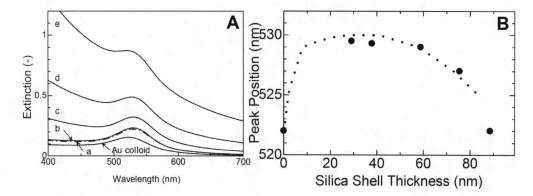

FIGURE 10.8 UV-VIS extinction spectra of shell-free metallic Au nanoparticle colloid solution (Au colloid) and Au/SiO$_2$ nanoparticle colloid solutions with various shell thicknesses Figure (A). Ethanol was added to the Au colloid to adjust water concentration of the dispersion medium to 10.7 M (1:4 (v/v) water/ethanol), which was the same concentration as that in the silica-coating. Figure (B) shows the SPR peak position and predictions by Mie theory. Refer to Figure 10.7 for symbols and TEOS concentrations. (Source: Originated from *J. Colloid Interface Sci.*, **2003**, *264*, 365, with kind permission from Elsevier.)

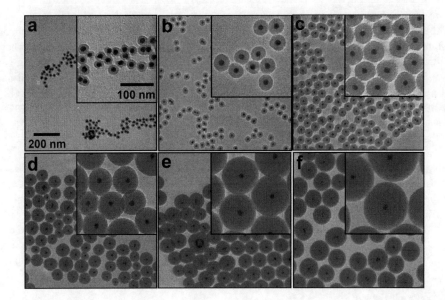

FIGURE 10.9   TEM images of Au/SiO$_2$ nanoparticles prepared at TEOS concentrations of (a) 0.3×10^{-3}, (b) 0.5×10^{-3}, (c) 1.0×10^{-3}, (d) 3.0×10^{-3}, (e) 5.0×10^{-3}, and (f) 10×10^{-3} M. (Source: Reprinted from *J. Colloid Interface Sci.*, **2011**, *358*, 329, with kind permission from Elsevier.)

aqueous solution was injected into a vigorously-stirred aqueous solution dissolving NaBH$_4$ and Na-cit cooled with ice water. The mixture turned to a colloid solution with a color of yellow in a few minutes after the injection, which implied the production of metallic Ag nanoparticles. Then, TEOS dissolved in ethanol was put into the metallic Ag nanoparticle colloidal solution. Thenceforth, a reaction on the silica coating of metallic Ag nanoparticles, or fabrication of Ag/SiO$_2$, was started by putting an amine aqueous solution into the Ag/TEOS colloidal solution. To thicken the silica shells, TEOS was further put into the colloidal solution.

Various amines were tested to obtain information on the effect of amines on the chemical stability of metallic Ag nanoparticles. Dissolution of the metallic Ag nanoparticles into water

containing ammonia or methylamine occurred by means of the production of amine complexes. Contrarily, dimethylamine (DMA) did not dissolve the metallic Ag nanoparticles. Thus, the DMA was regarded as a catalyst suitable to silica coating of metallic Ag nanoparticles, and it was used in all subsequent experiments. For the application of core-shell particles to photonic crystals (García-Santamaría et al. 2002), it is quite meaningful to accurately vary the shell thickness. Silica coating at different TEOS concentrations is very simple for the variation of shell thickness, as in the cases of Au/SiO$_2$. Figure 10.11 gives the TEM images of Ag/SiO$_2$ particles formed at various TEOS concentrations. At low TEOS concentrations, the majority of the particles had a nearly perfect core-shell structure with a single Ag core. The addition of more TEOS thickened the shell, though

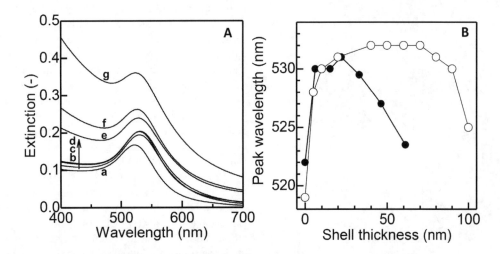

FIGURE 10.10   UV-VIS extinction spectra of (a) shell-free metallic Au nanoparticle colloid solution and Au/SiO$_2$ nanoparticle colloid solutions with various shell thicknesses Figure (A). Ethanol was added to the Au colloid to adjust water concentration of the dispersion medium to 10.7 M (1:4 (v/v) water/ethanol), which was the same concentration as that in the silica-coating. Symbols b–g stand for silica shell thicknesses of 6.1, 15.2, 22.2, 32.9, 46.3, and 61.0 nm, respectively. Figure (B) shows SPR peak position of Au/SiO$_2$ nanoparticle colloid solution vs. silica shell thickness. Experimental data and calculated positions are shown as ● and ○, respectively. A peak position of the shell-free Au colloid solution is also shown as a position at a thickness of 0 nm. (Source: Reprinted from *J. Colloid Interface Sci.*, **2011**, *358*, 329, with kind permission from Elsevier.)

production of many core-free particles also occurred. The shell was thickened in the thickness range of 28±1–76±4 nm with raising TEOS concentration. Figure 10.12 (A) gives the extinction spectra of the colloidal solutions of shell-free metallic Ag nanoparticles and of Ag/SiO₂ nanoparticles with different silica shell thicknesses. In the shell thickness range of 28–48 nm, the spectra having a single SPR band were recorded. The SPR wavelengths were located at approximately 408.5 nm, which meant that the SPR wavelengths were red-shifted compared to that of the uncoated metallic Ag nanoparticles (λ_{max} = 399 nm) (Figure 10.12 (B)). For the shell thickness range of 57–76 nm, a blue-shift of the SPR wavelength to approximately 400 nm and a decrease in the apparent extinction of the SPR took place. As in the case of the Au/SiO₂ nanoparticles (Mine et al. 2003), the experimentally obtained peak wavelengths were qualitatively matched with

the calculations performed by using Equations (10.1)–(10.3), as given in Figure 10.12 (B). This match supported successful silica coating that allowed the silica shell thickness to be varied. Thus, the colloidal stability of the metallic Ag nanoparticles was also found to be maintained by the silica coating.

10.4 Silica Coating of Other Metal-Related Nanoparticles

Our research on the silica coating of nanoparticles of mono-component metal systems was introduced in Section 10.3. Nanoparticles of multi-component metal systems, multilayered nanoparticles composed of multi-component system materials, and non-spherical nanoparticles are also expected to have many functions. This section describes our research on the silica-coating methods for metal-related nanoparticles not made from mono-component metal systems.

10.4.1 Platinum-Ruthenium Alloy Nanoparticles

Metallic Pt nanoparticles are a representative catalyst for direct methanol fuel cells (DMFCs), i.e. a type of cell that generates electricity via oxidation of methanol on the anode and reduction of oxygen on the cathode. A difficulty in the practical use of DMFCs is based on their low activity and instability, which lower the power density and shorten lifetime of the cell. A decrease in the size of the nanoparticle catalyst is obviously easy to make the catalytic activity increased, because the decrease raises surface area per unit volume. Yoshitake et al. synthesized a catalyst by impregnation of catalytic metallic Pt nanoparticles onto carbon nanohorns (CNHs), and presented that the catalytic activity was raised due to the effect of decrease in size based on the unique shape of CNHs (Yoshitake et al. 2002). However, by means of high surface energy and tendency toward aggregation and/or agglomeration for small nanoparticles, an increase in the surface area often worsens the catalytic activity, especially in the strong acidic medium. Our work revealed that nanoparticles consisted of alloy of metallic platinum (Pt) and metallic ruthenium (Ru) (PtRu), which were synthesized by reducing Pt and Ru ions with NaBH₄ in an aqueous solution of citric acid (capping agent), had an ability for methanol oxidation comparable with that of a commercial catalyst (Shimazaki et al. 2005). The formation of silica on the surface of PtRu nanoparticles is promising for solving the difficulty based on their low activity and instability. This section describes our research on the development of the silica coating of PtRu nanoparticles (PtRu/SiO₂) (Shimazaki et al. 2006).

The process for fabricating the catalyst is composed of three steps. The first is synthesis of PtRu nanoparticles. A NaBH₄ aqueous solution was put into a stirred aqueous solution of hexachloroplatinic (IV) acid (H₂PtCl₆), ruthenium (III) chloride (RuCl₃), citrate, and NaOH at room temperature. The placing of NaBH₄ made the solution a colloid solution with a color of brown, which implied formation of PtRu nanoparticles. The second is silica coating of the PtRu nanoparticles or immobilization of silica on the PtRu nanoparticles. Prior to the silica-immobilization, an APMS aqueous solution was put into the PtRu nanoparticle colloid solution. Then, for efficient silica coating, a sodium silicate

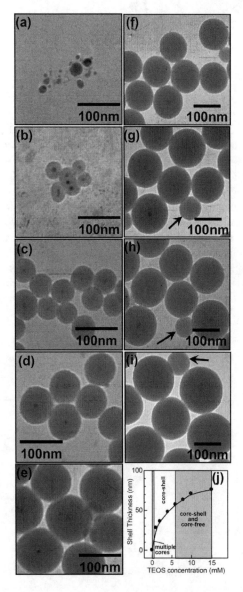

FIGURE 10.11 TEM images of Ag/SiO₂ nanoparticles prepared using total TEOS concentrations of 0.0001 (a), 0.0005 (b), 0.001 (c), 0.002 (d), 0.004 (e), 0.006 (f), 0.008 (g), 0.01 (h), and 0.015 M (i). Arrows stand for core-free silica particles. Figure (j) shows plot of silica shell thickness as a function of TEOS concentration. (Source: Reprinted from *J. Colloid Interface Sci.*, **2005**, *283*, 392, with kind permission from Elsevier.)

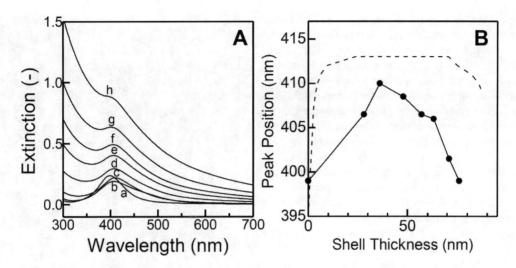

FIGURE 10.12 UV-VIS extinction spectra of colloid solutions of (a) shell-free metallic Ag nanoparticles and AgSiO$_2$ nanoparticles with shell thicknesses of (b) 28, (c) 36, (d) 48, (e) 57, (f) 63, (g) 71, and (h) 76 nm Figure (A). Figure (B) shows the SPR peak position (closed circles) and predictions by Mie theory (dashed line). (Source: Originated from *J. Colloid Interface Sci.*, **2005**, *283*, 392, with kind permission from Elsevier.)

solution, of which pH was adjusted to 10.0, was dipped slowly into the PtRu nanoparticle colloid solution. After the dipping, pH adjustment to 9.0 with a HCl solution was carried out. Third, adsorption of the silica-coated PtRu nanoparticles on a carbon support was done. A suspension of carbon powder (BP-2000) was mixed with the colloidal solution of silica-coated PtRu nanoparticles, and the mixture was sonicated to well disperse the carbon powder in the solution. During the sonication, a NaCl aqueous solution was dipped slowly into the colloidal solution to increase the amount of PtRu nanoparticles adsorbed on the carbon (C) powder (PtRu/SiO$_2$/C). The final powder suspension was obtained by performing a process composed of centrifugation, removal of the supernatant, and drying in vacuum at 70°C overnight.

XRD measurements confirmed the production of nanoparticles consisted of PtRu. Figure 10.13 presents a scanning transmittance electron microscope (STEM) image of the PtRu/SiO$_2$

nanoparticle catalyst on carbon powder. PtRu/SiO$_2$ nanoparticles with a particle size of ca. 2 nm, which were observed as black spots, were adsorbed on the carbon with no aggregation. Energy dispersive X-ray microscopic (EDX) measurements revealed that elements of silicone (Si) and oxygen (O) were present at the black spots as well as Pt and Ru, and neither Si nor O elements were on the carbon with no black spots. These results indicated that selective immobilization of silica occurred on the PtRu nanoparticle surfaces. The amino group of APMS acted on a Ru atom on the PtRu particle surface (Brayner et al. 2002), and the silanol group of APMS was toward the outer surface, which increased affinity between the silica and the PtRu particle surface. This selective immobilization was probably given with the increased affinity. Inductively coupled plasma emission spectroscopy (ICP) revealed the existence of ca. 1 wt % Si in the PtRu/SiO$_2$/C. Since the silica shell was too thin to be visualized by the TEM, the silica shell thickness was calculated to be ca. 0.2 nm by using the ICP data, on the assumption that the uniform shell of silica with a density of 2 g/cm^3 was on the PtRu nanoparticle. Accordingly, the PtRu nanoparticles were successfully silica coated by the developed method.

10.4.2 Cobalt–Platinum Alloy Nanoparticles

Pure metals, e.g. metallic Fe, metallic Co, and metallic Ni, have good magnetic features (Park et al. 2000; Racka et al. 2005; Osuna et al. 1996; Giersig and Hilgendorff 1999; Ely et al. 1999; Cordente et al. 2001). Generally, their nanoparticles tend to be oxidized in atmosphere, which spoils their magnetic features. Apart from the pure magnetic metals, chemical stability of magnetic metal alloys is high (Tyson et al. 1996). The magnetic metal alloys have other features, and one of them is magnetic anisotropy (Weller et al. 1993). The magnetic anisotropy is available for application to magneto-optical storage media. The alloy of metallic Co and metallic Pt (CoPt) is promising for high-density magnetic recording media by means of excellent properties on magnetic anisotropy and chemical stability against oxidation (Ely et al. 2000; Carpenter et al. 1999; Liou et al. 1999; Thielen et al.

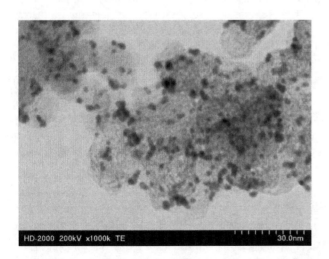

FIGURE 10.13 STEM image of catalyst fabricated by simply adding carbon powder to PtRu/SiO$_2$ nanoparticle colloidal solution. (Source: Reprinted from *J. Colloid Interface Sci.*, **2006**, *300*, 253, with kind permission from Elsevier.)

1998; Yamada et al. 1998). Various studies on the synthesis and characterization of CoPt nanoparticles have been performed (Park and Cheon et al. 2001; Yu et al. 2002; Shevchenko et al. 2003; Sobal et al. 2003; Gibot et al. 2005; Du et al. 2006). Magnetic nanoparticles are required to be separated to make each magnetic particle act as a single magnetic dipole for preservation of magnetic features. Silica coating of the magnetic nanoparticles is promising for separation of nanoparticles or prevention of aggregation. The silica coating also makes it possible to chemically stabilize the nanoparticles and to surface-modify the nanoparticles for the preparation of nanoparticle colloid solution dispersed in non-aqueous medium and for the change of interactions between particles. As mentioned in Section 10.2.1, we developed the silica coating technique for metallic Co nanoparticles produced by reducing Co ions with $NaBH_4$ in aqueous solutions (Kobayashi et al. 2003). In advance, citric acid was dissolved in the solution in order to stabilize the metallic Co nanoparticles. The obtained metallic Co nanoparticles were surface-modified with a silane coupling agent, and then coated with silica by a sol-gel method. In this section, we introduce our research on silica-coated CoPt nanoparticles ($CoPt/SiO_2$) fabricated by extending our method of producing Co/SiO_2 nanoparticles (Kobayashi et al. 2008a).

The CoPt nanoparticle colloidal solution was prepared by simultaneously injecting aqueous solutions of $CoCl_2$ and H_2PtCl_6 to an aqueous solution of $NaBH_4$ and citric acid. The CoPt nanoparticles were silica coated by putting an ethanolic solution of APMS and TEOS to the CoPt nanoparticle colloidal solution.

Figure 10.14 gives the TEM images of various $CoPt/SiO_2$ nanoparticles. Metals and silica are the darker and lighter parts of the particles, respectively. For the as-prepared $CoPt/SiO_2$ particles, multiple cores were coated with silica. For the $CoPt/SiO_2$ particles annealed at temperatures of 300 and 500°C, the clear difference between the cores and the shells could be recognized. The increase in an annealing temperature up to 700°C resulted in sintering among the particles and in no formation of core-shell structure. In our research concerning $CoPt/SiO_2$ nanoparticles (Kobayashi et al. 2006), 700°C-annealing spread a slightly darker area, which gave an implication in diffusion of cobalt toward the silica shell. A similar phenomenon also occurred in this study.

The $CoPt/SiO_2$ nanoparticles were further characterized by STEM. Figure 10.15 gives the light-field and dark-field STEM images of 500°C-annealed $CoPt/SiO_2$ nanoparticles. The core consisted of many nanoparticles. Most nanoparticles in the core were somewhat spaced out. EDX analysis was performed at the places indicated with arrows in the dark-field images (Figures 10.15 (b) and (d)). The entire place of the particle (arrow 1) had 84.5 atom%–Co and 15.5 atom%–Pt, which approximately agreed with the initial ratio of 70 atom%–Co and 30 atom%–Pt. The nanoparticles in the core were roughly categorized into a Co-rich group (arrows 4, 6, 7) and a Pt-rich one (arrows 2, 3, 5, 8). This result gave an implication that several productions of Co nanoparticles and Pt nanoparticles took place initially, and the CoPt formed next. The several productions resulted in the uneven composition in the core.

Figure 10.16 presents the XRD patterns of various CoPt/ SiO_2 nanoparticles. In the pattern of the as-prepared $CoPt/SiO_2$ nanoparticles, peaks were detected at 39.79, 46.51, and 68.01 degrees, of which d-values were 0.2263, 0.1950, and 0.1377 nm,

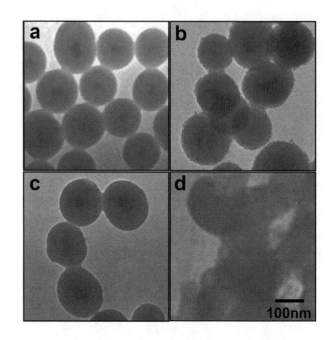

FIGURE 10.14 TEM images of $CoPt/SiO_2$ nanoparticles. The sample (a) stands for as-prepared $CoPt/SiO_2$ nanoparticles. The as-prepared CoPt/ SiO_2 nanoparticles were annealed at (b) 300, (c) 500, and (d) 700°C. (Source: Reprinted from *J. Sol-Gel Sci. Technol.*, **2008**, *47*, 16, with kind permission from Springer.)

respectively. Since the d-values of Pt are 0.2265 (111), 0.1962 (200), and 0.1387 (220) nm (JCPDS card No. 4–802), and those of fcc $CoPt_3$ are 0.2224 (111), 0.1927 (200), and 0.13627 (220) nm (JCPDS card No. 29–499), the d-values for the XRD peaks were positioned between those for Pt and fcc CoPt. This positioning indicated that the core was CoPt. These peak degrees tended to increase with raising the annealing temperature. The d-value decreased with increasing annealing temperature, which indicated that the formation of CoPt was promoted by the annealing. The increase in annealing temperature up to 500°C made a new XRD peak assigned to fcc metallic Co detected at 44.30 degrees in addition to the CoPt peaks (Figure 10.16 (c)). These tendencies on XRD peak positions suggested that the metallic Co that was amorphous or too small to be detected with XRD was produced in the as-prepared $CoPt/SiO_2$ nanoparticles, and that the production of metallic Co crystallized by annealing formed the CoPt. The presence of Co-rich and Pt-rich phases in the cores of the particles annealed at 500°C, which was confirmed by the EDX, was backed by the presence of two phases of CoPt and metallic Co in the particles, which was confirmed by the XRD. Thus, the formation of CoPt and the production of silica coating of CoPt/ SiO_2 nanoparticles were successfully performed by the proposed method.

10.4.3 Metallic Gold/Silica/Gadolinium Compound Multilayered Nanoparticles

Au is advantageous when using for X-ray imaging because Au shows high absorbance of X-rays. Nanoparticles of metallic Au have been examined as X-ray contrast agents for taking images of tissues in living bodies in the size of nanometer (Menk et al. 2011; Peng et al. 2012; Wang et al. 2013). Gadolinium (Gd)

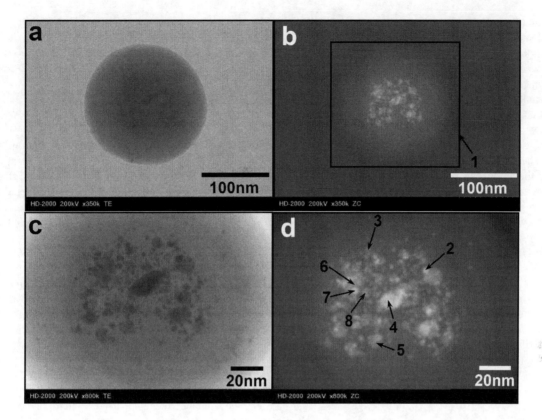

FIGURE 10.15 STEM images of CoPt/SiO$_2$ nanoparticles annealed at 500°C. Images (a) and (c) were taken in a bright-field, and (b) and (d) are dark-field images for (a) and (c), respectively. High magnification images for (a) and (b) are shown in (c) and (d), respectively. Arrows with numbers 1–8 stand for points analyzed with EDX. (Source: Reprinted from *J. Sol-Gel Sci. Technol.*, **2008**, *47*, 16, with kind permission from Springer.)

compounds (GdC) work as contrast agents by virtue of their paramagnetic features, and several kinds of Gd complexes are marketed (Secchi et al. 2011; Telgmann et al. 2013; Yu et al. 2013). Materials consisting of units with disparate characteristics should play numerous roles. Based on the perspective on the aim of making materials play numerous roles, particles comprising metallic

Au and GdC should serve as both an X-ray contrast agent and a magnetic resonance imaging (MRI) contrast agent. However, the toxicity of metallic nanoparticles (Schulz et al. 2012; Beer et al. 2012; Grosse et al. 2013) and adverse reactions to free Gd ions that might be released from Gd complexes (Telgmann et al. 2013; Thomsen 2011; Pietsch et al. 2011) have been suggested by several researchers. Coating the contrast agents with shells of silica, which is harmless to living bodies, is a good candidate as a method for controlling adverse reactions originated from contrast agents for the reason that the silica shell inhibits the contrast agents from being in touch with the living body. Our previous work has proposed methods for the GdC coating of particles such as SiO$_2$ (Kobayashi et al. 2007c) and silica-coated quantum dots (Kobayashi et al. 2010b, c). Those GdC-coated particles have been further coated with silica shells, producing multilayered core-shell particles. Accordingly, multilayered core-shell particles consisting of metallic Au nanoparticle (core), silica (first shell), GdC (second shell), and silica (third shell) (Au/SiO$_2$/GdC/SiO$_2$) should be fabricated by extending the method and may serve as both an X-ray contrast agent and an MRI contrast agent. This section familiarizes our research on the development of a method for synthesizing Au/SiO$_2$/GdC/SiO$_2$ nanoparticles (Kobayashi et al. 2013c).

Metallic Au nanoparticles having an average particle size of 16.9±1.2 nm were synthesized by reducing gold salt with Na-cit, according to TEM observation (see ref. Kobayashi et al. 2012a). Our work performed silica coating of the metallic Au nanoparticles by a modified Stöber method (Kobayashi et al. 2011), which is a step of formation of the first shell, as follows..

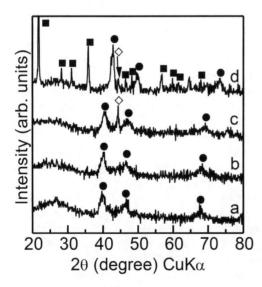

FIGURE 10.16 XRD patterns of CoPt/SiO$_2$ nanoparticles. The samples (a)–(d) were the same as those in Figure 10.14. ●: fcc CoPt, ■: cristobalite, ◇: fcc Co. (Source: Reprinted from *J. Sol-Gel Sci. Technol.*, **2008**, *47*, 16, with kind permission from Springer.)

The colloid solution of metallic Au nanoparticles, APMS/ethanol solution, and TEOS/ethanol solution were put into a water/ethanol solution in turn. Next, the reaction for silica coating was started by speedily putting NaOH aqueous solution into the Au/APMS/TEOS colloidal solution. As shown in a TEM image (see Kobayashi et al. 2012a), an average particle size of the Au/SiO$_2$ particles was 47.3±3.5 nm. The Au/SiO$_2$ nanoparticle colloidal solution was concentrated by executing a process comprising centrifugation and removal of the supernatant, followed by sonication. The Au concentration was increased by ten times by adjusting the amount of supernatant removed. The Au/SiO$_2$ nanoparticles were coated with the GdC shell (the second shell) by homogeneous precipitation, as follows. Into a 1-propanol/water solution were consecutively injected the concentrated Au/SiO$_2$ nanoparticle colloidal solution and polyvinylpyrrolidinone aqueous solution at 35°C. Then, into the mixture were consecutively put urea aqueous solution, nitric acid (for adjusting pH to 5) and gadolinium (III) nitrate (Gd(NO$_3$)$_3$) aqueous solution, followed by stirring at 80°C. Ions such as OH$^-$ and CO$_3^{2-}$ are generated in the homogeneous precipitation method using urea by heating urea in aqueous solution at 80–100°C, which precipitates metallic ions. According to Li et al. (Li et al. 2008), Gd(OH)CO$_3$·1.3H$_2$O is produced by the homogeneous precipitation reaction. Thus, the reaction equation should be as follows:

$$Gd(NO_3)_3 + (NH_2)_2CO + 4.3H_2O$$
$$\rightarrow Gd(OH)CO_3 \cdot 1.3H_2O \downarrow + 2NH_4NO_3 + HNO_3 \quad (10.4)$$

The Au/SiO$_2$/GdC nanoparticles in colloidal suspension were washed by centrifuging the colloidal suspension, removing the supernatant, adding water, and sonicating the suspension. The washing process was repeated. The solvent was changed into a water/ethanol solution by adding not water but water/ethanol solution at the last washing step. According to TEM observation (see Kobayashi et al. 2012a), the particle size increased to 55.0±3.8 nm after the GdC coating, which indicated the production of a multilayered core-shell particles with GdC shell with a thickness of 3.9 nm. The Au/SiO$_2$/GdC nanoparticles were then coated with silica shell (the third shell) by the modified Stöber method. Into the Au/SiO$_2$/GdC nanoparticle colloidal solution that was strenuously stirred were consecutively injected TEOS and NaOH aqueous solutions at 35°C. The Au/SiO$_2$/GdC/SiO$_2$ nanoparticles were washed by performing the same process as that for the Au/SiO$_2$/GdC nanoparticles. As shown in a TEM image (see Kobayashi et al. 2012a), multiple darker particles were coated with lighter-colored shells contrast, and the resulting particles were not spherical but distorted. The darker particles were considered to be Au/SiO$_2$/GdC nanoparticles because they were almost the same size as Au/SiO$_2$/GdC nanoparticles, though a multilayered core-shell structure was not clearly observed in the darker particles. The lighter parts around the darker parts were determined to be silica shells. The silica was considered to form the third shell on the GdC surface on the basis of the strong affinity between the GdC surface and the silica surface, by a mechanism similar to the GdC coating on the silica surface. The multilayered particles had a size of ca. 200 nm. Aggregates of Au/SiO$_2$/GdC nanoparticles that were probably produced during the silica coating were also coated with silica, which provided

the production of distorted particles incorporating several Au/SiO$_2$/GdC nanoparticles. Thus, multilayered Au/SiO$_2$/GdC/SiO$_2$ nanoparticles could be successfully fabricated by this method, despite the formation of multiple cores and the distortion of the particles.

10.4.4 Metallic Gold Nanorods

Metallic Au nanoparticles exhibit SPR absorption. The dependence of SPR peak position on the shape or aspect ratio of the nanoparticles has been acquainted with (Cao et al. 2014, Chandra et al. 2015; Kluczyk and Jacak 2016). An SPR band is detected at ca. 500 nm for metallic Au nanoparticles, of which an aspect ratio is 1, i.e. spherical metallic Au nanoparticles, based on the transverse oscillation of electrons. Contrastively, an SPR band is detected at wavelengths as long as 600–900 nm for metallic Au nanoparticles having larger aspect ratios, e.g. metallic Au nanorods (AuNR), on the basis of the longitudinal oscillation of electrons. Light with a wavelength in the range of 650–900 nm permeates into living bodies better than light with wavelengths shorter than 650 nm and longer than 900 nm. The energy absorbed based on SPR is partially converted into heat energy. Accordingly, this phenomenon is applied to produce hyperthermia by irradiation with light from outside living cells in the range of 650–900 nm. For reduction of the toxicity of metal ions released from metallic nanoparticles and for preventing nanoparticles from aggregating, silica coating is a promising approach, as mentioned above. Several research groups have also coated AuNRs with silica (Pastoriza-Santos et al. 2006; Huang et al. 2012; Son et al. 2014; Li et al. 2014; Wu and Tracy 2015). In most performances of silica coating, AuNRs are synthesized by concurrently reducing Ag ions and Au ions with a reductant in aqueous solution dissolving a surfactant and dispersing Au-seed nanoparticles. The surfactant used was n-hexadecyltrimethylammonium bromide (CTAB). The CTAB should be left on the AuNR surface, even after a washing process utilizing centrifugation. For the reason that quaternary ammonium surfactants like CTAB damage cells, the amount of CTAB contained in the nanoparticles must be decreased. Even so, removal of all the CTAB breaks the rod-like structure of AuNR. Thence, successful silica coating requires elimination of an appropriate amount of CTAB. This section presents our research on the development of a technique for synthesizing silica-coated AuNRs, in which much CTAB was removed from the AuNR colloidal solution to the extent that the rod-like structure was not broken in advance of silica coating (Inose et al. 2017).

A colloidal solution of Au-seed nanoparticles was synthesized by reducing Au ions, as described below. Aqueous solutions of HAuCl$_4$ and NaBH$_4$ were in turn put into a strenuously-stirred aqueous solution of CTAB at 40°C. The placing of NaBH$_4$ made the solution have a color of brown, implying the production of Au nanoparticles. A colloidal solution of AuNRs was then prepared by the growth of Au-seed nanoparticles, as described below. Aqueous solutions of HAuCl$_4$, AgNO$_3$, and L-ascorbic acid and the Au-seed nanoparticle colloidal solution were in turn injected into a CTAB aqueous solution at 40°C with strenuous stirring. The AuNRs were washed by repeating a process of centrifugation, removal of the supernatant by decantation, addition of water to the residue, and shaking the mixture in a vortex mixer to remove the CTAB from the AuNR colloidal solution. In advance of a subsequent sol-gel reaction of TEOS for silica coating, the washed AuNR colloidal solution and an APMS aqueous solution were in turn put into a water/ethanol solution in order to vitrify the AuNR surface, i.e. in order to increase affinity between AuNR surface and silica. A TEOS/ethanol solution was put into the mixture. Initiation of a reaction for silica coating was then performed by fast injection of a NaOH aqueous solution into the AuNR/TEOS colloidal solution at 35°C. Salting out of AuNR/SiO$_2$ particles was then carried out with the addition of a NaCl aqueous solution to the as-prepared AuNR/SiO$_2$ colloidal solution. The salted-out AuNR/SiO$_2$ particles were washed by a process involving centrifugation, removal of the supernatant by decantation, and addition of water to the residue, followed by shaking the mixture in a vortex mixer. The decrease in water content in the washing process increased the concentration of AuNR/SiO$_2$ colloidal solution.

A photograph of the AuNR colloidal solution is given in Figure 10.17 (a) (inset). The solution had a color of bluish-purple, which suggested the production of AuNRs. Neither precipitation nor flocculation occurred, which promised the colloidal-stability of AuNRs. The TEM image of the AuNRs is presented in Figure 10.17 (a) (inset). The structure of majority of nanoparticles was rod-like, though a small amount of spherical and oblong nanoparticles was also produced. Average particle sizes were 20.2±3.6 (lateral) and 41.9±5.7 nm (longitudinal). Figure 10.17 (a) also shows an extinction spectrum of the AuNR colloidal solution in the visible region. Two peaks were dominantly detected at 516.0 and 683.0 nm. They were assigned to the SPR bands of AuNRs based on the transverse and longitudinal oscillations of electrons, respectively, so that the production of AuNRs was confirmed. Figure 10.17 (b) (inset) gives a photograph of the AuNR/SiO$_2$ nanoparticle colloidal solution. The color of the solution was bluish-purple. The bluish-purple color was similar to that of the AuNRs, which implied that the rod-like structure of AuNR was still maintained even after the silica coating process. Neither precipitation nor flocculation took place, indicating that the silica

coating did not change the colloidal stability of AuNRs. Figure 10.17 (b) (inset) gives a TEM image of the AuNR/SiO$_2$. The successful coating of AuNRs with silica shells was achieved. The shell thickness was 36.8±7.3 nm. Aggregates of AuNRs were present as cores in several particles. Silanol groups were generated with the hydrolysis of TEOS, which provided an increase in the ionic strength of the solution. An increase in ionic strength thin double layer of colloidal particles (Dickson et al. 2012; Li et al. 2015; Dimic-Misic et al. 2015). The generation of silanol groups therefore diminished the double-layer repulsion between the nanoparticles due to thinning of the double layer. Thus, the increase in ionic strength principally brought about the aggregation of AuNRs in this system. Next, the aggregates were silica coated. The AuNRs appeared to aggregate at their tips. As particle growth of AuNRs at their tips form the final AuNRs, matters such as molecules, clusters, nuclei, and fine particles might be easily adsorbed on sites like surfaces near the tips. Based on the mechanism of adsorption, these sites on the AuNRs were speculated to approach the corresponding sites of other AuNRs, so that the approach formed the aggregates. Figure 10.17 (b) gives an extinction spectrum of the AuNR/SiO$_2$ colloidal solution in the visible region. The SPR band derived from the transverse oscillation of electrons was detected at 519.5 nm, and the SPR bands assigned to the longitudinal oscillation of electrons were recorded at two wavelengths of 713.0 and 793.5 nm. The 713.0 nm-SPR band was assigned to the longitudinal oscillation of electrons, because the wavelength of 713.0 nm was closer to 683.0 nm for the AuNRs. With an increase in an aspect ratio of metallic Au nanoparticles, the wavelength of SPR band due to the longitudinal oscillation of electrons shifts to a longer wavelength (Cao et al. 2014, Chandra et al. 2015; Kluczyk and Jacak 2016). Thence, AuNRs with a large aspect ratio were implied to be produced in this system by means of the detection of SPR bands at wavelengths as long as 793.5 nm. The TEM image (Figure 10.17 (a) (inset)) revealed aggregates composed of several AuNRs. The increase in apparent aspect ratio of AuNRs might have taken place with this aggregation, which might have

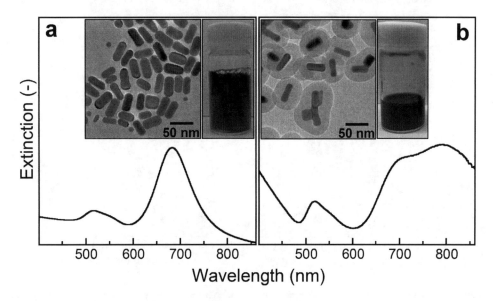

FIGURE 10.17 Visible extinction spectra of colloid solutions of (a) AuNRs and (b) AuNR/SiO$_2$ nanoparticles. Insets show their TEM images and colloid solutions. (Source: Reprinted from *Biochem. Biophys. Res. Commun.*, **2017**, *484*, 318, with kind permission from Elsevier.)

supported the detection of the SPR bands at the longer wavelength. Consequently, the section could conclude that the AuNRs were also successfully silica coated by the proposed method.

10.5 Properties

As explained in the introduction, the coating of nanoparticles with solid shells offers unique abilities such as the prevention of degradation, the control of the aggregation of nanoparticles, and the prevention of the dissolution of nanoparticles. These abilities are expected to allow the utilization of silica-coated nanoparticles as various functional materials. This section describes our studies on various properties such as magnetism, catalysis, MRI ability, X-ray imaging ability, and the photothermal conversion of silica-coated nanoparticles, aiming for a practical use of the nanoparticles.

10.5.1 Magnetism

The magnetic characterization of the Co/SiO$_2$ nanoparticles mentioned in Section 10.2.1 was executed at room temperature (Kobayashi et al. 2001). In order to discuss the magnetic characteristics and to compare the different samples, two kinds of plots were constructed. One was a plot of saturation magnetization as a function of annealing temperature, and the other was a plot of coercive field as a function of annealing temperature, as shown in Figure 10.18. The former plots for the particles fabricated at various [citrate]/[Co] ratios are given in Figure 10.18 (A). Expect for the particles fabricated at a [citrate]/[Co] ratio of 1, the maximum saturation magnetization values were obtained at annealing temperatures of 400 or 500°C. Those temperatures concurred with the annealing temperatures that gave the largest intensities of XRD peaks due to metallic Co for each sample (see Figure 10.4). However, for the particles fabricated at the ratio of 1, the saturation magnetization was the largest at the annealing temperature of 200°C, while no marked XRD peak was detected for 200°C. It has been demonstrated previously that the saturation magnetization decreases with an increase in surface/volume ratio, i.e. with a decrease in surface/volume ratio, on the basis of an increase in the disorder of the orientation of the magnetic moments at the various sites (Hochepied et al. 2000). A tendency similar to that in the research on the demonstration also reappeared in this work. Figure 10.18 (B) gives the relationships of the coercive field of Co/SiO$_2$ particles to the annealing temperature. For all the particles, the coercive fields increased as the annealing temperature was raised up to 400–600°C, and then decreased down to zero with raising the annealing temperature over 400–600°C. The annealing temperatures, at which the largest coercive fields of approximately 300 Oe were recorded, also concurred with the temperatures that gave the largest XRD peak of metallic Co.

Room-temperature magnetic characterization was also executed for the CoPt/SiO$_2$ nanoparticles mentioned in Section 10.4.2. Figure 10.19 shows the relationships of saturation magnetization and coercive field of the CoPt/SiO$_2$ nanoparticles to the annealing temperature. The saturation magnetization increased with increasing annealing temperature up to 300°C in consequence of the formation of CoPt during the annealing. The largest saturation magnetization of 11.4 emu/g-sample was recorded at the annealing temperature of 300°C. The saturation magnetization decreased with raising the annealing temperatures in a range above 300°C. The result on the decrease in saturation magnetization gave an implication for partial oxidation of CoPt alloy nanoparticles by annealing in atmosphere. Notwithstanding, the XRD revealed no detection of peaks assigned to oxide (see Figure 10.16), which might have meant that the oxide was not detected by XRD due to the production of too small oxide particles. The coercive field rose with increasing annealing temperature up to 500°C because of the acceleration of the formation of CoPt by the annealing. The largest coercive field of 365 Oe was recorded at the annealing temperature of 500°C. The coercive field decreased with raising the annealing temperatures in a range above 500°C. Such tendency for the coercive field might have been related to the partial oxidation of CoPt nanoparticles, notwithstanding that the reason giving the tendency has not been clarified as yet.

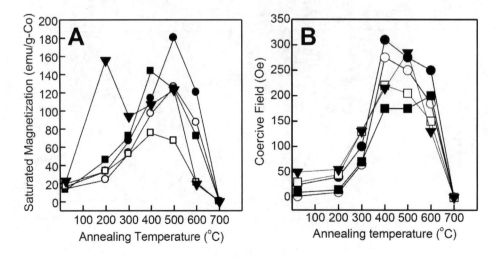

FIGURE 10.18 Room temperature saturation magnetization (A) and coercive field (B) vs. annealing temperature of Co/SiO$_2$ nanoparticles. Symbols ●, ○, ■, □, and ▼ stand for the samples (a)–(e) in Figure 10.4. (Source: Originated from *J. Phys. Chem. B*, **2003**, *107*, 7420, with kind permission from the American Chemical Society.)

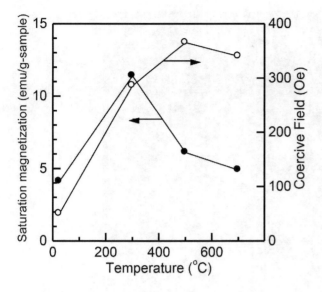

FIGURE 10.19 Room temperature saturation magnetization and coercive field of CoPt/SiO$_2$ nanoparticles as a function of annealing temperature. (Source: Reprinted from *J. Sol-Gel Sci. Technol.*, **2008**, *47*, 16, with kind permission from Springer.)

10.5.2 Catalysis

To evaluate the activity of the PtRu/SiO$_2$/C mentioned in Section 10.4.1, the current for methanol oxidation vs. electrode potential was measured for a mixture of sulfuric acid aqueous solution and methanol at 25°C (Shimazaki et al. 2006). A working electrode was fabricated by repeating a process composed of spreading of a Nafion® solution dispersing the catalyst onto carbon paper and evaporating the solvent by drying.

The current was similar to that for the electrode produced by using the commercially available catalyst (Shimazaki et al. 2005), which meant that active surface and electron paths to the electrode still remained even after the silica coating. The monolayers of silica were deduced to form incompetently on the PtRu surface, which granted that the PtRu surface attached to the carbon support. The incomplete formation of silica monolayers might have bared the surface of the PtRu nanoparticles partially and made the methanol oxidation reaction occur on the bare surface. There were no large differences in current among the unannealed and the catalysts annealed at 250 and 400°C, which meant that the PtRu/SiO$_2$ nanoparticles were thermally stable. The silica on surface probably inhibited the nanoparticles from aggregating and the active surface area from decreasing after annealing, which provided thermal stability. The electrochemical stability of an electrode with the PtRu/SiO$_2$ nanoparticle catalyst is exhibited in Figure 10.20. For the electrode with PtRu/SiO$_2$ nanoparticles, the oxidation current retained almost a constant value at immersion times of 300–1000 h. In contrast, for the electrode with the PtRu nanoparticles made according to our work (Shimazaki et al. 2005), the oxidation current became smaller with increasing immersion time. These results confirmed that the PtRu/SiO$_2$ nanoparticles were electrochemically stable. Possibly, the silica coating prevented the aggregation and/or agglomeration of PtRu nanoparticles, stabilizing the surface area of the catalyst. Additionally, the silica coating might have

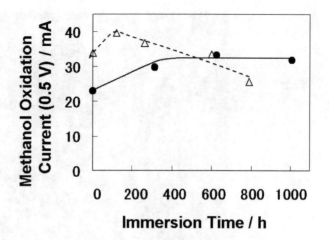

FIGURE 10.20 Plots of methanol oxidation current (0.5 V) against immersion time of electrodes to sulfuric acid solution; (filled circle) PtRu/SiO$_2$ nanoparticles, (unfilled triangle) silica-free citrate-stabilized PtRu nanoparticles. The lines are for the eyes. (Source: Reprinted from *J. Colloid Interface Sci.*, **2006**, *300*, 253, with kind permission from Elsevier.)

prevented the Ru from dissolving, which would control poisoning of the Pt surface by carbon monoxide.

10.5.3 Magnetic Resonance Imaging Ability

Section 10.4.3 shows the method for fabricating Au/SiO$_2$/GdC/SiO$_2$ particles (Kobayashi et al. 2013c). Our work proved that multilayered SiO$_2$/GdC/SiO$_2$ core-shell particles showed MRI ability (Kobayashi et al. 2007a). This result signifies that the Au/SiO$_2$/GdC/SiO$_2$ particles will act as an MRI contrast agent.

In addition, our other work revealed that silica-coated Au nanoparticles had X-ray imaging ability (Kobayashi et al. 2011). It was confirmed in a preliminary experiment that mice remained active even after injection of the silica-coated Au nanoparticle colloidal solution, which meant that the silica-coated Au nanoparticles were not strongly toxic to mice. Therefore, the multilayered Au/SiO$_2$/GdC/SiO$_2$ nanoparticles will be harmless because of their outer silica shell. Accordingly, the Au/SiO$_2$/GdC/SiO$_2$ nanoparticles will function in both MRI and X-ray imaging.

10.5.4 X-Ray Imaging Ability and Photothermal Conversion

Methods for fabricating Au/SiO$_2$ nanoparticles were developed as described in Section 10.3.1.2. This section introduces our research on an X-ray imaging process that uses the Au/SiO$_2$ nanoparticles (Kobayashi et al. 2014c). The immune system is convinced that hydrophobic materials are foreign, which controls the blood circulation of hydrophobic materials in vivo. PEGylation, which is a process for immobilizing poly(ethylene glycol) (PEG) on materials, is usually carried out to hydrophilize the surface (Niidome et al. 2010; Yoshino et al. 2012; Otsuka et al. 2012). For efficient PEGylation of Au/SiO$_2$ nanoparticles, surface-aminated Au/SiO$_2$ nanoparticles (Au/SiO$_2$-NH$_2$) were synthesized by injecting a NaOH aqueous solution into ethanol containing Au nanoparticles surface-modified with APMS, TEOS and water, and then putting an APMS/ethanol solution into the mixture. The as-prepared Au/SiO$_2$-NH$_2$ nanoparticle colloid

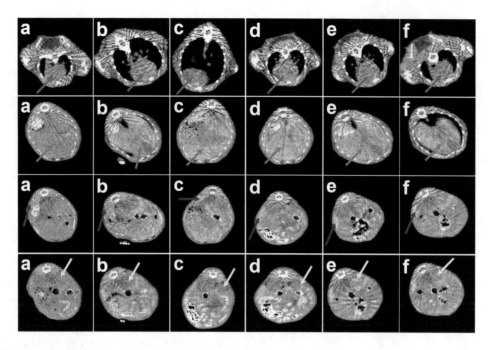

FIGURE 10.21 CT images of heart (red arrows), liver (green arrows), spleen (blue arrows), and kidney (yellow arrows) of mouse after injection of Au/SiO$_2$/ PEG nanoparticle colloid solution. The images were taken (a) prior to injection, and at (b) 5, (c) 60, (d) 180, (e) 360, and (f) 720 min after injection. (Source: Reprinted from *J. Nanopart. Res.*, **2014**, *16*, 1, with kind permission from Springer.)

solution was then concentrated by the washing process similar to that described in Section 10.4.4. Poly oxy-1 2-ethanediyl α-methyl-ω- {2-[2 5-dioxo-1-pyrrolidinyl oxy]-6-oxohexyloxy} (ME-050HS) was used for the PEGylation, because the activated carboxyl group of ME-050HS should react on the amino group on the Au/SiO$_2$-NH$_2$ nanoparticles. An aqueous solution of ME-050HS was put into the concentrated Au/SiO$_2$-NH$_2$ nanoparticle colloidal solution to cause the PEGylation (Au/SiO$_2$/PEG). X-ray imaging revealed that the computed tomography (CT) value per the Au concentration of the Au/SiO$_2$/PEG nanoparticle colloidal solution was 8.20×10^3 HU/M, which was higher than the value of 4.76×10^3 HU/M for Iopamiron 300 (Kobayashi et al. 2013b) because of the high X-ray absorption of Au. Figure 10.21 gives X-ray images of a mouse prior to and after injection of the Au/SiO$_2$/PEG nanoparticle colloid solution. Figure 10.22 presents the CT values of various tissues versus time after the injection, where the CT values at 0 min stands for those prior to the injection. In the kidney, there appeared to be no large difference in the contrast between prior to and after the injection; its CT value was a nearly fixed value of ca. 65 HU. After the injection, the images of heart, liver, and spleen appeared to become slightly brighter than those prior to the injection. With increasing the time after the injection from 0 to 5 min, the CT values drastically increased from 55 to 101 HU for the heart, from 83 to 112 HU for the liver, and from 95 to 105 HU for the spleen. In the liver and spleen, the CT values became nearly fixed after 5 min, though there was a little change. This result proved that the Au/SiO$_2$/PEG nanoparticles flew in the mouse during the whole of the measurement and were not remarkably trapped in the kidney, the liver, and the spleen. However, in the heart, the CT value began to fall off after 60 min and finally came back to its original value after 360 min. This fall-off gave that the retention time of the Au/SiO$_2$/PEG nanoparticles in a mouse was 360 min. This

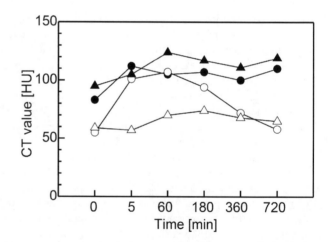

FIGURE 10.22 CT values of the heart (open circles), liver (closed circles), spleen (open triangles) and kidney (closed triangles) of a mouse after injection of Au/SiO$_2$/PEG nanoparticle colloid solution. (Source: Reprinted from *J. Nanopart. Res.*, **2014**, *16*, 1, with kind permission from Springer.)

demonstration indicated that tissues in a mouse could be imaged by X-ray up to 6 h after injection.

Section 10.4.4 introduced our research on the development of the method for fabricating AuNR/SiO$_2$ nanoparticles. This section introduces their photothermal conversion property (Inose et al. 2017). Figure 10.23 (a) presents the temperatures of the AuNR colloidal solutions with various Au concentrations versus the laser-irradiation time. The temperatures with the AuNRs were higher than that with the Au concentration of 0 mM, i.e. that for water. The temperatures recorded at 5 min rose with raising the Au concentration. This result proved affectual changeover of the light energy derived from absorption by the AuNRs to heat

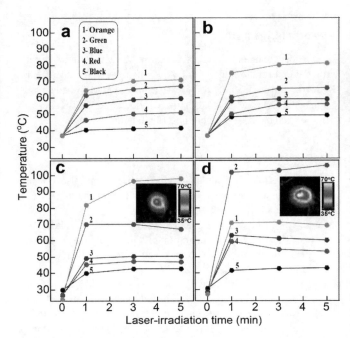

FIGURE 10.23 Temperatures vs. laser-irradiation time for (a) an AuNR colloid solution, (b) an AuNR/SiO$_2$ colloid solution, (c) a mouse injected with AuNR colloid solution, and (d) a mouse injected with AuNR/SiO$_2$ nanoparticle colloid solution. For (a) and (b), Au concentrations in the laser-irradiated colloid solutions were 0 (black, 5), 0.1 (red, 4), 0.3 (blue, 3), 1 (green, 2), and 3 (orange, 1) mM. For (c) and (d), the colloid solutions with Au concentrations of 0 (black, 5), 1 (red, 4), 3 (blue, 3), 10 (green, 2), and 30 (orange, 1) mM were injected into the mice. Insets of the Figures (c) and (d) show typical thermographs of the mice. They were taken at 5 min after the injection of the colloid solution with Au concentration of 1 mM. (Source: Reprinted from Biochem. Biophys. Res. Commun., 2017, 484, 318, with kind permission from Elsevier.)

energy. Figure 10.23 (b) gives the temperatures of the AuNR/SiO$_2$ colloidal solutions with various Au concentrations versus the laser-irradiation time. Dependences of temperature on laser-irradiation time were similar to those for the AuNR colloidal solutions, which proved that the silica coating did not change the behavior of photothermal conversion by AuNRs. Figure 10.23 (c) (inset) presents the thermograph of the skin of a mouse taken at 5 min after injection of the AuNR colloidal solution into the mouse and laser irradiation to the mouse. Thermographs were also taken at 0, 1, 2, 3, and 4 min after the injection and the laser irradiation. The temperature of mouse skin versus the laser-irradiation time was plotted by using such thermographs, as shown in Figure 10.23 (c) that gives the dependences of temperatures on laser-irradiation time for various amounts of injected AuNR colloidal solutions. The temperatures increased with increasing time for all the AuNR colloidal solutions. For the reason that the AuNR colloidal solutions were injected between the epidermis and the subcutaneous tissue, the light arrived at the AuNRs in spite of the conduct of the irradiation from outside the body. This arrival brought about the photothermal conversion followed by the increase in temperature. The temperature of the laser-irradiated skin increased with increasing the Au concentration. Thus, the skin temperature could be raised by the laser irradiation, and the rise could be accelerated by the injection of AuNR colloidal solution, even in vivo. For the AuNR/SiO$_2$ nanoparticle

colloidal solution, thermographs were also taken at 0–5 min after the injection and the laser irradiation. For example, the thermograph of the skin of a mouse taken at 5 min after injection of the AuNR/SiO$_2$ nanoparticle colloidal solution into the mouse and laser irradiation to the mouse is given in Figure 10.23 (d) (inset). The temperature of mouse skin versus the laser-irradiation time was also plotted by using such thermographs, as shown in Figure 10.23 (d) that gives the dependences of temperatures on laser-irradiation time for various amounts of injected AuNR/SiO$_2$ nanoparticle colloidal solutions. The temperatures also increased with increasing time for all the AuNR/SiO$_2$ nanoparticle colloidal solutions. Thus, the result confirmed the robust similarity in the dependence of the achieved temperature on time between the AuNR colloidal solutions and the AuNR/SiO$_2$ nanoparticle colloidal solutions, which meant that the AuNR/SiO$_2$ nanoparticle colloidal solution could also have an ability to raise the temperature in vivo upon laser irradiation.

Figure 10.24 (a) shows the dependences of temperatures of cancer cells incubated in AuNR colloidal solution on the laser-irradiation time for various Au concentrations. With no AuNR, i.e. at 0 mM Au, the dependence of temperature on time was quite similar to that with no cancer cells (Figure 10.23 (a)), which expected that the photothermal conversion ability of various nanoparticles would not be influenced with the presence of cancer cells. For other Au concentrations, the dependence of temperature on time was also close to that for the AuNR colloidal

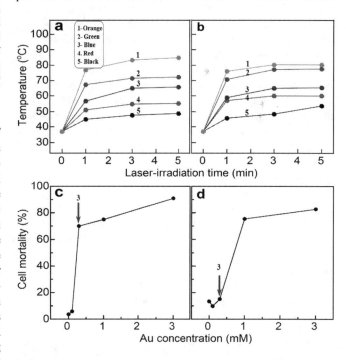

FIGURE 10.24 Results of the Trypan Blue viability test. Figure (a) and Figure (b) show temperature vs. laser-irradiation time for an AuNR colloid solution containing cancer cells and an AuNR/SiO$_2$ nanoparticle colloid solution containing cancer cells, respectively. Au concentrations in the laserirradiated colloid solutions were 0 (black, 5), 0.1 (red, 4), 0.3 (blue, 3), 1 (green, 2), and 3 (orange, 1) mM. Figures (c) and (d) show cell mortality vs. Au concentration in the AuNR colloid solution containing cancer cells and the AuNR/SiO2 nanoparticle colloid solution containing cancer cells, respectively. The arrows indicate a concentration of 0.3 mM AuNR or AuNR/SiO$_2$. (Source: Reprinted from Biochem. Biophys. Res. Commun., 2017, 484, 318, with kind permission from Elsevier.)

solutions with the corresponding Au concentrations and with no cancer cells (Figure 10.23 (a)). Figure 10.24 (b) gives the dependences of temperatures of the AuNR/SiO$_2$ nanoparticle colloidal solution containing the cancer cells on the laser-irradiation time for various Au concentrations. The dependences of temperatures on time were quite similar to those for the AuNR colloidal solutions containing cancer cells. Accordingly, both AuNRs and AuNR/SiO$_2$ were confirmed to exert photothermal conversion ability even together with cancer cells. Figure 10.24 (c) presents the dependence of mortality of cancer cells in AuNR colloidal solutions after laser-irradiation on the Au concentration. The cell mortality increased with an increase in Au concentration, and leveled out at 70–90% at 0.3 mM Au. The photothermal conversion by the AuNRs raised the temperature (Figure 10.24 (a)), which probably resulted in this increase in cell mortality. Figure 10.24 (d) gives the dependence of mortality of cancer cells in AuNR/SiO$_2$ nanoparticle colloidal solutions after laser irradiation on the Au concentration. The plot was approximately close to that for the AuNR colloidal solution. These results proved that the AuNR/SiO$_2$ nanoparticles had the photothermal conversion ability comparable to that of the AuNR colloidal solution (Figure 10.24 (c)), i.e. the AuNR/SiO$_2$ nanoparticles were also useful to kill cancer cells.

The cell mortality was ca. 70% at 0.3 mM Au for AuNR (Figure 10.24 (c) (blue arrow)), whereas the cell mortality was as small as 15% at 0.3 mM for AuNR/SiO$_2$ (Figure 10.24 (d) (blue arrow)). Because the AuNR colloidal solution should have contained much CTAB compared with that of AuNR/SiO$_2$, this difference in cell mortality between AuNR and AuNR/SiO$_2$ may be related to the high cell toxicity caused by the CTAB contained in the AuNR colloidal solution. Thus, the silica coating was proved to control the cell toxicity of nanoparticles. As a consequence, the results for cell mortality gave us the knowledge that the AuNR/SiO$_2$ nanoparticles not only could kill cancer cells in vivo by photothermal conversion but also might be harmless to living cells.

10.6 Conclusion

This book chapter has described the silica coating of various metallic nanoparticles by a liquid-phase process. The silica coating was effective for heightening both the chemical stability and colloidal stability of nanoparticles, as confirmed for nanoparticles of metallic Co, Cu, Au, and Ag. The developed silica-coating techniques could be applied to the silica coating of other metal-related nanoparticles, as confirmed for PtRu nanoparticles, CoPt nanoparticles, Au/SiO$_2$/GdC multilayered nanoparticles, and AuNRs. This chapter also mentioned that the resulting silica-coated metal-related nanoparticles exhibited or were expected to exhibit various unique properties such as magnetism for Co/SiO$_2$ and CoPt/SiO$_2$, catalysis for PtRu/SiO$_2$, MRI for Au/SiO$_2$/GdC/ SiO$_2$ and SiO$_2$/GdC/SiO$_2$, X-ray imaging for Au/SiO$_2$, and photothermal conversion for AuNR/SiO$_2$. Accordingly, the silica-coated metal-related nanoparticles can be strongly believed to be applied in various fields, e.g. optics, electronics, magnetism, catalysis, and medicine. Further studies of silica-coated metal-related nanoparticles are in progress, focusing on both fundamental research and practical use.

Acknowledgments

The authors are indebted to Prof. L.M. Liz-Marzán, Prof. M. Konno, Prof. S. Gu, Prof. D. Nagao, Prof. A. Kasuya, Prof. N. Ohuchi, Prof. N. Suzuki, Prof. M. Sato, Prof. K. Nakashima, Dr. M. Takeda, Dr. Y. Shimazaki, Dr. E. Mine, and many co-workers, who have energetically worked on the researches described in this chapter.

REFERENCES

Anžlovar, A., Z.C. Orel, and M. Žigon. (2007). Copper(I) oxide and metallic copper particles formed in 1,2-propane diol. *J. Eur. Ceram. Soc.*, 27, 987–991.

Ayame, T., Y. Kobayashi, T. Nakagawa, K. Gonda, M. Takeda, and N. Ohuchi. (2011). Preparation of silica-coated AgI nanoparticles by an amine-free process and their X-ray imaging properties. *J. Ceram. Soc. Jpn.*, 119, 397–401.

Bahadur, N.M., S. Watanabe, T. Furusawa et al. (2011). Rapid one-step synthesis, characterization and functionalization of silica coated gold nanoparticles. *Colloids Surf. A*, 392, 137–144.

Beer, C., R. Foldbjerg, Y. Hayashi, D.S. Sutherland, and H. Autrup. (2012). Toxicity of silver nanoparticles: Nanoparticle or silver ion? *Toxicol. Lett.*, 208, 286–292.

Brayner, R., G. Viau, and F. Bozon-Verduraz. (2002). Liquid-phase hydrogenation of hexadienes on metallic colloidal nanoparticles immobilized on supports via coordination capture by bifunctional organic molecules. *J. Mol. Catal. A*, 182–183, 227–238.

Brus, L. (1986). Electronic wave functions in semiconductor clusters: experiment and theory. *J. Phys. Chem.*, 90, 2555–2560.

Cao, J., T. Sun, and K.T.V. Grattan. (2014). Gold nanorod-based localized surface plasmon resonance biosensors: A review. *Sens. Actuators B*, 195, 332–351.

Carpenter, E.E., C.T. Seip, and C.J. O'Connor. (1999). Magnetism of nanophase metal and metal alloy particles formed in ordered phases. *J. Appl. Phys.*, 85, 5184–5186.

Chandra, S., J. Doran, and S.J. McCormack. (2015). Two step continuous method to synthesize colloidal spheroid gold nanorods. *J. Colloid Interface Sci.*, 459, 218–223.

Chou, K.-S., H.-L. Liu, L.-H. Kao, C.-M. Yang, and S.-H. Huang. (2014). A quick and simple method to test silica colloids' ability to resist aggregation. *Colloids Surf. A*, 448, 115–118.

Cong, L., M. Takeda, Y. Hamanaka et al. (2010). Uniform silica coated fluorescent nanoparticles: Synthetic method, improved light stability and application to visualize lymph network tracer. *PLoS ONE*, 5, e13167.

Cordente, N., M. Respaud, F. Senocq, M.-J. Casanove, C. Amiens, and B. Chaudret. (2001). Synthesis and magnetic properties of nickel nanorods. *Nano Lett.*, 1, 565–568.

Correa-Duarte, M.A., M. Giersig, N.A. Kotov, and L.M. Liz-Marzán. (1998b). Control of packing order of self-assembled monolayers of magnetite nanoparticles with and without SiO$_2$ coating by microwave irradiation. *Langmuir*, 14, 6430–6435.

Correa-Duarte, M.A., M. Giersig, and L.M. Liz-Marzán. (1998a). Stabilization of CdS semiconductor nanoparticles against photodegradation by a silica coating procedure. *Chem. Phys. Lett.*, 286, 497–501.

Correa-Duarte, M.A., Y. Kobayashi, R.A. Caruso, and L.M. Liz-Marzán. (2001). Photodegradation of SiO$_2$-coated CdS nanoparticles within silica gels. *J. Nanosci. Nanotechnol.*, 1, 95–99.

de Oliveira, A.M., L.E. Crizel, R.S. da Silveira, S.B. Pergher, and I.M. Baibich. (2007). NO decomposition on mordenite-supported Pd and Cu catalysts. *Catal. Commun.*, *8*, 1293–1297.

Dickson, D., G. Liu, C. Li, G. Tachiev, and Y. Cai. (2012). Dispersion and stability of bare hematite nanoparticles: Effect of dispersion tools, nanoparticle concentration, humic acid and ionic strength. *Sci. Total Environ.*, *419*, 170–177.

Dimic-Misic, K., M. Hummel, J. Paltakari, H. Sixta, T. Maloney, and P. Gane. (2015). From colloidal spheres to nanofibrils: Extensional flow properties of mineral pigment and mixtures with micro and nanofibrils under progressive double layer suppression. *J. Colloid Interface Sci.*, *446*, 31–43.

Doremus, R.H. (1964). Optical properties of small gold particles. *J. Chem. Phys.*, *40*, 2389–2396.

Du, X., M. Inokuchi, and N. Toshima. (2006). Preparation and characterization of Co-Pt bimetallic magnetic nanoparticles. *J. Magn. Magn. Mater.*, *299*, 21–28.

Ekimov, A.I., A.L. Efros, and A.A. Onushchenko. (1993). Quantum size effect in semiconductor microcrystals. *Solid State Commun.*, *88*, 947–950.

Ely, T.O., C. Amiens, B. Chaudret et al. (1999). Synthesis of nickel nanoparticles. Influence of aggregation induced by modification of poly(vinylpyrrolidone) chain length on their magnetic properties. *Chem. Mater.*, *11*, 526–529.

Ely, T.O., C. Pan, C. Amiens et al. (2000). Nanoscale bimetallic Co_xPt_{1-x} particles dispersed in poly(vinylpyrrolidone): Synthesis from organometallic precursors and characterization. *J. Phys. Chem. B*, *104*, 695–702.

Enüstün, B.V., and J. Turkevich. (1963). Coagulation of colloidal gold. *J. Am. Chem. Soc.*, *85*, 3317–3328.

Farbman, I., O. Lev, and S. Efrima. (1992). Optical response of concentrated colloids of coinage metals in the near-ultraviolet, visible, and infrared regions. *J. Chem. Phys.*, *96*, 6477–6485.

García-Santamaría, F., V. Salgueiriño-Maceira, C. López, and L.M. Liz-Marzán. (2002). Synthetic opals based on silica-coated gold nanoparticles. *Langmuir*, *18*, 4519–4522.

Gholami, T., M. Salavati-Niasari, M. Bazarganipour, and E. Noori. (2013). Synthesis and characterization of spherical silica nanoparticles by modified Stöber process assisted by organic ligand. *Superlattices Microstruct.*, *61*, 33–41.

Gibot, P., E. Tronc, C. Chanéac, J.P. Jolivet, D. Fiorani, and A.M. Testa. (2005). (Co, Fe) Pt nanoparticles by aqueous route; self-assembling, thermal and magnetic properties. *J. Magn. Magn. Mater.*, 290–291, 555–558.

Giersig, M., and M. Hilgendorff. (1999). The preparation of ordered colloidal magnetic particles by magnetophoretic deposition. *J. Phys. D*, *32*, L111.

Grosse, S., L. Evje, and T. Syversen. (2013). Silver nanoparticle-induced cytotoxicity in rat brain endothelial cell culture. *Toxicol. Vitro*, *27*, 305–313.

Gu, S., J. Onishi, E. Mine, Y. Kobayashi, and M. Konno. (2004). Preparation of multilayered gold-silica-polystyrene core-shell particles by seeded polymerization. *J. Colloid Interface Sci.*, *279*, 284–287.

Hall, S.R., S.A. Davis, and S. Mann. (2000). Cocondensation of organosilica hybrid shells on nanoparticle templates: A direct synthetic route to functionalized core-shell colloids. *Langmuir*, *16*, 1454–1456.

Hochepied, J.F., P. Bonville, and M.P. Pileni. (2000). Nonstoichiometric zinc ferrite nanocrystals: Syntheses and unusual magnetic properties. *J. Phys. Chem. B*, *104*, 905–912.

Huang, C., C. Huang, I. Kuo, L. Chau, and T. Yang. (2012). Synthesis of silica-coated gold nanorod as Raman tags by modulating cetyltrimethylammonium bromide concentration. *Colloids Surf. A*, *409*, 61–68.

Huang, H.H., F.Q. Yan, Y.M. Kek et al. (1997). Synthesis, characterization, and nonlinear optical properties of copper nanoparticles. *Langmuir*, *13*, 172–175.

Iler, R.K. (1979). *The Chemistry of Silica*. Wiley.

Inose, T., T. Oikawa, K. Shibuya et al. (2017). Fabrication of silica-coated gold nanorods and investigation of their property of photothermal conversion. *Biochem. Biophys. Res. Commun.*, *484*, 318–322.

Joshi, S., A. Rao, H.-J. Lehmler, B.L. Knutson, and S.E. Rankin. (2014). Interfacial molecular imprinting of Stöber particle surfaces: A simple approach to targeted saccharide adsorption. *J. Colloid Interface Sci.*, *428*, 101–110.

Khanna, P.K., S. Gaikwad, P.V. Adhyapak, N. Singh, and R. Marimuthu. (2007). Synthesis and characterization of copper nanoparticles. *Mater. Lett.*, *61*, 4711–4714.

Kluczyk, K., and W. Jacak. (2016). Damping-induced size effect in surface plasmon resonance in metallic nano-particles: Comparison of RPA microscopic model with numerical finite element simulation (COMSOL) and Mie approach. *J. Quant. Spectrosc. Radiat. Transfer*, *168*, 78–88.

Kobayashi, Y., T. Ayame, T. Nakagawa, K. Gonda, and N. Ohuchi. (2012a). X-ray imaging technique using colloid solution of AgI/silica/poly(ethylene glycol) nanoparticles. Mater. Focus, *1*,127–130.

Kobayashi, Y., T. Ayame, T. Nakagawa, Y. Kubota, K. Gonda, and N. Ohuchi. (2013a). Preparation of AgI/silica/poly(ethylene glycol) nanoparticle colloid solution and X-ray imaging using It. *ISRN Nanomater.*, 670402.

Kobayashi, Y., T. Ayame, K. Shibuya et al. (2016a). Stabilization of silica-coated silver iodide nanoparticles by ethanol-washing. *Pigm. Resin Technol.*, *45*, 99–105.

Kobayashi, Y., M.A. Correa-Duarte, and L.M. Liz-Marzán. (2001). Sol-gel processing of silica-coated gold nanoparticles. *Langmuir*, *17*, 6375–6379.

Kobayashi, Y., K. Gonda, and N. Ohuchi. (2014a). Imaging processes using core-shell particle colloid solutions for medical diagnosis. *Athens J. Nat. Form. Sci.*, *1*, 31–41.

Kobayashi, Y., M. Horie, M. Konno, B. Rodríguez-González, and L.M. Liz-Marzán. (2003). Preparation and properties of silica-coated cobalt nanoparticles. *J. Phys. Chem. B*, *107*, 7420–7425.

Kobayashi, Y., M. Horie, D. Nagao, Y. Ando, T. Miyazaki, and M. Konno. (2006). Preparation of silica-coated Co-Pt alloy nanoparticles. *Mater. Lett.*, *60*, 2046–2049.

Kobayashi, Y., J. Imai, D. Nagao et al. (2007a). Preparation of multilayered silica-Gd-silica core-shell particles and their magnetic resonance images. *Colloids Surf. A*, *308*, 14–19.

Kobayashi, Y., J. Imai, D. Nagao, and M. Konno. (2009a). Fabrication of monodispesed, multilayered silica-Y:Eu-silica core-shell particles and their photonic crystals. *J. Chem. Eng. Jpn.*, *42*, 47–50.

Kobayashi, Y., H. Inose, R. Nagasu et al. (2013b). X-ray imaging technique using colloid solution of Au/silica/poly(ethylene glycol) nanoparticles. *Mater. Res. Innov.* 17, 507–514.

Kobayashi, Y., H. Inose, T. Nakagawa et al. (2011). Control of shell thickness in silica-coating of Au nanoparticles and their X-ray imaging properties. *J. Colloid Interface Sci.*, *358*, 329–333.

Kobayashi, Y., H. Inose, T. Nakagawa et al. (2012b). Synthesis of Au-silica core-shell particles by a sol-gel process. *Surf. Eng.*, *28*, 129–133.

Kobayashi, Y., H. Inose, T. Nakagawa, Y. Kubota, K. Gonda, and N. Ohuchi. (2013c). Fabrication of multilayered Au/silica/gadolinium compound/silica core-shell particles. *Mater. Focus, 2,* 369–373.

Kobayashi, Y., H. Inose, T. Nakagawa, Y. Kubota, K. Gonda, and N. Ohuchi. (2013d). X-ray imaging technique using colloid solution of Au/silica core-shell nanoparticles. *J. Nanostruct. Chem., 3,* 62.

Kobayashi, Y. and T. Iwasaki. (2016b). Silica-coating of nitrogen-doped titanium oxide particles and their electrical conductivity. *Adv. Powder Technol. 27,* 819–824.

Kobayashi, Y., T. Iwasaki, K. Kageyama et al. (2014b). Fabrication of nitrogen-doped titanium oxide/silica core-shell particles and their electrical conductivity. *Colloids Surf. A, 457,* 244–249.

Kobayashi, Y., H. Kakinuma, D. Nagao, Y. Ando, T. Miyazaki, and M. Konno. (2008a). Synthesis and properties of Co–Pt alloy silica core-shell particles. *J. Sol-Gel Sci. Technol., 47,* 16–22.

Kobayashi, Y., H. Kakinuma, D. Nagao, Y. Ando, T. Miyazaki, and M. Konno. (2009b). Silica-coating of Co-Pt alloy nanoparticles prepared in the presence of poly (vinylpyrrolidone). *J. Nanopart. Res., 11,* 1787–1794.

Kobayashi, Y., H. Katakami, E. Mine, D. Nagao, M. Konno, and L.M. Liz-Marzán. (2005a). Silica coating of silver nanoparticles using a modified Stöber method. *J. Colloid Interface Sci., 283,* 392–396.

Kobayashi, Y., and L.M. Liz-Marzán. (2001). Preparation of silica-coated magnetic nanoparticles. *Stud. Surf. Sci. Catal., 132,* 363–366.

Kobayashi, Y., H. Matsudo, Y. Kubota, T. Nakagawa, K. Gonda, and N. Ohuchi. (2015a). Preparation of silica-coated quantum dot nanoparticle colloid solutions and their application in in-vivo fluorescence imaging. *J. Chem. Eng. Jpn., 48,* 112–117.

Kobayashi, Y., H. Matsudo, T. Li et al. (2016c). Fabrication of quantum dot/silica core-shell particles immobilizing Au nanoparticles and their dual imaging functions. *Appl. Nanosci., 6,* 301–307.

Kobayashi, Y., H. Matsudo, T. Nakagawa et al. (2013e). In-vivo fluorescence imaging technique using colloid solution of multiple quantum dots/silica/poly(ethylene glycol) nanoparticles. *J. Sol-Gel Sci. Technol., 66,* 31–37.

Kobayashi, Y., M. Minato, K. Ihara et al. (2010a). Synthesis of silica-coated AgI nanoparticles and immobilization of proteins on them. *J. Nanosci. Nanotechnol., 10,* 7758–7761.

Kobayashi, Y., M. Minato, K. Ihara et al. (2012c). Synthesis of high concentration colloid solution of silica-coated AgI nanoparticles. *J. Nanosci. Nanotechnol., 12,* 6741–6745.

Kobayashi, Y., K. Misawa, M. Kobayashi et al. (2004a). Silica-coating of fluorescent polystyrene microspheres by a seeded polymerization technique and their photo-bleaching property. *Colloids Surf. A, 242,* 47–52.

Kobayashi, Y., K. Misawa, M. Kobayashi et al. (2005b). Silica-coating of fluorescent polystyrene microspheres by a modified Stöber method and their stability against photobleaching. *e-Polym.,* no. 052.

Kobayashi, Y., K. Misawa, M. Takeda et al. (2004b). Silica-coating of AgI semiconductor nanoparticles. *Colloids Surf. A, 251,* 197–201.

Kobayashi, Y., K. Misawa, M. Takeda, N. Ohuchi, A. Kasuya, and M. Konno. (2007b). Control of shell thickness in silica-coating of AgI nanoparticles. *Adv. Mater. Res., 29–30,* 191–194.

Kobayashi, Y., K. Misawa, M. Takeda, N. Ohuchi, A. Kasuya, and M. Konno. (2008b). Preparation and properties of silica-coated AgI nanoparticles with a modified Stöber method. *Mater. Res. Soc. Symp. Proc., 1074E,* 110–117.

Kobayashi, Y., H. Morimoto, T. Nakagawa, K. Gonda, and N. Ohuchi. (2013f). Preparation of silica-coated gadolinium compound particle colloid solution and its application in imaging. *Adv. Nano Res., 1,* 159–169.

Kobayashi, Y., R. Nagasu, T. Nakagawa, Y. Kubota, K. Gonda, and N. Ohuchi. (2015b). Preparation of Au/silica/poly(ethylene glycol) nanoparticle colloid solution and its use in X-ray imaging process. *Nanocomposites, 2,* 83–88.

Kobayashi, Y., R. Nagasu, T. Nakagawa, Y. Kubota, K. Gonda, and N. Ohuchi. (2016d). Preparation of a colloid solution of Au/silica core-shell nanoparticles surface-modified with cellulose and its X-ray imaging properties. *J. Nanomater. Mol. Nanotechnol., 5,* 1–6.

Kobayashi, Y., R. Nagasu, K. Shibuya et al. (2014c). Synthesis of a colloid solution of silica-coated gold nanoparticles for X-ray imaging applications. *J. Nanopart. Res., 16,* 2551.

Kobayashi, Y., T. Nozawa, T. Nakagawa et al. (2010b). Direct coating of quantum dots with silica shell. *J. Sol-Gel Sci. Technol., 55,* 79–85.

Kobayashi, Y., T. Nozawa, T. Nakagawa, K. Gonda, M. Takeda, and N. Ohuchi. (2012d). Fabrication and fluorescence properties of multilayered core-shell particles composed of quantum dot, gadolinium compound and silica. *J. Mater. Sci., 47,* 1852–1859.

Kobayashi, Y., T. Nozawa, M. Takeda, N. Ohuchi, and A. Kasuya. (2010c). Direct silica-coating of quantum dots. *J. Chem. Eng. Jpn., 43,* 490–493.

Kobayashi, Y., S. Saeki, M. Yoshida, D. Nagao, and M. Konno. (2008c). Synthesis of spherical submicron-sized magnetite/silica nanocomposite particles. *J. Sol-Gel Sci. Technol., 45,* 35–41.

Kobayashi, Y., and T. Sakuraba. (2008). Silica-coating of metallic copper nanoparticles in aqueous solution. *Colloids Surf. A, 317,* 756–759.

Kobayashi, Y., K. Shibuya, M. Tokunaga, Y. Kubota, T. Oikawa, and K. Gonda. (2016e). Preparation of high-concentration colloidal solution of silica-coated gold nanoparticles and their application to X-ray Imaging. *J. Sol-Gel Sci. Technol., 78,* 82–90.

Kobayashi, Y., N. Shimizu, K. Misawa et al. (2008d). Preparation of amine-free silica-coated AgI nanoparticles with modified Stöber method. *Surf. Eng., 24,* 248–252.

Kobayashi, Y., Y. Shindo, T. Oikawa, M. Tokunaga, Y. Kubota, and K. Gonda. (2017). Fabrication of multilayered Au/silica/gadolinium compound core-shell particles and their imaging properties. *Mater. Sci. Technol., 33,* 963–970.

Kobayashi, Y., M. Yoshida, D. Nagao, Y. Ando, T. Miyazaki, and M. Konno. (2007c). Synthesis of SiO$_2$-coated magnetite nanoparticles and immobilization of proteins on them. *Ceram. Trans., 198,* 135–141.

Kreibig, U., and L. Genzel. (1985). Optical absorption of small metallic particles. *Surf. Sci., 156,* 678–700.

Li, J.-G., X. Li, X. Sun, T. Ikegami, and T. Ishigaki. (2008). Uniform colloidal spheres for $(Y_{1-x}Gd_x)_2O_3$ ($x = 0 - 1$): Formation mechanism, compositional impacts, and physicochemical properties of the oxides. *Chem. Mater., 20,* 2274–2281.

Li, Y., T. Wen, R. Zhao et al. (2014). Localized electric field of plasmonic nanoplatform enhanced photodynamic tumor therapy. *ACS Nano., 8,* 11529–11542.

Li, Z., J. Li, R. Xu, Z. Hong, and Z. Liu. 2015. Streaming potential method for characterizing the overlapping of diffuse layers of the electrical double layers between oppositely charged particles. *Colloids Surf. A*, *478*, 22–29.

Lin, X.M., C.M. Sorensen, K.J. Klabunde, and G.C. Hadjipanayis. (1998). Temperature dependence of morphology and magnetic properties of cobalt nanoparticles prepared by an inverse micelle technique. *Langmuir*, *14*, 7140–7146.

Liou, S.H., S. Huang, E. Klimek, R.D. Kirby, and Y.D. Yao. (1999). Enhancement of coercivity in nanometer-size CoPt crystallites. *J. Appl. Phys.*, *85*, 4334–4336.

Lipani, E., S. Laurent, M. Surin, L.V. Elst, P. Leclère, and R.N. Muller. (2013). High-relaxivity and luminescent silica nanoparticles as multimodal agents for molecular imaging. *Langmuir*, *29*, 3419–3427.

Lisiecki, I., F. Billoudet, and M.P. Pileni. (1996). Control of the shape and the size of copper metallic particles. *J. Phys. Chem.*, *100*, 4160–4166.

Lisiecki, I., and M.P. Pileni. (1993). Synthesis of copper metallic clusters using reverse micelles as microreactors. *J. Am. Chem. Soc.*, *115*, 3887–3896.

Liu, Q., Z. Xu, J.A. Finch, and R. Egerton. (1998). A novel two-step silica-coating process for engineering magnetic nanocomposites. *Chem. Mater.*, *10*, 3936–3940.

Liu, Z., and Y. Bando. (2003). A novel method for preparing copper nanorods and nanowires. *Adv. Mater.*, *15*, 303–305.

Liz-Marzán, L.M., M. Giersig, and P. Mulvaney. (1996a). Homogeneous silica coating of vitreophobic colloids. *Chem. Commun.*, 731–732.

Liz-Marzán, L.M., M. Giersig, and P. Mulvaney. (1996b). Synthesis of nanosized gold-silica core-shell particles. *Langmuir*, *12*, 4329–4335.

Lu, Y., Y. Yin, Z.Y. Li, and Y. Xia. (2002a). Synthesis and self-assembly of Au@SiO$_2$ core-shell colloids. *Nano Lett.*, *2*, 785–788.

Lu, Y., Y. Yin, B.T. Mayers, and Y. Xia. (2002b). Modifying the surface properties of superparamagnetic iron oxide nanoparticles through a sol-gel approach. *Nano Lett.*, *2*, 183–186.

Massé, P., S. Mornet, E. Duguet et al. (2013). Synthesis of size-monodisperse spherical Ag@SiO$_2$ nanoparticles and 3-D assembly assisted by microfluidics. *Langmuir*, *29*, 1790–1795.

Menk, R.H., E. Schultke, C. Hall et al. (2011). Gold nanoparticle labeling of cells is a sensitive method to investigate cell distribution and migration in animal models of human disease. *Nanomed. Nanotechnol. Biol. Med.*, *7*, 647–654.

Mine, E., and M. Konno. (2001). Secondary particle generation at low monomer concentrations in seeded growth reaction of tetraethyl orthosilicate. *J. Chem. Eng. Jpn.*, *34*, 545–548.

Mine, E., A. Yamada, Y. Kobayashi, M. Konno, and L.M. Liz-Marzán. (2003). Direct coating of gold nanoparticles with silica by a seeded polymerization technique. *J. Colloid Interface Sci.*, *264*, 385–390.

Morimoto, H., M. Minato, T. Nakagawa et al. (2011). X-ray imaging of newly-developed gadolinium compound/silica core-shell particles. *J. Sol-Gel Sci. Technol.*, *59*, 650–657.

Mulvaney, P. (1996), Surface plasmon spectroscopy of nanosized metal particles. *Langmuir*, *12*, 788–800.

Mulvaney, P., L.M. Liz-Marzán, M. Giersig, and T. Ung. (2000). Silica encapsulation of quantum dots and metal clusters. *J. Mater. Chem.*, *10*, 1259–1270.

Nagao, D., T. Satoh, and M. Konno. (2000). A generalized model for describing particle formation in the synthesis of monodisperse oxide particles based on the hydrolysis and condensation of tetraethyl orthosilicate. *J. Colloid Interface Sci.*, *232*, 102–110.

Niidome, T., A. Ohga, Y. Akiyama et al. (2010). Controlled release of PEG chain from gold nanorods: Targeted delivery to tumor. *Bioorg. Med. Chem.*, *18*, 4453–4458.

Ohmori, M., and E.Matijević. (1992). Preparation and properties of uniform coated colloidal particles. VII. Silica on hematite. *J. Colloid Interface Sci.*, *150*, 594–598.

Ohmori, M., and E. Matijević. (1993). Preparation and properties of uniform coated inorganic colloidal particles: 8. Silica on iron. *J. Colloid Interface Sci.*, *160*, 288–292.

Osuna, J., D. de Caro, C. Amiens et al. (1996). Synthesis, characterization, and magnetic properties of cobalt nanoparticles from an organometallic precursor. *J. Phys. Chem.*, *100*, 14571–14574.

Otsuka, H., Y. Nagasaki, and K. Kataoka. (2012). PEGylated nanoparticles for biological and pharmaceutical applications. *Adv. Drug Delivery Rev.*, *64*, 246–255.

Park, B.K., S. Jeong, D. Kim, J. Moon, S. Lim, and J.S. Kim. (2007). Synthesis and size control of monodisperse copper nanoparticles by polyol method. *J. Colloid Interface Sci.*, *311*, 417–424.

Park, J.-I., and J. Cheon. (2001). Synthesis of "solid solution" and "core-shell" type cobalt-platinum magnetic nanoparticles via transmetalation reactions. *J. Am. Chem. Soc.*, *123*, 5743–5746.

Park, S.-J., S. Kim, S. Lee, Z.G. Khim, K. Char, and T. Hyeon. (2000). Synthesis and magnetic studies of uniform iron nanorods and nanospheres. *J. Am. Chem. Soc.*, *122*, 8581–8582.

Park, Y.-S., L. M. Liz-Marzán, A. Kasuya et al. (2006). X-ray absorption of the gold nanoparticles with thin silica shell. *J. Nanosci. Nanotechnol.*, *6*, 3503–3506.

Parpaite, T., B. Otazaghine, A. Taguet, R. Sonnier, A.S. Caro, and J.M. Lopez-Cuesta. (2014). Incorporation of modified Stöber silica nanoparticles in polystyrene/polyamide-6 blends: Coalescence inhibition and modification of the thermal degradation via controlled dispersion at the interface. *Polymer*, *55*, 2704–2715.

Pastoriza-Santos, I., J. Pérez-Juste, and L.M. Liz-Marzán. (2006). Silica-coating and hydrophobation of CTAB-stabilized gold nanorods. *Chem. Mater.*, *18* 2465–2467.

Peng, C., L. Zheng, Q. Chen et al. (2012). PEGylated dendrimer-entrapped gold nanoparticles for in vivo blood pool and tumor imaging by computed tomography. *Biomaterials*, *33*, 1107–1119.

Philipse, A.P., M.P.B. van Bruggen, and C. Pathmamanoharan. (1994). Magnetic silica dispersions: Preparation and stability of surface-modified silica particles with a magnetic core. *Langmuir*, *10*, 92–99.

Pietsch, H., G. Jost, T. Frenzel et al. (2011). Efficacy and safety of lanthanoids as X-ray contrast agents. *Eur. J. Radiol.*, *80*, 349–356.

Plumeré, N., A. Ruff, B. Speiser, V. Feldmann, and H.A. Mayer. (2012). Stöber silica particles as basis for redox modifications: Particle shape, size, polydispersity, and porosity. *J. Colloid Interface Sci.*, *368*, 208–219.

Qi, L., J. Ma, and J. Shen. (1997). Synthesis of copper nanoparticles in nonionic water-in-oil microemulsions. *J. Colloid Interface Sci.*, *186*, 498–500.

Qu, H., S. Tong, K. Song et al. (2013). Controllable *in situ* synthesis of magnetite coated silica-core water-dispersible hybrid nanomaterials. *Langmuir*, *29*, 10573–10578.

Racka, K., M. Gich, A. Ślawska-Waniewska, A. Roig, and E. Molins. (2005). Magnetic properties of Fe nanoparticle systems. *J. Magn. Magn. Mater.*, *290–291*, 127–130.

Rodríguez-González, B., A. Sánchez-Iglesias, M. Giersig, and L.M. Liz-Marzán. (2004). AuAg bimetallic nanoparticles: formation, silica-coating and selective etching. *Faraday Discuss.*, *125*, 133–144.

Sakurai, Y., H. Tada, K. Gonda et al. (2012). Development of silica-coated silver iodide nanoparticles and their biodistribution. *Tohoku J. Exp. Med.*, *228*, 317–323.

Schulz, M., L. Ma-Hock, S. Brill et al. (2012). Investigation on the genotoxicity of different sizes of gold nanoparticles administered to the lungs of rats. *Mutat. Res.*, *745*, 51–57.

Secchi, F., G.D. Leo, G.D.E. Papini, F. Giacomazzi, M.D. Donato, and F. Sardanelli. (2011). Optimizing dose and administration regimen of a high-relaxivity contrast agent for myocardial MRI late gadolinium enhancement. *Eur. J. Radiol.*, *80*, 96–102.

Selvan, S.T., T. Hayakawa, M. Nogami et al. (2002). Sol-gel derived gold nanoclusters in silica glass possessing large optical nonlinearities. *J. Phys. Chem. B*, *106*, 10157–10162.

Shevchenko, E.V., D.V. Talapin, H. Schnablegger et al. (2003). Study of nucleation and growth in the organometallic synthesis of magnetic alloy nanocrystals: The role of nucleation rate in size control of $CoPt_3$ nanocrystals. *J. Am. Chem. Soc.*, *125*, 9090–9101.

Shimazaki, Y., Y. Kobayashi, M. Sugimasa et al. (2006). Preparation and characterization of long-lived anode catalyst for direct methanol fuel cells. *J. Colloid Interface Sci.*, *300*, 253–258.

Shimazaki, Y., Y. Kobayashi, S. Yamada, T. Miwa, and M. Konno. (2005). Preparation and characterization of aqueous colloids of Pt–Ru nanoparticles. *J. Colloid Interface Sci.*, *292*, 122–126.

Sobal, N.S., U. Ebels, H. Möhwald, and M. Giersig. (2003). Synthesis of core-shell PtCo nanocrystals. *J. Phys. Chem. B*, *107*, 7351–7354.

Son, M., J. Lee, and D. Jang. (2014). Light-treated silica-coated gold nanorods having highly enhanced catalytic performances and reusability. *J. Mol. Catal. A*, *385*, 38–45.

Song, X., S. Sun, W. Zhang, and Z. Yin. (2004). A method for the synthesis of spherical copper nanoparticles in the organic phase. *J. Colloid Interface Sci.*, *273*, 463–469.

Stöber, W., A. Fink, and E. Bohn. (1968). *J. Colloid Interface Sci.*, *26*, 62–69.

Sun, Y.-P., H.W. Rollins, and R. Guduru. (1999). Preparations of nickel, cobalt, and iron nanoparticles through the rapid expansion of supercritical fluid solutions (RESS) and chemical reduction. *Chem. Mater.*, *11*, 7–9.

Tago, T., T. Hatsuta, K. Miyajima, M. Kishida, S. Tashiro, and K. Wakabayashi. (2002). Novel synthesis of silica-coated ferrite nanoparticles prepared using water-in-oil microemulsion. *J. Am. Ceram. Soc.*, *85*, 2188–2194.

Telgmann, L., M. Sperling, and U. Karst. (2013). Determination of gadolinium-based MRI contrast agents in biological and environmental samples: A review. *Anal. Chim. Acta*, *764*, 1–16.

Thielen, M., S. Kirsch, A. Weinforth, A. Carl, and E.F. Wassermann. (1998). Magnetization reversal in nanostructured Co/Pt multilayer dots and films. *IEEE Trans. Magn.*, *34*, 1009–1011.

Thomsen, H.S. (2011). Contrast media safety-an update. *Eur. J. Radiol.*, *80*, 77–82.

Tyson, T.A., S.D. Conradson, R.F.C. Farrow, and B.A. Jones. (1996). Observation of internal interfaces in Pt_xCo_{1-x} ($x \approx 0.7$) alloy films: A likely cause of perpendicular magnetic anisotropy. *Phys. Rev. B*, *54*, R3702–R3705.

Underwood, S., and P. Mulvaney. (1994). Effect of the solution refractive index on the color of gold colloids. *Langmuir*, *10*, 3427–3430.

Ung, T., L.M. Liz-Marzán, and P. Mulvaney. (1998). Controlled method for silica coating of silver colloids. Influence of coating on the rate of chemical reactions. *Langmuir*, *14*, 3740–3748.

Wang, H., L. Zheng, C. Peng, M. Shen, X. Shi, and G. Zhang. (2013). Folic acid-modified dendrimer-entrapped gold nanoparticles as nanoprobes for targeted CT imaging of human lung adencarcinoma. *Biomaterials*, *34*, 470–480.

Weller, D., H. Brändle, and C. Chappert. (1993). Relationship between Kerr effect and perpendicular magnetic anisotropy in $Co_{1-x}Pt_x$ and $Co_{1-x}Pd_x$ alloys. *J. Magn. Magn. Mater. 121*, 461–470.

Wu, S.-H., and D.-H. Chen. (2004). Synthesis of high-concentration Cu nanoparticles in aqueous CTAB solutions. *J. Colloid Interface Sci.*, *273*, 165–169.

Wu, W., and J.B. Tracy. (2015). Large-scale silica overcoating of gold nanorods with tunable shell thicknesses. *Chem. Mater.*, *27*, 2888–2894.

Xu, J., and C.C. Perry. (2007). A novel approach to $Au@SiO_2$ core-shell spheres. *J. Non-Cryst. Solids.*, *353*, 1212–1215.

Yamada, Y., T. Suzuki, and E.N. Abarra. (1998). Magnetic properties of electron beam evaporated CoPt alloy thin films. *IEEE Trans. Magn.*, *34*, 343–345.

Ye, J., B. Van de Broek, R. De Palma et al. (2008). Surface morphology changes on silica-coated gold colloids. *Colloids Surf. A*, *322*, 225–233.

Yeh, M.-S., Y.-S. Yang, Y.-P. Lee, H.-F. Lee, Y.-H. Yeh, and C.-S. Yeh. (1999). Formation and characteristics of Cu colloids from CuO powder by laser irradiation in 2-propanol. *J. Phys. Chem. B*, *103*, 6851–6857.

Yin, Y., Y. Lu, Y. Sun, and Y. Xia. (2002). Silver nanowires can be directly coated with amorphous silica to generate well-controlled coaxial nanocables of silver/silica. *Nano Lett. 2*, 427–430.

Yoshino, K., K. Nakamura, Y. Terajima et al. (2012). Comparative studies of irinotecan-loaded polyethylene glycol-modified liposomes prepared using different PEG-modification methods. *Biochim. Biophys. Acta*, *1818*, 2901–2907.

Yoshitake, T., Y. Shimakawa, S. Kuroshima et al. (2002). Preparation of fine platinum catalyst supported on single-wall carbon nanohorns for fuel cell application. *Phys. B*, *323*, 124–126.

Yu, A.C.C., M. Mizuno, Y. Sasaki, H. Kondo, and K. Hiraga. (2002). Structural characteristics and magnetic properties of chemically synthesized CoPt nanoparticles. *Appl. Phys. Lett.*, *81*, 3768–3770.

Yu, S.-M., S.-H. Choi, S.-S. Kim, E.-H. Goo, Y.-S. Ji, and B.-Y. Choe. (2013). Correlation of the R1 and R2 values of gadolinium-based MRI contrast media with the ΔHounsfield unit of CT contrast media of identical concentration. *Curr. Appl. Phys.*, *13*, 857–863.

Zhang, X., H. Yin, X. Cheng, H. Hu, Q. Yu, and A. Wang. (2006). Effects of various polyoxyethylene sorbitan monooils (Tweens) and sodium dodecyl sulfate on reflux synthesis of copper nanoparticles. *Mater. Res. Bull.*, *41*, 2041–2048.

Zhou, X., Y. Kobayashi, V. Romanyuk et al. (2005). Preparation of silica encapsulated CdSe quantum dots in aqueous solution with the improved optical properties. *Appl. Surf. Sci.*, *242*, 281–286.

11

Silica Nanoparticles: Methods of Fabrication and Multidisciplinary Applications

Atul Dev, Mohammed Nadim Sardoiwala, and Surajit Karmakar

CONTENTS

11.1 Introduction

The field of nanotechnology is multidisciplinary, which provides the force of development and innovation. Silica, one of the most abundant materials on earth with very little application in its natural form, converted to different nanoscale structures using nanotechnology approaches.Variation in its nanoscale size and physical state resulted in new unique properties, which were not present in its natural state. These newly added properties have revolutionized its application window. Silica nanoparticles are prepared in solid to mesoporous form depending upon the applicability. This chapter focuses on the various fabrication and functionalization strategies of silica nanoparticles, which resulted in widespread applications of the material.

11.2 Strategy for Functionalization/ Fabrication of Silica Nanoparticles

Functionalization is useful in the advancement of the utility of silica nanoparticles. Functionalized nanoparticles are applicable in the food, agriculture, environment, and bio-medical sectors with the purpose of imaging, delivery, sensing, diagnosis, and treatment. There are various methods for the synthesis of silica nanoparticles, some of the most common methods are explained in detail.

11.2.1 Stöber Method

The Stöber method was first introduced in 1968 (Stöber et al. 1968). This procedure is used for the synthesis of solid colloidal silica particles with a large size distribution: a size range from

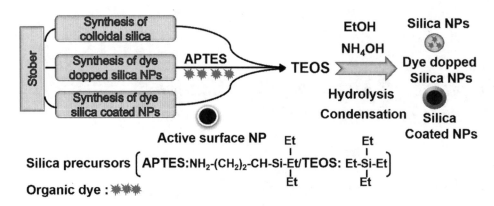

FIGURE 11.1 Stöber method for silica nanoparticle synthesis.

nanoscale to micron. In this method, a silica alkoxide precursor such as TEOS (tetraethoxysilane) is used which results in the formation of monodisperse silica particles upon hydrolysis and condensation in a mixture of ethanol and ammonium hydroxide (Figure 11.1).

11.2.2 Chemical Vapor Deposition (CVD)

CVD is a route to deposit a solid material onto a substrate via surface reaction in the gas phase. Various forms of CVD have been utilized, butmost commonly, firstly precursors are vaporized. Sonication, thermal heating, and pressure reduction are common methods to be used for vaporization (Licausi et al. 2011). In the CVD process, after vaporization, reactants are activated by using heating (Shi et al. 2011), electromagnetic radiation (Santucci et al. 2010), and plasma activation (Wang et al. 2011b). Hydrophobic material deposition with CVD is a challenging process. In general, the outcome of CVD provides a flat and chemically homogeneous deposition. So, surface roughness has been achieved by the further introduction of nanoparticles (Wang et al. 2011a). In the case of silica coating, surface chemical composition has been modified with post-treatment. Silica surface is generally hydrophilic due to the presence of hydroxyl groups on their surface that favors reaction with water molecules (Laskowski and Kitchener 1969). Lower surface energy and reduced interaction with water help to achieve hydrophobic silica. Commonly, FAS molecules are applied to obtain a superhydrophobic surface of silica (Xu et al. 2009), which is costly. Hence, new and cost-effective routes are required to deposit superhydrophobic material by using CVD. Generally, surface coating of SiO_2 is performed via CVD using $SiCl_4$. The reaction is bifurcated into two reactions as follows (Klaus et al. 1997):

$$SiOH*+SiCl_4 \rightarrow SiOSiCl_3*+HCl \qquad (11.1)$$

$$SiCl*+H_2O \rightarrow SiOH*+HCl(*\text{ indicates surface species}) \qquad (11.2)$$

The requirement of a high temperature around 600–800 K and HCl production during reaction limits the use of this reaction for coating of soft materials or biomaterials. However, the use of TEOS has overcome one of the limitations of this reaction by preventing the production of HCl. The reaction is as follows:

$$SiOH*+(OCH_2CH_3)_4Si \rightarrow SiOSi(OCH_2CH_3)_3 + CH_2CH_3OH \qquad (11.3)$$

$$SiOSi(OCH_2CH_3)_3 + H_2O \rightarrow SiOH + CH_2CH_3OH \qquad (11.4)$$

The use of the amine catalysts has been proven to reduce the higher temperature requirement, and also a modification of silica surface with vinyl alkoxysilanes helps in obtaining reaction at the optimum temperature (Effati and Pourabbas 2012). Further, advancesin the CVD process for the coating of complex geometries and substrates for the modification of nanoparticles has been studied recently. In the study, a new one-step CVD modification method for silica nanoparticles synthesis is demonstrated. In this method, deposition and modification of a silica coating was performed with an all-gas phase to overcome the limitations of CVD. Post-treatment of silica nanoparticles, using ammonia and the advancesin the CVD process facilitated the utilization of temperature-sensitive substrates or biomaterials.

11.2.3 Sol-Gel Method

In general, CVD is the primary method for synthesis of silica nanoparticles (Silva 2004). However, there is a limitation of CVD to control the particle size: its morphology and phase composition (Kempster 1992). As an alternative to CVD, the sol-gel method is applied for the production of silica-based materials. This process can form pure and homogenous products at mild reaction conditions. In this process, hydrolysis and the condensation of tetraethylorthosilicate (TEOS) or sodium silicate salt has been obtained with a mild acidic or basic condition. In the basic condition, ammonia (NH3) is widely used as a catalyst (Hench and West 1990; Stöber et al. 1968).

In the formation of silica structure, siloxane bridges are formed due to the hydrolysis of TEOS and condensation of silanol groups. Further, the formation of silica nanoparticles comprises two steps: nucleation and growth. The two models are proposed to understand the growth pattern of silica nanoparticles:

(1) monomer addition (Matsoukas and Gulari 1988)

(2) controlled aggregation (Bogush et al. 1988; Bogush and Zukoski 1991).

In monomer addition, hydrolyzed monomers are added to the reaction after completion of the nucleation process. Meanwhile, in controlled aggregation, the resulting nuclei are aggregated to form nanoparticles. Both models are applicable for the formation of spherical or gel structure. Researchers have worked widely to understand the size of nuclei and primary particle (Bailey and Mecartney 1992; Green et al. 2003). Rahman et al. (2007) synthesized 7.1 ± 1.9 nm of nearly monodispersed and stable silica nanoparticles. Many synthesis processes of nanosilica have evolved by following the Stöber method. An advantage of following the Stöber method is the ability to form homogeneous and spherical silica nanoparticles in comparison to the acid-catalyzed method that mostly produces a gel structure. In this method, fabrication or functionalization of silica nanoparticles has been performed by chemical modification of the silica surface. Surface modifications of silica nanoparticles have been demonstrated to improve the affinity between inorganic and organic phases and also enhance the dispersion ability of silica nanoparticles (Kickelbick 2003; Wei et al. 2011; Shu et al. 2008; Bailly et al. 2010). Generally, silane coupling agents are preferred to modify the silica surface. Silane coupling agents (Si(OR)₃R) facilitate bond formation between inorganic silica and other organic material like resins. In this reaction, Si(OR)₃ reacts with inorganic material and the R group reacts with the organic phase. This reaction generally conducted in aqueous or nonaqueous solvent systems that are known as post-modification. Nonaqueous systems are mostly used to functionalize APTS molecules onto the silica surface. The major advantage of a nonaqueous system is that it prevents hydrolysis. Silanes like APTS have the property to hydrolyze in an uncontrolled manner that leads to polycondensation in an aqueous system. Therefore, use of organic solvent facilitates a good control of reactions and it is preferred to utilize hydrolysis-prone silane coupling agents. The silane molecules are functionalized to silica through a direct condensation reaction under the nonaqueous system (Vansant 1995). In contrast, the aqueous system is preferred for large-scale production. In the aqueous system, silanes first hydrolyze and then condense before coating on the silica surface. The alkoxy silanes are hydrolyzed and self-condensed to form siloxane bonds with silanol groups of silica particles. Aminopropylmethydiethoxy silane (APMDS) and methacryloxypropyltriethoxysilane (MPTS) are also used as silane molecules to modify the surface of silica nanoparticles (Kang et al. 2001; Yu et al. 2003). The use of these silane molecules resulted in an increased size of particles. Besides that, amino groups carrying aminoethylaminopropyltrimethoxy-silane (AEAPTS) and 3-glycidyloxypropyltriethoxysilane (GPTS) with epoxy groups are also utilized to functionalize silica nanoparticles. Pre-treatment of nanosilica with lower silane molecules via sonication and posttreatment with epoxy silane for a longer time period resulted in monodispersed silica nanoparticles. In this reaction, both treatments were performed in an aqueous system. In addition, the advantage of epoxy silane is described as being that its enhanced dispersion of nanosilica in comparison to amino silanes is due to the absence of H-bonding between silica nanoparticles (Sun et al. 2005). Pharm et al. also modified nanosilica of 30 nm size by using 3-aminopropyltrimethoxysilane (APTS) and aminopropyldimethylmethoxysilane (APMS) molecules via aqueous route (Pham et al. 2007). Trimethoxy silane or monomethoxy silane molecules were utilized in less

concentration in respect to silica molecules, to avoid irreversible aggregation of nanosilica. Therefore, these studies suggest the use of a low concentration of silane agents with a longer reaction period for the synthesis of monodispersed nanosilica. Vejaykumaran et al. also fabricated 7 nm-sized silica particles with APTS (Vejaykumaran et al. 2008). They applied a one-pot synthesis approach that is an alternative method to reduce time, energy, and overcome the disadvantages of the post-modification method. In this approach, co-condensation is applied to modify the silica nanoparticles. However, co-condensation methods were mainly used to synthesis porous silica nanoparticles and very few reports are present with the modification of nanosilica (Kobler and Bein 2008; Suzuki et al. 2008). Kobler and Bein demonstrated a co-condensation reaction to synthesize ultrasmall nanosilica by using triethanolamine as a catalyst and phenylethoxysilane as a coupling agent (Kobler and Bein 2008). A one pot sol-gel method was also reported to prepare amino-functionalized nanosilica by using tetraethoxysilane and aminopropyltriethoxysilane (Chen et al. 2008). They prepared mixtures of both precursors in ethanol/water solutions and obtained a 200 nm size of silica nanoparticles that depends on mixing the ratio of precursors. A one-pot microemulsion method was also reported to functionalize silica nanoparticles by utilizing mixtures of TEOS and various organosilanes (Naka et al. 2010). The one-pot synthesis method and post-modification approach are both useful methods to functionalize silica nanoparticles. The one-pot method has an advantage over the post-modification approach in the synthesis of monodispersed and low aggregated particles. On the other hand, post-modification methods have a better ability to functionalize small-size particles with less effect on its size. In the one-pot method, the presence of NH2 group enhances hydrolysis rate that leads to an increase in particle size. Therefore, a low concentration of silane molecules is advantageous for control of the size of particles. Amino-functionalized nanosilica have potential bio-medical applications to be utilized as nanocarriers for drugs, DNA, and enzymes.

11.2.4 Microemulsion

The microemulsion method is also useful to synthesize silica nanoparticles. Arriagada and Osseo-Asare reported synthesis of silica nanoparticles by using a reverse microemulsion method. They used decane, ammonium hydroxide, and sodium bis(2-ethylhexyl)sulfosuccinate (AOT) in the formation of reverse microemulsion (Arriagada and Osseo-Asare 1995). Hollow silica nanospheres are also synthesized by using the microemulsion method. The report has demonstrated hollow silica synthesis by using a template of unilamellar vesicles. They synthesized it by hydrolysis and condensation of silicon alkoxides (Hubert et al. 2000). Zoldesi and Imhof have reported the synthesis of monodispersed, spherical nanosilica, silica capsules, and microballons using microemulsion methods. The properties of these different morphological silica particles could be tuned by modifying reaction conditions (Zoldesi and Imhof 2005). Microemulsion templates have recently gained attention due to the ease of control for tuning the particles' morphology and size. Ossero-Asare also synthesized nanosilica by using water-in-oil reverse microemulsion (Arriagada and Osseo-Asare 1999a). They also proposed a statistical nucleation model, which is based on the relationship of nanosilica size and water-to-sur-factant ratio. Nanosilica was also synthesized by controlling the hydrolysis of TEOS in the NP 5/cyclohexane/ammonium microemulsion solution (Arriagada and Osseo-Asare 1999b). They also showedthe advantage of the microemulsion method incontrolling the size of silica nanoparticles by analyzing the influence of ammonia concentration and the water-to-surfactant ration on the size of nanosilica. Micoemulsion methods are also preferred to synthesize various morphological silica nanoparticles. Tao and Li have demonstrated the synthesis of microskeletal silica nanosphere by using a reverse microemulsion method (Tao and Li 2005). Nevertheless, this method has limitations in synthesizing nanoparticles due to the use of the aqueous system. Hence, the room temperature ionic liquid (RTIL) method is preferred over a conventional microemulsion method. In this method, low melting points containing a series of organic salts are used. This method has wide applications in catalysis, electrochemical deposition, organic synthesis, and separation (Welton 1999; Mehnert et al. 2002; He et al. 2006). The major advantage of the RTIL is that it can be modified and designed via changing anions and cations. Hence, this microemulsion method has great potential applications. Han et al. have produced silica nanorods by using ionic liquid microemulsion in an aqueous system composed of water, 1-butyl-3-methylimidazolium hexafluorophosphate(bmimPF6), and Triton X-100 (TX-100) (Li et al. 2006). They investigated the importance of bmimPF6 in the synthesis of nanosilica and also showed that this method also shows similar limitations as a conventional method due to the presence of an aqueous system. Recently, nonaqueous ion liquid microemulsion gained interest and a few reports have shown the synthesis of nanosilica. The first report was for the synthesis of hollow silica spheres with the use of a nonaqueous ionic liquid microemulsion method. They prepared nonaqueous microemulsion with benzene, TX-100, and 1-butyl-3-methylimidazolium tetrafluoroborate (bmimBF4) (Zhao et al. 2008). However, catalysis of TEOS by the microemulsion without the addition of acid or alkali has a slow reaction rate. Hence, a recent report has studied the synthesis of silica nanoparticles via nonaqueous ionic liquid microemulsion method with the addition of acid or alkali. They show that the addition of acid resultedin ellipsoid silica nanostructures and the addition of alkali leads to the formation of hollow silica spheres. In addition, they also demonstrated that the use of acid or alkaliaccelerated the reaction rate. This microemulsion method has been proposed as a facile method to prepare nanosilica with advantages over a conventional,aqueous microemulsion method.

11.2.5 ESD

For the betterment of the deposition of silica nanoparticles, it is essential to develop methods for micropatterning (Innocenzi et al. 2008). Self-assembly patterning is one of the methods in which modification of wet surface can be processed (Masuda et al. 2005). However, there is a limitation to this wet-based patterning method. The resulting pattern may be compromised due to the surface tension and capillary flow during the drying process. Therefore, dry-based patterning methods are also developed (Lu et al. 2001; Malfatti et al. 2006). These methods include lithography process like electron-beam lithography(Wu et al. 2004) and

interferometric lithography (Xia et al. 2007). Lithography methods provide high resolution and uniform thickness. However these drying based patterning processes are costly due to the multiple fabrication steps. Recently, an electrospray deposition (ESD) method has been developed for the micropatterning of silica nanoparticles (Higashi et al. 2014). In this method, a stencil mask is used that is similar to the masks utilized for micropatterning during CVD (Kim et al. 2003; Brugger et al. 2000). ESD is suitable to perform at room temperature and atmospheric pressure that makes it cost-effective. This method has advantages in that it can be utilized for water sensitive and vacuum sensitive materials. In the process of ESD, a TEOS in water/ethanol/HCl is stirred at room temperature for 5 min. TEOS sol is heated to achievea viscosity of 7 mPa·s. In this system, an 8 mm diameter nozzle of the glass syringe is filled with the prepared TEOS sol. In the arrangement of components, the syringe is positioned horizontally at a height of 120 mm. A 0.5 mm thick 400 mm square silicon substrate is placed 40 mm far from the nozzle of the syringe.The silicon substrate is covered with the stencil mask to micropattern the nanosilica.

11.2.6 Thermal Plasma

In recent decades, nanoparticle synthesis by using thermal plasma has gained attention (Macwan et al. 2014; Kim and Kim 2019). In thermal plasma, a high temperature is achieved to evaporate all input materials. Then, the resulting vaporized materials can be condensed to form nanoparticles. These gas phase synthesis methods have some advantages over conventional sol-gel and other methods. Various experimental and modeling studies are reported to explain the effects of reactions parameters on synthesis of the nanoparticles by using a thermal plasma method (Girshick et al. 1993; Ishigaki et al. 2005; Shigeta et al. 2004; Suda et al. 2002). The reaction parameters mostly included are: pressure, feed rate, quench gas injection, and other reaction conditions. It is shown that a solid feed rate and quench gas injection majorly affect morphology and the size of nanoparticles. Sundstrom and DeMichiell have compared quenching methods by reviewing the literature and have shown that the direct quench gas mixing method is more effective (Sundstrom and DeMichiell 1971). However, it leads to a diluted product. Pratsini's research group also studied a comparison of two mechanisms for the quench gas injection: cup mixing and no-cup mixing (Wegner et al. 2002). They employed an aerosol flow reactor in the study and demonstrated that appropriate cup mixing of the quench gas with Bi vapors leads to the production of small-sized homogenous nanoparticles.

The report has shown that primary particle size is increased with increasing the feed rate(Girshick et al. 1993). They applied quench gas injection along the length of the radiofrequency plasma reactor. Similarly, the effect of quench configuration of RF plasma reactors on particle morphology and size has beeninvestigated (Leparoux et al. 2005). They employed two gas jets and eight nozzle ring configurations and demonstrated that eight nozzle configurations are likely to synthesize uniform nanoparticles. Ishigaki et al. have analyzed the effect of quench gas injection by applying it to an RF plasma tail and shown that it leads to narrow sized nanoparticles (Ishigaki et al. 2005). In advances in the field, the systematic study of the effects of different quenching configurations on the production of silica nanoparticles was performed (Mendoza-Gonzalez et al. 2007). In addition, they evaluated aggregation level and sintering in the produced nanopowder and differentiated nanostructure as highly aggregated, sintered nanospheres, and spherical nanoparticles. This study also showed the effect of temperature and velocity distributions inside the reactor on the particle size and structure. In details of this RF plasma-based study, three reactors (radical-top quench reactor, radical-bottom quench reactor, and alumna-wall reactor) are designed with the same set-up of feed injection system, RF torch, and filtering system. In the process of particle synthesis, quartz micrometer-sized silica particles in methanol are injected into an RF torch. The particles suddenly start to evaporate and form silica vapors and then it condenses to form nuclei due to the cold wall and quench gas injection. The nuclei grow throughout the length of plasma reactors and are finally collected from the filter. In a comparative analysis of all three reactors, they showed that a radial-top quench reactor has synthesized very fine primary nuclei with a 20 nm diameter. But a high degree of aggregation and nano-micrometer-sintered particles are also formed. In the case of a radical-bottom quench reactor, a similar kind of results are obtained,except for the lower level of sintered particles. On the other hand, an alumina wall reactor has prepared complete spherical and monodispersed nanoparticles (diameter = 90 nm) with a lower level of aggregation. Hence, the proper position of quench inside the plasma reactors and reactor wall temperature are both important factors in controlling the size and morphology of silica nanoparticles.

11.2.7 Thermal Spraying

To fabricate or functionalize ceramic/polymer nanocomposites, a low viscosity is required for the dispersion and coating of nanoparticles. In this process, either a higher temperature or large amounts of solvent are needed. The difficultyof proceeding with melt mixing of polymer with nanoparticles is due to the dramatic increase in the viscosity of the melt. Hence, a higher temperature of melt is essential to lower its viscosity with the addition of nanoparticles. In addition, thermal treatment is used to coat the surface of substrate with ceramic/polymer mixture and for that; parts of substrate are kept in an oven. This method applies to perform coat on a field or large surface. On the other hand, the use of solvent at a high temperature for coating is problematic due to the volatile nature of the solvent. Therefore, thermal spraying is used to overcome these limitations. During the thermal spray process, particles are melted in a thermal jet and heated via plasma or combustion. The vaporized particles are accelerated against substrate. The viscosity of particles is reduced in flight and powder spread against substrate. No requirement of solvent is needed to spread coating. In the beginning, a thermal spray method is reported to make nanoparticulate-silica-reinforced nylon 11 coating (Schadler et al. 1997). They prepared silica/nylon nanocomposite coating, which is scratch- and wear-resistant. In addition, they showed that a hydrophobic silica surface (methylated) provided better mechanical properties to nanocomposite than an hydrophilic surface. Further, it is utilized to coat silica nanoparticles to create a suitable substrate (Siegmann et al. 2005).

11.3 Applications of Functionalized Silica Nanoparticles

Colloidal and monodispersed silica nanoparticles are one of the most applied materials due to their intrinsic properties like functionality, surface structure, optical properties, and biocompatibility (Noll 1956). Surface-functionalized silica nanoparticles have a wide range of applications in biomedical imaging, food and agriculture, electronics, paints, pigments, sensors, catalysis (Lim et al. 2010; Wang et al. 2010)(Figure 11.2). Bio-compatible nanoparticles are useful in drug delivery systems and bio-imaging (Muhammad et al. 2011; Bonacchi and Zaccheron 2010). However, different applications of silica nanoparticles depend on surface properties and size (Noll 1956).

11.3.1 Environmental Application

Mesoporous silica has a high surface area and great shape selectivity, which is required in catalytic reactions. High surface area, better adsorption capacity, and fast transportation of molecules with a good thermal stability make porous material more useful for catalytic applications (Valtchev and Tosheva 2013; Walcarius and Mercier 2010). The silanol groups are easily functionalized to tune surface properties like hydrophilicity, hydrophobicity, biocompatibility, etc. The environmental applications of nano-silica are useful in air/water purification, pollution remediation, emission control, and biomass conversion to produce energy (Roy et al. 2009; Perego and Bosetti 2011; Taarning et al. 2011; Walcarius and Mercier 2010). The conversion of biomass has gained interest and development is reported across the globe (Huber et al. 2006; Van de Vyver et al. 2011; Zhou et al. 2011; Dapsens et al. 2012). Lignin-enriched cellulose biomass is utilized for biomass conversion. Cellulose are made up of glucose and it is important to break β-glycosidic bonds for hydrolysis and biomass conversion. Catalysts are used for hydrolyzing the cellulose. Liquid catalyst is difficult to recycle, therefore a solid

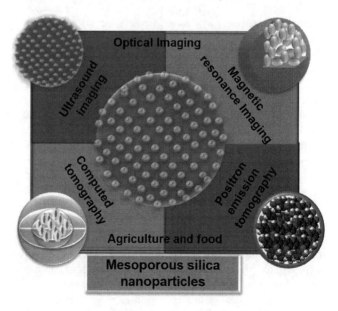

FIGURE 11.2 Interdisciplinary application of mesoporous silica nanoparticles.

catalyst is used to convert bio-mass due to their easy removal and recycling (Van de Vyver et al. 2011; Zhou et al. 2011). The use of a solid catalyst also overcomes the limited solubility of cellulose. Silica-based zeolites are mostly employed to convert biomass (Perego and Bosetti 2011; Taarning et al. 2011). Recently, zeolites are reported to transform biomass pyrolysis oil into hydrocarbons (Gayubo et al. 2004; Gayubo et al. 2005; Onda et al. 2008), olefin from ethanol (Nikolla et al. 2011; Bermejo-Deval et al. 2012), and glucose from cellulose (Onda et al. 2008). The study demonstrated that a higher ratio of Si/Al provides better selectivity for glucose and higher conversion. Research is on the way to establish better technology using silica-based zeolites for biomass conversion and energy production. Air and water purification and pollution remediation are also remarkable applications of silica nanoparticles. Surface functionalization of porous nanosilica endows their adsorption ability by enhancing electrostatic interactions and providing a binding site for heavy metal chelation. Functionalized silica nanoparticles are widely used in heavy metal adsorption (Chen et al. 2009; Li et al. 2011). Fryxell et al. demonstrated the recovery of radioactive elements from contaminated water by using silica-based materials (Fryxell et al. 2005). Magnetic mesoporous silica nanoparticles are also employed to adsorb metal elements. The use of magnetic silica enhances the separation of metal elements. There are innumerous reports that showed water purification by removing heavy and transition metal elements (Chen et al. 2009; Arruebo et al. 2006). Development in magnetic silica has reported the use of iron oxide nanoparticles as a core for mesoporous silica nanoparticle. Thus, a permanent magnet is utilized to separate magnetic solid adsorbents (Chen et al. 2009; Li et al. 2011). Thiol functionalized hybrid silica nanoparticles were reported to adsorb and remove Hg2+ because of a higher binding affinity of the Hg–S. They also showed that functionalized silica has better selectivity for Hg2+ ions over other metal ions (e.g., Pb2+, Ni2+, Zn2+, Fe3+, Co3+, and Cu2+) (Delacôte al. 2009; Antochshuk et al. 2003). On the other hand, N-donor functionalized silica materials are capable of binding with acids (Pb2+, Ni2+ Zn2+, Co2+, Cu2+ Cr3+, etc.) (Benhamou et al. 2009). Fabrication of carbamoylphosphonic acid was also studied to quench heavy and transition metal ions (Co2+, Cd2+, Cu2+, Pb2+, Cr3+, Ni2+, Mn2+, and Zn2+) (Yantasee et al. 2003). Amino- and carboxyl-functionalized nanoparticles are reported to adsorb organic compounds like methylene blue, phenosafranine, and rhodamine B (Deka et al. 2014). Photoactive compound-functionalized silica materials are widely studied as catalysis for environmental remediation (Zaccariello et al. 2014; Corma and Garcia 2004). Photocatalyst-functionalized hydrophobic silica nanoparticles are reported to be efficient adsorbent and also remediate impurities of water(Kuwahara et al. 2009).

11.3.2 Application in Agri-Food Industry

Silica has been utilized for many decades in the agri-food industry. No environmental risks have been associated with silica according to safety, toxicity, and physio-chemical and epidemiology data (Fruijtier-Pölloth 2012). Further, Peters et al.(2012) have shown that nanosilica is a food additive during in vitro digestion of foods (Peters et al. 2012). In a study, they explained that higher amounts (5–40%) of silica were present in the saliva digestion stage, while it disappeared after successive gastric digestion. They also noted that in the intestine where pH comes to neutral, nanosilica particlesreappeared in higher amounts than present at the saliva stage. Hence, this study clearly shows that human intestines are majorly exposed to nanosilica. Applications of silica in the food sector include catalysis, packaging, and sensing.Initially, Diaz et al. (2000) and Márquez-Alvarez et al. (2004) showedthe application of silica in food (Díaz et al. 2000; Márquez-Alvarez et al. 2004). They used silica nanoparticles for catalysis of fatty acids. The product of catalysis, and the mono-esters of glycerol, are utilized as anemulsifier in the food and cosmetic sectors. The use of open-structured nanoporous solids are reported by Thomas and Raja (2006) to transform organic compounds (Thomas and Raja 2006). They described that these materials have better centers and regioselectivity. One of the products (nylon-6) synthesized by a green and clean method is in demand for the production of textiles, plastics,and films for food packaging. Moelans et al. (2005) has demonstrated a new area of immobilization of molecules to porous materials (Moelans et al. 2005). Initially, the use of nanoparticles to immobilize enzymes was reported (Wang 2009), and later, functionalized and modified nanostructures have beenpreferred to catalyze reactions. Xylitol dehydrogenase (XDH) can be used to produce sugars enzymatically. Hence, Zhang et al. (2011a) havereported the use of silica nanoparticles to immobilize recombinant *rhizobium etli* CFN42 xylitol dehydrogenase (ReXDH) to produce 1-xylulose sugar that is used in the diagnosis and treatment of hepatitis (Zhang et al. 2011a). Besides that, silica materials are useful in the synthesis of nutritional compounds. Kisler et al. (2003) firstly described this possibility to be utilized as molecular sieves (Kisler et al. 2003). Mesoporous silica nanostructures have promising applications for the separation of larger molecules like proteins, which have importance in the food industries. However, mesoporous structures have limited stability in aqueous solutions. To overcome this, hexamethyldisilazane was used to functionalize silica particles that provide hydrophobicity and stability to nanostructures in aqueous nanostructures.

11.3.3 Application in Sensing

The emergence of nanotechnology has opened new horizons for the development of a sensitive platform for sensing different biological, environmental, targets. Silica-based nanoparticles provide the advantage of optical transparency over other developed sensors. It allows for the surface functionalization of key biomarkers for the detection of substrates; higher surface area provides a better sensitivity and stability of the assays. Utilizing this property, a number of sensors are developed which detect a variety of target molecules. (Yang et al. 2003, 2004; Rossi et al. 2006; Wang et al. 2006; Dyba and Hell 2003; Kneuer et al. 2000; Ashtari et al. 2005). Silica nanoparticles are also currently used in a wide range of applications as sensors in the food industry. Sudan I (diazo-conjugate dye with the chemical formula of 1-phenylazo- 2-naphthol) is carcinogenic and its use in the food industry was banned,therefore an electrochemical method was developed to sense Sudan I (Yang and He 2010). This rapid and sensitive technique is based on the property of mesoporous silica. A sensitive oxidation of Sudan I is observed and the peak current goes higher at a mesoporous silica-coated electrode. This method was successfully applied to sense and quantify the presence of Sudan I in juices and hot chili powder. Liu et al. (2011) also developed an immunosensor by using nanogold-assembled mesoporous silica (GMSNs) to determine streptomycin residues (STR) in food (Liu et al. 2011). The developed method was validated by using STR containing milk, honey, and kidney. Further, Zhao et al (2012) developed a chemiluminescence-molecular imprinting (CL–MI) sensor with the use of mesoporous silica to detect fenpropathrin, an insecticide (Zhao et al. 2012). This sensor is also applied to determine fenpropathrin in the food samples. Further, in agriculture, mesoporous nanosilica bound to photosystem II was reported to increase photosynthetic oxygen evolution (Noji et al. 2011). Yao et al. (2009) used antibody-functionalized fluorescent silica nanoparticles in the detection of microorganisms to determine plant disease (Yao 2009). Further, porous silica nanoparticles were used to deliver pesticides. Liu et al. (2006) demonstrated the delivery of validamycin (pesticide) by using mesoporous silica nanoparticles (Liu et al. 2006). They show this as an efficient nanocarrier system for the controlled release of water-soluble pesticides. Barik et al. (2016) also reviewed the application of nanosilica as a nano-insecticide (Barik et al. 2016). Insects generally developed resistance to insecticide by the presence of various cuticular lipids to protect their bodies. But nanosilica has the ability to be adsorbed at cuticular lipids and that makes them effective against insect pests. A review study has shown that functionalized hydrophobic silica nanoparticles are efficient to be implemented to control various ectoparasites of agricultural and animal pests (Sekhon 2014). Nanosilica was used to increase salinity tolerance in plants by Mushtaq et al. (2018). They synthesized chitosan and sodium alginate-coated nanosilica and showed a slow release rate of nutrients to control salinity during drought or high saline conditions. Suriyaprabha et al. (2014) also applied silica nanoparticle to study the growth of maize plants and showed that the phytochemical response of nanosilica is better than bulk silica (Suriyaprabha et al. 2014). Thus, bio-compatibility, better catalytic activity, high surface functionalization, and rapid sensitivity make silica nanoparticles highly important materials to be utilized in the environmental and agri-food sectors. Nanosilica helps to manage an eco-friendly environment and agri-food products.

11.3.4 Application in Other Industries

Silica nanoparticles haveversatile properties to employ in diverse application fields. Applications of silica nanoparticles are useful in drilling and fracturing fluids (Sensoy et al. 2009). The use of silica nanoparticles reducesthe invasion of water in the shake on addition of nanoparticles to water-based muds. Shale is sedimentary rock having clays, quartz, carbonate, and silicate minerals. Further, a study has tested two fluid samples, one containing silica nanoparticles and only brine solutions on shales and noticed that silica nanoparticles control the penetration of water to shales (Sensoy et al. 2009; Al-Bazali et al. 2008). Hence, nanofluid is useful to control lubricity, rheology, and shale stability. Silica nanoparticles are also employed in oil well cementing (Madani et al. 2012; Patil and Deshpande 2012). Cement is required for the provision of mechanical support and protection from corrosive fluids (Omosebi et al. 2015, 2016). Wait-on cementing is time allowed to cement properly and ensure prevention of negative

impact. However, less time is desired for cost saving (Santra et al. 2012). Therefore, silica nanoparticles were studied and showed that the presence of nanosilica enhanced the hydration of cement by developing an early and final compressive strength of cement (Zhang et al. 2009, 2011b). In addition, silica nanoparticles provided the control of fluid loss and compatibility with other cement composition. Silica nanoparticles are also reported to enhance oil recovery (Onyekonwu and Ogolo 2010). The tiny size of nanoparticles allows them to cross the pore throat of reservoir formation that makes nanoparticles stabilized emulsions applicable to enhance oil recovery. Polyethylene glycol-modified silica nanoparticles have been reported to enhance the stability of emulsion (Zhang et al. 2009). Here, hexane, tolune, decane, crude, and mineral oil were also used to study the recovery of oil. Further studies also reported the stabilization of emulsion by surface-coated nanosilica and showed stability for a longer time with a higher temperature (Zhang et al. 2011b). The study also confirmed the use of silica nanoparticles in the enhancement of oil recovery in water-wet formation (Onyekonwu and Ogolo 2010). Recently, fabricated wrinkled silica nanostructures have been demonstrated for use in the chemical mechanical planarization process (Ryu et al. 2018). In a study, they remarked that the cooling step is an important point to differentiate nucleation and growth stages. They further concluded that rapid cooling stabilized seeds and allowed seeds to grow further instead of forming new particles. Hence, it is useful to fabricate and form monodisperse nanoparticles to be utilized in a chemical mechanical planarization process.Textile application ofsilica nanoparticles was also reported to fabricate cotton fabric. The study showed that nanosilica-fabricated cotton is superhydrophobic, less air-permeable, and tear-proof in comparisonto water repellent agent-fabricated cotton. Hence, silica nanoparticle isa widely used nano system, and functionalization of silica nanoparticles endows the potential of silica nanoparticles to be applied in every industrial sector, including the bio-medical, environmental, health, agricultural, food, textile, and petroleum industries.

11.3.5 Application in Biomedical Imaging

The biomedical field is developing at an enormous speed with regard to imaging techniques. Early and exact detection of the disease is the current need of treatments. Currently available imaging tools suffer from various limitations thatdo not allow the progress of the technique at a rapid pace. Nanoparticle-based imaging technology provides the edge over conventional forms. Considering the limits of current imaging contrast agents and the potential advantages of nanoparticles for early diagnosis and microstructure visualization, interest in nanotechnology for biomedical imaging is rapidly increasing.

11.3.5.1 Fluorescence Imaging

Fluorescence molecular imaging is a crucial part of the biomedical field. Early detection and diagnosis of the disease provide a treatment window that helps in improving the survival rate. The urgent need for early detection pushes the limit of imaging technologies to the next level. Currently available imaging agents suffer from various limitations that are vigorously challenged by currently developed nanomaterials. These nano agents provide improved biomedical detection and imaging due to their unique size, shape, active, passive, and physical targeting abilities along with intrinsic imaging capabilities, such as superparamagnetism, fluorescence, and absorption(Doane and Burda 2012; Kim et al. 2009). The smaller size of nanoparticles allows enhanced permeability and retention (EPR) effects in lesions, which allows increased concentrations of fluorescent agent at the local site. (Oh et al. 2013) Surface labeling of nanoparticles with various target specific ligands helps increase the imaging agent localization through specific binding to target receptors in lesions (Das et al. 2015; Ke et al. 2010). In the series of nanoparticles, silica nanoparticles have shown tremendous ability to be utilized in different fluorescent-based bioimaging applications. In general, silica bulk form or its nanoform do not possess any specific imaging abilities until the tailoring of surface properties. The surface of silica NPs can be structured and functionalized with a different imaging agent and biomolecules (Kim et al. 2009), which can control the interaction with biological environments, tune binding to target surfaces and provide a long circulation time. Mesoporous silica nanoparticles are vigorously modified to achieve imaging abilities in the biological system.

11.3.5.1.1 SNPs Conjugated with Organic Dye

Covalent conjugation of organic dyes is the easiest chemical method to attach a fluorescent molecule to the silica nanoparticles (SNPs) and provide imaging abilities. The organic dyes are conjugated on the surface through chemical modification of the intrinsic silanol groups (Heidegger et al. 2016; Pan et al. 2013), or chemically linked to the silica matrix through a silane agent to form an organic dye and silane agent complex which finally forms the dye-incorporated SNPs (Huang et al. 2011; Liong et al. 2008; Lu et al. 2009, 2010; Slowing et al. 2006). Except for these two methods, physical adsorption of the dyes inside the mesoporous silica nanoparticles was also utilized for imaging application, but this method suffers from leeching issues, therefore require a pore protection mechanism (Ma et al. 2012).

11.3.5.1.2 SNPs Conjugated with QuantumDots (QDs)

Inorganic quantum dots provide added imaging advantage over classical organic dyes. Size-dependable controllable emission of light from visible to infrared wavelengths, wider excitation spectra, high quantum yield of fluorescence, high absorption coefficients, better brightness, and photostability, along with minimized photobleaching, are a few properties of quantum dotsthat provide a better coverage of imaging requirements. Quantum dots are made up of heavy metals, therefore toxicity is a concern in the biological system; encapsulation of mesoporous silica layer on the surface has resolved this concern and become a method of conjugation of quantum dots with SNPS (Kim et al. 2006; Lai et al. 2003; Pan et al. 2011; Sathe et al. 2006).

11.3.5.1.3 SNPs Conjugated with Carbon Dots (CDs)

Carbon dots provide an excellent chemiluminescence property as quantum dots but differ in the mechanism of fluorescence generation. Chemical functionalization of carbon dots via sol-gel methods provide an attractive synthetic route. Their water solubility and the property of surface functionalization with different organic species makes them highly compatible with sol-gel chemistry, which allows the synthesis of carbon dot-conjugated self-assembled mesoporous silica nanoparticles. (Fu et al. 2015; Lai et al. 2012; Lei et al. 2015; Pandey et al. 2014; Sahu et al. 2014).

11.3.5.1.4 SNPs Conjugated with Upconversion Nanoparticles (UCNPs)

Conventional fluorescence molecules like organic dye, quantum dots, and carbon dots absorb a shorter wavelength light and emit a higher wavelength light which results in poor in vivo penetration and limit imaging abilities. Up-conversion nanoparticles absorb a higher wavelength light and emit a lower wavelength light which provides the advantage of better penetrance in tissue and less background noise. These particles are synthesized and protected by hydrophobic moieties which allow conjugation with silica nanoparticles. These particles are synthesized via phase transfer with a cationic detergent, followed by a silica sol-gel reaction (Fan et al. 2014; Gai et al. 2010; Li et al. 2013; Liu et al. 2012; Qian et al. 2009).

11.3.5.2 Magnetic Resonance Imaging (MRI)

MRI is a high-resolution imaging tool thatis based on a hydrogen nuclei spin in the presence of an external magnetic field when excited with a radio frequency pulse. MRI requires a contrast agent for imaging which provides better resolution and interpretation of an image taken. MRI images can be classified as T1 contrast and T2 contrast images, depending on the use of contrast agent. The details provided by MRI imaging from the lesion has accelerated the development of contrast agents. The spatial resolution of the MRI is the advancement in optical imaging which provides a new dimension to its imaging. Nevertheless, the low sensitivity of MRI is still a concern thatneeds to be addressed. MSN-based contrast agents show more sensitivity compared to their conventional counterparts, due to enhanced relaxivity and a large surface area, with a high payload of active magnetic centers(Rieter et al. 2007). The ability of mesopores of MSNs to provide easy access of protons into the magnetic center results in a significant reduction of T1 and T2 decay relaxation times (Wartenberg et al. 2013) which allows improved resolution and sensitivity. Silica-based T2 contrast agents were developed through surface conjugation of magnetic nanoparticles on MSNs (Chen et al. 2011; Kim et al. 2008; Liu et al. 2008; Peng et al. 2014; Taboada et al. 2009; Lee et al. 2010), while T1 contrast agents were developed through conjugation of Gd and Mn-chelates (Hsiao et al. 2008; Yu-Shen Lin and Mou 2004; Palmai et al. 2017; Taylor et al. 2008).

11.3.5.3 Positron Emission Tomography (PET)

In the field of nuclear medicine, positron emission tomography (PET) is a powerful and widely used technique. It provides the advantage of improved sensitivity with high tissue penetration, and real-time quantitation of images. It requires positron-emitting specific molecular probes to visualize in vivo biological processes. This requirementleads to the development of MSN-based positron-emitting radionuclides with longer half-lives (Chen et al. 2013a, 2014a, b, 2015; Miller et al. 2014; Tang et al. 2012).

11.3.5.4 Computed Tomography (CT)

Computed tomography (CT) is anadvanced technique thatprovides three-dimensional images using differential tissue X-ray attenuation. To create CT images, contrast agents play a crucial role. It distinguishes the tissues with similar attenuation coefficients from the background. Currently, iodine- and bismuth-based CT contrast agents are used intravenously but they suffer from fast clearance, nonspecific distribution, potential toxicity, and anaphylaxis. To overcome these limitations, nanosized contrast agents have been designed, which are more efficient compared to conventional contrast agents and provide a long circulation time with a high degree of specificity to the tissues. Silica-based nano agents were specifically developed with coating of magnetic and superparamagnetic molecules to broaden the imaging application (Feng et al. 2014; Luo et al. 2011; Song et al. 2015; Chou et al. 2010).

11.3.5.5 Ultrasound Imaging

Ultrasound (US) imaging is one of the most widely used medical diagnostic imaging setups, which requires a handheld probe over the imaging site, and uses a water-based gel to provide clear acoustic coupling (Chen et al. 2013b; Shi et al. 2013). Its portability, noninvasiveness, high spatial resolution, low cost, and real-time imaging properties make it a highly rated imaging tool. It requires contrast agents whichprovide the acoustic signal differences between normal tissues and the lesions to generate contrast.

TABLE 11.1

Bio-Medical Applications of Silica Nanoparticles

Sr. No.	Application	References
1.	**Bioimaging application**	
a.	**SNPs coated with organic dye**	
I.	Amine functionalized FITC MSN	Heidegger et al. (2016)
II.	APTES functionalized FITC MSN	Huang et al. (2011)
III.	folic acid functionalized FITC-MSNs	Lu et al. (2010)
IV.	ICG tagged MSN	Lee et al. (2009)
V.	Squaraine tagged graphene oxide (GO)-enwrapped MSN	Sreejith et al. (2012)
b.	**SNPs Conjugated with Quantum Dots**	
I.	CdSe/ZnS conjugated CTAB stabilized MSN	Kim et al. (2006)
II.	CdSe/ZnS conjugated PEG liposome stabilized MSN	Pan et al. (2011)
III.	CdSe/ZnS QDs and iron oxide loaded MSN	Sathe et al. (2006)
c.	**SNPs Conjugatedwith Carbon Dots**	
I.	Citric acid Si-CDs	Lei et al. (2015)
II.	Rice husk-induced MSN capped CDs	Pandey et al. (2014)
d.	**SNPs Conjugatedwith UCNPs**	
I.	NaYF4:Tm/Yb/Gd@MSNs	Liu et al. (2012)
II.	NaYF4:Yb,Er@NaYF4@mSiO2-Ru	Xu et al. (2018)
III.	NaYF4:Yb,Er@SiO2	Zhou et al. (2019)
2.	**Magnetic Resonance Imaging**	
I.	Fe3O4@MSNs	Kim et al. (2008)
II.	Fe3O4@MSNs with bromo-terminated ligands	Lee et al. (2010)
III.	Gd-MSNs-RGD	Hu et al. (2016a)
IV.	Fe3O4@SiO2@MSNs	Ma et al. (2012)
V.	Mn-DTPA-doped MSNs	Palmai et al. (2017)
VI.	MnO@MSNs	Kim et al. (2011)
VII.	Fe3O4/MnO@MSNs	Peng et al. (2017)
3.	**PositronEmission Tomography**	
I.	^{64}Cu-NOTA-MSNs-PEG-TRC105	Chen et al. (2013a)
II.	^{89}Zr-MSNs	Chen et al. (2015)
III.	8F-DBCOT-PEG-MSNs	Lee et al. (2013)
4.	**Computed Tomography**	
I.	i-fmSiO4@SPIONs	Xue et al. (2014)
II.	Au@MSNs	Song et al. (2015)
III.	FePt@MSN@PDA	Chen et al. (2017c)
5.	**Ultrasound Imaging**	
I.	MSN-PFH	Wang et al. (2012)
6.	**Multimodal Imaging**	
I.	(ZGOCS@MSNs@Gd2O3)	Zou et al. (2017)
II.	MnOx-doped hollow MSNs	Chen et al. (2012)
III.	^{64}Cu/GdIII/ ZW800	Huang et al. (2012)

The current contrast agents comprise microbubbles with 1 to 8 μm diameter(Hu et al. 2016b). These agents only provide blood pool contrast but fail to provide contrast at a cellular level due to the larger size. Silica-based nano-contrast agents have been developed which provide improved contrast and control in tissues (Liu et al. 2017; Wang et al. 2012, 2013; Zhang et al. 2014; Chen et al. 2017a; Yildirim et al. 2016).

11.3.5.6 Multimodal Imaging

Multimodal imaging techniques have been developed to overcome the limitations of single model imaging systems. This technique laid the development of multifunctional imaging agents which provided flexibility in imaging applications, along with the added advantages of a minimum payload of contrast agents inside the tissue and a detailed knowledge of the lesions. Silica-based nanoformulations were developed using MnOx, Copper (Cu), Gadolinium (Gd3+) doping, along with many other contrast agents (Chen et al. 2012; Fan et al. 2014; Gai et al. 2010; Nakamura et al. 2017; Zou et al. 2017).

11.3.6 Therapeutic Delivery Application

Silica-based nano delivery systems are well developed because of the unique properties of loading hydrophobic and hydrophilic cargos with a very high percentage of loading. Tunability of the sizeand easy surface functionalization allows rapid development of tissue-specific cargo delivery systems. Delivery of anticancer drugs, antibiotics, enzymes are well characterized and established using mesoporous silica nanoparticles. It can be functionalized by different moieties to provide advantages in triggered (light, magnet, temperature, enzyme, pH) and responsive applications in controlled drug delivery. The property of sustained release and minimal toxicity of the delivery vehicle makes it a preferable method for therapeutic delivery application(Chen et al. 2014a, 2017b; Zhu et al. 2005; Qu et al. 2006; Doadrio et al. 2006; Wang et al. 2016; Khosravian et al. 2016; Liu et al. 2015)(Table 11.1).

11.4 Conclusion

This chapter described in detail the different fabrication roots and application of silica nanoparticles in various fields. Silica nanoparticles show promising application in therapeutic delivery and biomedical imaging due to their tailorable and well-defined surface chemistry, which allows conjugation of various imaging agents on the surface. The ability of silica nanoparticles to work in a hydrophilic and hydrophobic environment further increases its usage in multidomain areas. Its biocompatibility makes a huge difference in comparison to its other inorganic nanomaterials to be used in biological application.

REFERENCES

Al-Bazali T., Zhang J, Chenevert ME, Sharma MM. (2008). Factors controlling the compressive strength and acoustic properties of shales when interacting with water-based fluids. *International Journal of Rock Mechanics and Mining Sciences*, *45*, 729–738. doi:10.1016/j.ijrmms.2007.08.012.

Antochshuk V, Olkhovyk O, Jaroniec M, Park I-S, Ryoo R. (2003). Benzoylthiourea-modified mesoporous silica for mercury(II) removal. *Langmuir*, *19*(7), 3031–3034. doi:10.1021/la026739z.

Arriagada FJ, Osseo-Asare K. (1995). Synthesis of nanosize silica in aerosol OT reverse microemulsions. *Journal of Colloid and Interface Science*, *170*(1), 8–17. doi:https://doi.org/10.1006/jcis.1995.1064.

Arriagada FJ, Osseo-Asare K. (1999a). Controlled hydrolysis of tetraethoxysilane in a nonionic water-in-oil microemulsion: A statistical model of silica nucleation. *Colloids and Surfaces A*, *154*, 311–326. doi:10.1016/S0927-7757(98)00870-X.

Arriagada FJ, Osseo-Asare K. (1999b). Synthesis of nanosize silica in a nonionic water-in-oil microemulsion: Effects of the water/surfactant molar ratio and ammonia concentration. *Journal of Colloid and Interface Science*, *211*(2), 210–220. doi:https://doi.org/10.1006/jcis.1998.5985.

Arruebo M, Galán M, Navascués N, Téllez C, Marquina C, Ibarra MR, Santamaría J. (2006). Development of magnetic nanostructured silica-based materials as potential vectors for drug-delivery applications. *Chemistry of Materials*, *18*(7), 1911–1919. doi:10.1021/cm051646z.

Ashtari P, He X, Wang K, Gong P. (2005). An efficient method for recovery of target ssDNA based on amino-modified silica-coated magnetic nanoparticles. *Talanta*, *67*(3), 548–554. doi:10.1016/j.talanta.2005.06.043.

Bailey JK, Mecartney ML. (1992). Formation of colloidal silica particles from alkoxides. *Colloids and Surfaces*, *63*(1), 151–161. doi:https://doi.org/10.1016/0166-6622(92)80081-C.

Bailly M, Kontopoulou M, El Mabrouk K. (2010). Effect of polymer/filler interactions on the structure and rheological properties of ethylene-octene copolymer/nanosilica composites. *Polymer*, *51*, 5506–5515. doi:10.1016/j.polymer.2010.09.051.

Barik R, Sarkar R, Biswas P, Bera R, Sharma S, Nath S, Karmakar S, Sen T. (2016). 5,7-dihydroxy-2-(3-hydroxy-4, 5-dimethoxy-phenyl)-chromen-4-one-a flavone from Bruguiera gymnorrhiza displaying anti-inflammatory properties. *Indian Journal of Pharmacology*, *48*(3), 304–311. doi:10.4103/0253-7613.182890.

Benhamou A, Baudu M, Derriche Z, Basly JP. (2009). Aqueous heavy metals removal on amine-functionalized Si-MCM-41 and Si-MCM-48. *Journal of Hazardous Materials*, *171*(1), 1001–1008. doi:https://doi.org/10.1016/j.jhazmat.2009.06.106.

Bermejo-Deval R, Gounder R, Davis ME. (2012). Framework and extraframework tin sites in zeolite beta react glucose differently. *ACS Catalysis*, *2*(12), 2705–2713. doi:10.1021/cs300474x.

Bogush GH, Tracy MA, Zukoski CF. (1988). Preparation of monodisperse silica particles: Control of size and mass fraction. *Journal of Non-Crystalline Solids*, *104*(1), 95–106. doi:https://doi.org/10.1016/0022-3093(88)90187-1.

Bogush GH, Zukoski CF. (1991). Studies of the kinetics of the precipitation of uniform silica particles through the hydrolysis and condensation of silicon alkoxides. *Journal of Colloid and Interface Science*, *142*(1), 1–18. doi:https://doi.org/10.1016/0021-9797(91)90029-8.

Bonacchi S., Genovese D., Juris R., Montalti M., Prodi L., Rampazzo E., Sgarzi M., Zaccheron N.. (2010). Luminescent chemosensors based on silica nanoparticles. In: Prodi L., Montalti M., Zaccheroni N. (Eds.), *Luminescence Applied in Sensor Science* (Vol. 300). Berlin: Springer. doi:10.1007/128_2010_104.

Brugger J, Berenschot JW, Kuiper S, Nijdam W, Otter B, Elwenspoek M. (2000). Resistless patterning of sub-micron structures by evaporation through nanostencils. *Microelectronic Engineering.* *53*(1), 403–405. doi:https://doi.org/10.1016/S 0167-9317(00)00343-9.

Chen F, Goel S, Valdovinos HF, Luo H, Hernandez R, Barnhart TE, Cai W. (2015). In vivo integrity and biological fate of chelator-free zirconium-89-labeled mesoporous silica nanoparticles. *ACS Nano, 9*(8), 7950–7959. doi:10.1021/acsnano.5b00526.

Chen F, Hong H, Shi S, Goel S, Valdovinos HF, Hernandez R, Theuer CP, Barnhart TE, Cai W. (2014a). Engineering of hollow mesoporous silica nanoparticles for remarkably enhanced tumor active targeting efficacy. *Scientific Reports, 4,* 5080. doi:10.1038/srep05080.

Chen F, Hong H, Zhang Y, Valdovinos HF, Shi S, Kwon GS, Theuer CP, Barnhart TE, Cai W. (2013a). In vivo tumor targeting and image-guided drug delivery with antibody-conjugated, radio-labeled mesoporous silica nanoparticles. *ACS Nano, 7*(10), 9027–9039. doi:10.1021/nn403617j.

Chen F, Ma M, Wang J, Wang F, Chern SX, Zhao ER, Jhunjhunwala A, Darmadi S, Chen H, Jokerst JV. (2017a). Exosome-like silica nanoparticles: A novel ultrasound contrast agent for stem cell imaging. *Nanoscale, 9*(1), 402–411. doi:10.1039/c6nr08177k.

Chen F, Nayak TR, Goel S, Valdovinos HF, Hong H, Theuer CP, Barnhart TE, Cai W. (2014b). In vivo tumor vasculature targeted PET/NIRF imaging with TRC105(Fab)-conjugated, dual-labeled mesoporous silica nanoparticles. *Molecular Pharmaceutics, 11*(11), 4007–4014. doi:10.1021/mp500306k.

Chen P-J, Hu S-H, Hsiao C-S, Chen Y-Y, Liu D-M, Chen S-Y. (2011). Multifunctional magnetically removable nanogated lids of Fe3O4–capped mesoporous silica nanoparticles for intracellular controlled release and MR imaging. *Journal of Materials Chemistry, 21*(8), 2535. doi:10.1039/c0jm02590a.

Chen S, Osaka A, Hayakawa S, Tsuru K, Fujii E, Kawabata K. (2008). Novel one-pot sol–gel preparation of amino-functionalized silica nanoparticles. *Chemistry Letters, 37*(11), 1170–1171. doi:10.1246/cl.2008.1170.

Chen X, Lam KF, Zhang Q, Pan B, Arruebo M, Yeung KL. (2009). Synthesis of highly selective magnetic mesoporous adsorbent. *Journal of Physical Chemistry C, 113*(22), 9804–9813. doi:10.1021/jp9018052.

Chen X, Sun H, Hu J, Han X, Liu H, Hu Y. (2017b). Transferrin gated mesoporous silica nanoparticles for redox-responsive and targeted drug delivery. *Colloids and Surfaces B, 152,* 77–84. doi:10.1016/j.colsurfb.2017.01.010.

Chen Y, Chen H, Shi J. (2013b). In vivo bio-safety evaluations and diagnostic/therapeutic applications of chemically designed mesoporous silica nanoparticles. *Advanced Materials, 25*(23), 3144–3176. doi:10.1002/adma.201205292.

Chen Y, Yin Q, Ji X, Zhang S, Chen H, Zheng Y, Sun Y, Qu H, Wang Z, Li Y, Wang X, Zhang K, Zhang L, Shi J. (2012). Manganese oxide-based multifunctionalized mesoporous silica nanoparticles for pH-responsive MRI, ultrasonography and circumvention of MDR in cancer cells. *Biomaterials, 33*(29), 7126–7137. doi:10.1016/j.biomaterials.2012.06.059.

Chen Y-W, Peng Y-K, Chou S-W, Tseng Y-J, Wu P-C, Wang S-K, Lee Y-W, Shyue J-J, Hsiao J-K, Liu T-M, Chou P-T. (2017c). Mesoporous silica promoted deposition of bioinspired polydopamine onto contrast agent: A universal strategy to achieve both biocompatibility and multiple scale molecular imaging. *Particle & Particle Systems Characterization, 34*(6), 1600415. doi:10.1002/ppsc.201600415.

Chou SW, Shau YH, Wu PC, Yang YS, Shieh DB, Chen CC. (2010). In vitro and in vivo studies of FePt nanoparticles for dual modal CT/MRI molecular imaging. *Journal of the American Chemical Society, 132*(38), 13270–13278. doi:10.1021/ja1035013.

Corma A, Garcia H. (2004). *Zeolite-based photocatalysts. Chemical Communications,* (13), 1443–1459. doi:10.1039/B400147H.

Dapsens PY, Mondelli C, Pérez-Ramírez J. (2012). Biobased chemicals from conception toward industrial reality: Lessons learned and to be learned. *ACS Catalysis, 2*(7), 1487–1499. doi:10.1021/cs300124m.

Das M, Duan W, Sahoo SK. (2015). Multifunctional nanoparticle-EpCAM aptamer bioconjugates: A paradigm for targeted drug delivery and imaging in cancer therapy. *Nanomedicine: Nanotechnology, Biology, and Medicine, 11*(2), 379–389. doi:10.1016/j.nano.2014.09.002.

Deka JR, Lin Y-H, Kao H-M. (2014). Ordered cubic mesoporous silica KIT-5 functionalized with carboxylic acid groups for dye removal. *RSC Advances, 4*(90), 49061–49069. doi:10.1039/C4RA08819K.

Delacôte C, Gaslain FOM, Lebeau B, Walcarius A. (2009). Factors affecting the reactivity of thiol-functionalized mesoporous silica adsorbents toward mercury(II). *Talanta, 79*(3), 877–886. doi:10.1016/j.talanta.2009.05.020.

Díaz I, Márquez-Alvarez C, Mohino F, Pérez-Pariente J, Sastre E. (2000). Combined alkyl and sulfonic acid functionalization of MCM-41-type silica. *Journal of Catalysis, 193,* 283–294. doi:10.1006/jcat.2000.2899

Doadrio JC, Sousa EMB, Izquierdo-Barba I, Doadrio AL, Perez-Pariente J, Vallet-Regí M. (2006). Functionalization of mesoporous materials with long alkyl chains as a strategy for controlling drug delivery pattern. *Journal of Materials Chemistry, 16*(5), 462–466. doi:10.1039/b510101h.

Doane TL, Burda C. (2012). The unique role of nanoparticles in nanomedicine: Imaging, drug delivery and therapy. *Chemical Society Reviews, 41*(7), 2885–2911. doi:10.1039/c2cs15260f.

Dyba M, Hell SW. (2003). Photostability of a fluorescent marker under pulsed excited-state depletion through stimulated emission. *Applied Optics, 42*(25), 5123–5129.

Effati E, Pourabbas B. (2012). One-pot synthesis of sub-50nm vinyl- and acrylate-modified silica nanoparticles. *Powder Technology, 219,* 276–283. doi:https://doi.org/10.1016/j.powt ec.2011.12.062.

Fan W, Shen B, Bu W, Chen F, He Q, Zhao K, Zhang S, Zhou L, Peng W, Xiao Q, Ni D, Liu J, Shi J. (2014). A smart upconversion-based mesoporous silica nanotheranostic system for synergetic chemo-/radio-/photodynamic therapy and simultaneous MR/UCL imaging. *Biomaterials, 35*(32), 8992–9002. doi:10.1016/j.biomaterials.2014.07.024.

Feng J, Chang D, Wang Z, Shen B, Yang J, Jiang Y, Ju S, He N. (2014). A FITC-doped silica coated gold nanocomposite for both in vivo X-ray CT and fluorescence dual modal imaging. *RSC Advances4*(94), 51950–51959. doi:10.1039/c4ra09392e.

Fruijtier-Pölloth C. (2012). The toxicological mode of action and the safety of synthetic amorphous silica: A nanostructured material. *Toxicology, 294*(2), 61–79. doi:https://doi.org/10.1016/j .tox.2012.02.001.

Fryxell GE, Lin Y, Fiskum S, Birnbaum JC, Wu H, Kemner K, Kelly S. (2005). Actinide sequestration using self-assembled monolayers on mesoporous supports. *Environmental Science & Technology*, 39(5), 1324–1331. doi:10.1021/es049201j.

Fu C, Qiang L, Liang Q, Chen X, Li L, Liu H, Tan L, Liu T, Ren X, Meng X. (2015). Facile synthesis of a highly luminescent carbon dot@silica nanorattle for in vivo bioimaging. *RSC Advances*, 5(57), 46158–46162. doi:10.1039/c5ra04311e.

Gai S, Yang P, Li C, Wang W, Dai Y, Niu N, Lin J. (2010). Synthesis of magnetic, up-conversion luminescent, and mesoporous core-shell-structured nanocomposites as drug carriers. *Advanced Functional Materials*, 20(7), 1166–1172. doi:10.1002/adfm.200902274.

Gayubo AG, Aguayo AT, Atutxa A, Aguado R, Olazar M, Bilbao J. (2004). Transformation of oxygenate components of biomass pyrolysis oil on a HZSM-5 zeolite. II. Aldehydes, ketones, and acids. *Industrial & Engineering Chemistry Research*, 43(11), 2619–2626. doi:10.1021/ie030792g.

Gayubo AG, Aguayo AT, Atutxa A, Valle B, Bilbao J. (2005). Undesired components in the transformation of biomass pyrolysis oil into hydrocarbons on an HZSM-5 zeolite catalyst. *Journal of Chemical Technology & Biotechnology*, 80(11), 1244–1251. doi:10.1002/jctb.1316.

Girshick SL, Chiu CP, Muno R, Wu CY, Yang L, Singh SK, McMurry PH. (1993). Thermal plasma synthesis of ultrafine iron particles. *Journal of Aerosol Science*. 24(3), 367–382. doi:https://doi.org/10.1016/0021-8502(93)90009-X.

Green DL, Lin JS, Lam Y-F, Hu MZC, Schaefer DW, Harris MT. (2003). Size, volume fraction, and nucleation of Stober silica nanoparticles. *Journal of Colloid and Interface Science*, 266(2), 346–358. doi:https://doi.org/10.1016/S0021-9797(03)00610-6.

He Y, Li Z, Simone P, Lodge TP. (2006). Self-assembly of block copolymer micelles in an ionic liquid. *Journal of the American Chemical Society*, 128(8), 2745–2750. doi:10.1021/ja058091t.

Heidegger S, Gossl D, Schmidt A, Niedermayer S, Argyo C, Endres S, Bein T, Bourquin C. (2016). Immune response to functionalized mesoporous silica nanoparticles for targeted drug delivery. *Nanoscale*, 8(2), 938–948. doi:10.1039/c5nr06122a.

Hench LL, West JK. (1990). The sol-gel process. *Chemical Reviews*, 90(1), 33–72. doi:10.1021/cr00099a003.

Higashi K, Uchida K, Hotta A, Hishida K, Miki N. (2014). Micropatterning of silica nanoparticles by electrospray deposition through a stencil mask. *Journal of Laboratory Automation*, 19(1), 75–81. doi:10.1177/2211068213495205.

Hsiao JK, Tsai CP, Chung TH, Hung Y, Yao M, Liu HM, Mou CY, Yang CS, Chen YC, Huang DM. (2008). Mesoporous silica nanoparticles as a delivery system of gadolinium for effective human stem cell tracking. *Small*, 4(9), 1445–1452. doi:10.1002/smll.200701316.

Hu H, Arena F, Gianolio E, Boffa C, Di Gregorio E, Stefania R, Orio L, Baroni S, Aime S. (2016a). Mesoporous silica nanoparticles functionalized with fluorescent and MRI reporters for the visualization of murine tumors overexpressing alphavbeta3 receptors. *Nanoscale*, 8(13), 7094–7104. doi:10.1039/c5nr08878j.

Hu Y, Wang Y, Jiang J, Han B, Zhang S, Li K, Ge S, Liu Y. (2016b). Preparation and characterization of novel perfluorooctyl bromide nanoparticle as ultrasound contrast agent via layer-by-layer self-assembly for folate-receptor-mediated tumor imaging. *BioMed Research International*, *2016*, 6381464. doi:10.1155/2016/6381464.

Huang X, Li L, Liu T, Hao N, Liu H, Chen D, Tang F. (2011). The shape effect of mesoporous silica nanoparticles on biodistribution, clearance, and biocompatibility in vivo. *ACS Nano*, 5(7), 5390–5399. doi:10.1021/nn200365a.

Huang X, Zhang F, Lee S, Swierczewska M, Kiesewetter DO, Lang L, Zhang G, Zhu L, Gao H, Choi HS, Niu G, Chen X. (2012). Long-term multimodal imaging of tumor draining sentinel lymph nodes using mesoporous silica-based nanoprobes. *Biomaterials*, 33(17), 4370–4378. doi:10.1016/j.biomaterials.2012.02.060.

Huber GW, Iborra S, Corma A. (2006). Synthesis of transportation fuels from biomass: Chemistry, catalysts, and engineering. *Chemical Reviews*, 106(9), 4044–4098. doi:10.1021/cr068360d.

Hubert DHW, Jung M, Frederik PM, Bomans PHH, Meuldijk J, German AL. (2000). Vesicle-directed growth of silica. *Advanced Materials*, 12(17), 1286–1290. doi:10.1002/1521-4095(200009)12:17<1286::AID-ADMA1286>3.0.CO;2-7.

Innocenzi P, Kidchob T, Falcaro P, Takahashi M. (2008). Patterning techniques for mesostructured films. *Chemistry of Materials*, 20(3), 607–614. doi:10.1021/cm071784j.

Ishigaki T, Oh S-M, Li J-G, Park D-W. (2005). Controlling the synthesis of TaC nanopowders by injecting liquid precursor into RF induction plasma. *Science and Technology of Advanced Materials*, 6(2), 111–118. doi:10.1016/j.stam.2004.11.001.

Kang S, Il Hong S, Rim Choe C, Park M, Rim S, Kim J. (2001). Preparation and characterization of epoxy composites filled with functionalized nanosilica particles obtained via sol-gel process. *Polymer*, 42, 879–887. doi:10.1016/S0032-3861(00)00392-X.

Ke R, Yang W, Xia X, Xu Y, Li Q. (2010). Tandem conjugation of enzyme and antibody on silica nanoparticle for enzyme immunoassay. *Analytical Biochemistry*, 406(1), 8–13. doi:10.1016/j.ab.2010.06.039.

Kempster A. (1992). The principles and applications of chemical vapour deposition. *Transactions of the IMF*, 70(2), 68–75. doi:10.1080/00202967.1992.11870945.

Khosravian P, Shafiee Ardestani M, Khoobi M, Ostad SN, Dorkoosh FA, Akbari Javar H, Amanlou M. (2016). Mesoporous silica nanoparticles functionalized with folic acid/methionine for active targeted delivery of docetaxel. *OncoTargets and Therapy*, 9, 7315–7330. doi:10.2147/OTT.S113815.

Kickelbick G. (2003). Concepts for the incorporation of inorganic building blocks into organic polymers on a nanoscale. *Progress in Polymer Science*, 28(1), 83–114. doi:https://doi.org/10.1016/S0079-6700(02)00019-9.

Kim GM, Kim B, Brugger J. (2003). All-photoplastic microstencil with self-alignment for multiple layer shadow-mask patterning. *Sensors and Actuators A*, *107*, 132–136. doi:10.1016/S0924-4247(03)00298-X.

Kim J, Kim HS, Lee N, Kim T, Kim H, Yu T, Song IC, Moon WK, Hyeon T. (2008). Multifunctional uniform nanoparticles composed of a magnetite nanocrystal core and a mesoporous silica shell for magnetic resonance and fluorescence imaging and for drug delivery. *Angewandte Chemie*, 47(44), 8438–8441. doi:10.1002/anie.200802469.

Kim J, Lee JE, Lee J, Yu JH, Kim BC, An K, Hwang Y, Shin CH, Park JG, Kim J, Hyeon T. (2006). Magnetic fluorescent delivery vehicle using uniform mesoporous silica spheres embedded with monodisperse magnetic and semiconductor nanocrystals. *Journal of the American Chemical Society*, 128(3), 688–689. doi:10.1021/ja0565875.

Kim J, Piao Y, Hyeon T. (2009). Multifunctional nanostructured materials for multimodal imaging, and simultaneous imaging and therapy. *Chemical Society Reviews*, *38*(2), 372–390. doi:10.1039/b709883a.

Kim KS, Kim TH. (2019). Nanofabrication by thermal plasma jets: From nanoparticles to low-dimensional nanomaterials. *Journal of Applied Physics*, *125*(7), 070901. doi:10.1063/1.5060977.

Kim T, Momin E, Choi J, Yuan K, Zaidi H, Kim J, Park M, Lee N, McMahon MT, Quinones-Hinojosa A, Bulte JW, Hyeon T, Gilad AA. (2011). Mesoporous silica-coated hollow manganese oxide nanoparticles as positive T1 contrast agents for labeling and MRI tracking of adipose-derived mesenchymal stem cells. *Journal of the American Chemical Society*, *133*(9), 2955–2961. doi:10.1021/ja1084095.

Kisler JM, Gee ML, Stevens GW, O'Connor AJ. (2003). Comparative study of silylation methods to improve the stability of silicate MCM-41 in aqueous solutions. *Chemistry of Materials*, *15*(3), 619–624. doi:10.1021/cm0116018.

Klaus JW, Ott AW, Johnson JM, George SM. (1997). Atomic layer controlled growth of SiO2 films using binary reaction sequence chemistry. *Applied Physics Letters*, *70*(9), 1092–1094. doi:10.1063/1.118494.

Kneuer C, Sameti M, Haltner EG, Schiestel T, Schirra H, Schmidt H, Lehr CM. (2000). Silica nanoparticles modified with aminosilanes as carriers for plasmid DNA. *International Journal of Pharmaceutics*, *196*(2), 257–261.

Kobler J, Bein T. (2008). Porous thin films of functionalized mesoporous silica nanoparticles. *ACS Nano*, 2(11), 2324–2330. doi:10.1021/nn800505g.

Kuwahara Y, Maki K, Matsumura Y, Kamegawa T, Mori K, Yamashita H. (2009). Hydrophobic modification of a mesoporous silica surface using a fluorine-containing silylation agent and its application as an advantageous host material for the TiO2 photocatalyst. *Journal of Physical Chemistry C*, *113*(4), 1552–1559. doi:10.1021/jp809191v.

Lai C-W, Hsiao Y-H, Peng Y-K, Chou P-T. (2012). Facile synthesis of highly emissive carbon dots from pyrolysis of glycerol; gram scale production of carbon dots/mSiO2 for cell imaging and drug release. *Journal of Materials Chemistry*, *22*(29), 14403. doi:10.1039/c2jm32206d.

Lai CY, Trewyn BG, Jeftinija DM, Jeftinija K, Xu S, Jeftinija S, Lin VS. (2003). A mesoporous silica nanosphere-based carrier system with chemically removable CdS nanoparticle caps for stimuli-responsive controlled release of neurotransmitters and drug molecules. *Journal of the American Chemical Society*, *125*(15), 4451–4459. doi:10.1021/ja0286501.

Laskowski J, Kitchener JA. (1969). The hydrophilic–hydrophobic transition on silica. *Journal of Colloid and Interface Science*, *29*(4), 670–679. doi:https://doi.org/10.1016/0021-9797(69)90219-7.

Lee C-H, Cheng S-H, Wang Y-J, Chen Y-C, Chen N-T, Souris J, Chen C-T, Mou C-Y, Yang C-S, Lo L-W. (2009). Near-infrared mesoporous silica nanoparticles for optical imaging: Characterization and in vivo biodistribution. *Advanced Functional Materials* *19*(2), 215–222. doi:10.1002/adfm.200800753.

Lee JE, Lee N, Kim H, Kim J, Choi SH, Kim JH, Kim T, Song IC, Park SP, Moon WK, Hyeon T. (2010). Uniform mesoporous dye-doped silica nanoparticles decorated with multiple magnetite nanocrystals for simultaneous enhanced magnetic resonance imaging, fluorescence imaging, and drug delivery. *Journal of the American Chemical Society*, *132*(2), 552–557. doi:10.1021/ja905793q.

Lee SB, Kim HL, Jeong HJ, Lim ST, Sohn MH, Kim DW. (2013). Mesoporous silica nanoparticle pretargeting for PET imaging based on a rapid bioorthogonal reaction in a living body. *Angewandte Chemie*, *52*(40), 10549–10552. doi:10.1002/anie.201304026.

Lei J, Yang L, Lu D, Yan X, Cheng C, Liu Y, Wang L, Zhang J. (2015). Carbon dot-incorporated PMO nanoparticles as versatile platforms for the design of ratiometric sensors, multichannel traceable drug delivery vehicles, and efficient photocatalysts. *Advanced Optical Materials*, *3*(1), 57–63. doi:10.1002/adom.201400364.

Leparoux M, Schreuders C, Shin J-W, Siegmann S. (2005). Induction plasma synthesis of carbide nano-powders. *Advanced Engineering Materials*, *7*(5), 349–353. doi:10.1002/adem.200500046.

Li C, Yang D, Ma P, Chen Y, Wu Y, Hou Z, Dai Y, Zhao J, Sui C, Lin J. (2013). Multifunctional upconversion mesoporous silica nanostructures for dual modal imaging and in vivo drug delivery. *Small*, *9*(24), 4150–4159. doi:10.1002/smll.201301093.

Li G, Zhao Z, Liu J, Jiang G. (2011). Effective heavy metal removal from aqueous systems by thiol functionalized magnetic mesoporous silica. *Journal of Hazardous Materials*, *192*(1), 277–283. doi:10.1016/j.jhazmat.2011.05.015.

Li Z, Zhang J, Du J, Han B, Wang J. (2006). Preparation of silica microrods with nano-sized pores in ionic liquid microemulsions. *Colloids and Surfaces A*, *286*(1), 117–120. doi:https://doi.org/10.1016/j.colsurfa.2006.03.011.

Licausi N, Rao S, Bhat I. (2011). Low-pressure chemical vapor deposition of CdS and atomic layer deposition of CdTe films for HgCdTe surface passivation. *Journal of Electronic Materials*, *40*(8), 1668–1673. doi:10.1007/s11664-011-1640-y.

Lim HM, Lee J, Jeong J-H, Oh S-G, Lee S-H. (2010). Comparative study of various preparation methods of colloidal silica. *Engineering*, *2*(12), 998–1005. doi:10.4236/eng.2010.212126.

Liong M, Lu J, Kovochich M, Xia T, Ruehm SG, Nel AE, Tamanoi F, Zink JI. (2008). Multifunctional inorganic nanoparticles for imaging, targeting, and drug delivery. *ACS Nano*, *2*(5), 889–896. doi:10.1021/nn800072t.

Liu B, Zhang B, Cui Y, Chen H, Gao Z, Tang D. (2011). Multifunctional gold–silica nanostructures for ultrasensitive electrochemical immunoassay of streptomycin residues. *ACS Applied Materials & Interfaces*, *3*(12), 4668–4676. doi:10.1021/am201087r.

Liu F, Wen L-X, Li Z-Z, Yu W, Sun H-Y, Chen J. (2006). Porous hollow silica nanoparticles as controlled delivery system for water-soluble pesticide. *Materials Research Bulletin*, *41*, 2268–2275. doi:10.1016/j.materresbull.2006.04.014.

Liu HM, Wu SH, Lu CW, Yao M, Hsiao JK, Hung Y, Lin YS, Mou CY, Yang CS, Huang DM, Chen YC. (2008). Mesoporous silica nanoparticles improve magnetic labeling efficiency in human stem cells. *Small*, *4*(5), 619–626. doi:10.1002/smll.200700493.

Liu J, Bu W, Zhang S, Chen F, Xing H, Pan L, Zhou L, Peng W, Shi J. (2012). Controlled synthesis of uniform and monodisperse upconversion core/mesoporous silica shell nanocomposites for bimodal imaging. *Chemistry*, *18*(8), 2335–2341. doi:10.1002/chem.201102599.

Liu K, Wang ZQ, Wang SJ, Liu P, Qin YH, Ma Y, Li XC, Huo ZJ. (2015). Hyaluronic acid-tagged silica nanoparticles in colon cancer therapy: Therapeutic efficacy evaluation. *International Journal of Nanomedicine*, *10*, 6445–6454. doi:10.2147/IJN. S89476.

Liu T, Zhang N, Wang Z, Wu M, Chen Y, Ma M, Chen H, Shi J. (2017). Endogenous catalytic generation of O2 bubbles for in situ ultrasound-guided high intensity focused ultrasound ablation. *ACS Nano*, *11*(9), 9093–9102. doi:10.1021/ acsnano.7b03772.

Lu F, Wu SH, Hung Y, Mou CY. (2009). Size effect on cell uptake in well-suspended, uniform mesoporous silica nanoparticles. *Small*, *5*(12), 1408–1413. doi:10.1002/smll.200900005.

Lu J, Liong M, Li Z, Zink JI, Tamanoi F. (2010). Biocompatibility, biodistribution, and drug-delivery efficiency of mesoporous silica nanoparticles for cancer therapy in animals. *Small*, *6*(16), 1794–1805. doi:10.1002/smll.201000538.

Lu Y, Yang Y, Sellinger A, Lu M, Huang J, Fan H, Haddad R, Lopez G, Burns AR, Sasaki DY, Shelnutt J, Brinker CJ. (2001). Self-assembly of mesoscopically ordered chromatic polydi-acetylene/silica nanocomposites. *Nature*, *410*(6831), 913–917. doi:10.1038/35073544.

Luo T, Huang P, Gao G, Shen G, Fu S, Cui D, Zhou C, Ren Q. (2011). Mesoporous silica-coated gold nanorods with embedded indo-cyanine green for dual mode X-ray CT and NIR fluorescence imaging. *Optics Express*, *19*(18), 17030–17039. doi:10.1364/ OE.19.017030.

Ma M, Chen H, Chen Y, Wang X, Chen F, Cui X, Shi J. (2012). Au capped magnetic core/mesoporous silica shell nanoparticles for combined photothermo-/chemo-therapy and multimodal imaging. *Biomaterials*, *33*(3), 989–998. doi:10.1016/j.biomaterials.2011.10.017.

Macwan DP, Balasubramanian C, Dave PN, Chaturvedi S. (2014). Thermal plasma synthesis of nanotitania and its characterization. *Journal of Saudi Chemical Society*, *18*(3), 234–244. doi:https://doi.org/10.1016/j.jscs.2011.07.009.

Madani H, Bagheri A, Parhizkar T. (2012). The pozzolanic reactivity of monodispersed nanosilica hydrosols and their influence on the hydration characteristics of Portland cement. *Cement and Concrete Research*, *42*(12), 1563–1570. doi:https://doi.org /10.1016/j.cemconres.2012.09.004.

Malfatti L, Kidchob T, Costacurta S, Falcaro P, Schiavuta P, Amenitsch H, Innocenzi P. (2006). Highly ordered self-assembled mesostructured hafnia thin films: An example of rewritable mesostructure. *Chemistry of Materials*, *18*(19), 4553–4560. doi:10.1021/cm060236n.

Márquez-Alvarez C, Sastre E, Pérez-Pariente J. (2004). Solid catalysts for the synthesis of fatty esters of glycerol, polyglycerols and sorbitol from renewable resources. *Topics in Catalysis*, *27*(1), 105–117. doi:10.1023/B:TOCA.0000013545.81809.bd.

Masuda Y, Itoh T, Koumoto K. (2005). Self-assembly patterning of silica colloidal crystals. *Langmuir*, *21*(10), 4478–4481. doi:10.1021/la050075m.

Matsoukas T, Gulari E. (1988). Dynamics of growth of silica particles from ammonia-catalyzed hydrolysis of tetra-ethylorthosilicate. *Journal of Colloid and Interface Science*, *124*(1), 252–261. doi:https://doi.org/10.1016/0021-9797(88) 90346-3.

Mehnert CP, Cook RA, Dispenziere NC, Afeworki M. (2002). Supported ionic liquid catalysis: A new concept for homogeneous hydroformylation catalysis. *Journal of the American Chemical Society*, *124*(44), 12932–12933. doi:10.1021/ ja0279242.

Mendoza-Gonzalez NY, Goortani BM, Proulx P. (2007). Numerical simulation of silica nanoparticles production in an RF plasma reactor: Effect of quench. *Materials Science and Engineering: C*, *27*(5), 1265–1269. doi:https://doi.org/10.1016/j.msec.2006 .09.042.

Miller L, Winter G, Baur B, Witulla B, Solbach C, Reske S, Linden M. (2014). Synthesis, characterization, and biodistribution of multiple 89Zr-labeled pore-expanded mesoporous silica nanoparticles for PET. *Nanoscale*, *6*(9), 4928–4935. doi:10.1039/c3nr06800e.

Moelans D, Cool P, Baeyens J, Vansant EF. (2005). Immobilisation behaviour of biomolecules in mesoporous silica materials. *Catalysis Communications*, *6*(9), 591–595. doi:https://doi.org /10.1016/j.catcom.2005.05.007.

Muhammad F, Guo M, Qi W, Sun F, Wang A, Guo Y, Zhu G. (2011). pH-triggered controlled drug release from mesoporous silica nanoparticles via intracelluar dissolution of ZnO nanolids. *Journal of the American Chemical Society*, *133*(23), 8778–8781. doi:10.1021/ja200328s.

Mushtaq A, Jamil N, Rizwan S, Mandokhel F, Riaz M, Hornyak GL, Najam Malghani M, Naeem Shahwani M. (2018). Engineered silica nanoparticles and silica nanoparticles containing controlled release fertilizer for drought and saline areas. IOP Conference Series, *414*, 012029. doi:10.1088/1757-899x/414/1/012029.

Naka Y, Komori Y, Yoshitake H. (2010). One-pot synthesis of organo-functionalized monodisperse silica particles in W/O microemulsion and the effect of functional groups on addition into polystyrene. *Colloids and Surfaces A*, *361*, 162–168. .doi:10.1016/j.colsurfa.2010.03.034

Nakamura M, Hayashi K, Kubo H, Kanadani T, Harada M, Yogo T. (2017). Relaxometric property of organosilica nanoparticles internally functionalized with iron oxide and fluorescent dye for multimodal imaging. *Journal of Colloid and Interface Science*, *492*, 127–135. doi:10.1016/j.jcis.2017.01.004.

Nikolla E, Román-Leshkov Y, Moliner M, Davis ME. (2011). "One-Pot" Synthesis of 5-(Hydroxymethyl)furfural from Carbohydrates using Tin-Beta Zeolite. *ACS Catalysis*, *1*(4), 408–410. doi:10.1021/cs2000544.

Noji T, Kamidaki C, Kawakami K, Shen J-R, Kajino T, Fukushima Y, Sekitoh T, Itoh S. (2011). Photosynthetic oxygen evolution in mesoporous silica material: Adsorption of photosystem II reaction center complex into 23 nm nanopores in SBA. *Langmuir*, *27*(2), 705–713. doi:10.1021/la1032916.

Noll W. (1956). The colloid chemistry of silica and silicates, von Ralph K. Iler. Cornell University Press, Ithaca, New York. 1. Aufl. XII, 324 S., 20 Tab., 62 Abb. Gebd. $ 5.50. *Angewandte Chemie*, *68*(8), 312–312. doi:10.1002/ange.19560680824.

Oh IH, Min HS, Li L, Tran TH, Lee YK, Kwon IC, Choi K, Kim K, Huh KM. (2013). Cancer cell-specific photoactivity of pheophorbide a-glycol chitosan nanoparticles for photodynamic therapy in tumor-bearing mice. *Biomaterials*, *34*(27), 6454–6463. doi:10.1016/j.biomaterials.2013.05.017.

Omosebi O, Maheshwari H, Ahmed R, Shah S, Osisanya S, Hassani S, DeBruijn G, Cornell W, Simon D. (2016). Degradation of well cement in HPHT acidic environment: Effects of CO2 concentration and pressure. *Cement and Concrete Composites*, *74*, 54–70. doi:https://doi.org/10.1016/j.cemconcomp.2016.09.006.

Omosebi O, Maheshwari H, Ahmed R, Shah S, Osisanya S, Santra A, Saasen A. (2015). Investigating temperature effect on degradation of well cement in HPHT carbonic acid environment. *Journal of Natural Gas Science and Engineering*, *26*, 1344–1362. doi:https://doi.org/10.1016/j.jngse.2015.08.018.

Onda A, Ochi T, Yanagisawa K. (2008). Selective hydrolysis of cellulose into glucose over solid acid catalysts. *Green Chemistry*, *10*(10), 1033–1037. doi:10.1039/B808471H.

Onyekonwu MO, Ogolo NA. (2010). Investigating the use of nanoparticles in enhancing oil recovery. Paper presented at the Nigeria Annual International Conference and Exhibition, Tinapa - Calabar, Nigeria, 2010/1/1.

Palmai M, Petho A, Nagy LN, Klebert S, May Z, Mihaly J, Wacha A, Jemnitz K, Veres Z, Horvath I, Szigeti K, Mathe D, Varga Z. (2017). Direct immobilization of manganese chelates on silica nanospheres for MRI applications. *Journal of Colloid and Interface Science*, *498*, 298–305. doi:10.1016/j.jcis.2017.03.053.

Pan J, Wan D, Gong J. (2011). PEGylated liposome coated QDs/mesoporous silica core-shell nanoparticles for molecular imaging. *Chemical Communications*, *47*(12), 3442–3444. doi:10.1039/c0cc05520d.

Pan L, Liu J, He Q, Wang L, Shi J. (2013). Overcoming multidrug resistance of cancer cells by direct intranuclear drug delivery using TAT-conjugated mesoporous silica nanoparticles. *Biomaterials*, *34*(11), 2719–2730. doi:10.1016/j.biomaterials.2012.12.040.

Pandey S, Mewada A, Thakur M, Pillai S, Dharmatti R, Phadke C, Sharon M. (2014). Synthesis of mesoporous silica oxide/C-dot complex (meso-SiO2/C-dots) using pyrolysed rice husk and its application in bioimaging. *RSC Advances*, *4*(3), 1174–1179. doi:10.1039/c3ra45227a.

Patil RC, Deshpande A. (2012). Use of nanomaterials in cementing applications. Paper presented at the SPE International Oilfield Nanotechnology Conference and Exhibition, Noordwijk, The Netherlands, 2012/1/1/.

Peng YK, Lui CN, Lin TH, Chang C, Chou PT, Yung KK, Tsang SC. (2014). Multifunctional silica-coated iron oxide nanoparticles: A facile four-in-one system for in situ study of neural stem cell harvesting. *Faraday Discussions*, *175*, 13–26. doi:10.1039/c4fd00132j.

Peng Y-K, Lui CNP, Chen Y-W, Chou S-W, Raine E, Chou P-T, Yung KKL, Tsang SCE. (2017). Engineering of single magnetic particle carrier for living brain cell imaging: A tunable T1-/T2-/dual-modal contrast agent for magnetic resonance imaging application. *Chemistry of Materials*, *29*(10), 4411–4417. doi:10.1021/acs.chemmater.7b00884.

Perego C, Bosetti A. (2011). Biomass to fuels: The role of zeolite and mesoporous materials. *Microporous and Mesoporous Materials*, *144*(1), 28–39. doi:https://doi.org/10.1016/j.micromeso.2010.11.034.

Peters R, Kramer E, Oomen AG, Herrera Rivera ZE, Oegema G, Tromp PC, Fokkink R, Rietveld A, Marvin HJP, Weigel S, Peijnenburg AACM, Bouwmeester H. (2012). Presence of nano-sized silica during in vitro digestion of foods containing silica as a food additive. *ACS Nano*, *6*(3), 2441–2451. doi:10.1021/nn204728k.

Pham KN, Fullston D, Sagoe-Crentsil K. (2007). Surface modification for stability of nano-sized silica colloids. *Journal of Colloid and Interface Science*, *315*(1), 123–127. doi:10.1016/j.jcis.2007.06.064.

Qian HS, Guo HC, Ho PC, Mahendran R, Zhang Y. (2009). Mesoporous-silica-coated up-conversion fluorescent nanoparticles for photodynamic therapy. *Small*, *5*(20), 2285–2290. doi:10.1002/smll.200900692.

Qu F, Zhu G, Huang S, Li S, Sun J, Zhang D, Qiu S. (2006). Controlled release of Captopril by regulating the pore size and morphology of ordered mesoporous silica. *Microporous and Mesoporous Materials*, *92*(1–3), 1–9. doi:10.1016/j.micromeso.2005.12.004.

Rahman IA, Vejayakumaran P, Sipaut CS, Ismail J, Bakar MA, Adnan R, Chee CK. (2007). An optimized sol–gel synthesis of stable primary equivalent silica particles. *Colloids and Surfaces A*, *294*(1), 102–110. doi:https://doi.org/10.1016/j.colsurfa.2006.08.001.

Rieter WJ, Kim JS, Taylor KM, An H, Lin W, Tarrant T, Lin W. (2007). Hybrid silica nanoparticles for multimodal imaging. *Angewandte Chemie*, *46*(20), 3680–3682. doi:10.1002/anie.200604738.

Rossi LM, Shi L, Rosenzweig N, Rosenzweig Z. (2006). Fluorescent silica nanospheres for digital counting bioassay of the breast cancer marker HER2/neu [correction of HER2/nue]. *Biosensors & Bioelectronics*, *21*(10), 1900–1906. doi:10.1016/j.bios.2006.02.002.

Roy S, Hegde MS, Madras G. (2009). Catalysis for NOx abatement. *Applied Energy*, *86*(11), 2283–2297.

Ryu J, Kim W, Yun J, Lee K, Lee J, Yu H, Kim JH, Kim JJ, Jang J. (2018). Fabrication of uniform wrinkled silica nanoparticles and their application to abrasives in chemical mechanical planarization. *ACS Applied Materials & Interfaces*, *10*(14), 11843–11851. doi:10.1021/acsami.7b15952.

Sahu S, Sinha N, Bhutia SK, Majhi M, Mohapatra S. (2014). Luminescent magnetic hollow mesoporous silica nanotheranostics for camptothecin delivery and multimodal imaging. *Journal of Materials Chemistry B*, *2*(24), 3799–3808. doi:10.1039/c3tb21669a.

Santra AK, Boul P, Pang X. (2012). Influence of nanomaterials in oilwell cement hydration and mechanical properties. Paper presented at the SPE International Oilfield Nanotechnology Conference and Exhibition, Noordwijk, The Netherlands, 2012/1/1/.

Santucci V, Maury F, Senocq F. (2010). Vapor phase surface functionalization under ultra violet activation of parylene thin films grown by chemical vapor deposition. *Thin Solid Films*, *518*(6), 1675–1681. doi:https://doi.org/10.1016/j.tsf.2009.11.064.

Sathe TR, Agrawal A, Nie S. (2006). Mesoporous silica beads embedded with semiconductor quantum dots and iron oxide nanocrystals: dual-function microcarriers for optical encoding and magnetic separation. *Analytical Chemistry*, *78*(16), 5627–5632. doi:10.1021/ac0610309.

Schadler LS, Laut KO, Smith RW, Petrovicova E. (1997). Microstructure and mechanical properties of thermally sprayed silica/nylon nanocomposites. *Journal of Thermal Spray Technology*, *6*(4), 475–485. doi:10.1007/s11666-997-0034-4.

Sekhon BS. (2014). Nanotechnology in agri-food production: An overview. *Nanotechnology Science and Applications*, *7*, 31–53. doi:10.2147/NSA.S39406.

Sensoy T, Chenevert ME, Sharma MM. (2009). Minimizing water invasion in shales using nanoparticles. Paper presented at the SPE Annual Technical Conference and Exhibition, New Orleans, Louisiana, 2009/1/1/.

Shi G, Franzke T, Xia W, Sanchez MD, Muhler M. (2011). Highly dispersed MoO3/Al2O3 shell-core composites synthesized by CVD of Mo(CO)6 under atmospheric pressure. *Chemical Vapor Deposition, 17*(4–6), 162–169. doi:10.1002/cvde.201106909.

Shi S, Chen F, Cai W. (2013). Biomedical applications of functionalized hollow mesoporous silica nanoparticles: Focusing on molecular imaging. *Nanomedicine, 8*(12), 2027–2039. doi:10.2217/nnm.13.177.

Shigeta M, Watanabe T, Nishiyama H. (2004). Numerical investigation for nano-particle synthesis in an RF inductively coupled plasma. *Thin Solid Films, 457*(1), 192–200. doi:https://doi.org/10.1016/j.tsf.2003.12.020.

Shu H, Li X, Zhang Z. (2008). Surface modified nano-silica and its action on polymer. *Progress in Chemistry, 20*, 1509.

Siegmann S, Leparoux M, Rohr L. (2005). The role of nano-particles in the field of thermal spray coating technology. In *Opto-Ireland 2005: Nanotechnology and Nanophotonics* (Vol. 5824). doi:10.1117/12.605225.

Silva GA. (2004). Introduction to nanotechnology and its applications to medicine. *Surgical Neurology, 61*(3), 216–220. doi:10.1016/j.surneu.2003.09.036.

Slowing I, Trewyn BG, Lin VS. (2006). Effect of surface functionalization of MCM-41-type mesoporous silica nanoparticles on the endocytosis by human cancer cells. *Journal of the American Chemical Society, 128*(46), 14792–14793. doi:10.1021/ja0645943.

Song JT, Yang XQ, Zhang XS, Yan DM, Wang ZY, Zhao YD. (2015). Facile synthesis of gold nanospheres modified by positively charged mesoporous silica, loaded with near-infrared fluorescent dye, for in vivo x-ray computed tomography and fluorescence dual mode imaging. *ACS Applied Materials & Interfaces, 7*(31), 17287–17297. doi:10.1021/acsami.5b04359.

Sreejith S, Ma X, Zhao Y. (2012). Graphene oxide wrapping on squaraine-loaded mesoporous silica nanoparticles for bioimaging. *Journal of the American Chemical Society, 134*(42), 17346–17349. doi:10.1021/ja305352d.

Stöber W, Fink A, Bohn E. (1968). Controlled growth of monodisperse silica spheres in the micron size range. *Journal of Colloid and Interface Science, 26*(1), 62–69. doi:https://doi.org/10.1016/0021-9797(68)90272-5.

Suda Y, Ono T, Akazawa M, Sakai Y, Tsujino J, Homma N. (2002). Preparation of carbon nanoparticles by plasma-assisted pulsed laser deposition method: Size and binding energy dependence on ambient gas pressure and plasma condition. *Thin Solid Films, 415*(1), 15–20. doi:https://doi.org/10.1016/S0040-6090(02)00532-1.

Sun Y, Zhang Z, Wong CP. (2005). Study on mono-dispersed nano-size silica by surface modification for underfill applications. *Journal of Colloid and Interface Science, 292*(2), 436–444. doi:10.1016/j.jcis.2005.05.067.

Sundstrom DW, DeMichiell RL. (1971). Quenching processes for high temperature chemical reactions. *Industrial & Engineering Chemistry Process Design and Development, 10*(1), 114–122. doi:10.1021/i260037a021.

Suriyaprabha R, Karunakaran G, Yuvakkumar R, Rajendran V, Kannan N. (2014). Foliar application of silica nanoparticles on the phytochemical responses of maize (Zea mays L.) and its toxicological behavior. *Synthesis and Reactivity in Inorganic Metal-Organic and Nano-Metal Chemistry, 44*(8), 1128–1131. doi:10.1080/15533174.2013.799197.

Suzuki TM, Nakamura T, Fukumoto K, Yamamoto M, Akimoto Y, Yano K. (2008). Direct synthesis of amino-functionalized monodispersed mesoporous silica spheres and their catalytic activity for nitroaldol condensation. *Journal of Molecular Catalysis A, 280*(1–2), 224–232. doi:10.1016/j.molcata.2007.11.012.

Taarning E, Osmundsen CM, Yang X, Voss B, Andersen SI, Christensen CH. (2011). Zeolite-catalyzed biomass conversion to fuels and chemicals. *Energy & Environmental Science, 4*(3), 793–804. doi:10.1039/C004518G.

Taboada E, Solanas R, Rodríguez E, Weissleder R, Roig A. (2009). Supercritical-fluid-assisted one-pot synthesis of biocompatible core(γ-Fe2O3)/shell(SiO2) nanoparticles as high relaxivity T2-contrast agents for magnetic resonance imaging. *Advanced Functional Materials, 19*(14), 2319–2324. doi:10.1002/adfm.200801681.

Tang L, Yang X, Dobrucki LW, Chaudhury I, Yin Q, Yao C, Lezmi S, Helferich WG, Fan TM, Cheng J. (2012). Aptamer-functionalized, ultra-small, monodisperse silica nanoconjugates for targeted dual-modal imaging of lymph nodes with metastatic tumors. *Angewandte Chemie, 51*(51), 12721–12726. doi:10.1002/anie.201205271.

Tao C, Li J. (2005). Morphosynthesis of microskeletal silica spheres templated by W/O microemulsion. *Colloids and Surfaces A, 256*, 57–60. doi:10.1016/j.colsurfa.2004.09.020.

Taylor KM, Kim JS, Rieter WJ, An H, Lin W, Lin W. (2008). Mesoporous silica nanospheres as highly efficient MRI contrast agents. *Journal of the American Chemical Society, 130*(7), 2154–2155. doi:10.1021/ja710193c.

Thomas JM, Raja R. (2006). The advantages and future potential of single-site heterogeneous catalysts. *Topics in Catalysis, 40*(1), 3–17. doi:10.1007/s11244-006-0105-7.

Valtchev V, Tosheva L. (2013). Porous nanosized particles: Preparation, properties, and applications. *Chemical Reviews, 113*(8), 6734–6760. doi:10.1021/cr300439k.

Van de Vyver S, Geboers J, Jacobs PA, Sels BF. (2011). Recent advances in the catalytic conversion of cellulose. *ChemCatChem, 3*(1), 82–94. doi:10.1002/cctc.201000302.

Vansant E.F., Van Der Voort P., Vrancken K.C. . (1995). *Characterization and Chemical Modification of the Silica Surface* (Vol. 93). 1st ed. Elsevier Science.

Vejayakumaran P, Rahman IA, Sipaut CS, Ismail J, Chee CK. (2008). Structural and thermal characterizations of silica nanoparticles grafted with pendant maleimide and epoxide groups. *Journal of Colloid and Interface Science, 328*(1), 81–91. doi:10.1016/j.jcis.2008.08.054.

Walcarius A, Mercier L. (2010). Mesoporous organosilica adsorbents: Nanoengineered materials for removal of organic and inorganic pollutants. *Journal of Materials Chemistry, 20*(22), 4478–4511. doi:10.1039/B924316J.

Wang J, Liu G, Engelhard MH, Lin Y. (2006). Sensitive immunoassay of a biomarker tumor necrosis factor-alpha based on poly(guanine)-functionalized silica nanoparticle label. *Analytical Chemistry, 78*(19), 6974–6979. doi:10.1021/ac060809f.

Wang P. (2009). Multi-scale features in recent development of enzymic biocatalyst systems. *Applied Biochemistry and Biotechnology*, *152*(2), 343–352. doi:10.1007/s12010-008-8243-y.

Wang S, Li Y, Fei X, Sun M, Zhang C, Li Y, Yang Q, Hong X. (2011a). Preparation of a durable superhydrophobic membrane by electrospinning poly (vinylidene fluoride) (PVDF) mixed with epoxy-siloxane modified SiO2 nanoparticles: A possible route to superhydrophobic surfaces with low water sliding angle and high water contact angle. *Journal of Colloid and Interface Science*, *359*(2), 380–388. doi:10.1016/j.jcis.2011.04.004.

Wang X, Chen H, Chen Y, Ma M, Zhang K, Li F, Zheng Y, Zeng D, Wang Q, Shi J. (2012). Perfluorohexane-encapsulated mesoporous silica nanocapsules as enhancement agents for highly efficient high intensity focused ultrasound (HIFU). *Advanced Materials*, *24*(6), 785–791. doi:10.1002/adma.201104033.

Wang X-D, Shen Z-X, Sang T, Cheng X-B, Li M-F, Chen L-Y, Wang Z-S. (2010). Preparation of spherical silica particles by Stöber process with high concentration of tetra-ethyl-orthosilicate. *Journal of Colloid and Interface Science*, *341*(1), 23–29. doi:https://doi.org/10.1016/j.jcis.2009.09.018.

Wang Y, Sun Y, Wang J, Yang Y, Li Y, Yuan Y, Liu C. (2016). Charge-reversal APTES-modified mesoporous silica nanoparticles with high drug loading and release controllability. *ACS Applied Materials & Interfaces*, *8*(27), 17166–17175. doi:10.1021/acsami.6b05370.

Wang Z, Shoji M, Ogata H. (2011b). Carbon nanosheets by microwave plasma enhanced chemical vapor deposition in CH4–Ar system. *Applied Surface Science*, *257*(21), 9082–9085. doi:https://doi.org/10.1016/j.apsusc.2011.05.104.

Wartenberg N, Fries P, Raccurt O, Guillermo A, Imbert D, Mazzanti M. (2013). A gadolinium complex confined in silica nanoparticles as a highly efficient T1/T2 MRI contrast agent. *Chemistry*, *19*(22), 6980–6983. doi:10.1002/chem.201300635.

Wegner K, Walker B, Tsantilis S, Pratsinis S. (2002). Design of metal nanoparticle synthesis by vapor flow condensation. *Chemical Engineering Science*, *57*, 1753–1762. doi:10.1016/S0009-2509(02)00064-7.

Wei B, Song S, Cao H. (2011). Strengthening of basalt fibers with nano-SiO2–epoxy composite coating. *Materials & Design*, *32*(8), 4180–4186. doi:https://doi.org/10.1016/j.matdes.2011.04.041.

Welton T. (1999). Room-temperature ionic liquids. Solvents for synthesis and catalysis. *Chemical Reviews*, *99*(8), 2071–2084. doi:10.1021/cr980032t.

Wu C-W, Aoki T, Kuwabara M. (2004). Electron-beam lithography assisted patterning of surfactant-templated mesoporous thin films. *Nanotechnology*, *15*(12), 1886–1889. doi:10.1088/0957-4484/15/12/035.

Xia D, Li D, Ku Z, Luo Y, Brueck SR. (2007). Top-down approaches to the formation of silica nanoparticle patterns. *Langmuir*, *23*(10), 5377–5385. doi:10.1021/la7005666.

Xu QF, Wang JN, Smith IH, Sanderson KD. (2009). Superhydrophobic and transparent coatings based on removable polymeric spheres. *Journal of Materials Chemistry*, *19*(5), 655–660. doi:10.1039/B812659C.

Xu S, Yu Y, Gao Y, Zhang Y, Li X, Zhang J, Wang Y, Chen B. (2018). Mesoporous silica coating NaYF4:Yb,Er@NaYF4 upconversion nanoparticles loaded with ruthenium(II) complex nanoparticles: Fluorometric sensing and cellular imaging

of temperature by upconversion and of oxygen by downconversion. *Mikrochimica Acta*, *185*(10), 454. doi:10.1007/s00604-018-2965-5.

Xue S, Wang Y, Wang M, Zhang L, Du X, Gu H, Zhang C. (2014). Iodinated oil-loaded, fluorescent mesoporous silica-coated iron oxide nanoparticles for magnetic resonance imaging/computed tomography/fluorescence trimodal imaging. *International Journal of Nanomedicine*, *9*, 2527–2538. doi:10.2147/IJN.S59754.

Yang H-H, Qu H-Y, Lin P, Li S-H, Ding M-T, Xu J-G. (2003). Nanometer fluorescent hybrid silica particle as ultrasensitive and photostable biological labels. *Analyst*, *128*(5), 462–466. doi:10.1039/b210192k.

Yang X, He D. (2010). Rapid determination of banned Sudan I in foodstuffs using a mesoporous SiO2 modified electrode. *Journal of AOAC International*, *93*(5), 1537–1541.

Yantasee W, Lin Y, Fryxell GE, Busche BJ, Birnbaum J. (2003). Removal of heavy metals from aqueous solution using novel nanoengineered sorbents: Self-assembled carbamoylphosphonic acids on mesoporous silica. *Separation Science and Technology*, *38*, 3809–3825. doi:10.1081/SS-120024232.

Yao KS, Li SJ, Tzeng KC, Cheng TC, Chang CY, Chiu CY, Liao CY, Hsu JJ, Lin ZP. (2009). Fluorescence silica nanoprobe as a biomarker for rapid detection of plant pathogens. Paper presented at the International Conference on Multifunctional Materials and Structures, Qingdao, China.

Yildirim A, Chattaraj R, Blum NT, Goodwin AP. (2016). Understanding acoustic cavitation initiation by porous nanoparticles: Toward nanoscale agents for ultrasound imaging and therapy. *Chemistry of Materials*, *28*(16), 5962–5972. doi:10.1021/acs.chemmater.6b02634.

Yu YQ, Chen C-Y, Chen W-C. (2003). Synthesis and characterization of organic-inorganic hybrid thin films from poly(acrylic) and monodispersed colloidal silica. *Polymers*, *44*(3), 593–601. doi:10.1016/S0032-3861(02)00824-8.

Yu-Shen Lin YH, Jen-KuanSu, Rain Lee, ChenChang,Meng-LiangLin, and, Mou C-Y. (2004). Gadolinium(III)-incorporated nanosized mesoporous silica as potential magnetic resonance imaging contrast agents. *Journal of Physical Chemistry B*, *108*, 15608–15611.

Zaccariello G, Moretti E, Storaro L, Riello P, Canton P, Gombac V, Montini T, Rodríguez-Castellón E, Benedetti A. (2014). TiO2–mesoporous silica nanocomposites: Cooperative effect in the photocatalytic degradation of dyes and drugs. *RSC Advances*, *4*(71), 37826–37837. doi:10.1039/C4RA06767C.

Zhang J, Yu M, Yuan P, Lu G, Yu C. (2011a). Controlled release of volatile (–)-menthol in nanoporous silica materials. *Journal of Inclusion Phenomena and Macrocyclic Chemistry*, *71*(3), 593–602. doi:10.1007/s10847-011-9996-4.

Zhang K, Chen H, Li F, Wang Q, Zheng S, Xu H, Ma M, Jia X, Chen Y, Mou J, Wang X, Shi J. (2014). A continuous triphase transition effect for HIFU-mediated intravenous drug delivery. *Biomaterials*, *35*(22), 5875–5885. doi:10.1016/j.biomaterials.2014.03.043.

Zhang T, Espinosa D, Yoon KY, Rahmani AR, Yu H, Caldelas FM, Ryoo S, Roberts M, Prodanovic M, Johnston KP, Milner TE, Bryant SL, Huh C. (2011b). Engineered nanoparticles as harsh-condition emulsion and foam stabilizers and as novel sensors. Paper presented at the Offshore Technology Conference, Houston, Texas, USA, 2011/1/1/.

Zhang T, Roberts M, Bryant SL, Huh C. (2009). Foams and emulsions stabilized with nanoparticles for potential conformance control applications. Paper presented at the SPE International Symposium on Oilfield Chemistry, The Woodlands. Texas, 2009/1/1/.

Zhao M, Zheng L, Li N, Yu L. (2008). Fabrication of hollow silica spheres in an ionic liquid microemulsion. *Materials Letters, 62,* 4591–4593. .doi:10.1016/j.matlet.2008.08.047

Zhao P, Yu J, Liu S, Yan M, Zang D, Gao L. (2012). One novel chemiluminescence sensor for determination of fenpropathrin based on molecularly imprinted porous hollow microspheres. *Sensors and Actuators B, 162,* 166–172. doi:10.1016/j.snb.2011.12.062.

Zhou C-H, Xia X, Lin C-X, Tong D-S, Beltramini J. (2011). Catalytic conversion of lignocellulosic biomass to fine chemicals and fuels. *Chemical Society Reviews, 40*(11), 5588–5617. doi:10.1039/C1CS15124J.

Zhou M, Ge X, Ke DM, Tang H, Zhang JZ, Calvaresi M, Gao B, Sun L, Su Q, Wang H. (2019). The bioavailability, biodistribution, and toxic effects of silica-coated upconversion nanoparticles in vivo. *Frontiers in Chemistry, 7,* 218. doi:10.3389/fchem.2019.00218.

Zhu Y, Shi J, Chen H, Shen W, Dong X. (2005). A facile method to synthesize novel hollow mesoporous silica spheres and advanced storage property. *Microporous and Mesoporous Materials, 84*(1–3), 218–222. doi:10.1016/j.micromeso.2005.05.001.

Zoldesi CI, Imhof A. (2005). Synthesis of monodisperse colloidal spheres, capsules, and microballoons by emulsion templating. *Advanced Materials, 17*(7), 924–928. doi:10.1002/adma.200401183.

Zou R, Gong S, Shi J, Jiao J, Wong K-L, Zhang H, Wang J, Su Q. (2017). Magnetic-NIR persistent luminescent dual-modal ZGOCS@MSNs@Gd2O3 core–shell nanoprobes for in vivo imaging. *Chemistry of Materials, 29*(9), 3938–3946. doi:10.1021/acs.chemmater.7b00087.

12

Biosynthetic Route for the Functionalization of Nanomaterials

Faisal Ahmad and Shamim Ahmad

CONTENTS

12.1 Introduction

The progress made in developing a large variety of nanostructured materials involving inorganic, organic, polymeric, and biomolecular species have clearly established them as possessing distinctly different physicochemical properties in contrast to their bulk counterparts. The associated quantum confinement effects have provided some special features like increased electron density, strong optical absorptions, fluorescent, and phosphorescent behavior of quantum dots and nanoparticulate-doped oxides, respectively, along with magnetic behavior of the iron oxide and cobalt in their colloidal suspensions. In addition, the application-specific surface functionalization schemes have also been explored extensively in parallel for enabling them to participate in site-specific interactions with the molecular complexes based on the phenomenon of molecular recognition in the surrounding environment. Such highly reactive features of these nanostructured species are currently being put to use in biomedicine and bioimaging applications besides many others, as discussed later (Table 12.1).

In continuation, appropriate syntheses protocols were also developed successfully to grow nanorods/nanotubes, and nanosheets in varying configurations, resulting in quantum confinement of the electron states in 1- and 2-D, respectively. Such confinements in lower dimensions, in contrast to those of their 3-D-confined nanoparticulate material states, could offer more additional features for their applications in active electron devices and circuits subsequently. Moreover, a large variety of nanocomposites involving these nanostructured material species in 1-, 2-, and 3-D have also been examined for preparing their synergistic combinations with the different intrinsic properties of each constituent. Of late, these nanostructured species have been studied for their biomimetic conjugations with numerous types of complex biomolecules, in which the weak chemical bonds established are expected to incorporate some possible intelligent behavior capable of detecting the presence of the external species in minute quantity, besides participating in biochemical

actions as well. Thus, in the last two decades, it has already been established theoretically and experimentally that quantum confinement in 1-, 2-, and 3-D would certainly be of significant help in realizing programmable features involving these engineered synthetic material building blocks and the numerous types of molecular species as linkers to incorporate unprecedented properties in these nanomaterials in different configurations including superlattice structure – a purely synthetic materials family.

A large number of nanoparticulate species of varying elemental compositions, morphologies, and physical/chemical properties are currently available commercially, with their hydrophilic and hydrophobic surfaces rendering them stable colloidal forms after their dispersion in aqueous and organic solvents, respectively. However, the amphiphilic nanospecies form stable dispersions in both types of solvents. In contrast to the nanoparticulate species as mentioned above, even clusters comprising of smaller size particles of a few to a few hundreds of atoms/molecules have also been realized similar to quasi-molecules. These topologically different nanoparticulate species comprising of different constituents along with their application-specific surface functionalizations are currently being explored as 'engineered building blocks' for future nanomaterial syntheses with an advantage. Generally, creating electrostatic repulsive forces during bottom-up type synthesis of nanoparticles involving some suitable surfactant molecules present there on the surfaces of the growing nuclei try to stabilize them against further aggregation. Various parameters that control the overall growth of the nan-crystallites resulting in to a specific type of morphology in the presence of surfactant molecules used are, in general, not only influenced by the constituent species but also the behavior of surfactant molecules as a solvent or reducing agent, particularly in the case of inorganic nanoparticles. A fairly large number of nanoparticulate material species have already been extensively studied and reviewed, particularly including nanoparticles of gold, silver, and magnetic nanoparticles, along with semiconducting quantum dots as described in the enclosed references covering all the aspects mentioned above in more detail (Murray et al. 2000;

TABLE 12.1

Polymeric Nanoparticulate Formulations of Nanomedicines

Compositions	Name	Manufacturer	Year
PEGylated Adenosine Deaminase Enzyme	Adagen®	Sigma-Tau Pharma	**1990**
Conjugate PEGylated L-Asparaginase	Oncaspar®	Enzon Pharma	**1994**
Copolymer - L-glutamate/alanine/lysine/tyrosine	Copaxone®	Teva	**1996**
Poly(allylamine hydrochloride)	Renagel®	Sanofi	**2000**
PEGylated IFN alpha-2b Protein	PegIntron®	Merck	**2001**
Leuprolide Acetate, PLGH [poly(dilactidecoglycolide)]	Eligard®	Tolmar	**2002**
PEGylated GCSF Protein	Neulasta®	Amgen	**2002**
PEGylated IFN Alpha-2a Protein	Pegasys®	Genentech	**2002**
PEGylated HGH Receptor Antagonist	Somavert®	Pfizer	**2003**
PEGylated Anti-VEGF Aptamer	Macugen®	Bausch & Lomb	**2004**
Synthesized ESA Methoxy PEG Glycol-epoetin β	Mircera®	Hoffman-LaRoche	**2007**
PEGylated Porcine-Likeuricase	Krystexxa®	Horizon	**2010**
PEGylated Antibody Fragment	Cimzia®	UCB	**2013**
Polymer/Protein/Conjugate (PEGylated IFNbeta-1a)	Pelegridy	Biogen	**2014**
Polymer/Protein/Conjugate (PEGylated factor VII	ADYNOVATE	Baxalta	**2015**

Masala and Seshadri 2004; Daniel and Astruc 2004; van Embden et al. 2007; Rogach et al. 2007; Schmid 2008; Reiss et al. 2009; Sperling and Parak 2010; Da et al. 2012; Murphy et al. 2015; Upadhyay and Verma 2015; Ahmad 2015; Wu et al. 2015; Ahmad 2016; Sengani et al. 2016; Payne et al. 2016; Shan et al. 2016; Pulit-Prociak and Banach 2016).

The possibilities of oriented growths were also confirmed experimentally by involving strong binding of certain ligands to some specific crystal facets of the nanoparticulate materials, with the help of appropriate surfactant molecules used in their syntheses. It has been noted that composite types of nanoparticulate species having the distinct domains of different materials were synthesized successfully in the case of pairs of constituents like CdS/FePt, Co/CdSe, and similar others having fluorescent and magnetic properties, as studied in-depth by the several groups involved in these investigations, as seen in the enclosed references (Sun and Xia 2002; Wiley et al. 2005; Xia et al. 2012, 2015; Niu et al. 2017; Chen et al. 2017; Chung et al. 2017; Kumar et al. 2017; Ahmad et al. 2017; Zhan et al. 2011; Sharma et al. 2017; Siddiqi and Husen 2017; Duan and Wang 2013; Banadaki and Kajbafvala 2014; Gilroy et al. 2016; Wang et al. 2016a; Boles et al. 2016; Jin et al. 2016; Tan et al. 2017).

Different kinds of naturally derived bio-membranes were found highly promising while exploring for some suitable surface functionalization of the nanoparticulate species. Especially, the chemical moieties intrinsically present on such membranes, e.g. those derived from the cell surfaces, were found extremely useful in imparting long-duration blood circulation, evading the immune system, along with targeting-affinity features without requiring any additional surface engineering for the required functionalities during their bottom-up syntheses. This strategy demonstrated a number of additional features including biodetoxification, antibacterial vaccination, antibiotic delivery, photo-thermal therapy, and cancer immunotherapy, to name a few cases. It was found interesting to conclude from such experiments that different types of bio-membranes prepared from a number of cell types could be used in bio-functionalizing the nanoparticulate surfaces with specific advantages associated

therein, as discussed by several research groups (Yoo et al. 2011; Hu et al. 2011, 2012, 2013a–c, 2015; Fang et al. 2012, 2014; Gao et al. 2013b, 2015; Parodi et al. 2013; Luk et al. 2014; Copp et al. 2014; Piao et al. 2014; Pang et al. 2015; Zhang et al. 2015). In this context, red blood cells (RBCs) were found as good sources for membrane coating, offering themselves as natural long-circulating carriers. Surface functionalization of NPs with RBC membrane was thus found very useful in promoting immune evasion, resulting in enhanced circulation time. However, in all such NP-based chemotherapies, the major emphasis was on safety and immune-compatibility. For example, membranes derived from RBC-cells were coated onto poly (lactic-*co*-glycolic acid) (PLGA)-NP-cores were found providing the improved drug loading facilitated by biocompatible cell membranes. Subsequently, RBC-membrane coated nanoparticles were successfully employed in the targeted delivery of doxorubicin (DOX) in a lymphoma disease model of a mouse, as discussed by several research groups in detail, and reflected in the enclosed references (Oldenborg et al. 2000; Rodriguez et al. 2013; Davis et al. 2008; Peer et al. 2007; Ishida et al. 2003; Knop et al. 2010; Aryal et al. 2013; Luk et al. 2016).

Features like prolonged blood circulation, along with targeting capability to tumor tissues, proper binding to the cancer cells, and their easier entry into the cell cytosol are found necessary in the cancer treatments employing nanoparticulate drug delivery carriers. For achieving the above-mentioned features in an identified nanoparticulate drug delivery carrier, it became necessary to have proper control over their morphology and surface properties to take into consideration while deciding the functionalization scheme. For instance, poly (ethylene glycol) (PEG) functionalized nanoparticles were not only found to be hydrophilic and flexible but also possessing prolonged circulations, which facilitated their tumor-accumulation via enhanced penetration and retention. Such long-living circulating nanocarriers, when decorated with functional moieties involving targeting ligands and cell penetrating peptides, indeed enhanced their adhesion and cell uptake. For incorporating multiple functionalities in the currently synthesized nanoparticulate drug delivery systems, it requires using complex chemical synthesis routes for optimizing their final properties. It was further observed that even PEGylated nanocarriers with intrinsic neutral behavior, presumed in earlier experiments, were not found to be compatible biologically, resulting in their accelerated clearance from blood circulation, as observed by the several researchers involved in these exhaustive studies covered in the cited publications mentioned together (Tanaka et al. 2004; Gupta et al. 2005; Ishida et al. 2006; Peer et al. 2007; Yu et al. 2007; Davis et al. 2008; Li and Huang 2008; Gratton et al. 2008; Jain and Stylianopoulos 2010; Knop et al. 2010; Danhier et al. 2010; Wong et al. 2011; Fang et al. 2011; Albanese et al. 2012; Dubey et al. 2012; Iversen et al. 2013; Abu Lila et al. 2013; Garcia et al. 2015; Sun et al. 2015; Liu et al. 2016; Wang et al. 2016b; Bose et al. 2016; Palivan et al. 2016).

Deploying the natural material sources for preparing nanoparticulate drug delivery carriers was, accordingly, found much easier for preparing their engineered surface-coatings, especially when their surfaces employed bio-membranes derived from cells or biovesicles. These bio-membrane-coated nanoparticulate drug carriers were found to be not only biocompatible, but also imparting prolonged circulation and/or tumor targeting properties depending on the nature of their specific cell-membranes as reported elsewhere in detail by several research groups worldwide (Langer and Tirrell 2004; Yu et al. 2007; Davis et al. 2008; Li and Huang 2008; Gratton et al. 2008; Jain and Stylianopoulos 2010; Knop et al. 2010; Danhier et al. 2010; Hu et al. 2011; Yoo et al. 2011; Wong et al. 2011; Fang et al. 2011; Albanese et al. 2012; Dubey et al. 2012; Gao et al. 2013a; Iversen et al. 2013; Gao[l] and Zhang 2015a; Hsieh et al. 2015; Luk and Zhang 2015; Wegst et al. 2015; Bose et al. 2016; Palivan et al. 2016; Dehaini et al. 2017; Kroll et al. 2017).

A number of successful attempts were made in developing vaccinations and detoxification processes involving cell-camouflaged nanoparticulate systems besides using them in cardiovascular treatment and cancer management. Various aspects of this important area of drug development were reviewed recently by discussing the techniques of membrane collection, core-particle coatings, and their detailed characterizations meant for their proper applications in tumor treatment (Hu et al. 2011; Yoo et al. 2011; Gao et al. 2013a; Luk and Zhang 2015; Gao and Zhang 2015b; Dehaini et al. 2017; Kroll et al. 2017; Zhai et al. 2017).

The basic principle involved in surface modification of nanoparticulate species of organic, inorganic, and biological origin has been schematically described in Figure 12.1 using biomembrane coating, along with various attachments of the other ligands and chemical moieties in brief.

12.2 Nanoparticulate Species in the Surrounding Environment

For understanding the nature of the biosynthesis route of surface functionalization of a nanoparticulate species, it is necessary to examine various interactions taking place at the interface between the NP-surface and the ionic, and the neutral molecular species present in the surrounding biofluid. The basic parameters are those particularly involved and related to the surface hydrophobicity, hydrophilicity, charge distributions, screening of the surface charges by oppositely charged ionic species drawn from the immediate surrounding, disturbing the pattern of ionic

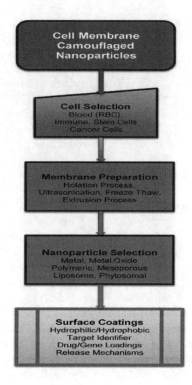

FIGURE 12.1 Principle involved in surface modification of nanoparticulate species of organic, inorganic, and biological origin.

concentration in the environment. Such interactions are bound to influence the physical and chemical behavior of both the constituents – the nanoparticulate system and the surrounding biofluid. Some of the specific examples of nanoparticulate systems with and without ligands are highlighted in brief based on the detailed discussions mentioned elsewhere (Pfeiffer et al. 2014).

Various kinds of influences anticipated under the practical circumstances might include repulsive/attractive interactions among the protein molecules and charged NPs or electrostatic attraction due to oppositely charged NPs and protein molecules after adding the nanoparticulate assemblies in the biofluids. Locally conjugated hydrophilic and hydrophobic molecules to the surfaces of NPs give rise to protein adsorptions known as protein corona affecting both the NPs and local environment. Introducing such proteins causes simultaneous changes in the hydrodynamic size of the NPs and their colloidal dispersions. For instance, any increase in local protein concentration on the surfaces of the NPs via protein adsorption might induce structural modifications resulting in protein destructions. This would eventually lead to protein-depleted solution around the NPs due to local charges and hydrophobicity, and thus further protein adsorption would increase the size of the NPs accordingly. However, the situation becomes different in the case of mono-dispersed colloidal NPs, in which a monolayer of protein-like human bovine serum or transferrin created a saturated condition by reducing the adsorptions of the unwanted proteins. PEGylated NPs, however, showed diminished non-specific protein binding. Similarly, introducing solution-dissociating salts was found to be disturbing to the total system of surface functionalized-NPs in the biofluid. Nanoparticulate assemblies were stabilized either by electrostatic or steric repulsions to prevent their agglomerations by van der Waal's attraction. However, the screening of the charges present on the NPs by the ions produced by the salt dissociation could induce colloidal instability at high concentrations, followed by agglomeration as discussed in detail elsewhere (Pellegrino et al. 2005; Verma et al. 2008; Lynch and Dawson 2008; Jiang et al. 2010; He et al. 2010; Zhang et al. 2010, 2011, 2012; Gebauer and Treuel 2011; Huynh and Chen 2011; Monopoli et al. 2011; Amin et al. 2012; Deng et al. 2012; Gebauer et al. 2012; Goy-Lopez et al. 2012; Shemetov et al. 2012; Caballero-Díaz et al. 2013; Hühn et al. 2013; Mahmoudi et al. 2013; Rehbock et al. 2013).

This kind of phenomenon was corroborated in the case of the typical examples of producing ligand-free gold and silver NPs while conducting laser assisted synthesis in NaCl electrolytes. These effects occur only with anions like Cl^-, Br^-, I^-, and SCN^-, whereas those including F^-, and SO_4^{2-} did not influence the stability. This could be understood better by considering that the ion-adsorptions significantly modify the NP's nano-environment due to the changes in the ion-concentrations in the vicinity of NP-surfaces and hence increase electrostatic stabilization. The accumulation of anions like Cl^-, Br^-, I^-, and SCN^- in the vicinity of the NPs did not only influence the colloidal stability, but also affected the growth mechanism of ligand-free NPs. Consequently, such ions could be employed in size quenching during the growth of gold and silver-NPs. It was noted recently that the NP-size reductions in the diluted electrolytes was found to be dependent on the NP-surface area, which is electrostatically stabilized by the available anions. The situation becomes more complex in the presence of ligand-coated NPs because of their additional influences. The situations, in the presence of the carboxylic acid ligands, and complex carboxylic acids, were found to possess a pH-dependent equilibrium and several pKa values, respectively. For pH < pKa, the NP-surfaces could lose their charges, hence reducing the colloidal stability and inducing agglomerations. Otherwise, for pH >> pKa, the NP-surfaces were ultimately saturated with negative charges resulting in colloidal stability. However, the situations might vary with the different ligands. It could thus be summarized that local pH could determine the surface charges of the NPs decided finally by the nature of the ligands used. These NPs can be stabilized also with macromolecular ligands, such as PEG, which provides colloidal stability via steric repulsions. It is accordingly concluded that the local ion-concentrations in the close vicinity of the NP-surfaces would certainly be different from the bulk concentration as a consequence of their direct influence on the immediately surrounding environment. In the same way, the screening caused by the adsorption of the counter-ions, together with the pH-dependent surface charge of the ligands would consequently determine the overall colloidal stability of the NPs (Bae et al. 2002; Sylvestre et al. 2004; Amendola and Meneghetti 2009; Zhang et al. 2012; Rehbock et al. 2013; Deka et al. 2015; Roth et al. 2016; Xu et al. 2016; Darr et al. 2017; de Espinosa et al. 2017; Farka et al. 2017; Kinnear et al. 2017; Liu and Liu 2017; Zhang et al. 2017).

12.3 Some Recent Results

The experimental investigations carried out in the field of biofunctionalized nanoparticulate systems for their numerous applications in the field of biomedicines are well known. Various types of nanomaterials used for nanomedicine use earlier are listed in Table 12.2.

Despite having an incomplete understanding of the phenomenon involved in protein and peptide-template-assisted one-pot-synthesis of biocompatible gold-NPs, Roth et al. could throw more light on the binding of the consensus tetratricopeptide repeat protein (CTPR) to gold ions and NPs for clarifying the issues related to bio-inorganic interfaces during gold-NP synthesis by using fluorescence spectroscopy and heteronuclear single quantum coherence NMR spectroscopic measurements. This study provided useful information regarding the rational-protein designs for synthesizing tailored functional nanomaterials for their biological, medical, and optical applications (Roth et al. 2016).

A depletion-based stabilization-strategy was implemented by Zhang, et al. in the presence of polyethylene glycol (PEG) to stabilize a diverse range of nanoparticulate species, including gold-NPs (i.e. 10 to 100nm in diameter), graphene oxide, quantum dots, silica NPs, and liposomes in the presence of Mg^{2+} (>1.6 M), heavy metal ions, extreme pH (pH 1–13), organic solvents, and adsorbed nucleosides and drugs. However, these NP-surfaces still remained accessible for the adsorption of both small and macromolecules as reported in the higher loading of thiolated DNA in one step, with just 2% PEG 20,000 in 2h (Zhang et al. 2012).

Xu, et al. succeeded in developing a method involving the oligoethylene glycol (OEG) as spacer to attach thiolated-DNA onto Au-NPs at physiological pH without the need for surfactants. An uncharged OEG spacer was found shielding against the repulsion between Au-NPs and DNAs, which could substantially enhance both the adsorption kinetics and thermodynamics of the thiolated DNAs. Subsequently, various kinds of thiolated DNAs were attached to large Au-NPs. This could also enable the direct immobilization of thiolated molecular beacons and eliminated the aggregations due to hydrogen bonding. These functionalized AuNPs were successfully used for the fluorescent detection of target DNA at nano-molar concentrations. Introducing the OEG spacer could be used for tuning DNA adsorption kinetics and thermodynamics besides pH and salt, which provided a novel method of controlled functionalization of AuNPs as reported (Xu et al. 2016).

Liu and Liu could offer better fundamental insights into the processes involved in colloidal stability by minimizing the electrostatic repulsions among the negatively charged DNAs and AuNPs after incorporating appropriate modifications in the DNA molecules. Accordingly, it became easier to use their useful optical properties with high extinction coefficients, distance-dependent color, strong fluorescence quenching, and localized surface plasmon resonance. Programming the DNA structures with suitable molecular recognition properties was thus put to use in producing a diverse range of useful biosensors and stimuli-responsive materials. The shortcomings of various methods developed for minimizing aggregations in the recent past including the salt-aging along with surfactants and sonication, and depletion stabilization with a concentrated polymer solution were better taken care of in this new method, as discussed in the enclosed reference (Liu and Liu 2017).

A novel approach of rapid synthesis of highly stable single-stranded DNA (ssDNA) functionalized Au-NPs was reported by

TABLE 12.2

Table Showing Liposomal, Micellar, Protein, and Nanocrystalline-Based Nanomedicines

Compositions	Trade Names	Manufacturer	Year
Liposomal Preparations			
Liposomal Amphotericin B Lipid	Abelcet®	Sigma-tau	1995
Liposomal Cytarabine	DepoCyt©	Sigma-Tau	1996
Liposomal Daunorubicin	DaunoXome	Galen	1996
Liposomal Amphotericin B	AmBisome®	Gilead Sciences	1997
Liposome-proteins SP-band SP-C	Curosurf®	Chieseifarmaceutici	1999
Liposomal Verteporfin	Neulasta®	Bausch & Lomb	2000
Liposomal Morphine Sulphate	Visudyne®	Pacira Pharma	2004
Liposomal Doxorubicin	Doxil®/Caelyx™	Janssen	2008
Liposomal Vincristine	Marqibo®	Onco TCS	2012
Liposomal Irinotecan	Onivyde®	Merrimack	2015
Micellar Formulation			
Micellar Estradiol	Mircera®	Novavax	2003
mPEG-PLA micelle + Paclitaxel	Genexol-PM	Samyang	2003
Protein Nanoparticulate Formulations			
Iron Dextran (Low MW)	INFed ®	Sanofi Aventis	1957
Engineered Protein + L-2 + Diphtheria Toxin	Ontak®	Horizon	1999
Albumin-bound paclitaxel NP	Abraxane®/ABI-007	Eisai Inc.	2013
Nanocrystalline Formulations			
Sirolimus	Rapamune®	Wyeth Pharma	2000
Iron Sucrose	Venofer	Lupitold Pharma	2000
Megestrol Acetate	MegaceES®	Par Pharma	2001
Aprepitant	Emend®	Merck	2003
Fenofibrate	Tricor®	Lupin Atlantis	2004
Dextran Coated SPION	Feridex/Endorem	AMAG Pharma	2008
Silicone on SPION	GastroMARK Umirem	AMAG Pharma	2009
Iron Oxide	NanoTherm	MagForce	2010

Deka et al. by using the combined effect of surface-passivation by (1-mercaptoundec-11-yl)hexa(ethylene glycol) and low pH-conditions with no salt pretreatment or excess of ssDNA. This technique was found applicable in the cases of oligonucleotides of any length or base sequence for preparing stabilized ssDNA-coated-Au-NPs conjugates at salt concentrations up to 3M, and causing DNA–DNA hybridization, as confirmed experimentally. The method could also be implemented in preparing ssDNA-Au-NP conjugates with a predefined number of ssDNA strands per particle, which became widely applicable in fabricating diverse families of biosensors involving ssDNA functionalized AuNPs (Deka et al. 2015).

The basic requirement of evolving properly engineered nanocomponents with highly sensitive bio-recognition ligands for site specific selective binding, and seamless integration for realizing biosensors with substantially better performances operating against the relatively noisier bio-environments are yet to be realized in the current devices. Farka and others reviewed the recent advances made in the areas including material synthesis, assembly, and applications of nanoengineered reporting and transducing components by examining several kinds of inorganic reporters and organic transducers. The design aspects of the required surface functionalizations leading to the desirable transducing paths, the characterization of interfacial architectures, and the integration of the multiple nanocomponents for

preparing multifunctional nanostructured materials were also studied. Various examples of the currently developing biosensors made using hybrid nanomaterials were also examined with a distinct emphasis on tailoring the nanosensor designs for specific operating environments, as discussed in the enclosed reference (Zhang et al. 2017).

The progress made during the last five years in developing immunochemical biosensors involving nanoparticulate material species was reviewed by Farka and associates, especially with a view to suggest ways and means for improving their sensitivity. Employing the antibodies as recognition elements in this approach, realization of optical sensors was proposed based on the phenomena of fluorescence, luminescence, and surface plasmon resonances. The electrochemical transducing techniques of amperometry, voltammetry, and impedance spectroscopy could be supplemented with the electrochemiluminescence and photoelectric conversions. Various kinds of nanoparticulate species including metals and metal oxides, magnetic NPs, carbon-based nanotubes, graphene and the other related nanomaterials, luminescent carbon dots, nanocrystals as quantum dots, and photon up-converting particles were examined as possible transducer elements. These sources, once integrated properly, were expected to provide extreme variability in the existing immune-sensors. Applications were also cited in the clinical analysis by employing markers, tumor cells, and pharmaceuticals for detecting the pathogenic microorganisms, toxic agents, and pesticides in the environmental field and food products in the reference enclosed (Farka et al. 2017).

De Espinosa et al. reviewed the ongoing R&D activities related to the development of bioinspired materials capable of responding to the external stimuli by changing their various mechanical features such as stiffness, shape, porosity, density, or hardness. The switchable mechanical properties of the living organisms were examined in detail to replicate some of them in hierarchically designed superstructures with adequate robustness and responsive behavior involving a number of events for translating the external stimuli into the subsequent changes occurring in the structure and/or connectivity of the constituent building blocks at various levels. Having acquired better awareness about the underlying principles operating in the living organisms, efforts are continuing to mimic some such structures and their associated functionalities in case of the synthetic materials using the available library of nanostructured material building blocks. Many practical examples are now available to demonstrate the feasibility as discussed in the enclosed reference (de Espinosa et al. 2017).

State-of-the-art manufacturing nanotechnology based on inorganic NPs prepared by a continuous hydrothermal flow synthesis (CHFS) process was reviewed by Darr and colleagues, giving comprehensive coverage of the current applications of the CHFS-made nanomaterials in optical, healthcare, electronics (including sensors, information, and communication technologies), catalysis, devices (including energy harvesting/conversion/fuels), and energy storage applications. Various aspects of selecting the right kind of precursors were taken into consideration for producing a variety of products, as well as materials or structures such as surface-functionalized hybrids, nanocomposites, nanograined coatings, and monoliths, and metal-organic frameworks. Some innovative equipment was also examined, in this context, such as those involving *in situ* flow/rapid heating setup, reactors supporting high-throughput flow syntheses, as well as the scale-up of hydrothermal processes (Darr et al. 2017).

Kinnear et al. examined the influence of the associated morphologies of the nanoparticulate entities on the biological system by starting from the simple case of single cell description and gradually extending to the whole organisms with special reference to the target specific cellular uptake mechanisms, bio distribution patterns, and pharmacokinetics. Two typical examples of spherical lipid-NPs introducing the marked changes in the field of chemotherapy and the negative impact of the asbestos-induced lung related ailments caused by fibrous materials, are sufficient to highlight the impact of NP morphology quantified in terms of many physicochemical parameters, namely, size, shape, elasticity, surface chemistry, and bio-persistence. Some of the morphologies observed in nature as well as those available as nanoparticulate assemblies synthesized in the laboratories were discussed prior to understanding the basic mechanisms involved in their cellular uptake through various theoretical models before comparing them with the *in vitro* and *in vivo* experimental observations in terms of their targeting efficacies, cytotoxicity, and cellular mechanics (Kinnear et al. 2017).

Having recognized the importance of a fast-growing number of applications involving a large-variety of peptide molecules with different structures in synthesizing a whole lot of

functional nanomaterials, Walsh and Knecht reviewed this area in-depth with the main emphasis on assessing the current status and predicting the future possibilities of creating complex molecular assemblies for their applications in the fields like catalysis, energy, and medicine. Despite having insufficient data on their structural details, the peptide molecules were found to influence in establishing collective bio-interfaces at the material surfaces. Peptide molecules were found endowed with the capability of molecular recognitions based on site-specific binding to the different nanomaterial surfaces, resulting in the screening and identification of hundreds of peptides involving a wide range of metals, metal-oxides, minerals, and polymer substrates. Formation of a complex bio-interface between the living and non-living components aided by complex biomolecules are known to bind to the materials with relatively higher affinity. Because of a very large number of material-binding sequences being anticipated, the quantitative material-binding characterization of these peptides has been accomplished only in a relatively small number of sequences after taking care of the challenges involved in determining the molecular-level structure(s) of these peptides in an adsorbed state on a material. A comprehensive overview of the possible applications was also included separately to confirm the versatility of peptide-mediated routes for growing, organizing, and activating the nanomaterials for specific applications (Walsh and Knecht 2017).

Zhao et al. could succeed in rapidly synthesizing the DNA-functionalized-Au-NPs as labeling agents in the molecular diagnostics and building blocks in nanotechnology. However, the existing methods of synthesis running overnight, along with the specific requirement of precise control of the ionic strength to compensate for the charge repulsion between the NPs' surface and the DNA strands, was taken care of by a mononucleotide-mediated conjugation method for synthesizing such DNA-functionalized Au-NPs within 4h in an environment with higher ionic strength. Au-NPs covered with a thermally tunable stabilization layers through mononucleotide adsorption were shown to readily conjugate with thiol-DNAs in heated 0.1M NaCl solutions. The resulting conjugated nano-entities from this experiment were loaded by ~80 strands per particle, comparable to the DNA-loading density of the current methods. The general applicability of this approach was further verified in a nanoparticle-bound DNA hybridization test confirming that the mononucleotide-mediated thermal conjugation was an attractive alternative that allowed temperature-controlled and salt-enhanced functionalization of gold-NPs with DNAs in just a few hours (Zhao et al. 2009).

Xu et al. attempted successful surface functionalizations of Au-NPs with DNA offering alternative platforms for their broader applications in biosensors, medical diagnostic, and biological analysis by evolving a novel and rapid approach of conjugating DNAs to Au-NPs to form functional DNA/AuNPs in 2–3h by employing Tween-80 as a protective agent. The DNA/Au-NPs, so synthesized, exhibited excellent stability as a function of temperature, pH, and freeze-thaw cycles besides confirming functionality of DNA/AuNPs conjugates (Xu et al. 2011).

Li et al. succeeded in conjugating DNA to the gold nanorods (AuNRs) for their extensive use in nanoassembly, gene therapy, biosensing, and drug delivery. However, it is a challenge to attach thiolated DNA on AuNRs, because the positively charged

AuNRs readily aggregate in the presence of negatively charged DNAs. This problem could be taken care of in a mPEG-SH/Tween-20-assisted method to load thiolated DNA on AuNRs in 1h via synergistic effect of Tween 20 and mPEG-SH in displacing the CTAB species on the surface of AuNRs by repeated centrifugation and re-suspension resulting in the attachment of thiolated DNA to the AuNRs in the presence of 1M NaCl, 100 mM MgCl$_2$, or 100 mM citrate. AuNRs with different sizes and aspect ratios were surface functionalized with DNA using this technique. The number of DNA loaded on each AuNRs could easily be controlled by the concentrations of mPEG-SH and Tween 20 or the ratio between DNA and AuNRs. Functionalized AuNRs were used for nanoparticle assembly and cancer cell imaging to confirm that DNA anchored on the surface of AuNRs retaining its hybridization and molecular recognition capability. This method turned out to be easy, rapid, and robust for the production of DNA functionalized AuNRs for a variety of applications such as cancer therapy, drug delivery, self-assembly, and imaging (Li et al. 2015).

An easy and efficient method of stabilizing a wide range of nanoparticulate metallic species was proposed by Chateau et al. using surfactants, long-chain polymers, and silica shells. For instance, nanospheres, nano bipyramids, and nanorods of gold and silver metals were surface-modified by optimized silicon oligomers prior to mixing them in monolithic silica-based sol-gel materials, which were efficiently used in preparing composite sol-gel materials for preparing plasmonic nanostructures for their optical applications (Chateau et al. 2017).

Wang et al. reported a cost-effective method appropriate for co-functionalization of gold-NPs with DNA and PEG polymers simultaneously by using a small quantity of thiolized-DNA in comparison to that based on pure-DNA-based functionalization but exhibiting similar binding efficacy of nanoparticulate gold species with DNA-based origami nanostructures. This protocol, in essence, offered scale-up potentials for preparing DNA-NP-conjugates involving much lower consumption of DNA material accordingly (Wang et al. 2017).

Baumann, and colleagues succeeded in surface-functionalizing the gold-nanorods with different DNA sequences using a ligand-exchange protocol by following sequential aqueous–organic–aqueous transfers followed by oligonucleotide grafting that exhibited very high and long-term colloidal stability in high ionic strength media as well as exhibiting adequate biocompatibility in the cell culture media. These two types of conjugated clusters of Au-NR–DNAs and Au-NP-DNAs could be selectively addressed using a laser beam in a binary mixture. This process could be used for their selective melting by exposing them with their respective plasmon resonance frequencies in the form of microsecond laser pulses at 532, and 1064 nm for spherical Au-NPs and Au-NRs, respectively. This was the first report of selective DNA-melting of the different sequences in one solution independent of their respective melting temperatures (Baumann et al. 2016).

Despite having recognized the importance of introducing the programmability features in the molecular-recognition based biopolymers combined with thiolated-DNAs-Au-NPs conjugates for their useful applications as nanomaterials possessing optical, thermal, and catalytic properties, the time-consuming and complex nature of the currently available protocols made them of limited use, barring the deployment of the larger size DNA-AuNPs. In response to these limitations, Li et al. evolved a rapid and facile method of conjugating the thiolated-DNAs onto the Au-NPs based on the stabilizing influence of the mPEG-SH. The Au-NPs were first coated with mPEG-SH in the presence of Tween 20 to provide excellent stability in the environment of higher ionic strengths and with extreme pH values. Adding NaCl solution to the mixture of DNAs + Au-NPs directly, did allow very efficient DNA-attachments onto the Au-NP-surfaces by reducing the electrostatic repulsions. The entire DNA loading process was thus completed only in 1.5h. These DNA-conjugated-Au-NPs retained their stability for more than 2 weeks at room temperature. Precise hybridizations experiments were accomplished with the complementary sequence to synthesize core-satellite nanostructures. The DNA-Au-NP-conjugates from this method could exhibit reduced cytotoxicity compared to those prepared by the standard methods (Li et al. 2014).

The strategy of achieving selective binding of the NPs to the required species being of utmost importance in synthesizing superstructures for detecting various biomarkers and targets. Seo, et al. could realize such binding by controlling the surface properties of the particles involved. Accordingly, solid/liquid interfacial layer comprising of multiple components on nanoparticle surfaces was established to have the required binding properties. In addition, the chemical structures of the nonfunctional diluent ligands were noted to influence the overall binding properties of the NPs. Taking gold-NPs and single-stranded oligonucleotides for synthesizing core nanoparticles and functional ligands, respectively, the influence of the various kinds of chemical and biological diluents were studied in detail as reported (Seo et al. 2017).

In another study, King et al. examined the capabilities of a number of compounds for stabilizing the Au-NPs against thermal sintering and compared them with and without functional groups that helped in conjugating the molecules onto the NP-surfaces. The conjugated stabilizing compounds with high thermal stability were also found to prevent the sintering of Au-NPs until the decomposition of the compound, as observed in the case of 1-pyrenebutanethiol stabilizer dissociating @ 390°C. The unanchored stabilizing compounds when applied to butanethiol-capped Au-NPs were found particularly effective. Out of oleyl amine ($T_{SE} \approx 300$ °C) and a perylene dicarboximide derivative ($T_{SE} \approx 540$°C), the latter could impart a thermal stability of an unprecedented kind on the ligand-stabilized Au-NPs. These experimental results amply demonstrated the importance of choosing the stabilizing compounds having an affinity with the capping ligands on the AuNPs to ensure a uniform mixture of AuNPs and stabilizer within a film (King et al. 2017).

The problem of measurement uncertainty was addressed to in a facile and rapid SERS quantification protocol by Yan et al., by deploying the target-induced nanoparticle self-assembly at oil/water interfaces for copper ion analysis. In response to copper, the core-molecule-shell-NPs were found migrating to the interface and organizing into densely-packed arrays with stronger plasmonic coupling. This could be further used in sensitive and selective measurements using Raman as well as visual detections. These kinds of 'target induced nanoparticle self-assembled interfaces' (TINSAIs) were accordingly deployed in fabricating visual and Raman detection platforms for high-throughput on-site screening with higher sensitivity and selectivity (Yan et al. 2016).

Despite having success with molecules likes polystyrene sulfonate (PSS), phosphine, DNA, and polyethylene glycol (PEG) in stabilizing the Au-NPs, larger size Au-NPs were found very difficult to stabilize. Consequently, biomedical applications of larger size Au-NPs (i.e. 30–100 nm range) possessing more favorable optical properties and easier cellular uptake were found constrained against the smaller Au-NPs. A novel method of preparing larger size Au-NPs was accordingly developed by Heo et al. to produce highly stabilized colloidal dispersions with facile chemical or biological functionality via surface passivation with an amphiphilic polymer polyvinylpyrrolidone (PVP). This kind of passivation could produce highly stable colloidal dispersion of Au-NPs in the range of 13 to 100nm in PBS for at least three months. These PVP-capped Au-NPs were found resistant to protein-adsorptions in the presence of serum containing media exhibiting negligible cytotoxicity. The Au-NP-PVPs functionalized with a DNA-aptamer (AS1411) was found adequately bioactive resulting in significant increase in the cell-uptake of these Au-NPs (~12 200 AuNPs per cell), in comparison with Au-NPs capped by a control DNA of the same length. The novel method would allow much wider application of these larger size Au-NPs in biomedical applications involving cellular imaging, molecular diagnosis, and targeted therapy (Heo et al. 2015).

The thermodynamic and kinetic studies of the interactions taking place among the DNAs and the NPs were carried out in detail by Carnerero et al. to conclude that knowing about the smaller ligands was necessary for better understanding of these interactions responsible for their control and modulation, which would offer newer avenues of conducting further research in nanomedicine (Carnerero et al. 2017).

Solvothermal synthesis of HDA-capped nanoparticulate ZnO was reported by Khoza et al. using different solvents like water, ethanol, and acetone with different polarities and maintained at constant temperature, pressure, and pH. Using ethanol and acetone could grow nanorods with higher aspect ratios. Water and ethanol with hydroxyl groups interacted with the NPs right from their nucleation, growth, and termination giving rise to non-spherical shapes. The hydroxyl groups were found to promote a nonuniform growth, resulting in nanostars and nanorods. The optical absorption spectra of these nanoparticulate-ZnO species exhibited the excitonic peaks in the range 368 to 374 nm, as reported elsewhere (Khoza et al. 2012).

In an attempt to alleviate the major concerns raised in connection with the inadequate availability of the experimental data and to establish the cytotoxicity of the nanoparticulate species, a detailed investigation was carried out by Najafi-Hajivar and others establishing that the nanoparticulate-material-species were observed either to stimulate and/or suppress the immune responses, and that their compatibility with the immune system was largely guided by their surface properties. Parameters of the nanoparticulate species involving their size, shape, composition, protein binding, and administration routes were noted to be the main factors that contributed to the overall interactions of the nanoparticulate species with the immune system (Figure 12.1) (Najafi-Hajivar et al. 2016).

12.4 Discussion and Conclusions

In the context of putting an extremely large variety of nanoparticulate materials species available in different shapes and sizes because of their inherently modified physicochemical properties to proper uses, it is necessary to take into account the general features that are invariably present in them. It is intriguing to note that various kinds of nano-sized material species developed during the last few decades and as such, have been found highly prone to aggregation/agglomeration in their respective media. In turn, they slowly start losing their nano characteristic properties and ultimately return to their bulk counterparts. One of the reasons behind this tendency is the presence of a large number of dangling bonds present on the surface of each NP due to their extremely large surface-to-volume ratios. Such dangling bonds are readily available for chemical conjugation via strong and weak type chemical bond formations, including inter-particle interactions among themselves. This novel phenomenon of higher chemical reactivity associated with nanoparticulate species makes them difficult to handle without adequate surface modifications. This includes surface stabilization, followed by controlled surface activation generated by attaching chemical moieties for site-specific attachments of the molecular species introduced subsequently. In order to get rid of the impeding features coming in the way, in doing so attempts were made using their novel topology and specific physical, chemical, and biological properties in preparing newer kinds of synthetic materials. A number of surface functionalization schemes have already been explored using numerous types of conjugation schemes with inorganic, organic, polymeric, and a whole host of biomolecular species. During these explorations, it has been noted in general that most of the synthetic methods developed for preparing inorganic NPs employ extremely hazardous combinations of chemicals as well as processing chemistries that are highly toxic to living organisms, and the environment invariably. However, the simplest solution for taking care of the associated toxicity issues conjugating with the polymeric/bio-polymeric molecules instead of depending mostly on inorganic moieties was preferred. Especially, one particular example of using RBC-derived membrane coatings on NP-core is there to highlight the extraordinary importance of such approaches of conjugations that are extended to a larger number of naturally derived molecular structures. Particularly, the rich source of a large variety of chemical moieties present there on the membrane surface makes their conjugation very similar to the natural entities that stay in blood circulation for longer durations, avoiding interactions with immune systems, besides providing a number of extremely useful features already mentioned. It is worth mentioning that liposomal entities derived from natural resources are alternately expected to play a very significant role in such conjugation.

Sailaja, et al. observed that bio-inspired self-assembled nano-species with adequate binding affinities to the numerous kinds of nanoparticulate species turned out to be a rich source of bio-functionalization useful in numerous applications. A detailed study of such fabrication protocols and their sensor applications was presented in this study involving the relevant nanospecies like DNAs, proteins, peptides, and viruses, as well as micro-organisms like bacteria and plant leaves. Their applications were further explored in detail with the help of characterization techniques like colorimetry, fluorescence spectroscopy, surface-plasmon-resonance properties, surface-enhanced Raman scattering, electrical resistance measurement, and electrochemistry. This exercise provided useful input for a better understanding of the interactions between the biomolecules/microorganisms and functional nanomaterial species, leading to their improved design and synthesis with unique properties (Sailaja et al. 2016).

Simpler material building blocks, like collagen fibrils employed by nature in producing a variety of hierarchical and complex superstructures, which are still very difficult to imitate in top-down nanofabrication techniques, required in biological tissues and structures possess remarkable physicochemical properties. Work has already started building such complex biomimetic structures with diverse optical, mechanical, and electrical properties using a bottom-up approach. In this context, some filamentous viruses were assembled into colorful films by Wang et al. and transferred onto the solid substrates, in which their colors were found to be changing when exposed to organic solvents and volatile organic compounds. These films, when exposed to different mechanical pressures, could exhibit piezoelectric properties for fabricating force sensors (Wang et al. 2016b).

In yet another study of smart host-tissue biomaterial interfaces, it was found to possess stimuli-responsive properties for participating in appropriate biochemical interactions. The basic requirement of a 'smart tissue-interface' has been found to exert a positive influence on healing and promoting bone-tissue regeneration as reviewed by Lee et al. recently. The associated healing process was found to accelerate through the development of a smart interface where biomaterial surface interacted synergistically with the extracellular matrix. The interface functionality was found dependent equally on the bound functional groups and the conjugated molecules belonging to the biomaterial and the biological tissues (Lee et al. 2017).

The study carried out by Ma and colleagues noted that the smart micelles being sensitive to a specific biological environment could be deployed in target-site-triggered drug-release actions by the reversible stabilization of micelle structure that were found to be very promising. A biocompatible and pH-sensitive copolymer was, accordingly, synthesized involving a bridge of poly (2-methacryloyloxyethyl phosphorylcholine) (PMPC) and poly (D-L-lactide) (PLA) blocks by a benzoyl-imine-linkage (*Blink*). These micelles were used in paclitaxel (PTX) delivery with excellent biocompatibility based on the PLA-*Blink*-PMPC copolymer. Due to their rapidly breaking linkages under acidic conditions, the micelle structure disruption could accelerate the PTX release. Such pH-triggered drug release involving acidic environment at tumor sites was found helpful in improving the utilization of drug by enhancing antitumor efficacy, as discussed in the cited publication (Ma et al. 2017).

A novel siRNA-nanocarrier was developed by Wang et al. for curing pancreatic cancer, in which biomimetic Au@BSA-nanoflowers were used as potential theranostic nanoplatforms. Gold-NPs with a higher X-ray absorption coefficient were found suitable for computed tomography and optical multimodal imaging when combined with the fluorescent markers. The Au@BSA-nanoflowers were obtained from the bovine serum albumin bio template for synthesizing the Au@BSA-nanocomposites after conjugating them with siRNA-FAM to form Au@BSA–siRNA-FAM through electrostatic layer-by-layer assembly. The resultant delivery system showed enhanced biocompatibility due to BSA coatings. The *in vitro* experiments demonstrated that the delivered siRNA could efficiently suppress the growth of BXPC-3 cells through gene silencing. In general, the biomimetic Au@BSA nanocarriers might be offering a new modality of cancer therapy in future, as has been reported (Wang et al. 2016d).

In another review, carried out by Ahmad and Hashim, the fast emerging prospects of using the nanoparticulate species prepared from the whole extracts of several naturally available herbs for their efficacious applications in treating a number of human ailments based on the experience gained through actual animal trials, could also be extended for their applications in drug and gene deliveries. It is quite possible to attach a number of nano-herbal micelles prepared from the herbal extracts onto the cellular-membrane coated Au-NPs or DNA-decorated metal/metal-oxide-NPs by their chemical conjugations for achieving a better targeted delivery with improved cellular uptakes and enhanced circulation lifetimes. This scheme could possibly be explored further for preparing natural phytochemical-based nanoparticulate species for treating a number of diseases which are not possible to handle using conventional molecular medicines already in use. Moreover, it is also possible to functionalize the surfaces of the specific metal nanoparticulate species with the phytochemical-based micelles along with some effective molecular drugs for treating the diseases for which pharmaceutical formulations are not yet available. This kind of unique synergistic strategy of using targeted delivery of molecular and natural herbal drugs might prove to be more effective in providing rather fast treatment of the human ailments with minimal side effects (Ahmad and Hashim 2012)

Taking all these possibilities into consideration, already explored through *ab initio* quantum calculations and experimental investigations, the possibility of realizing intelligent features in synthesized nanomaterial species is slowly inching towards fruitful results in the near future. Employing these nanoparticulate species synthesized by design, as fundamental building blocks as already discussed in the text, would certainly pave the way for reaching the stipulated goals in times to come.

Based on the above-mentioned schemes of bio-functionalization of the nanoparticulate species, it may be safely concluded that such surface-engineered nanoparticulate species might offer almost unlimited possibilities of incorporating intelligent features in the nanomaterials by design. Biomimetic route of surface engineering of the NPs is certainly going to be almost green in nature. Using naturally available material resources in the different forms would not only imbibe all the advantages of their constituent components synergistically but also acquire renewable characteristics, which will stop depleting the scarce materials from the surface of Earth in the coming times. Such hybrid schemes of improving the quality of human lives would not only facilitate providing better treatment for deadly diseases at an affordable cost, but also offer a better and more sustainable management of human health-related problems as a whole, in a sustainable manner.

Some of the FDA-approved nanoformulations, available commercially, are compiled in Table I and II, to highlight the level of technology development that has already been acquired in this context and which reflect the translation of basic ideas into usable products, based on clearing all the stringent steps of clinical trials before introducing them to the market.

Acknowledgments

The authors are grateful to their colleagues, graduate students, and especially Prof. Dinesh Kumar, VC, YMCA University of Science & Technology, Faridabad, Haryana, for providing a very congenial research environment through very fruitful interactions at different levels in connection with preparing this chapter. The authors are also thankful to the research workers in the Interdisciplinary Innovative Research Center, YMCA, for providing all necessary supports and encouragement throughout.

REFERENCES

Abu Lila, A.S., H. Kiwada, and T. Ishida. (2013). The accelerated blood clearance (ABC) phenomenon: Clinical challenge and approaches to manage. *Journal of Controlled Release, 172,* 38–47.

Ahmad, B., N. Hafeez, S. Bashir, A. Rauf, and Mujeeb-ur-Rehman. (2017). Phyto fabricated gold nanoparticles and their biomedical applications. *Biomedicine & Pharmacotherapy, 89,* 414–25.

Ahmad, S. (2015). Engineered nanomaterials for drug and gene deliveries: A review. *Journal of Nanopharmaceutics and Drug Delivery, 3,* 1–50.

Ahmad, S. (2016). Band-structure-engineered materials synthesis: Nano crystals and hierarchical superstructures: Current status and future trend. *International Journal of Material Science, 6*(1), 1–34.

Ahmad, S. and U. Hashim. (2012). Nano herbals in human healthcare: A proposed research and development roadmap, part I, and part II. *ASEAN Journal on Science and Technology for Development, 29*(1), 55–75.

Albanese, A., P.S. Tang, and W.C. Chan. (2012). The effect of nanoparticle size, shape, and surface chemistry on biological systems. *Annual Review of Biomedical Engineering, 14,* 1–16.

Amendola, V. and M. Meneghetti. (2009). Laser ablation synthesis in solution and size manipulation of noble metal nanoparticles. *Physical Chemistry Chemical Physics, 11,* 3805–21.

Amin, F., D.A. Yushchenko, J.M. Montenegro, and W.J. Parak. (2012). Integration of organic fluorophores in the surface of polymer-coated colloidal nanoparticles for sensing the local polarity of the environment. *ChemPhysChem, 13,* 1030–5.

Aryal, S., C.M. Hu, R.H. Fang, D. Dehaini, C. Carpenter, D.E. Zhang, and L. Zhang. (2013). Erythrocyte membrane-cloaked polymeric nanoparticles for controlled drug loading and release. *Nanomedicine, 8,* 1271–80.

Bae, C.H., S.H. Nam, and S.M. Park. (2002). Formation of silver nanoparticles by laser ablation of a silver target in NaCl solution. *Applied Surface Science, 197,* 628–34.

Banadaki, A.D., and A. Kajbafvala. (2014). Recent advances in facile synthesis of bimetallic nanostructures: An overview. *Journal of Nanomaterials, 2014,* 985948.

Baumann, V., P.J.F. Röttgermann, F. Haase, K. Szendrei, P. Dey, K. Lyons, R. Wyrwich, M. Gräßel, J. Stehr, L. Ullerich, F. Bürsgens, and J. Rodríguez-Fernández. (2016). Highly stable and biocompatible gold nanorod–DNA conjugates as NIR probes for ultrafast sequence-selective DNA melting. *RSC Advances, 6,* 103724–39.

Boles, M.A., M. Engel, and D.V. Talapin. (2016). Self-assembly of colloidal nanocrystals: from intricate structures to functional materials. *Chemical Reviews, 116*(18), 11220–89.

Bose, R.J.C., S.H. Lee, and H. Park. (2016). Bio-functionalized nanoparticles: An emerging drug delivery platform for various disease treatments. *Drug Discovery Today, 21,* 1303–12.

Caballero-Díaz, E., C. Pfeiffer, L. Kastl, P. Rivera-Gil, B. Simonet, M. Valcárcel, J. Jiménez-Lamana, F. Laborda, and W.J. Parak. (2013). The toxicity of silver nanoparticles depends on their uptake by cells and thus on their surface chemistry. *Particle and Particle System Characterization, 30,* 1079–85.

Carnerero, J.M., A. Jimenez-Ruiz, P.M. Castillo, and R. Prado-Gotor. (2017). Covalent and non-covalent DNA–gold-nanoparticle interactions: New avenues of research. *ChemPhysChem, 18*(1), 17–33.

Chateau, D., A. Liotta, D. Gregori, F. Lerouge, F. Chaput, A. Desert, and S. Parola. (2017). Controlled surface modification of gold nanostructures with functionalized silicon polymers. *Journal of Sol-Gel Science and Technology, 81*(1), 147–53.

Chen, J., X.-J. Wu, Y. Gong, Y. Zhu, Z. Yang, B. Li, Q. Lu, Y. Yu, S. Han, Z. Zhang, Y. Zong, Y. Han, L. Gu, and H. Zhang. (2017). Edge epitaxy of two-dimensional $MoSe_2$ and MoS_2 nanosheets on one-dimensional nanowires. *Journal of American Chemical Soceity, 139*(25), 8653–60.

Chung, D.Y., S.W. Jun, G. Yoon, H. Kim, J.M. Yoo, K.-S. Lee, T. Kim, H. Shin, A.K. Sinha, S.G. Kwon, K. Kang, T. Hyeon, and Y.-E. Sung. (2017). Large-scale synthesis of carbon-shell-coated FeP nanoparticles for robust hydrogen evolution reaction electro catalyst. *Journal of American Chemical Society, 139*(19), 6669–74.

Copp, J.A., R.H. Fang, B.T. Luk, C.M. Hu, W. Gao, K. Zhang, and L. Zhang. (2014). Clearance of pathological antibodies using biomimetic nanoparticles. *Proceedings of the National Academy of Sciences of the United States of America, 111,* 13481–6.

Da, L., S.Y. Tao, F.Y. Yu, and F. Wei. (2012). Recent progress in the fields of tuning the band gap of quantum dots. *Science China, 55*(4), 903–12.

Danhier, F., O. Feron, and V. Prea. (2010). To exploit the tumor microenvironment: Passive and active tumor targeting of nanocarriers for anti-cancer drug delivery. *Journal of Controlled Release, 148,* 135–46.

Daniel, M.-C. and D. Astruc. (2004). Gold nanoparticles: Assembly, supramolecular chemistry, quantum-size-related properties, and applications toward biology, catalysis, and nanotechnology. *Chemical Reviews, 104*(1), 293–346.

Darr, J. A., J. Zhang, N.M. Makwana, and X. Weng. (2017). Continuous hydrothermal synthesis of inorganic nanoparticles: Applications and future directions. *Chemical Reviews, 117*(17), 11125–38.

Davis, M.E., Z.G. Chen, and D.M. Shin. (2008). Nanoparticle therapeutics: An emerging treatment modality for cancer. *Nature Reviews Drug Discovery, 7,* 771–82.

de Espinosa, L.M., W. Meesorn, D. Moatsou , and C. Weder. (2017). Bioinspired polymer systems with stimuli-responsive mechanical properties. *Chemical Reviews,* DOI: 10.1021/acs. chemrev.7b00168.

Dehaini, D., X. Wei, R.H. Fang, S. Masson, P. Angsantikul, B.T. Luk, Y. Zhang, M. Ying, Y. Jiang, A.V. Kroll, W. Gao, and L. Zhang. (2017). Erythrocyte-platelet hybrid membrane coating for enhanced nanoparticle functionalization. *Advanced Materials*, 29, 1606209.

Deka, J., R. Měch, L. Ianeselli, H. Amenitsch, F. Cacho-Nerin, P. Parisse, and L. Casalis. (2015). Surface passivation improves the synthesis of highly stable and specific DNA-functionalized gold nanoparticles with variable DNA density. *ACS Applied Materials & Interfaces*, 7(12), 7033–40.

Deng, Z.J., M. Liang, I. Toth, M.J. Monteiro, and R.F. Minchin. (2012). Molecular interaction of poly(acrylic acid) gold nanoparticles with human fibrinogen. *ACS Nano*, 6, 8962–9.

Duan, S., and R. Wang. (2013). Bimetallic nanostructures with magnetic and noble metals and their physicochemical applications. *Progress in Natural Science: Materials International*, 23(2), 113–26.

Dubey, N., R. Varshney, J. Shukla, A. Ganeshpurkar, P.P. Hazari, G.P. Bandopadhaya, A.K. Mishra, and P. Trivedi. (2012). Synthesis and evaluation of biodegradable PCL/PEG nanoparticles for neuroendocrine tumor targeted delivery of somatostatin analog. *Drug Delivery*, 19, 132–42.

Fang, J., H. Nakamura, and H. Maeda. (2011). The EPR effect: Unique features of tumor blood vessels for drug delivery, factors involved, and limitations and augmentation of the effect. *Advanced Drug Delivery Reviews*, 63, 136–51.

Fang, R.H., C.M. Hu, B.T. Luk, W. Gao, J.A. Copp, Y. Tai, D.E. O'Connor, and L. Zhang. (2014). Cancer cell membrane-coated nanoparticles for anticancer vaccination and drug delivery. *Nano Letters*, 14, 2181–8.

Fang, R.H., C.M. Hu, and L. Zhang. (2012). Nanoparticles disguised as red blood cells to evade the immune system. *Expert Opinion on Biological Therapy*, 12, 385–9.

Farka, Z., T. Juřík, D. Kovář, L. Trnková, and P. Skládal. (2017). Nanoparticle-based immunochemical biosensors and assays: Recent advances and challenges. *Chemical Reviews*, 117(15), 9973–10042.

Gao, W., R.H. Fang, S. Thamphiwatana, B.T. Luk, J. Li, P. Angsantikul, Q. Zhang, C.M. Hu, and L. Zhang. (2015). Modulating antibacterial immunity via bacterial membrane-coated nanoparticles. *Nano Letters*, 15, 1403–9.

Gao, W., C.M. Hu, R.H. Fang, B.T. Luk, J. Su, and L. Zhang. (2013a). Surface functionalization of gold nanoparticles with red blood cell membranes. *Adv. Mater.*, 25, 3549–53.

Gao, W.W., C.M.J. Hu, R.H. Fang, and L.F. Zhang. (2013b). Liposome-like nanostructures for drug delivery. *Journal of Materials Chemistry B*, 1, 6569–85.

Gao, W.W. and L.F. Zhang. (2015a). Coating nanoparticles with cell membranes for targeted drug delivery. *Journal of Drug Targeting*, 23, 619–26.

Gao, W.W. and L.F. Zhang. (2015b). Engineering red-blood-cell-membrane-coated nanoparticles for broad biomedical applications. *AICHE, J.*, 61, 738–46.

Garcia, I., A. Sanchez-Iglesias, M. Henriksen-Lacey, M. Grzelczak, S. Penades, and L.M. Liz-Marzan. (2015). Glycans as biofunctional ligands for gold nanorods: Stability and targeting in protein-rich media. *J. Am. Chem. Soc.*, 137, 3686–92.

Gebauer, J.S., M. Malissek, S. Simon, S.K. Knauer, M. Maskos, R.H. Stauber, W. Peukert, and L. Treuel. 2012. Impact of the nanoparticle–protein corona on colloidal stability and protein structure. *Langmuir*, 28, 9673–9.

Gebauer, J.S. and L. Treuel. (2011). Influence of individual ionic components on the agglomeration kinetics of silver nanoparticles. *J. Colloid Interface Sci.*, 354, 546–54.

Gilroy, K.D., A. Ruditskiy, H.-C. Peng, D. Qin, and Y. Xia. (2016). Bimetallic nanocrystals: syntheses, properties, and applications. *Chem. Rev.*, 116(18), 10414–72.

Goy-Lopez, S., J. Juarez, M. Alatorre-Meda, E. Casals, V.F. Puntes, P. Taboada, and V. Mosquera. (2012). Physicochemical characteristics of protein–NP bioconjugates: The role of particle curvature and solution conditions on human serum albumin conformation and fibrillogenesis inhibition. *Langmuir*, 28, 9113–26.

Gratton, S.E.A., P.A. Ropp, P.D. Pohlhaus, J.C. Luft, V.J. Madden, M.E. Napier, J.M. DeSimone. (2008). The effect of particle design on cellular internalization pathways. *Proc. Natl. Acad. Sci U.S.A.*, 15, 11613–8.

Guo, Y., D. Wang, Q. Song, T. Wu, X. Zhuang, Y. Bao, M. Kong, Y. Qi, S. Tan, and Z. Zhang. (2015). Erythrocyte membrane-enveloped polymeric nanoparticles as nanovaccine for induction of antitumor immunity against melanoma. *ACS Nano*, 9, 6918–33.

Gupta, B., T.S. Levchenko, and V.P. Torchilin. (2005). Intracellular delivery of large molecules and small particles by cell-penetrating proteins and peptides. *Adv. Drug Delivery Rev.*, 57, 637–51.

He, Q., J. Zhang, J. Shi, Z. Zhu, L. Zhang, W. Bu, L. Guo, and Y. Chen. (2010). The effect of PEGylation of mesoporous silica nanoparticles on nonspecific binding of serum proteins and cellular responses. *Biomaterials*, 31, 1085–92.

Heo, J.H., K.-I. Kim, H.H. Cho, J.W. Lee, B.S. Lee, S. Yoon, K.J. Park, S. Lee, J. Kim, D. Whang, and J.H. Lee. (2015). Ultra-stable-stealth large gold nanoparticles with DNA directed biological functionality. *Langmuir*, 31(51), 13773–82.

Hsieh, C.C., S.T. Kang, Y.H. Lin, Y.J. Ho, C.H. Wang, C.K. Yeh, and C.W. Chang. (2015). Biomimetic acoustically-responsive vesicles for theragnostic applications. *Theranostics*, 5, 1264–74.

Hu, C.M., R.H. Fang, J. Copp, B.T. Luk, and L. Zhang. (2013a). A biomimetic nanosponge that absorbs pore-forming toxins. *Nat. Nanotechnol.*, 8, 336–40.

Hu, C.M., R.H. Fang, B.T. Luk, K.N. Chen, C. Carpenter, W. Gao, K. Zhang, and L. Zhang. (2013c). Marker-of-self functionalization of nano scale particles through a top-down cellular membrane coating approach. *Nanoscale*, 5, 2664–8.

Hu, C.M.J., R.H. Fang, B.T. Luk, and L. Zhang. (2013b). Nanoparticle-detained toxins for safe and effective vaccination. *Nat. Nanotechnol.*, 8, 933–8.

Hu, C.M.J, R.H. Fang, K.C. Wang, B.T. Luk, S. Thamphiwatana, D. Dehaini, P. Nguyen, P. Angsantikul, C.H. Wen, A.V. Kroll, C. Carpenter, M. Ramesh, V. Qu, S.H. Patel, J. Zhu, W. Shi, F.M. Hofman, T.C. Chen, W. Gao, K. Zhang, S. Chien, and L. Zhang. (2015). Nanoparticle bio interfacing by platelet membrane cloaking. *Nature*, 526, 118–21.

Hu, C.M.J., R.H. Fang, and L. Zhang. (2012). Erythrocyte-inspired delivery systems. *Adv. Healthcare. Mater.*, 1, 537–47.

Hu, C.M.J., L. Zhang, S. Aryal, C. Cheung, R.H. Fang, and L.F. Zhang. (2011). Erythrocyte membrane-camouflaged polymeric nanoparticles as a biomimetic delivery platform. *Proc. Natl. Acad. Sci. U.S.A.*, 108, 10980–5.

Hühn, D., K. Kantner, C. Geidel, S. Brandholt, I. De Cock, S.J. Soenen, P.R. Gil, J.M. Montenegro, K. Braeckmans, K. Müllen, G.U. Nienhaus, M. Klapper, and W.J. Parak. (2013). Polymer-coated nanoparticles interacting with proteins and cells: Focusing on the sign of the net charge. *ACS Nano*, 7, 3253–63.

Huynh, K.A. and K.L. Chen. (2011). Aggregation kinetics of citrate and polyvinylpyrrolidone coated silver nanoparticles in monovalent and divalent electrolyte solutions. *Environ. Sci. Technol.*, *45*, 5564–71.

Ishida, T., M. Ichihara, X. Wang, K. Yamamoto, J. Kimura, E. Majima, and H. Kiwada. (2006). Injection of PEGylated liposomes in rats elicits PEG-specific IgM, which is responsible for rapid elimination of a second dose of PEGylated liposomes. *J. Controlled Release*, *112*, 15–25.

Ishida, T., R. Maeda, M. Ichihara, K. Irimura, and H. Kiwada. (2003). Accelerated clearance of PEGylated liposomes in rats after repeated injections. *J. Controlled Release*, *88*, 35–42.

Iversen, F., C.X. Yang, F. Dagnaes-Hansen, D.H. Schaffert, J. Kjems, and S. Gao. (2013). Optimized siRNA-PEG conjugates for extended blood circulation and reduced urine excretion in mice. *Theranostics*, *3*, 201–9.

Jain, R.K. and T. Stylianopoulos. (2010). Delivering nanomedicine to solid tumors. *Nat. Rev. Clin. Oncol.*, *7*, 653–64.

Jiang, X., S. Weise, M. Hafner, C. Röcker, F. Zhang, W. J. Parak, and G. U. Nienhaus. (2010). Quantitative analysis of the protein corona on FePt nanoparticles formed by transferrin binding. *J. R. Soc. Interface*, *7*, S5–13.

Jin, R., C. Zeng, M. Zhou, and Y. Chen. (2016). Atomically precise colloidal metal nano clusters and nanoparticles: Fundamentals and opportunities. *Chem. Rev.*, *116*(18), 10346–413.

Khoza, P.B., M.J. Moloto, and L.M. Sikhwivhilu. (2012). The effect of solvents, acetone, water, and ethanol, on the morphological and optical properties of ZnO nanoparticles prepared by microwave. *J. Nanotechnology*, *2012*, 195106.

King, S.R., S. Shimmon, D.D. Totonjian, and A.M. McDonagh. (2017). Influence of bound versus non-bound stabilizing molecules on the thermal stability of gold nanoparticles. *J. Phys. Chem. C*, *121*(25), 13944–51.

Kinnear, C., T.L. Moore, L. Rodriguez-Lorenzo, B. Rothen-Rutishauser, and A. Petri-Fink. (2017). Form follows function: Nanoparticle shape and its implications for nanomedicine. *Chem. Rev.*, *117*(17), 11476–521.

Knop, K., R. Hoogenboom, D. Fischer, and U.S. Schubert. (2010). Poly (ethylene glycol) in drug delivery: Pros and cons as well as potential alternatives. *Angew. Chem. Int. Ed.*, *49*, 6288–308.

Kroll, A.V., R.H. Fang, and L. Zhang. (2017). Bio-interfacing and applications of cell membrane-coated nanoparticles. *Bioconjugate. Chem.*, *28*, 23–32.

Kumar, J.S., S.R. Kumar, and S.V. Kumar. (2017). Phyto-assisted synthesis, characterization and applications of gold nanoparticles: A review. *Biochem. Biophys. Rep.*, *11*, 46–57.

Langer, R., and D.A. Tirrell. (2004). Designing materials for biology and medicine. *Nature*, *428*, 487–92.

Lee, D.H., W. Song, and B.Y. Lee. (2017). Biomimetic materials and structures for sensor applications. In, Kyung, C.M., Yasuura, H., Liu, Y., and Lin, Y.L. (Eds.), *Smart Sensors and Systems*. Cham: Springer. https://doi.org/10.1007/978-3-319-33201-7_1.

Li, J., B. Zhu, X. Yao, Y. Zhang, Z. Zhu, S. Tu, S. Jia, R. Liu, H. Kang, and C.J. Yang. (2014). Synergetic approach for simple and rapid conjugation of gold nanoparticles with oligonucleotides. *ACS Appl. Mater. Interfaces*, *6*(19), 16800–7.

Li, J., B. Zhu, Z. Zhu, Y. Zhang, X. Yao, S. Tu, R. Liu, S. Jia, and C.J. Yang. (2015). Simple and rapid functionalization of gold nanorods with oligonucleotides using an mPEG-SH/tween 20-assisted approach. *Langmuir*, *31*(28), 7869–76.

Li, S.D., and L. Huang. (2008). Pharmacokinetics and bio distribution of nanoparticles. *Mol. Pharmacol.*, *5*, 496–504.

Liu, B. and J. Liu. (2017). Methods for preparing DNA-functionalized gold nanoparticles, a key reagent of bioanalytical chemistry. *Anal. Methods*, *9*, 2633–43.

Liu, Y.Y., L. Mei, C.Q. Xu, Q.W. Yu, K.R. Shi, L. Zhang, Q. Zhang, H. Gao, Z. Zhang, and Q. He. (2016). Dual receptor recognizing cell penetrating peptide for selective targeting, efficient intra tumoral diffusion and synthesized anti-glioma therapy. *Theranostics*, *6*, 177–91.

Luk, B.T., R.H. Fang, C.-M. J. Hu, J.A. Copp, S. Thamphiwatana, D. Dehaini, W. Gao, K. Zhang, S. Li, and L. Zhang. (2016). Safe and immunocompatible nanocarriers cloaked in RBC membranes for drug delivery to treat solid tumors. *Theranostics*, *6*(7), 1004–11.

Luk, B.T., C.M. Hu, R.H. Fang, D. Dehaini, C. Carpenter, W. Gao, and L. Zhang. (2014). Interfacial interactions between natural RBC membranes and synthetic polymeric nanoparticles. *Nanoscale*, *6*, 2730–7.

Luk, B.T. and L.F. Zhang. (2015). Cell membrane-camouflaged nanoparticles for drug delivery. *J. Controlled Release*, *220*, 600–7.

Lynch, I., and K.A. Dawson. (2008). Protein–nanoparticle interactions. *Nano Today*, *3*, 40–7.

Ma, B., W. Zhuang, G. Liu, and Y. Wang. (2017). A biomimetic and pH-sensitive polymeric micelle as carrier for paclitaxel delivery. *Regener. Biomater.*, *5*(1), 15–24.

Mahmoudi, M., A.M. Abdelmonem, S. Behzadi, J.H. Clement, S. Dutz, M.R. Ejtehadi, R. Hartmann, K. Kantner, U. Linne, P. Maffre, S. Metzler, M.K. Moghadam, C. Pfeiffer, M. Rezaei, P. Ruiz-Lozano, V. Serpooshan, M.A. Shokrgozar, G.U. Nienhaus, and W.J. Parak. (2013). Temperature: The 'ignored' factor at the nanobiointerface. *ACS Nano 7*, 6555–62.

Masala, O. and R. Seshadri. (2004). Synthesis routes for large volumes of nanoparticles. *Annu. Rev. Mater. Res.*, *34*, 41–81.

Monopoli, M.P., D. Walczyk, A. Campbell, G. Elia, I. Lynch, F.B. Bombelli, and K.A. Dawson. (2011). Physical–chemical aspects of protein corona: Relevance to in vitro and in vivo biological impacts of nanoparticles. *J. Am. Chem. Soc.*, *133*, 2525–34.

Murphy, M., K. Ting, X. Zhang, C. Soo, and Z. Zheng. (2015). Current development of silver nanoparticle preparation, investigation, and application in the field of medicine. *J. Nanomaterials*, *2015*, 696918.

Murray, C.B., C.R. Kagan, and M.G. Bawendi. (2000). Synthesis and characterization of monodisperse nanocrystals and close-packed nanocrystal assemblies. *Annu. Rev. Mater. Sci. 30*, 545–610.

Najafi-Hajivar, S., P. Zakeri-Milani, H. Mohammadi, M. Niazi, M. Soleymani-Goloujeh, B. Baradaran, and H. Valizadeh. (2016). Overview on experimental models of interactions between nanoparticles and the immune system. *Biomed. Pharmacother.*, *83*, 1365–78.

Niu, Z., F. Cui, Y. Yu, N. Becknell, Y. Sun, G. Khanarian, D. Kim, L. Dou, A. Dehestani, K. Schierle-Arndt, and P. Yang. (2017). Ultrathin epitaxial Cu@Au core–shell nanowires for stable transparent conductors. *J. Am. Chem. Soc.*, *139*(21), 7348–54.

Oldenborg, P.A., A. Zhelezynak, Y.F. Fang, C.F. Lagenaur, H.D. Gresham, and F.P. Lindberg. (2000). Role of CD47 as a marker of self on red blood cells. *Science*, *288*, 2051–4.

Palivan, C.G., R. Goers, A. Najer, X.Y. Zhang, A. Car, and W. Meier. (2016). Bio inspired polymer vesicles and membranes for biological and medical applications. *Chem. Soc. Rev.*, *45*, 377–411.

Pang, Z., C.M. Hu, R.H. Fang, B.T. Luk, W. Gao, F. Wang, E. Chuluun, P. Angsantikul, S. Thamphiwatana, W. Lu, X. Jiang, and L. Zhang. (2015). Detoxification of organophosphate poisoning using nanoparticle bio-scavengers. *ACS Nano*, *9*, 6450–8.

Parodi, A., N. Quattrocchi, A.L. van de Ven, C. Chiappini, M. Evangelopoulos, J.O. Martinez, B.S. Brown, S.Z. Khaled, I.K. Yezdi, M.V. Enzo, L. Eisenhart, M. Ferrari, and E. Tasciotti. (2013). Synthetic nanoparticles functionalized with biomimetic leukocyte membranes possess cell-like functions. *Nat. Nanotechnol.*, *8*, 61–8.

Payne, J.N., H.K. Waghwani, M.G. Connor, W. Hamilton, S. Tockstein, H. Moolani, F. Chavda, V. Badwaik, M.B. Lawrenz, and R. Dakshinamurthy. (2016). Novel *Synthesis of Kanamycin Conjugated Gold Nanoparticles with Potent Antibacterial Activity*. *Front. Microbiol.*, *7*, 607.

Peer, D, J.M. Karp, S. Hong, O.C. Farokhzad, R. Margalit, and R. Langer. (2007). Nanocarriers as an emerging platform for cancer therapy. *Nat. Nanotechnol.*, *2*, 751–60.

Pellegrino, T., S. Kudera, T. Liedl, A.M. Javier, L. Manna, and W.J. Parak. (2005). On the development of colloidal nanoparticles towards multifunctional structures and their possible use for biological applications. *Small*, *1*, 48–63.

Pfeiffer, C., C. Rehbock, D. Hu¨hn, C. Carrillo-Carrion, D.J. de Aberasturi, V. Merk, S. Barcikowski, and W.J. Parak. (2014). Interaction of colloidal nanoparticles with their local environment: The (ionic) nanoenvironment around nanoparticles is different from bulk and determines the physico-chemical properties of the nanoparticles. *J. R. Soc. Interface*, *11*, 20130931.

Piao, J.G., L. Wang, F. Gao, Y.Z. You, Y. Xiong, and L. Yang. (2014). Erythrocyte membrane is an alternative coating to polyethylene glycol for prolonging the circulation lifetime of gold nanocages for photothermal therapy. *ACS Nano*, *8*, 10414–25.

Pulit-Prociak, J., and M. Banach. (2016). Silver nanoparticles: A material of the future? *Open Chem.*, *14*, 76–91.

Rehbock, C., V. Merk, L. Gamrad, R. Streubel, and S. Barcikowski. (2013). Size control of laser-fabricated surfactant-free gold nanoparticles with highly diluted electrolytes and their subsequent bioconjugation. *Phys. Chem. Chem. Phys.*, *15*, 3057–67.

Reiss, P., M. Protière, L. Li. (2009). Core/shell semiconductor nanocrystals. *Small*, *5*, 154–68.

Rodriguez, P.L., T. Harada, D.A. Christian, D.A. Pantano, R.K. Tsai, and D.E. Discher. (2013). Minimal 'self' peptides that inhibit phagocytic clearance and enhance delivery of nanoparticles. *Science 339*, 971–5.

Rogach, A.L., A. Eychmüller, S.G. Hickey, and S.V. Kershaw. (2007). Infrared-emitting colloidal nanocrystals: Synthesis, assembly, spectroscopy, and applications. *Small*, *3*(4), 536–57.

Roth, K.L., X. Geng, and T.Z. Grove. (2016). Bioinorganic interface: Mechanistic studies of protein-directed nanomaterial synthesis. *J. Phys. Chem. C*, *120*(20), 10951–60.

Sailaja, G.S., P. Ramesh, S. Vellappally, S. Anil, and H.K. Varma. (2016). Biomimetic approaches with smart interfaces for bone regeneration. *J. Biomed. Sci.*, *23*, 77.

Schmid, G. (2008). Ionically cross-linked gold clusters and gold nanoparticles. *Angew. Chem., Int. Ed. 47*(19), 3496–8.

Sengani, M., A.M. Grumezescu, and V.D. Rajeswari. (2016). Recent trends and methodologies in gold nanoparticle synthesis: A prospective review on drug delivery aspect. *Open Nano*, *2*, 37–46.

Seo, S., J.H. Joo, D.H. Park, and J.-S. Lee. (2017). Functionality of nonfunctional diluent ligands within bicomponent layers on nanoparticles. *J. Phys. Chem. C*, *121*(25), 13906–15.

Shan, J., L. Wang, H. Yu, H. Ji, W.A. Amer, Y. Chen, G. Jing, H. Khalid, M. Akram, and N.M. Abbasi. (2016). Recent progress in Fe_3O_4 based magnetic nanoparticles: From synthesis to application. *J. Mater. Sci. Technol.*, *32*(6). doi.org/10.1179 /1743284715Y.0000000122.

Sharma, G., A. Kumar, S. Sharma, Mu. Naushad, R.P. Dwivedi, Z.A. Al Othman, and G.T. Mola. (2017). Novel development of nanoparticles to bimetallic nanoparticles and their composites: A review. *J. King Saud Univ. Sci.*, *31*, 257–69. http://dx .doi.org/ 10.1016 /j.jksus.2017.06.012.

Shemetov, A.A., I. Nabiev, and A. Sukhanova. (2012). Molecular interaction of proteins and peptides with nanoparticles. *ACS Nano*, *6*, 4585–602.

Siddiqi, K.S. and A. Husen. (2017). Recent advances in plant-mediated engineered gold nanoparticles and their application in biological system. *J. Trace Elem. Med. Biol.*, *40*, 10–23.

Sperling, R.A., and W.J. Parak. (2010). Surface modification, functionalization and bioconjugation of colloidal inorganic nanoparticles. *Phil. Trans. R. Soc. A*, *368*, 1333–83.

Sun, Y., and Y. Xia. (2002). Shape-controlled synthesis of gold and silver nanoparticles. *Science*, *298*(5601), 2176–9.

Sun, Z.C., G. Tong, T.H. Kim, N. Ma, G. Niu, F. Cao, and X. Chen. (2015). PEGylated exendin-4, a modified GLP-1 analog exhibits more potent cardioprotection than its unmodified parent molecule on a dose to dose basis in a murine model of myocardial infarction. *Theranostics*, *5*, 240–50.

Sylvestre, J.P., S. Poulin, A.V. Kabashin, E. Sacher, M. Meunier, and J.H.T. Luong. (2004). Surface chemistry of gold nanoparticles produced by laser ablation in aqueous media. *J. Phys. Chem. B*, *108*(16), 864–9.

Tan, C., X. Cao, X.-J. Wu, Q. He, J. Yang, X. Zhang, J. Chen, W. Zhao, S. Han, G.-H. Nam, M. Sindoro, and H. Zhang. (2017). Recent advances in ultrathin two-dimensional nanomaterials. *Chem. Rev.*, *117*(9), 6225–331.

Tanaka, T., S. Shiramoto, M. Miyashita, Y. Fujishima, and Y. Kaneo. (2004). Tumor targeting based on the effect of enhanced permeability and retention (EPR) and the mechanism of receptor-mediated endocytosis (RME). *Int. J. Pharm.*, *277*, 39–61.

Upadhyay, L.S.B. and N. Verma. (2015). Recent developments and applications in plant-extract mediated synthesis of silver nanoparticles. *Anal. Lett.*, *48*(17), 2676–92.

van Embden, J., J. Jasieniak, D.E. Gómez, P. Mulvaney, and M. Giersig. (2007). Review of the synthetic chemistry involved in the production of core/shell semiconductor nano crystals. *Aust. J. Chem.*, *60*(7), 457–71.

Verma, A., O. Uzun, Y.H. Hu, Y. Hu, H.S. Han, N. Watson, S.L. Chen, D.J. Irvine, and F. Stellacci. (2008). Surface structure-regulated cell-membrane penetration by monolayer-protected nanoparticles. *Nat. Mater.*, *7*, 588–95.

Walsh, T.R. and M.R. Knecht. (2017). Bio interface structural effects on the properties and applications of bioinspired peptide-based nanomaterials. *Chem. Rev.*, *117*, 12641–704. DOI: 10.1021/acs.chemrev.7b00139.

Wang, L., Y. Sun, Z. Li, A. Wu, and G. Wei. (2016a). Bottom-up synthesis and sensor applications of biomimetic nanostructures. *Materials, 9*, 53.

Wang, R., I. Bowling and W. Liu. (2017). Cost effective surface functionalization of gold nanoparticles with a mixed DNA and PEG monolayer for nanotechnology applications. *RSC Adv., 7*, 3676–9.

Wang, X., J. Feng, Y. Bai, Q. Zhang, and Y. Yin. (2016b). Synthesis, properties, and applications of hollow micro-/nanostructures. *Chem. Rev., 116*(18), 10983–1060.

Wang, X., H. Tang, C.Z. Wang, J.L. Zhang, W. Wu, and X.Q. Jiang. (2016c). Phenylboronic acid-mediated tumor targeting of chitosan nanoparticles. *Theranostics, 6*, 1378–92.

Wang, Z., H. Wu, H. Shi, M. Wang, C. Huang, and N. Jia. (2016d). A novel multifunctional biomimetic Au@BSA nanocarrier as a potential siRNA theranostic nanoplatform. *J. Mater. Chem. B, 4*, 2519–26.

Wegst, U.G.K., H. Bai, E. Saiz, A.P. Tomsia, and R.O. Ritchie. (2015). Bioinspired structural materials. *Nat. Mater., 14*, 23–36.

Wiley, B., Y. Sun, B. Mayers, and Y. Xia. (2005). Shape-controlled synthesis of metal nanostructures: The case of silver., *Chem. Eur. J., 11*, 454–63.

Wong, C., T. Stylianopoulos, J. Cui, J. Martin, V.P. Chauhan, W. Jiang, Z. Popovic, R.K. Jain, M.G. Bawendi, and D. Fukumura. (2011). Multistage nanoparticle delivery system for deep penetration into tumor tissue. *Proc. Natl. Acad. Sci. U.S.A., 108*, 2426–31.

Wu, W., Z. Wu, T. Yu, C. Jiang, W.-S. Kim. (2015). Recent progress on magnetic iron oxide nanoparticles: Synthesis, surface functional strategies and biomedical applications. *Sci. Technol. Adv. Mater., 16*, 023501 (43pp).

Xia, X., J. Zeng, Q. Zhang, C.H. Moran, and Y. Xia. (2012). Recent developments in shape-controlled synthesis of silver nanocrystals. *J. Phys. Chem. C, 116*(41), 21647–56.

Xia, Y., X. Xia, and H.-C. Peng. (2015). Shape-controlled synthesis of colloidal metal nanocrystals: Thermodynamic versus kinetic products. *J. Am. Chem. Soc., 137*(25), 7947–66.

Xu, Q., X. Lou, L. Wang, X. Ding, H. Yu, and Y. Xiao. (2016). Rapid, surfactant-free, and quantitative functionalization of gold nanoparticles with thiolated DNA under physiological pH and its application in molecular beacon-based biosensor. *ACS Appl. Mater. Interfaces, 8*(40): 27298–304.

Xu, S., H. Yuan, A. Xu, J. Wang, and L. Wu. (2011). Rapid synthesis of stable and functional conjugates of DNA/gold nanoparticles mediated by tween 80. *Langmuir, 27*(22), 13629–34.

Yan, L., K. Zhang, H. Xu, J. Ji, Y. Wang, B. Liu, and P. Yang. (2016). Target induced interfacial self-assembly of nanoparticles: A new platform for reproducible quantification of copper ions. *Talanta, 158*, 254–61.

Yoo, J.W., D.J. Irvine, D.E. Discher, and S. Mitragotri. (2011). Bio-inspired, bio-engineered and biomimetic drug delivery carriers. *Nat. Rev. Drug Discovery., 10*, 521–35.

Yu, J.J., H.A. Lee, J.H. Kim, W.H. Kong, Y. Kim, Z.Y. Cui, K.G. Park, W.S. Kim, H.G. Lee, and S.W. Seo. (2007). Bio-distribution and anti-tumor efficacy of PEG/PLA nano particles loaded doxorubicin. *J. Drug Targeting., 15*, 279–84.

Zhai, Y., J. Su, W. Ran, P. Zhang, Q. Yin, Z. Zhang, H. Yu, and Y. Li. (2017). Preparation and application of cell membrane-camouflaged nanoparticles for cancer therapy. *Theranostics, 7*(10), 2575–92.

Zhan, G., J. Huang, M. Du, I. Abdul-Rauf, Y. Ma, and Q. Li. (2011). Green synthesis of Au-Pd bimetallic nanoparticles: Single-step bio-reduction method with plant extract. *Mater. Lett. 63*, 2989–91.

Zhang, F., Z. Ali, F. Amin, A. Feltz, M. Oheim, and W.J. Parak. (2010). Ion and pH sensing with colloidal nanoparticles: Influence of surface charge on sensing and colloidal properties. *ChemPhysChem, 11*, 730–5.

Zhang, F., E. Lees, F. Amin, P. Rivera Gil, F. Yang, P. Mulvaney, and W.J. Parak. (2011). Polymer-coated nanoparticles: A universal tool for bio labelling experiments. *Small, 7*, 3113–27.

Zhang, J., W. Gao, R.H. Fang, A. Dong, and L. Zhang. (2015). Synthesis of nanogels via cell membrane-templated polymerization. *Small, 11*, 4309–13.

Zhang, S., R. Geryak, J. Geldmeier, S. Kim, and V.V. Tsukruk. (2017). Synthesis, assembly, and applications of hybrid nanostructures for biosensing. *Chem. Rev.* 117(20):12942–13038.

Zhang, X., M.R. Servos, and J. Liu. (2012). Ultrahigh nanoparticle stability against salt, pH, and solvent with retained surface accessibility via depletion stabilization. *J. Am. Chem. Soc. 134*, 9910–13.

Zhao, W., L. Lin, and I.-M. Hsing. (2009). Rapid synthesis of DNA-functionalized gold nanoparticles in salt solution using mono-nucleotide-mediated conjugation. *Bioconjugate Chem., 20*(6), 1218–22.

13

Nanoferrite Composites: Synthesis, Characterization, and Catalytic Action on Thermal Decomposition of Ammonium Perchlorate

Jalpa A. Vara and Pragnesh N. Dave

CONTENTS

13.1 Introduction

Nanoscience is an interdisciplinary field of science which is closely connected with the research of the properties of matters at the atomic, molecular, and macromolecular ranges. 'Nanotechnology' denotes the technology that performs at a nanoscale level by controlling the shape and size at nanometer scale for design, production, characterization, and application. Nanotechnology was first described by the renowned physicist Richard Feynman in a meeting held at the American Physical Society, California Institute of Technology in 1959, entitled

'There's plenty of room at the bottom: An invitation to enter a new field of physics'. The term 'nano' comes from the Greek word 'nano' which means 'dwarf'. One nanometer is designated as 1nm and is equal to 10–9 m, which means one nanometer in length is approximately equivalent to the width of 6 carbon atoms or 10 water molecules (Mansoori 2005; Roukes 2009).

Nanomaterials are commonly classified as materials with an average grain size of less than 100 nanometers (Das and Ansari 2009); the materials include nanoparticles, and can are appreciated due to having improved properties like lower weight and higher strength. Nanomaterials comprise a branch involving atomic, molecular, and bulk structures. The novel and increased shape and size under such properties are exposed in comparison to their equivalent bulk materials. Nanomaterials are of attention as, at this range, the unique and relatively different magnetic, electrical, and optical properties emerge. Nanomaterials contain a much larger surface area-to-volume ratio than their bulk materials, which can cause better chemical reactivity and influence their strength. Also, at the nano range, quantum effects can happen to be highly important in determining the properties of materials and qualities, or most important to novel electrical, magnetic, and optical behaviors.

Ferrites are a huge group of oxides with amazing magnetic properties, which have been studied and proven useful throughout the last approximate 50 years (Valenzuela 2005). Ferrites are familiar magnetic nanomaterials effectively studied because of their greater physical properties. The properties of ferrites create a best applicant for technical uses like catalysis, magnetic resonance imaging enhancement, pigments, and sensors (Mathew and Juaug 2007). Mixed ferrites are considered effective for the coming years, owing to their potential utilization. The chemical formula of ferrites has MFe_2O_4 in which M can be any divalent metal cations.

Ferrites are fundamentally ferromagnetic oxide substances having high permeability and resistivity, while the diffusion magnetization of ferrite is under partially that of ferromagnetic alloys. The ferrite has benefits as an application due to its high resistivity, high frequency, higher resistance of heat, superior decomposition resistance, and low cost. Although its vast application is as a bulk material, the source of magnetism is a nano range occurrence (Beringer and Heald 1954). The improvement of magnetic nanostructure materials is a topic of concern, and equally for the scientific importance of considering the exclusive functional properties of materials, and for the technological importance in increasing the act of obtaining substances.

Composite solid propellants (CSPs) are the major source of chemical energy in space vehicles and missiles. Ammonium perchlorate (AP) is widely used as an oxidizer in composite solid propellants (Said 1991; Shen et al. 1993; Gao et al. 2001; Nema et al. 2004; Chaturvedi and Dave 2011). It is commonly observed that since catalytic activity is primarily a surface phenomenon, the size reduction of the catalysts increases their catalytic activity (Singh et al. 2009b). The ballistics of a composite propellant can be improved by adding a catalyst such as ferric oxide (Fe_2O_3), copper oxide (CuO), copper chromite ($CuO.Cr_2O_3$), nickel oxide (NiO), etc., which accelerates the rate of decomposition of AP (Said 1991; Shen et al. 1993; Gao et al. 2001; Nema et al. 2004; Chaturvedi and Dave 2011). Recent investigations have shown that nanoparticles of transition metal oxides without any agglomeration can increase the burning rate (Jacobs and Whitehead 1969). The efficiency of catalytic action increases more sharply in nanometer-size oxide particles than microscale oxide particles (Boldyrev 2006). The size distribution, morphology, and nanostructure of particles are very important characteristics and affect the kinetics of decomposition of ammonium perchlorate.

13.2 What is Nanoferrite?

Ferrite is an important class of material that has potential applications in integrated circuitry, transformer cores, and magnetic recording (O'Connor 1999). The study of ferrites has attracted immense attention from the scientific community because of their novel properties and technological applications, especially when the size of the particles approaches a nanometer scale.

The study of nano-crystalline materials is an active area of research in physics, chemistry, medical sciences, and material engineering as well as in multidisciplinary fields. Nowadays, material science research is focused on the invention of new materials with enhanced properties and novel synthesis techniques to cope with the increased technological demand. The study of ferrite nanoparticles is of interest due to the fundamental difference in their magnetic and electronic properties compared with the bulk counterparts. Nano-oxides, due to their electrical properties, have been found to be good catalysts, as compared to their corresponding normal oxides (Willard et al. 2004).

Nanomaterials are the center of attention due to their tremendous applications and interesting properties. Ferrites, a class of ferromagnetic compounds with the chemical formula of MFe_2O_4, where M represents any one of several divalent metallic elements such as Mg, Fe, Co, Ni, Cu, etc., commonly use magnetic nanoparticles. Nanoferrites have been the up-and-coming focus of attention of recent scientific research both from a synthesis and an application perspective. The properties of nanoferrites are very sensitive to the method of preparation and the sintering condition. Therefore, the selection of an appropriate process is the key to obtaining high-quality ferrites. The synthesis of various nanoparticles within the sol-gel auto-combustion technique, specifically ferrites, has demonstrated the ability to control the particle size, size distribution (Sugimoto 1999; Suzuki 2001; Nayak and Jena 2014), chemical stoichiometry, and cation occupancy.

13.2.1 Crystal Structure of Ferrites

Crystal structure and chemical compositions being the basis, ferrites are divided into three different types: (i) Spinel Ferrites, (ii) Garnets, and (iii) Hexaferrites.

13.2.1.1 Spinel Ferrites

Spinel ferrites have the general formula MFe_2O_4 or AB_2O_4 (where M or A is divalent metal cations such as Co^{2+}, Ni^{2+}, Zn^{2+}, etc., and B is Fe^{3+}) and belong to the space group Fd3m, that is derived from the crystal structure of natural mineral spinel $MgAl_2O_4$, where M is divalent metal cations, such as Fe^{2+}, Mn^{2+}, Co^{2+}, Ni^{2+}, Cu^{2+}, etc. (Gopalan and Anantharaman 2009). The crystal lattice structure of spinel ferrites is shown in Figure 13.1 (Issa et al. 2013). In spinel ferrites, 32 oxygen atoms form a unit cell having a face-centered cubic (FCC) structure arrangement, leaving two kinds of sites: tetrahedral sites (A) and octahedral sites (B), that is surrounded by four and six oxygen atoms respectively. The FCC structure of spinel ferrites contains 64 tetrahedral sites and 32 octahedral sites (B). To maintain the electrical neutrality of the lattice, one-eighth of the tetrahedral sites (A) and one-half of the octahedral sites (B) are occupied by the cations in a unit cell. The crystal structure of spinel ferrites has eight molecules per formula unit cell (Sugimoto 1999). On the basic of cation distribution in tetrahedral (A) and octahedral (B) sites, spinel ferrites are divided into three categories: normal spinel, inverse spinel, and mixed spinel ferrite structure. When tetrahedral sites

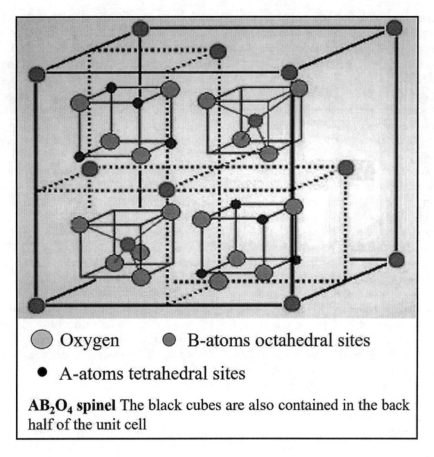

FIGURE 13.1 Crystal structure of spinel ferrites (Issa et al. 2013).

(A) are occupied by divalent metal cations and octahedral sites (B) are occupied by trivalent metal cations, then the structure is known as normal spinel. The normal spinel ferrites are defined by the formula $(M_1-\delta Fe\delta)$ A $[M\delta Fe_2-\delta]$ B O_4, where the first term in brackets describes tetrahedral (A) and the second term in square brackets describes the octahedral (B) sites components, while δ is the degree of inversion parameter of cations having zero values. $ZnFe_2O_4$ is the example of normal spinel ferrite. In an inverse spinel ferrite system, all divalent cations are distributed over half of the octahedral sites (B) and trivalent cations distributed over tetrahedral (A), whereas the remaining in octahedral sites. The value of δ in inverse spinel ferrites is one, and in the case of mixed spinel ferrites the value of δ lies at $0 < \delta < 1$ (Harris et al. 1996; Sickafus and Hughes 1999).

13.2.1.2 Garnet Ferrites

Garnet ferrites have a crystal structure of garnet minerals $Mn_3Al_2Si_3O_{12}$. Garnet ferrites are formed when Al and Si are replaced by Fe^{3+} ions and Mn is replaced by rare earth cation (R) to form a magnetic garnet having the general formula $R_3^{3+}Fe_5^{3+}O_{12}$. The garnet ferrite has a body-centered cubic structure with eight formula units. The garnet ferrites crystal has cubic symmetry and consists of three sub-lattices that are 24 tetrahedral (A), 16 octahedral (B) and 24 dodecahedral (C) sites. The largest dodecahedral sites are occupied by rare earth cations such as Y, La, Er, Gd, Sm, Eu, etc., and tetrahedral (A) and octahedral sites (B) are occupied by Fe^{3+} cations (Geller and Gilleo 1957).

13.2.1.3 Hexagonal Ferrites

Hexagonal ferrites and their magnetic property identification were firstly achieved by Went, Rathenau, Gorter, Van Oostershout, Jonker, Wijn, and Braun (Zijlstra 1982). Hexagonal ferrites are the class of permanent magnets having a general formula $MFe_{12}O_{19}$ where M is Ba, Sr, Ca, and Pb. The crystal structure of the hexagonal ferrites is made by oxygen ions having a close-packing hexagonal arrangement unit cell which contains two molecules of $MFe_{12}O_{19}$. The high magneto-crystalline anisotropy energy and coercivity of hexagonal ferrites makes it possible to form a permanent magnet. On the basis of the chemical formula and crystal structure, they are divided into five categories such as M-type or $SrFe_{12}O_{19}$, W-type or $SrMe_2Fe_{16}O_{27}$, Y-type or $SrMe_2Fe_{12}O_{22}$, X-type or $Sr_2Me_2Fe_{28}O_{46}$, and Z-type or $Sr_2Me_2Fe_{24}O_{41}$ (Stuijts et al. 1954; Standley 1972).

13.2.2 Properties of Nanoferrites

The properties of magnetic nanoparticles depend on particle size, surface effect, cations distribution in tetrahedral (A) and octahedral (B) sites, synthesis techniques, amount and types of dopant, and annealing temperature. As the particle size decreases from a bulk to nanometer range, there is a significant change in the physical, chemical, and mechanical properties of the system because of the increase in the surface-to-volume ratio of atoms in particles (Mathew and Juang 2007; William and Rethwisch 2012). The magnetic properties such as magnetic moment or magnetization and magnetic anisotropy of a nanoparticle system are different from those of a bulk specimen (Stoner and Wohlfarth 1948). The magnetic properties of magnetic materials which are size-dependent are observed in the range of a few microns to a few nanometers. In bulk form, particles exist in a multi-domain structure. Individually, domain is the region where all the spins' magnetic moments of atoms are aligned in a particular direction and this causes uniform magnetization. The magnetization of the multi-domain structure can be reversed by the moment of domain walls. When the particle size decreases in a nanometer range, the domain walls of the system become energetically unfavorable, leading to the formation of a single domain state where all spins' moments are aligned in one direction.

The Curie temperature of the spinel ferrite nanoparticles also depends on a super exchange interaction. The stronger the super exchange interaction, the higher the value of the Curie temperature. Spinel ferrites are good candidates for application from a commercial point of view because of their significant properties. The unique properties of spinel ferrites such as high magnetic permeability, high thermal stability, high electrical resistivity, high electrolytic activity, low magnetic losses, and resistance to corrosion make them a promising candidate in electronic and magnetic device applications (Olsen and Thonstad 1999; Suzuki et al. 2001). The chemical, structural, electrical, and magnetic properties of spinel type ferrite nanoparticles also depend on the distribution of cations in different lattices having different ionic radii, their composition, and synthesis method (Gubbala et al. 2004).

13.3 Synthesis Method of Nanoferrites

Synthesis of nanoferrites is an attractive and interesting research field due to their technical value, and magnetic property ferrite nanoparticles are used in various fields such as: biomedical (Mulens et al. 2013; Manikandan et al. 2014; Lee et al. 2015; Obaidat et al. 2015; Chinen et al. 2015); adsorbents and catalysts (Senapati et al. 2011; Flores et al. 2012; Wei et al. 2014; Zhao et al. 2014; Faria et al. 2014; Mahfouz et al. 2015; Tan et al. 2015; Ibrahim et al. 2016); manufacturing of electronic materials (Mohamed et al. 2010; Tang and Lo 2013); and wastewater treatment (Brar et al. 2010; Qu et al. 2013a, b).

A reported number of methods have been used in the synthesis of nanoferrites. These synthesis methods are divided into two wide categories that are 'bottom-up' and 'top-down'. Throughout bottom-up, the ions are chemically joined as one to form the particles, while in top-down, route materials are pulverized to form tiny particles. There are various synthesis techniques that incorporate the bottom-up alternative, which includes hydrothermal, co-precipitation, thermal decomposition, sol-gel, solvothermal, sonochemical, flame spray pyrolysis, microwave-assisted, vapor deposition, polyol techniques, and microemulsion; among them the initial four techniques are more prevalent. Mechanical milling and pulsed laser ablation are well-known for the top-down synthesis techniques.

13.3.1 Co-Precipitation

The development of nanoferrites using the co-precipitation method is between the most typical functional procedures used to obtain identical nanoparticles (Jain et al. 2005; Zhao et al. 2008; Drbohlavova et al. 2009; Amiri and Shokrollahi 2013; Rahimi et al. 2014; Li et al. 2015; Xing et al. 2015). In this process, aqueous solutions of trivalent and divalent transition metals are mixed together in the mole ratio of 1:2, respectively. Soluble salts having Fe (III) are usually used as a basis of trivalent metal ions and the process is generally conducted in an alkaline medium (Lopez et al. 2010; Ghandoor et al. 2012; Karimi et al. 2014). The synthesis procedure needs careful adjustment and control of pH in arrangements to prepare for top quality nanoferrites. The solution pH is generally adjusted by an ammonium solution or alkali solution. Then forceful stirring or agitation of the solution is required with or without heat under inert conditions. Considerable research studies have adopted this, and the route of synthesis for nanoferrites has been reported and the most recent ones are recognized herein (Zi et al. 2009; Xing et al. 2015; Kefeni et al. 2015; Thakur et al. 2015). These nanoferrites were synthesized in an aqueous solution in which the pH was adjusted by ammonium and alkali solution respectively. The subsequent yield of Fe_3O_4 nanoparticles was around 68% (6.5 g). Also, one of the striking characteristics of the technique is that it is more environmentally benign because salt dissolution is carried out in water. The final waste can be easily discharged after being treated for anion removal.

13.3.2 Hydrothermal

In the hydrothermal synthesis method, soluble salts having transition metals are dissolved individually and mixed in the mole ratio of 1:2, correspondingly (Peng et al. 2011; Zhang and Zhu 2012; Wang et al. 2013; Tadic et al. 2014; Chen et al. 2014; Tan et al. 2015; Wongpratat et al. 2015; Deepak et al. 2015; Georgiadou et al. 2015). This is followed by the drop-wise inclusion of organic solvents such as ethanol or ethylene glycol to the mixtures of aqueous solutions under continuous strong stirring in order to homogenize the solution made. Then the solution is heated under a pressure with high intensity in an autoclave. The time and temperature of heating are reliant on the kind of the target nanoferrites desired to be obtained (Pauline and Amaliya 2011; Zhang et al. 2015). The hydrothermal procedure is vital, since it is a capable synthesis procedure for large-scale ferrite nanoparticle production. Construction of better quality of nanoparticles with size distribution and controlled size can be accomplished by picking the proper mixture of solvents and varying variables such as pressure, reaction time, and temperature (Wang et al. 2005; Sharifi et al. 2012).

13.3.3 Sol-Gel

The sol-gel procedure usually comprises the aid of metal alkoxide solutions and it undergoes hydrolysis and condensation polymerization reactions to form gels at room temperature. Further heat treatments are required to eliminate any volatile by-products in order to get the final crystalline state (Dobre et al. 2009; Niu et al. 2010; Mulens et al. 2013; Masthoff et al. 2014; Singampalli et al. 2014).

A few of the benefits of the sol-gel synthesis process include the fact that it functions in low temperature, is low cost, and does not require a particular apparatus. The temperature of reaction in sol-gel differs between 25 and 200°C, and a narrow size distribution and controlled shaped ferrite nanoparticles are obtained using this technique.

These benefits, coupled with its cleanness in terms of the synthesis procedures of nanoferrites make the sol-gel method very attractive. Furthermore, the variable likes, stirring rate, sol concentration, and annealing temperature can be utilized to maintain the control composition, purity, microstructure, and shape of nanoparticles (Reda 2010). The main disadvantage of the procedure is that the nanoferrites produced high impurity because of contamination from reaction by-products; thus, purification is necessary in order to obtain pure nanoferrites.

13.3.4 Thermal Decomposition

The thermal decomposition method is among the easiest techniques for the formation of nanoferrites. This process involves thermal decomposition of organo-metallic precursors such as metallic acetylacetonates and carbonyls in the presence of organic solvents and surfactants (oleic acid and hexadecyl amine) for the synthesis of nanoferrites (Byun et al. 2009). Based on the type of precursor of metals, the low or high temperature can be used like a temperature of about 500°C is employed for the calcinations of maghemite (c-Fe_2O_3) in the air in order to form a-Fe_2O_3 (Darezereshki et al. 2012). On the other hand, during synthesis of Fe_3O_4, a low temperature of 165°C is used to decompose $Fe_3(CO)_{12}$ complex in diethylene glycol diethyl ether, with the oleic acid used as a stabilizer. The heating rate, temperature, or concentrations of precursors are the adjustable variable to get a controlled size and shape of ferrite nanoparticles, with nanoparticles with a highly monodisperse, uniform texture, and narrow distribution particle size (Dong et al. 2015). The oleic acid and iron oleate precursor's ratios are changing, and the thermal decomposition time changed from 2 and 10 h used in order to get spherical and cubic c-Fe_2O_3, respectively (Salazar-Alvarez et al. 2008). The production at a high scale with controlled size and shape can be carried out by employing this technique (Park et al. 2004; Wu et al. 2015; Fantechi et al. 2015). The procedure can be employed for the growth of ferrite nanoparticles required for medical as well as industrial applications.

13.3.5 Solvothermal

In the solvothermal procedure, both non-aqueous and aqueous solvents can be used to produce nanoparticles with specific switch over the shape, size distribution, and crystalline phases (Wu et al. 2015). These physical characteristics can be modified by altering certain experimental variables such as the reaction temperature, solvent, reaction time, surfactant, and precursors. A number of ferrite nanoparticles and their corresponding composites have been prepared using the solvothermal synthesis method. Usually, the solvothermal technique is conveniently applied for the development of ferrite nanoparticles required with developed physical and chemical characteristics and it is applicable to both industrial and biomedical areas, as required.

13.3.6 Sonochemical

Formation of nanoferrites by the sonochemical procedure has been reported as a suitable method, especially to form Fe_3O_4

and c-Fe_2O_3 nanoferrites (Shafi et al. 2001, 2002). During ultrasonic irradiation, bubbles are produced in the solvent medium and can effectively accumulate the diffuse energy of ultrasound; upon extreme collapse, high energy is released to heat the content of the bubble. It produces a transient localized hot spot with an actual temperature and pressure esoteric of the bubbles of approximately 5000 K and 1000 bars respectively with heating and cooling rates >1010 K S^{-1}. These extraordinary circumstances allow the use of a range of chemical reactions that are normally not accessible (Bang and Suslick. 2010). The composition of the systems expected to be synthesized using the sonochemical method is identical to the composition of the vapor in the bubbles; this assists in controlling the purity of nanoparticles (Tartaj et al. 2003). It is exclusively vital in crystal growth reduction, enables control over the particle size distribution, and uniformity of mixing but less significant for the formation of nanoferrites with controllable shapes and disparity (Wu et al. 2015). During the development of industrially vital non-ferrite NPs such as a-Fe_2O_3, the dependency of particle size on the intensity of ultrasound wave and reaction temperature has also been observed (Hassanjani-Roshan et al. 2011). In this study, deviation in a-Fe_2O_3 nanoparticle size was noticed, with a change in temperature and ultrasonic intensity. This indicates that with the sonochemical technique, utilized temperature, and ultrasonic intensity are the major determining factors that influence the particle size of nanoparticles. Commonly, it is attractive for the formation of magnetic nanoparticles due to the comfort of controlling reaction conditions and the opportunity for obtaining magnetic nanoparticles with high crystalline, coupled with low working temperature conditions.

13.3.7 Microwave-Assisted

The microwave-assisted synthesis procedure is a current technique which is used for the formation of versatile nonmaterial. In a microwave method, energy is conveyed directly to materials by using molecular interaction with the EMR. Heat is generated as the result of electromagnetic energy conversion to thermal energy from 100 to 200°C with a shorter reaction time (Mondal et al. 2015).

The exhaust drain connected to the Teflon vessel is used to drain any vapor produced during operation (Bhatt et al. 2011; Wu et al. 2015; Gonzalez-Moragas et al. 2015). The reasonable cost production of nanoferrites is possible with narrow size distribution in an instant, and is able to produce a high quality yield (Ding et al. 2008; Hu and Yu, 2008; Wu et al. 2015). The improved multi-mode equipment of the microwave-assisted method for the synthesis of nanoferrites has recently been reported (Gonzalez-Moragas et al. 2015). The large-scale production of nanoferrites can be carried out in equipment in multi-mode with exposure of several vessels in parallel mode. It is important to overcome hindrances faced in microwave-assisted synthesis with a single vessel. It offers the formation of nanoferrites on a higher scale with low yield, as compared to other methods.

13.3.8 Microemulsion

In the microemulsion synthesis technique, surfactant is used to stabilize the dispersion of two relatively immiscible liquids.

The different surfactant such as n-butanol and polyoxyethylene is used as a co-surfactant and non-ionic surfactant with n-hexane as an oil phase (Foroughi et al. 2015). It is a beneficial technique and provides diversity in the preparation of ferrites by varying the doses of oil water, co-surfactant, reaction conditions, and the nature of the surfactant (Hasany et al. 2013). Therefore, controlling the size of particles produced is one of the major benefits of the microemulsion method (Langevin 1992). Furthermore, it is a technique with a low temperature. As a low temperature reaction, it carries a drawback for the formation of more polydispersed and low crystallinity, owing to a slow nucleation rate (Lin and Samia 2006). The method can be classified into two major categories, namely, 'reverse water-in-oil' (Zhi et al. 2006; Zhou et al. 2013; Foroughi et al. 2015), and 'normal oil-in water' (Moumen et al. 1995; Pemartin et al. 2014). The mono-disperse droplet exists in the size of 2–100 nm in both cases. In both cases, the dispersed phase consists of mono-disperse droplets in the size range of 2–100 nm. This dispersed phase offers a limited environment for the synthesis of nano range particles (Lin and Samia 2006). The attractive aspect of microemulsion is that stable nanoferrites can be produced.

13.3.9 Polyol

Recently, the polyol technique has received more consideration in the preparation of nanoferrites. In this technique, the role of ethylene glycol is both the solvent and reducing agent; in particular, synthesis of mixed nanoferrites such as $Zn_{0.7}Ni_{0.3}Fe_2O_4$ (Flores-Arias et al. 2015), ($Mn_{0.48}Zn_{0.12}Fe_{2.4}O_4$) (Gaudisson et al. 2015), etc. were recently reported.

The different six nanoferrites have been synthesized by utilizing this technique (Solano et al. 2015). In the modified polyol method, mixtures of three different constituents were used in the starting materials such as MCl_2 (M = Mn, Fe, Co, Ni, Cu, Zn), Iron (III) acetylacetonate as an iron source and triethylene glycol, which acts as a solvent and coating legend at the same time.

The system is designed for synthesis of 13 mM MFe_2O_4 per run, and the heater is prepared with a magnetic stirring and reflux system. The mixture is heated until 280°C with a controlled ramp of 1°C min⁻¹ starting from the room temperature, after a 2.5 hr annealing time at 280°C, the mixture is cooled to room temperature and separated from the solution by an external magnet. Irrespective of different sizes and type of divalent metal ions, all the synthesized ferrites had an average size of about 10 nm. From the experimental results, it was observed that a lower heating ramp resulted in narrow size distribution, while increasing annealing time resulted in a larger particle size. The synthesized nanoferrites were of high quality, high saturation magnetization (Ms), and polar dispersible properties.

13.3.10 Electrochemical

The electrochemical procedure of synthesis is similar to co-precipitation, the difference is that the ion sources are due to scarification of anode electrodes through the oxidation process. Once the anode electrode is oxidized and released into the surfactant solution, ions interact between the electrode–electrolyte interface and form the corresponding nanoferrites (Mazario et al. 2012). This technique provides the advantage of a highly pure and regulated particle size over other methods. It can be achieved by adjusting the current, or optimizing the space between the electrodes (Wu et al. 2015).

The size and crystallinity of NPs have been influenced by the space between electrodes and density of current. The enhancement in the particle size is observed when reducing the space between electrode and increase current density (Fajaroh et al. 2012). This is because the closer the electrodes, the faster the reduction oxidation happens, releasing more oxidized metal ions into the solution; this in turn results in a fast rate of nanoparticle formation. It is simple, inexpensive, cost-effective, and environmentally benign (Ramimoghadam et al. 2014). However, factors such as pH, current density, electrode choice, electrolyte concentration, and space between the electrodes should be properly estimated in order to achieve the necessary quality of nanoferrites intended to be synthesized.

13.3.11 Mechanical Milling

This is a typical technique recommended for the development of nanoparticles utilizing planetary balls, high-energy shaker, or tumbler mills. It forms nanoferrites by a 'top-down' approach. The formation of NPs in tonnage scale can be carried out by this technique but it carries drawbacks like the contamination of nanoparticles throughout milling (Manova et al. 2004). A number of nanoferrites have been synthesized using this technique. Therefore, advances in the technique are necessary in order to get nanoferrites of higher crystallinities and super paramagnetic (SPM) properties.

13.3.12 Laser Ablation

In the laser ablation synthesis method, small-size particles are produced by the irradiation solid surface with a laser beam in liquid. The solvent can be a organic solvent or water. The preparation of an ultra-size particle takes place when diethylene glycol (DEG) is used as a liquid medium because large viscosity of solvent retards the growth of particles (Luo et al. 2013). The three main stages of this technique are: initial stage; absorption of laser energy; generation of heat waves which proliferate inside the target material and liquid. This results in material ejection and plume formation. In limiting the liquid environment, the laser column becomes bigger.

The aggregation of nanoparticles takes place when liquid is mixed with plume. The tiny nanoparticles are formed with the beneficial effect of ultra-short and short laser pulses (Itina 2011). This technique is supportive with forming size-controlled nanoparticles by the laser-induced reduction of size. The highly pure nanoparticles can be synthesized with the technique due to free use of a reducing agent (Luo et al. 2013; Chen et al. 2015). The technique is an auspicious one for the development of nanoferrites with high Ms (Franzel et al. 2012). In particular, the synthesized nanoferrites created by this technique are appropriate for imaging and cancer treatment applications.

13.4 Characterization of Nanoferrites

Many techniques are used for the characterization of nanoferrites such as X-ray diffraction (XRD), transmission electron microscopy (TEM), scanning electron microscopy (SEM), Raman spectroscopy (RS), X-ray photoelectron spectroscopy (XPS), Fourier transform infrared (FT-IR), energy dispersive X-ray spectroscopy (EDS), vibrating sample magnetometer (VSM), atomic force microscopy (AFM), superconducting quantum interference device (SQUID), Mössbauer spectroscopy (MS), electron paramagnetic resonance (EPR), thermo gravimetric analysis (TGA), and magnetic force microscopy (MFM).

13.4.1 Size and Shape of Nanoferrites

Stability and magnetic properties are governed through the shape and size of nanoferrites. XRD techniques are used to determinate the size of nanoferrites. X-ray diffraction patterns show the nature of nanoferrites; the sharp peaks show crystalline materials; and broad peak or no sharp peaks indicate amorphous materials (Chaturvedi et al. 2014). XRD is a universal, simple, comparatively inexpensive and non-destructive technique. It is generally utilized for assessment of percentage crystallinity, assessment of sample purity, and unit cell dimensions (Chekli et al. 2016). The size of nanoferrites is evaluated from XRD diffraction peaks utilizing the Scherrer equation. Shape determination of nanoferrites is mostly utilized through scanning electron microscopy (SEM) and transmission electron microscopy (TEM). In these methods, the atomic arrangement of nanoparticle is determined. TEM applies an accelerated electron beam to elucidate particles. HRTEM can image particles given through photographic data about shape, size, crystallinity and lattice spacing, and aggregation state (Chekli et al. 2016). The measurements of XRD were not capable to resolve the size of extremely small or no crystalline nanoferrites by reason of broad peak or not well recognized peaks, while TEM gives an enhanced size resolve (Gabbasov et al. 2015).

Nanoferrites can provide analysis of elemental composition with good sensitivity and precision through utilizing inductively coupled plasma mass spectroscopy (ICP–MS) (Galindo et al. 2012). ICP–OES (inductively coupled plasma – optical emission spectrometry) and atomic absorption spectroscopy (AAS) are commonly used for elemental estimation of nanoferrites and are dissolved in appropriate acids or bases. SEM and TEM are also used for the estimation of the sample elemental composition when coupled with EDS (Farre et al. 2011). XRD is used for the bulk mineral composition of crystalline nanoparticles. Many other techniques are used for the estimation of average size of particle and size distributions, such as Mössbauer spectroscopy, photon correlation spectroscopy, and dynamic light scattering (DLS).

13.4.2 Surface Morphology of Nanoferrites

The surface morphology of nanoferrites can be considered through SEM, TEM, and AFM. These techniques are commonly utilized to visualize the morphology of nanoferrites and to determine their diameter. SEM is a surface-sensitive imaginative technique; it uses the electrons that are created when the beam interacts with the surface of the sample. Therefore, the samples require conductive samples, and non-conductive samples need surface coating before analysis through SEM. It is generally used to study surface morphology and the aggregation state of nanoparticles. TEM is an extra imaginative system, given the morphological and constitutional aspects of the crystallite of nanoferrites (Lee et al. 2015). It can give a good resolution behind the atomic scale. When attached with other instruments like EDS and electron energy loss (EEL), TEM can produce atomic-scale elemental diagrams for composition analysis. The topography of nanoparticles is analyzed through AFM and also gives quantitative data on the morphology, particle size, and its distribution over a flat surface. In SEM and TEM, only dry samples are applied but AFM can be used in a broad range of conditions (air, liquid or moist, vacuum conditions) (Farre et al. 2011; Mahfouz et al. 2015).

13.4.3 Structure and Bonding of Nanoferrites

The most familiar methods are FT–IR, Raman spectroscopy (RS), XPS, X-ray absorption spectroscopy (XAS), and TGA. FT–IR and XPS are utilized to prove the arrangement of metal and oxygen bonds, recognizing the determination of the inverse spinel structure of ferrites and the occurrence of other chemical compounds, which are adsorbed on the surface of particles with organic functional groups existing. XPS has frequently been applied for the identification of the surface composition of a range of different nanoparticles (Wang et al. 2011). XPS are capable of providing data on binding energy, oxidation state, and elemental constitution of materials on the surface (Xing et al. 2015). Determining the unknown structure of substances and for investigation of the spinel lattice is carried out by using RS (Li et al. 2015; Georgiadou et al. 2015), and to estimate any organic content bound to the surface of nanoferrites through utilizing TGA (Douvas et al. 2014). XAS is extremely vital for data about the bond length, oxidation state, coordination number, identity of nearest neighbors, and electronic configuration of an element of interest (Reddy et al. 2012; Grafe et al. 2014).

13.4.4 Surface Area Nanoferrites

Surface area determination of nanoferrites uses the Brunauer–Emmett–Teller (BET) standard method. In this method, the dry powder of nanoferrites in a vacuum is used and then determines the adsorption of nitrogen gas on the surface and micropores (Hassellov et al. 2008). The basic premise in the BET system is that nitrogen gas has to acquire the complete surface of the particles.

13.4.5 Magnetic Property Nanoferrites

Magnetic property nanoferrites are evaluated through using the VSM, EPR, and SQUID techniques. VSM are applied when highly sensitive magnetization estimates are needed (Graham 2000). EPR is also a sensitive technique in the detection and identification of free radicals and paramagnetic centers such as F-center in chemicals and chemical reactions. It is a method to study the physical properties of magnetic nanoparticles and their actions on the effect of external magnetic fields. It is based on the interaction of an external magnetic field with magnetic moments of unpaired electrons in a sample (Dobosz et al. 2016). SQUID measures different types of samples; for example, crystals, powders, liquids, gases, and thin films (Levy et al. 2011a, b). The samples are characteristically filled into a gelatin capsule. Utilizing these systems, there is potential to estimate corresponding coercivity (Hs), saturation magnetization (Ms), and residual magnetization of nanoferrites at a utilized constant external magnetic field and temperature. In addition, the structural and magnetic properties, bonding, oxidation and spin state, covalence and electro negativity of an assumed element in the nanoferrites can be attained utilizing a Mössbauer spectroscopy. It is a multipurpose analytical method that is based on estimating nuclear energy level transitions with γ rays (Hurley et al. 2015). Usually, these methods give the entire image of nanoferrite dipole interactions and magnetic properties.

13.5 Catalytic Efficiency on Thermal Decomposition of Ammonium Perchlorate

Ferrites are generally ferromagnetic oxide constituents comprising high resistivity and high permeability. Ferrite-carried saturation magnetization is less compared to moderate ferromagnetic alloys but it has benefits, such as: it can used at a high resistivity, has a greater heat resistivity, is low-cost, has a high frequency, and has an advanced corrosion resistance. Commercial use of ferrites hiked after 1950 in television, computer circuitry, carrier telephony, radio, and microwave devices. Despite its vast utilization as a bulk material for the origin of magnetism, it is a nanosize aspect (Beringer and Heald 1954). Ferrite comprises the divalent metal ions mixture of two and its ratio may differ. The catalytic proficient ability of material is contingent on the surface properties of it. The chemical reaction rate can be enhanced by nano-size ferrite owing to the huge number of cations and the nano size.

Ammonium perchlorate (AP) is mainly used as an energetic material. It is employed as oxidizer in propellants recognized as ammonium perchlorate composite propellants (APCP). Unfortunately, AP is also one of the energetic materials which are only slightly understood. Over the course of some decades, a lot of research found the structural properties and thermal decomposition mechanism of AP. The AP possesses tough thermal decay, thanks to being comprised of four different components N, H, Cl, and O. If one compares all the probable oxidation states of these four components, more than 1,000 probable chemical reactions can be written for the thermal decomposition of AP (Brill and Budenz 2000). The burning tendency of the propellants are affected by the thermal decay proficiency of AP (Singh and Felix 2003; Singh et al. 2009a–c). To require a combustion catalyst for greater burning rates, conventionally used transition metals act as a catalyst in the AP-based composite solid propellants (Kishore and Sunitha 1979a, b; Singh and Felix 2003; Carnes and Klabunde 2003). Preceding studies recommended that catalysts are dynamic mostly in the condensed phase and therefore activity of the catalysts in the condensed phase thermolysis of AP can be a better main additive of the catalytic proficiency of the ingredient in combustion of CSPs (Singh and Pandey 2002).

P N Dave et al. (Dave et al. 2015) reported that nanoferrites act as a catalyst for decomposition of AP. The five different types of nanoferrite composite such as NiF, ZnF, CoF, CuF, and MnF were synthesized through a chemical co-precipitation method. The XRD and SEM techniques were applied to define their nature, shape, and crystallinity and the average particle size is given in Table 13.1. The catalytic efficiency of nanoferrites on the decomposition of ammonium perchlorate was measured utilizing TGA-DSC techniques. In the thermal analysis of pure AP, there are occurrences in two steps: the 25% weight loss at around 300°C, which is a low temperature decomposition (LTD) and 80% of weight loss at around 450°C, which is a high temperature decomposition (HTD). It can be concluded from a TGA curve of AP with nanoferrite composite that the catalysts affect both the LTD and HTD of AP. A TGA curve shows higher percentage of weight loss of AP in thermal decomposition route with nanoferrite, as compared with pure AP.

Binary nanoferrites are good catalysts for decomposition of AP, as reported by Singh et al. (2009c). Binary nanoferrites are developed by employing the co-precipitation method. The average particle size was calculated (Table 13.1) through using the Scherrer equation for CoF, CuF, and NiF from XRD data. The catalytic proficiency of binary nanoferrites on the thermal decomposition of AP were studied by using the TGA and DTA techniques. Nanoferrites were established to enhance the thermal decomposition of AP to a greater range owing to the nanosize of the ferrites. They evaluated the order of the catalytic proficiency of catalysts for thermal decomposition of AP based on the above experiments, as below: CoF > CuF > NiF

Furthermore, the author has explored the effect of nanoferrites on the decay of AP by changing their content. In the isothermal

TABLE 13.1

Particle Size of Nanoferrites

Sample	Particle size by XRD	Reference
NiF	1.1	Dave et al. (2015)
ZnF	6.8	
CoF	5.2	
CuF	3.9	
MnF	2.7	
CoF	39.9	Singh et al. (2009)c
CuF	27.3	
NiF	43.8	
$CuFe_2O_4$	26.0	Liu et al. (2008)
$CoFe_2O_4$	24.0	Zhao and Ma (2010)
$CdFe_2O_4$	14.9	Singh et al. (2010)
NiZnF	7.2	Singh et al. (2008)
CuCoF	3.0	
NiCuF	3.0	
CuZnF	6.8	
CoNiF	3.9	
CoZnF	5.8	
CuCoNiF	40.3	Srivastava et al. (2009)
NiCuZnF	50.0	
CuCoZnF	19.9	
MnF	27.4	Singh et al. (2013)
CoF	26.4	
NiF	31.4	

TG curve of AP in the existence of nanoferrites with different contents, the experimental results clearly point out that the catalytic activity enhances with increases in the amount of binary nanoferrites in AP.

Singh et al. (2008) reported that ternary nanoferrites can be used as modifiers in decomposition of AP through a thermal route. The ternary transition nanoferrite was prepared by employing co-precipitation method X-ray diffraction (XRD) and a BET equation was used to establish its characterization. The average particle size of nanoparticle by using XRD is shown in Table 13.1. The non-isothermal thermogravimetry (TG) curves in flowing inert air and N_2 atmosphere for pure AP are shown in the Figure 13.2. It is clearly pointed out that thermal decomposition of AP takes place in two steps (Bircomshaw and Newman 1955; Rosser et al. 1968). In the TG curves for AP with nanoferrites, it clearly shows that the catalyst influences both LTD and HTD processes and additional gasification of AP in the attendance of catalyst throughout the HTD process. It not only starts early but is also entirely at a low temperature (30–60°C). This is the reason that the temperature variance detected in the static air and inert atmosphere is because of experimental conditions. The last exothermic peak of AP was shifted at a low temperature, as shown in Figure 13.3, which is attributable to the presence of nanoferrites. IR spectra results of pure AP and AP with nanoferrites, which display increased intensity of νCl–O (1,087cm^{-1}) and the mode of ClO^-_4 (625cm^{-1}) with temperature. Up to 523K, νas (N–H) 3,179cm^{-1} and (N–H) bending ,1407cm^{-1}, intensity reduces slowly; but above 523K, a marked reduction happens. The ClO^-_4 modes are the maximum temperature sensitive ones attributable to the enhanced rotational motion at high temperatures. These variations in the action and anion internal modes equally, could be associated with the enhanced ion release along with LTD decomposition. The catalytic proficiency was found to be in the order: CoZnF > CoNiF> CuZnF> CuCoF > NiCuF > NiZnF

With the addition of nanoferrite catalyst into the AP, it increased the rate of heterolysis of N–H bond in NH_4^+ and O–H bond creating in $HClO_4$. The AP decomposition reaction frequently includes both bond breaking and making steps. Approximately one thousand reactions may be intricate in the decomposition and burning of AP (Rosser et al. 1968; Ermolin et al. 1982), attributable to the four elements and the complete range of oxidation states applied through nitrogen, chlorine, and supports to proton transfer mechanism. Concurrently, a sequence of reaction occurs to procedure gaseous NH_3 and $HClO_4$. The products are Cl_2, O_2, N_2O, H_2O, and a small quantity of NO. The breaking of an N–H bond, proton transfer from NH_4^+ to ClO^-_4 and formation of an O–H bond shows in the making of NH_3 and $HClO_4$ is an initial step in condense phases. The final step decomposition of AP is gas phase reaction, and the products are Cl_2, O_2, NO, and H_2O (Rosser et al. 1968; Boldyrev 2006).

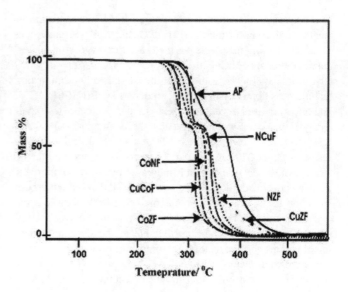

FIGURE 13.2 TG curve of pure AP and AP with Nanoferrite. (Singh et al. 2008).

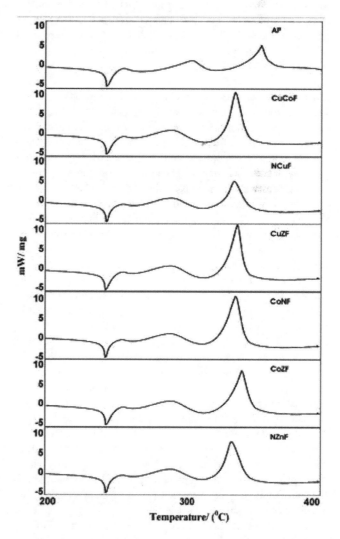

FIGURE 13.3 DSC curve of pure AP and AP with Nanoferrite (Singh et al. 2008).

$$NH_4^+ + ClO_4^- \leftrightarrow NH_3\text{–}H\text{–}ClO_4 \leftrightarrow NH_3\text{–}HClO_4$$
$$\text{(i)}$$

$$\leftrightarrow NH_{3(a)} + HClO_{4(a)} \leftrightarrow NH_{3(g)} + HClO_{4(g)}$$
$$\text{(ii)} \qquad\qquad\qquad \text{(iii)}$$

The quaternary nanoferrite is enhanced by the catalytic proficiency on decomposition of AP, HTPB, and CSPs (Srivastava et al. 2009). In this investigation, there are three types of quaternary nanoferrite (NiCuZnF, CuCoNiF, and CuCoZnF), which have been synthesized utilizing the sol-gel auto-combustion technique. The sophisticated techniques like FE-SEM, TEM, and XRD are used to characterize quaternary nanoferrites. Their catalytic behavior was explored by the decomposition of AP, CSPs, and HTPB with TGA and DTA thermal techniques. The result of TGA–DTA on AP and AP with quaternary nanoferrite stated that the decomposition is increased by adding quaternary nanoferrites. In addition to nanoferrites in AP that influence both LTD and HTD, it is not alone in enhancing the weight loss but also lowers the HTD range of AP. The AP with catalyst decomposition increased may be attributable to the development of metal perchlorates or metal perchlorate amines and the exothermic catalyzed decomposition of AP. They have good catalytic proficiency on the high temperature decomposition of AP and the order as: NiCuZnF > CuCoZnF > CuCoNiF.

Zhao and Ma (2010) studied the cobalt ferrites (CoFe$_2$O$_4$) synthesized via the polyol medium solvothermal method and it was characterized through TEM and XRD. For cobalt ferrites, the average particle size was predicted from XRD and TEM is given in Table 13.1. The catalytic efficiency of cobalt ferrites (1, 2, and 5 wt %) on the thermal decomposition of AP was utilized through TGA and DSC. The TGA and DSC curve for decomposition of AP in the existence of cobalt ferrites with a different ratio has revealed a direct variation in the decomposition pattern. The catalytic efficiency of cobalt ferrite nanoparticles is important not only for the high temperature decomposition but also on the low temperature decomposition process, and particularly through the HTD process; it not only starts early, but also completes with only one-step weight loss at a low temperature. Its situation depends dynamically on the content of the catalyst, and the catalytic efficiency is found to be enhanced with enhancing the quantity of the catalyst. The catalytic efficiency is improved in the ratio of 5 wt %. Kinetic parameters for HTD process of AP in the existence of 5wt. % cobalt ferrite different heating rates have also been described by authors. It was noticed that the activation energy (*Ea*) of AP in the occurrence of 5wt. % for HTD was considerably lower than pure AP. This lowering in *Ea* is in concurrence with the commonly noticed trend of lowering of Ea for a reaction whose rate is enhanced through a catalyst.

Nano-MnF$_2$O$_4$ is also an excellent catalyst for the decomposition of AP, as reported by Han et al. (2011). Synthesizing of nano-MnFe$_2$O$_4$ occurred by applying two procedures: the co-precipitation phase inversion process, and low temperature combustion process. Characterizing the structure and morphology occurred through utilizing TEM and XRD, while TG, DTA and DSC were applied to study the catalytic efficiency of nano-MnFe$_2$O$_4$ on the thermal decomposition of AP. Nano-MnFe$_2$O$_4$ has an excellent catalytic efficiency over AP decomposition. Nano-MnFe$_2$O$_4$ was synthesized through a low temperature combustion technique, and displayed excellent catalytic efficiency attributable to its smaller particle size.

Liu et al. (2008) synthesized copper nanoferrite CuFe$_2$O$_4$ through an auto-combustion technique. There occurred characterization of all precursors and an as-burnt sample by XRD and TEM. The catalytic efficiency of copper nanoferrite (2, 5, and 6 wt. %) on the thermal behavior of AP was studied through DTA. The copper nanoferrite has high catalytic efficiency on the decomposition temperature of AP transfer 105°C descending with the presence of 2wt% copper nano ferrite. The best results were achieved when adding 5% of copper nanoferrite to the catalytic efficiency. The authors have also proposed the catalytic mechanism of the copper nanoferrite on the AP decomposition. There are lattice defects in the as-burnt copper nanoferrite, and numerous positive holes and electrons exist in it. In the decomposition process of AP, copper nanoferrite produces a bridge for the transfer of electrons from the perchlorate ions to the ammonium ions. Copper nanoferrite endorses the transfer of electrons, which produces the activation energy of thermal decomposition of AP reduce; significantly, the decomposition temperature of AP reduces and the decomposition rate of AP accelerates.

The synthesis of nanoferrites of Mn, Co, and Ni such as nano-rods, nano-spheres, and nano-cubes through a wet-chemical technique under dissimilar synthesis conditions was reported by Singh et al. (2013), and also their characterization of nanoferrites by using FE-SEM, HR-TEM, XRD, TEM, and EDS techniques. The average size of nanoferrites, determined from XRD, is revealed in Table 13.1. The catalytic efficiency of nanoparticles on decomposition of AP was investigated through DSC. Noticeable change of the first and second exothermic peak significantly shifted in the DSC curve. MnF$_2$O$_4$ rods are predictable at having excellent catalytic efficiency on decomposition of AP through mainly revealing the reactive planes and superior number of well-defined active sites. The burning rate data of CSPs with these nanoferrites is enhanced when MnF$_2$O$_4$, Co F$_2$O$_4$, and Ni F$_2$O$_4$ are used as catalysts.

CdFe$_2$O$_4$ nanoferrites were utilized as catalysts for the decomposition of AP and CSPs. The synthesis of CdFe$_2$O$_4$ nanoferrites occurred through a wet chemical technique (Singh et al. 2010). CdFe$_2$O$_4$ nanoferrites were characterized through XRD and morphology via TEM. The catalytic efficiency was investigated through decomposition of AP with and without CdFe$_2$O$_4$ nanoferrites. AP has three peaks in DSC curve; the first peak is an endothermic peak at 242°C that recognized its transition from an orthorhombic to cubic form (Boldyrev 2006). The second peak is a first exothermic peak at 310°C that recognized the incomplete decomposition of AP and the development of intermediate products. The last exothermic peak displays at a high temperature of around 430°C, demonstrating the whole decomposition of AP. While AP with CdFe$_2$O$_4$ nanoparticles were the endothermic peak displays at a similar temperature, demonstrating that catalysts have no effect on the crystallographic transition temperature but obvious variations were noticed in first and second exothermic peaks. The two exothermic peaks were merged into one, and showed at about 320°C. The burning rate of CSPs is increased when adding CdF$_2$O$_4$ as a catalyst. The enhanced burning rate might be attributable to the increased thermal decomposition of AP and CSPs.

13.6 Conclusion

Nanoferrites are an important class of ferromagnetic compounds which are used as a catalyst in the thermal decomposition of ammonium perchlorate. The synthesis and characterization of nanoferrites are an attractive research field because of the high interest in the best quality nanoferrites. Good synthesis methods produce nanoferrites which increase their performance such as the size, shape, and properties of nanoferrites which require advanced technology. Nanoferrite particle sizes are reliant on the nature of transition metal and the synthesis method used. Nanoferrites give best results as a catalyst for the thermal decomposition of ammonium perchlorate; they enable a decrease in the decomposition temperature of ammonium perchlorate.

ABBREVIATIONS

AP	Ammonium perchlorate
CSPs	Composite solid propellants
XRD	X-ray diffraction
TEM	Transmission electron microscopy
SEM	Scanning electron microscopy
RS	Raman spectroscopy
XPS	X-ray photoelectron spectroscopy
FT-IR	Fourier transform infrared
EDS	Energy dispersive X-ray spectroscopy
VSM	Vibrating sample magnetometer
AFM	Atomic force microscopy
SQUID	Superconducting quantum interference device
MS	Mössbauer spectroscopy
EPR	Electron paramagnetic resonance
TGA	Thermo gravimetric analysis
MFM	Magnetic force microscopy
ICP-MS	plasma mass spectroscopy
LTD	Low temperature decomposition
HTD	High temperature decomposition
DSC	Differential scanning calorimetry
DTA	Differential thermal analysis
HTPB	Hydroxyl-terminated polybutadiene

REFERENCES

Amiri, S., and H. Shokrollahi. (2013). Magnetic and structural properties of RE doped Co-ferrite (REåNd, Eu, and Gd) nano-particles synthesized by co-precipitation. *J. Magn. Magn. Mater.*, *345*, 18–23.

Bang, J.H., and K.S. Suslick. (2010). Applications of Ultrasound to the Synthesis of Nanostructured Materials. *Adv. Mater.*, *22*, 1039–59.

Beringer, R., and M.A. Heald. (1954). Electron spin magnetic moment in atomic hydrogen. *Phys. Rev.*, *95*, 1474–81.

Bhatt, A.S., D.K. Bhat, C. Tai, and M.S. Santosh. (2011). Microwave-assisted synthesis and magnetic studies of cobalt oxide nanoparticles. *Mater. Chem. Phys.*, *125*, 347–50.

Bircomshaw, L., and B. Newman. (1955). Thermal decomposition of ammonium perchlorate. *Proc. R. Soc. A*, *227*, 228–37.

Boldyrev, V.V. (2006). Thermal decomposition of ammonium perchlorate. *J. Thermochim. Acta*, *443*, 1–36.

Brar, S.K., M. Verma, R.D. Tyagi, and R.Y. Surampalli. (2010). Engineered nanoparticles in wastewater and wastewater sludge--evidence and impacts. *Waste Manage.*, *30*, 504–20.

Brill, T.B., and B.T. Budenz. (2000). Progress in aeronautics and astronautics. In Yang, V., Brill, T.B., and Ren, W.Z. (Eds.), *Solid Propellant Chemistry, Combustion and Motor Interior Ballistics* (vol. 185, Chapter 1.1, pp. 3–32). American Institute of Aeronautics and Astronautics.

Byun, M., J. Wang, and Z. Lin. (2009). Massively ordered microstructures composed of magnetic nanoparticles. *J. Phys. Condens. Matter.*, *21*, 264014.

Carnes, C.L., and K.J. Klabunde. (2003). The catalytic methanol synthesis over nanoparticle metal oxide catalysts. *J. Mol. A: Catal. Chem.*, *194*(2003), 227–36.

Chaturvedi, S., and P.N. Dave. (2011). Review: Nano metal oxide: Potential catalyst on thermal decomposition of ammonium perchlorate. *J. Exp. Nanosci.*, *7*(2),1–27.

Chaturvedi, S., P.N. Dave and N.N. Patel. (2014). Nano-alloys: Potential catalyst for thermal decomposition of ammonium perchlorate. *Synth. React. Inorg. Met.-Org. Nano-Met. Chem.*, *42*, 258–62.

Chekli, L., B. Bayatsarmadi, R. Sekine, B. Sarkar, A.M. Shen, K.G. Scheckel, W. Skinner, R. Naidu, H.K. Shon, E. Lombi, and E. Donner. (2016). Analytical characterisation of nanoscale zero-valent iron: A methodological review. *Anal. Chim. Acta*, *903*, 13–35.

Chen, F., M. Chen, C. Yang, J. Liu, N. Luo, G. Yang, D. Chen, and L. Li. (2015). Terbium-doped gadolinium oxide nanoparticles prepared by laser ablation in liquid for use as a fluorescence and magnetic resonance imaging dual-modal contrast agent. *Phys. Chem. Chem. Phys.*, *17*, 1189–96.

Chen, R., W. Wang, X. Zhao, Y. Zhang, S. Wu, and F. Li. (2014). Rapid hydrothermal synthesis of magnetic CoxNi1−xFe2O4 nanoparticles and their application on removal of Congo red. *Chem. Eng. J.*, *242*, 226–33.

Chinen, A.B., C.M. Guan, J.R. Ferrer, S.N. Barnaby, T.J. Merkel, and C.A. Mirkin. (2015). Nanoparticle probes for the detection of cancer biomarkers, cells, and tissues by fluorescence. *Chem. Rev.*, *115*, 10530–74.

Darezereshki, E., F. Bakhtiari, M. Alizadeh, A. Behrad vakylabad, and M. Ranjbar. (2012). Direct thermal decomposition synthesis and characterization of hematite (α-Fe2O3) nanoparticles. *Mater. Sci. Semicond. Process.*, *15*, 91–97.

Das, I., and S.A. Ansari. (2009). Nanomaterials in science and technology. *J. Sci. Ind. Res.*, *68*, 657–67.

Dave, P.N., P.N. Ram, and S. Chaturvedi. (2015). Nanoferrites: Catalyst for thermal decomposition of ammonium perchlorate. *Part. Sci. Technol.*, *33*, 677–81.

Deepak, F.L., M. Banobre-Lopez, E. Carbo-Argibay, M.F. Cerqueira, Y. Pineiro- Redondo, J. Rivas, C.M. Thompson, S. Kamali, C. Rodríguez-Abreu, K. Kovnir, and Y.V. Kolen'ko. (2015). A systematic study of the structural and magnetic properties of Mn-, Co-, and Ni-doped colloidal magnetite nanoparticles. *J. Phys. Chem. C*, *119*, 11947–57.

Ding, Y., L. Xu, C. Chen, X. Shen, and S.L. Suib. (2008). Syntheses of nanostructures of cobalt hydrotalcite like compounds and Co3O4 via a microwave-assisted reflux method. *J. Phys. Chem. C*, *112*, 8177–83.

Dobosz, B., R. Krzyminiewski, G. Schroeder, and J. Kurczewska. (2016). Diffusion of functionalized magnetite nanoparticles forced by a magnetic field studied by EPR method. *Curr. Appl. Phys.*, *16*, 562–67.

Dobre, T., O.C. Pârvulescu, G. Iavorschi, A. Stoica, and M. Stroescu. (2009). Analysis of sol evolution in sol-gel synthesis by use of rheological measurements. *UPB Sci. Bull. Ser. B*, *71*, 55–64.

Dong, H., S.-R. Du, X.-Y. Zheng, G.-M. Lyu, L.-D. Sun, L.-D. Li, P.-Z. Zhang, C. Zhang, and C.-H. Yan. (2015). Lanthanide nanoparticles: From design toward bioimaging and therapy. *Chem. Rev.*, *115*, 10725–815.

Douvas, A.M., Vasilopoulou, M., Georgiadou, D.G., Soultati, A., Davazoglou, D., et al. (2014). Sol-gel synthesized, low-temperature processed, reduced molybdenum peroxides for organic optoelectronics applications. *J. Mater. Chem. C*, *2*, 6290–6300.

Drbohlavova, J., R. Hrdy, V. Adam, R. Kizek, O. Schneeweiss, and J. Hubalek. (2009). Preparation and Properties of various magnetic nanoparticles. *Sensors*, *9*, 2352–62.

Ermolin, N.E., O.P. Korobeinichev, A.G. Tereshenk, and V.M. Foomin. (1982). Kinetic calculations and mechanism definition for reactions in an ammonium perchlorate flame. *Combust. Explos. Shock Waves*, *18*, 180–89.

Fajaroh, F., H. Setyawan, W. Widiyastuti, and S. Winardi. (2012). Synthesis of magnetite nanoparticles by surfactant-free electrochemical method in an aqueous system. *Adv. Powder Technol.*, *23*, 328–33.

Fantechi, E., C. Innocenti, M. Albino, E. Lottini, and C. Sangregorio. (2015). Influence of cobalt doping on the hyperthermic efficiency of magnetite nanoparticles. *J. Magn. Magn. Mater.*, *380*, 365–71.

Faria, M.C.S., R.S. Rosemberg, C.A. Bomfeti, D.S. Monteiro, F. Barbosa, L.C.A. Oliveira, M. Rodriguez, M.C. Pereira, and J.L. Rodrigues. (2014). Arsenic removal from contaminated water by ultrafine δ-FeOOH adsorbents. *Chem. Eng. J.*, *237*, 47–54.

Farre, M., J. Sanchis, and D. Barcelo. (2011). Analysis and assessment of the occurrence, the fate and the behavior of nanomaterials in the environment. *TrAC Trends Anal. Chem.*, *30*, 517–27.

Flores, R.G., S.L. Andersen, L.K. Maia, H.J. José, and R F. Moreira. (2012). Recovery of iron oxides from acid mine drainage and their application as adsorbent or catalyst. *J. Environ. Manage.*, *111*, 53–60.

Flores-Arias, Y., G.V. Azquez-Victorio, R. Ortega-Zempoalteca, U. Acevedo- Salas, S. Ammar, and R. Valenzuelaa. (2015). Magnetic phase transitions in ferrite nanoparticles characterized by electron spin resonance. *J. Appl. Phys.*, *117*, 17A503.

Foroughi, F., S.A. Hassanzadeh-Tabrizi, J. Amighian, and A. Saffar-Teluri. (2015). A designed magnetic CoFe2O4–hydroxyapatite core–shell nanocomposite for Zn (II) removal with high efficiency. *Ceram. Int.*, *41*, 6844–50.

Franzel, L., M.F. Bertino, Z.J. Huba, and E.E. Carpenter. (2012). Synthesis of magnetic nanoparticles by pulsed laser ablation. *Appl. Surf. Sci.*, *261*, 332–36.

Gabbasov, R., M. Polikarpov, V. Cherepanov, M. Chuev, I. Mischenko, A. Lomov, A. Wang, and V. Panchenko. (2015). Mössbauer, magnetization and X-ray diffraction characterization methods for iron oxide nanoparticles. *J. Magn. Magn. Mater.*, *380*, 111–16.

Galindo, R., E. Mazario, S. Gutiérrez, M.P. Morales, and P. Herrasti. (2012). Electrochemical synthesis of NiFe2O4 nanoparticles: Characterization and their catalytic applications. *J. Alloys Compd.*, *536*, S241–44.

Gao, J., F. Guan, Y. Zhao, and W. Yang. (2001). Preparation of ultrafine nickel powder and its catalytic dehydrogenation activity. *Mat. Chem. Phys.*, *71*, 215–19.

Gaudisson, T., Z. Beji, F. Herbst, S. Nowak, S. Ammar, and R. Valenzuela. (2015). Ultrafine grained high-density manganese zinc ferrite produced using polyol process assisted by spark plasma sintering. *J. Magn. Magn. Mater.*, *387*, 90–95.

Geller, S., and M. A. Gilleo. (1957). The crystal structure and ferrimagnetism of yttrium-iron garnet, Y 3 Fe 2 (FeO4) 3. *J. Phys. Chem. Solids*, *3*, 30–36.

Georgiadou, V., V. Tangoulis, I. Arvanitidis, O. Kalogirou, and C. Dendrinou-Samara. (2015). Unveiling the physicochemical features of CoFe2O4 nanoparticles synthesized via a variant hydrothermal method: NMR relaxometric properties. *J. Phys. Chem. C*, *119*, 8336–48.

Ghandoor, H.E., H.M. Zidan, M.M.H. Khalil, and M.I.M. Ismai. (2012). Synthesis and some physical properties of magnetite (Fe3O4) nanoparticles. *Int. J. Electrochem. Sci.*, *7*, 5734–45.

Gonzalez-Moragas, L., S.-M. Yu, N. Murillo-Cremaes, A. Laromaine, and A. Roig. (2015). Scale-up synthesis of iron oxide nanoparticles by microwave-assisted thermal decomposition. *Chem. Eng. J.*, *281*, 87–95.

Gopalan, E.V., and M.R. Anantharaman. (2009). On the synthesis and multifunctional properties of some nanocrystalline spinel ferrites and magnetic nanocomposites (PhD dissertation, Cochin University of Science and Technology).

Grafe, M., E. Donner, R.N. Collins, and E. Lombi. (2014). Speciation of metal(loid)s in environmental samples by X-ray absorption spectroscopy: A critical review. *Anal. Chim. Acta*, *822*, 1–22.

Graham, C.D. (2000). High-sensitivity magnetization measurements. *Mater. Sci. Technol.*, *16*, 97–101.

Gubbala, S., H. Nathani, K. Koizol, and R.D.K. Misra. (2004). Magnetic properties of nanocrystalline Ni-Zn, Zn-Mn, and Ni-Mn ferrites synthesized by reverse micelle technique. *Phys. B*, *348*, 317–28.

Han, A., L.J. Liao, M. Ye, Y. Li, and X. Peng. (2011). Preparation of nano-MnFe2O4 and its catalytic performance of thermal decomposition of ammonium perchlorate. *Chin. J Chem. Eng.*, *19*, 1047–51.

Harris, V. G., N.C. Koon, C.M. Williams, Q. Zhang, M. Abe, and J.P. Kirkland. (1996). Cation distribution in NiZn-ferrite films via extended X-ray absorption fine structure. *Appl. Phys. Lett.*, *68*, 2082–84.

Hasany, S.F., N.H. Abdurahman, A.R. Sunarti, and R. Jose. (2013). Magnetic iron oxide nanoparticles: Chemical synthesis and applications review. *Curr. Nanosci.*, *9*, 561–75.

Hassanjani-Roshan, A., M.R. Vaezi, A. Shokuhfar, and Z. Rajabali. (2011). Synthesis of iron oxide nanoparticles via sonochemical method and their characterization. *Particuology*, *9*, 95–99.

Hassellov, M., J.W. Readman, J.F. Ranville, and K. Tiede. (2008). Nanoparticle analysis and characterization methodologies in environmental risk assessment of engineered nanoparticles. *Ecotoxicology*, *17*, 344–61.

Hu, X.L., and J.C. Yu. (2008). Continuous aspect-ratio tuning and fine shape control of monodisperse α-Fe2O3 nanocrystals by a programmed microwave–hydrothermal method. *Adv. Funct. Mater.*, *18*, 880–87.

Hurley, K.R., H.L. Ring, H. Kang, N.D. Klein, and C.L. Haynes. (2015). Characterization of magnetic nanoparticles in biological matrices. *Anal. Chem.*, *87*, 11611–19.

Ibrahim, I., I.O. Ali, T.M. Salama, A.A. Bahgat, and M.M. Mohamed. (2016). Synthesis of magnetically recyclable spinel ferrite (MFe2O4, M = Zn, Co, Mn) nanocrystals engineered by sol gel-hydrothermal technology: High catalytic performances for nitroarenes reduction. *Appl. Catal. B: Environ.*, *181*, 389–402.

Issa, B., I.M. Obaidat, B.A. Albiss, and Y. Haik. (2013). Magnetic nanoparticles: Surface effects and properties related to biomedicine applications. *Int. J. Mol. Sci.*, *14*, 21266–305.

Itina, T.E. (2011). On nanoparticle formation by laser ablation in liquids. *J. Phys. Chem. C*, *115*, 5044–48.

Jacobs, P.W., and H.M. Whithead. (1969). Thermal decomposition and combustion of ammonium perchlorate. *Chem. Rev.*, *4*, 551–90.

Jain, T.K., M.A. Morales, S.K. Sahoo, D.L. Leslie-Pelecky, and V. Labhasetwar. (2005). Iron oxide nanoparticles for sustained delivery of anticancer agents. *Mol. Pharm.*, *2*, 194–205.

Karimi, Z., Y. Mohammadifar, H. Shokrollahi, S.K. Asl, G. Yousefi, and L. Karimi. 2014. Magnetic and structural properties of nano sized Dy-doped cobalt ferrite synthesized by co-precipitation. *J. Magn. Magn. Mater.*, *361*, 150–56.

Kefeni, K.K., T.M. Msagati, and B.B. Mamba. (2015). Metals and sulphate removal from acid mine drainage in two steps via ferrite sludge and barium sulphate formation. *Chem. Eng. J.*, *276*, 222–231.

Kishore, K., and M.R. Sunitha. (1979a). Comprehensive view of the combustion models of composite solid propellants. *AIAA J.*, *17*, 1216–24.

Kishore, K., and M.R. Sunitha. (1979b). Effect of transition metal oxides on decomposition and deflagration on composite solid propellant systems: A survey. *AIAA J.*, *17*, 1118–25.

Langevin, D. (1992). Micelles and microemulsions. *Annu. Rev. Phys Chem.*, *43*, 341–369.

Lee, H., T.H. Shin, J. Cheon, and R. Weissleder. 2015. Recent developments in magnetic diagnostic systems. *Chem. Rev.*, *115*, 10690–724.

Lee, N., D. Yoo, D. Ling, M.H. Cho, T. Hyeon, and J. Cheon. 2015. Iron oxide based nanoparticles for multimodal imaging and magnetoresponsive therapy. *Chem. Rev.*, *115*(19), 10637–89.

Levy, M., N. Luciani, D. Alloyeau, D. Elgrabli, V. Deveaux, C. Pechoux, S. Chat, G. Wang, N. Vats, F. Gendron, C. Factor, S. Lotersztajn, A. Luciani, C. Wilhelm, and F. Gazeau. (2011a). Long term in vivo biotransformation of iron oxide nanoparticles. *Biomaterials*, *32*, 3988–99.

Levy, M., C. Wilhelm, N. Luciani, V. Deveaux, F. Gendron, A. Luciani, M. Devaud, and F. Gazeau. (2011b). Nanomagnetism reveals the intracellular clustering of iron oxide nanoparticles in the organism. *Nanoscale*, *3*, 4402–10.

Li, H., L. Qin, Y. Feng, L. Hu, and C. Zhou. 2015. Preparation and characterization of highly water-soluble magnetic Fe3O4 nanoparticles via surface double-layered self-assembly method of sodium alpha-olefin sulfonate. *J. Magn. Magn. Mater.*, *384*, 213–18.

Lin, X.-M., and A.C.S. Samia. (2006). Synthesis, assembly and physical properties of magnetic nanoparticles. *J. Magn. Magn. Mater.*, *305*, 100–109.

Liu, T., L. Wang, P. Yang, and B. Hu. (2008). Preparation of nanometer CuFe2O4 by auto-combustion and its catalytic activity on the thermal decomposition of ammonium perchlorate. *Mater. Lett.*, *62*, 4056–58.

Lopez, J.A., F. González, F.A. Bonilla, G. Zambrano, and M.E. Gomez. (2010). Synthesis and characterization of Fe3O4 magnetic nanofluid. *Rev. Latinoam. Metal. Mater.*, *30*, 60–66.

Luo, N., X. Tian, J. Xiao, W. Hu, C. Yang, L. Li, and D. Chen. (2013). High longitudinal relaxivity of ultra-small gadolinium oxide prepared by microsecond laser ablation in diethylene glycol. *J. Appl. Phys.*, *113*, 164306.

Mahfouz, M.G., A.A. Galhoum, N.A. Gomaa, S.S. Abdel-Rehem, A.A. Atia, T. Vincent, and E. Guibal. (2015). Uranium extraction using magnetic nano-based particles of diethylenetri-amine-functionalized chitosan: Equilibrium and kinetic studies. *Chem. Eng. J.*, *262*, 198–209.

Manikandan, A., R. Sridhar, S. Arul Antony, and S. Ramakrishna. (2014). A simple aloe vera plant-extracted microwave and conventional combustion synthesis: Morphological, optical, magnetic and catalytic properties of CoFe2O4 nanostructures. *J. Mol. Struct.*, *1076*, 188–200.

Manova, E., B. Kunev, D. Paneva, I. Mitov, L. Petrov, C. Estournes, C. D'Orlean, J.-L. Rehspringer, and M. Kurmoo. (2004). Mechano-synthesis, characterization, and magnetic properties of nanoparticles of cobalt ferrite, CoFe2O4. *Chem. Mater.*, *16*, 5689–96.

Mansoori, G.A. (2005). *Principles of naNotechnology: Molecular-based Study of Condensed Matter in Small Systems*. World Scientific.

Masthoff, I.C., M. Kraken, D. Mauch, D. Menzel, J.A. Munevar, E. Baggio Saitovich, F.J. Litterst, and G. Garnweitner. (2014). Study of the growth process of magnetic nanoparticles obtained via the non-aqueous sol–gel method. *J. Mater. Sci.*, *49*, 4705–14.

Mathew, D.S., and R.S. Juang. (2007). An overview of the structure and magnetism of spinel ferrite nanoparticles and their synthesis in micro emulsions. *Chem. Eng. J.*, *129*, 51–65.

Mazario, E., M.P. Morales, R. Galindo, P. Herrasti, and N. Menendez. (2012). Influence of the temperature in the electrochemical synthesis of cobalt ferrites nanoparticles. *J. Alloys Compd.*, *536*, S222–25.

Mohamed, R.M., M.M. Rashad, F.A. Haraz, and W. Sigmund. (2010). Structure and magnetic properties of nanocrystalline cobalt ferrite powders synthesized using organic acid precursor method. *J. Magn. Magn. Mater.*, *322*, 2058–64.

Mondal, A.K., S. Chen, D. Su, K. Kretschmer, H. Liu, and G. Wang. (2015). Microwave synthesis of α-Fe2O3 nanoparticles and their lithium storage properties: A comparative study. *J. Alloys Compd.*, *648*, 732–39.

Moumen, N., P. Veillet, and M.P. Pileni. (1995). Controlled preparation of nanosize cobalt ferrite magnetic particles. *J. Magn. Magn. Mater.*, *149*, 67–71.

Mulens, V., M.P. Morales, and D.F. Barber. (2013). Development of magnetic nanoparticles for cancer gene therapy: A comprehensive review. *ISRN Nanomater.*, *2013*, 1–14.

Nayak, H., and A.K. Jena. (2014). Catalyst effect of transition metal nano oxides on the decomposition of lanthanum oxalate hydrate: A thermogravimetric study. *Int. J. Sci. Res.*, *3*, 381–88.

Nema, A.K., S. Jain, S.K. Sharma, S.K. Nema, and S.K. Verma. (2004). Mechanistic aspect of thermal decomposition and burn rate of binder and oxidiser of AP/HTPB composite propellants comprising HYASISCAT. *Int. J. Plast. Technol.*, *8*, 344–54.

Niu, D., Y.S. Li, Z. Ma, H. Diao, J.L. Gu, H.R. Chen, W.R. Zhao, M.L. Ruan, Y.L. Zhang, and J.L. Shi. (2010). Preparation of uniform, water-soluble, and multifunctional nanocomposites with tunable sizes. *Adv. Funct. Mater.*, *20*, 773–780.

Obaidat, I.M., B. Issa, and Y. Haik. (2015). Magnetic properties of magnetic nanoparticles for efficient hyperthermia. *Nanomaterials*, *5*, 63–89.

O'Connor, C.J., C.T. Seip, E.E. Carpenter, S.C. Li, and V.T. John. (1999). Synthesis and reactivity of nanophase ferrites in reverse micellar solution. *Nanostruct. Mater.*, *12*, 65–70.

Olsen, E., and J. Thonstad. (1999). Nickel ferrite as inert anodes in aluminium electrolysis: Part I. Material fabrication and preliminary testing. *J. Appl. Electrochem.*, *29*, 293–99.

Park, J., K. An, Y. Hwang, J.-G. Park, H.-J. Noh, J.-Y. Kim, J.-H. Park, N.-M. Hwang, and T. Hyeon. (2004). Ultra-large-scale syntheses of monodisperse nanocrystals. *Nat. Mater.*, *3*(2004),891–95.

Pauline, S., and A.P. Amaliya. (2011). Synthesis and characterization of highly monodispersive CoFe2O4 magnetic nanoparticles by hydrothermal chemical route. *Arch. Appl. Sci. Res.*, *3*, 213–23.

Pemartin, K., C. Solans, J. Alvarez-Quintana, and M. Sanchez-Dominguez. (2014). Synthesis of Mn–Zn ferrite nanoparticles by the oil-in-water microemulsion reaction method. *Colloids Surf. A Physicochem. Eng. Asp.*, *451*, 161–71.

Peng, J.H., M. Hojamberdiev, Y.H. Xu, B.W. Cao, J. Wang, and H. Wu. (2011). Hydrothermal synthesis and magnetic properties of gadolinium-doped CoFe2O4 nanoparticles. *J. Magn. Magn. Mater.*, *323*, 133–37.

Qu, X., P.J.J. Alvarez, and Q. Li. (2013a). Applications of nanotechnology in water and wastewater treatment. *Water Res.*, *47*, 3931–46.

Qu, X., J. Brame, Q. Li, and P.J.J. Alvarez. (2013b). Nanotechnology for a safe and sustainable water supply: Enabling integrated water treatment and reuse. *Acc. Chem. Res.*, *46*, 834–43.

Rahimi, R., A. Maleki, S. Maleki, A. Morsali, and M.J. Rahimi. (2014). Synthesis and characterization of magnetic dichromate hybrid nanomaterials with triphenylphosphine surface modified iron oxide nanoparticles (Fe3O4@SiO2@PPh3@Cr2O72–). *Solid State Sci.*, *28*, 9–13.

Ramimoghadam, D., S. Bagheri, S. Bee, and A. Hamid. (2014). Progress in electrochemical synthesis of magnetic iron oxide nanoparticles. *J. Magn. Magn. Mater.*, *368*, 207–29.

Reda, S.M. 2010. Synthesis of ZnO and Fe2O3 nanoparticles by sol–gel method and their application in dye-sensitized solar cells. *Mater. Sci. Semicond. Process.*, *13*, 417–425.

Reddy, L.H., J.L. Arias, J. Nicolas, and P. Couvreur. (2012). Magnetic nanoparticles: Design and characterization, toxicity and biocompatibility, pharmaceutical and biomedical applications. *Chem. Rev.*, *112*, 5818–78.

Rosser, W.A., S.H. Inami, and H. Wise. (1968). Thermal decomposition of ammonium perchlorate. *Combust. Flame*, *12*(5), 427–35.

Roukes, M. (2009). Plenty of room, plenty of history. *Nat. Nanotechnol.*, *4*, 783–84.

Said, A.A. (1991). Thermal decomposition of ammonium metavanadate doped with iron, cobalt or nickel hydroxides. *Thermal Analysis*, *37*, 959–62.

Salazar-Alvarez, G., J. Qin, V. Sepelak, I. Bergmann, M. Vasilakaki, K.N. Trohidou, J.D. Ardisson, W.A.A. Macedo, M. Mikhaylova, M. Muhammed, M.D. Bar, and J. Nogues. (2008). Cubic versus spherical magnetic nanoparticles: The role of surface anisotropy. *Am. Chem. Soc. 130*, 13234–39.

Senapati, K.K., C. Borgohain, and P. Phukan. (2011). Synthesis of highly stable CoFe$_2$O$_4$ nanoparticles and their use as magnetically separable catalyst for Knoevenagel reaction in aqueous medium. *J. Mol. Catal. A: Chem.*, *339*, 24–31.

Shafi, K.V.P., A. Ulman, X. Yan, N.L. Yang, C. Estournes, H. White, and M. Rafailovich. (2001). Sonochemical synthesis of functionalized amorphous iron oxide nanoparticles. *Langmuir, 17*, 5093–97.

Shafi, K.V.P.M., A. Ulman, A. Dyal, X. Yan, N. Yang, C. Estournes, L. Fournes, A. Wattiaux, H. White, and M. Rafailovich. (2002). Magnetic enhancement of γ-Fe2O3 nanoparticles by sonochemical coating. *Chem. Mater.*, *14*, 1778–87.

Sharifi, I., H. Shokrollahi, and S. Amiri. (2012). Ferrite-based magnetic nanofluids used in hyperthermia applications. *J. Magn. Magn. Mater.*, *324*, 903–15.

Shen, S.M., S. I. Chen, and B.B. Wu. (1993). The thermal decomposition of ammonium perchlorate (AP) containing a burning-rate modifier. *Thermochim. Acta*, *223*, 135–43.

Sickafus, E.K., and R. Hughes. (1999). Spinel compounds: structure and property relations. *Journal of the American ceramic society*, *82*(12), 3277–78.

Singampalli, R., C.S. Beera, P.S.V. Subba Rao, and B.P. Rao. (2014). Microstructural and magnetic behavior of mixed Ni–Zn–Co and Ni–Zn–Mn ferrites. *Ceram. Int.*, *40*, 8729–35.

Singh, G., and S.P. Felix. (2003). Prem, studies on energetic compounds: Part 36: Evaluation of transition metal salts of NTO as burning rate modifiers for HTPB-AN composite solid propellant. *Combust. Flame*, *132*, 145–50.

Singh, G., I.P.S. Kapoor, and S. Dubey. (2009a). Bimetallic nanoalloys: Preparation, characterization and their catalytic activity. *J. Alloys Compd.*, *480*, 270–74.

Singh, G., I.P.S. Kapoor, S. Dubey, and P.F. Siril. (2009b). Preparation, characterization and catalytic activity of transition metal oxide nanocrystals. *J. Sci. Conf. Proc.*, *1*, 11–17.

Singh, G., I.P.S. Kapoor, S. Dubey, and P.F. Siril. (2009c). Kinetics of thermal decomposition of ammonium perchlorate with nanocrystals of binary transition metal ferrites. *Propellants Explos. Pyrotech.*, *34*, 78–83.

Singh, G., I.P.S. Kapoor, S. Dubey, P.F. Siril, Yi.J. Hua, F.Q. Zhao, and R.Z. Hu. (2008). Effect of mixed ternary transition metal ferrite nanocrystallites on thermal decomposition of ammonium perchlorate. *Thermochem. Acta*, *477*, 42–47.

Singh, G., I.P.S. Kapoor, R. Dubey, and P. Srivastava. (2010). Praparation, characterization and catalytic behavior of CdFe2O4 and Cd nanocrystals on AP, HTPB and composite solid propellants, Part: 79. *Thermochim. Acta*, *51*, 112–18.

Singh, G., and D.K. Pandey. (2002). Studies on energetic compound: Part 24-haxamine metal perchlorate as high energetic burning rate catalysts. *J. Energy Mater.*, *20*, 223–44.

Singh, S., P. Srivastava, and G Singh. (2013). Nanorods, nanospheres, nanocubes: Synthesis, characterization and catalytic activity of nanoferrites of Mn,Co,Ni, Part-89. *Mater. Res. Bul.*, *48*, 739–46.

Solano, E., R. Yáñez, S. Ricart, and J. Ros. (2015). New approach towards the polyol route to fabricate MFe2O4 magnetic nanoparticles: The use of MCl2 and Fe(acac)3 as chemical precursors. *J. Magn. Magn. Mater.*, *382*, 380–85.

Srivastava, P., I.P.S. Kapoor, and G. Singh. (2009). Nano ferrites: Preparation characterization and catalytic properties *J. Alloys Compd.*, *485*(1–2), 88–92.

Standley, K.J. (1972). *Oxide Magnetic Materials*. Oxford University Press.

Stoner, E.C., and E.P. Wohlfarth. (1948). A mechanism of magnetic hysteresis in heterogeneous alloys. *Philos. Trans. R. Soc. London. Ser. A, 240,* 599–642.

Stuijts, A.L., G.W. Rathenau, and G.H. Weber. (1954). Ferroxdure ii and iii, anisotropic permanent magnet materials. *Philips Tech. Rev., 16,* 141–80.

Sugimoto, M. (1999). The past, present, and future of ferrites. *J. Am. Ceram. Soc., 82,* 269–80.

Suzuki, T., T. Tanaka, and K. Ikemizu. (2001). High density recording capability for advanced particulate media. *J. Magn. Magn. Mater., 235,* 159–64.

Suzuki, Y. (2001). Epitaxial spinel ferrite thin films. *Annual Rev. Mater. Res., 31,* 265–89.

Tadic, M., S. Kralj, M. Jagodic, D. Hanzel, and D. Makovec. 2014. Magnetic properties of novel superparamagnetic iron oxide nanoclusters and their peculiarity under annealing treatment. *Appl. Surf. Sci.* 322:255–64.

Tan, L., Q. Liu, X. Jing, J. Liu, D. Song, S. Hu, L. Liu, and J. Wang. (2015). Removal of uranium (VI) ions from aqueous solution by magnetic cobalt ferrite/multiwalled carbon nanotubes composites. *Chem. Eng. J., 273,* 307–15.

Tang, S.C.N., and I.M.C. Lo. (2013). Magnetic nanoparticles: Essential factors for sustainable environmental applications. *Water Res., 47,* 2613–32.

Tartaj, P., M.D.P. Morales, S. Veintemillas-Verdaguer, T. González-Carreno, and C.J. Serna. (2003). The preparation of magnetic nanoparticles for applications in biomedicine. *J. Phys. D. Appl. Phys., 36,* R182–97.

Thakur, S., R. Rai, and S. Sharma. (2015). Structural characterization and magnetic study of NiFexO4 synthesized by co-precipitation method. *Mater. Lett., 139,* 368–72.

Valenzuela, R. (2005). *Magnetic Ceramics.* Cambridge University Press.

Wang, F., X.F. Qin, Y.F. Meng, Z.L. Guo, L.X. Yang, and Y.F. Ming. (2013). Hydrothermal synthesis and characterization of α-Fe2O3 nanoparticles. *Mater. Sci. Semicond. Process.,* 16:802–6.

Wang, H., R. Wang, L. Wang, X. Tian. (2011). Preparation of multi-core/single-shell OA-Fe3O4/PANI bifunctional nanoparticles via miniemulsion polymerization. *Colloids Surf. A, 384,* 624–29.

Wang, J., W.B. White, and J.H. Adair. (2005). Optical properties of hydrothermally synthesized hematite particulate pigments. *J. Am. Chem. Soc., 88,* 3449–54.

Wei, J., X. Zhang, Q. Liu, Z. Li, L. Liu, and J. Wang. (2014). Magnetic separation of uranium by CoFe2O4 hollow spheres. *Chem. Eng. J., 241,* 228–34.

Willard, M.A., L.K. Kurihara, E.E. Carpenter, S. Calvin, and V.G. Harris. (2004). Chemically prepared magnetic nanoparticles. *Int. Mater. Rev., 49,* 125–70.

William, D.C., and D.G. Rethwisch. (2012). *Fundamentals of Materials Science and Engineering: An Integrated Approach.* Wiley.

Wongpratat, U., S. Maensiri, and E. Swatsitang. (2015). EXAFS study of cations distribution dependence of magnetic properties in Co1−xZnxFe2O4 nanoparticles prepared by hydrothermal method. *Microelectron. Eng., 146,* 68–75.

Wu, W., Z. Wu, T. Yu, C. Jiang, and W.S. Kim. (2015). Recent progress on magnetic iron oxide nanoparticles: Synthesis, surface functional strategies and biomedical applications. *Sci. Technol. Adv. Mater., 16,* 23501.

Xing, Y., Y. Y. Jin, J.-C. Si, M.-L. Peng, X. F. Wang, C. Chen, and Y.-L. Cui. (2015). Controllable synthesis and characterization of Fe3O4/Au composite Nanoparticles. *J. Magn. Magn. Mater., 380,* 150–56.

Zhang, H., and G. Zhu. (2012). One-step hydrothermal synthesis of magnetic Fe3O4 nanoparticles immobilized on polyamide fabric. *Appl. Surf. Sci., 258,* 4952–59.

Zhang, J., J.-M. Song, H.-L. Niu, C.-J. Mao, S.-Y. Zhang, and Y.-H. Shen. (2015). ZnFe2O4 nanoparticles: Synthesis, characterization, and enhanced gas sensing property for acetone. *Sens. Actuators B Chem., 221,* 55–62.

Zhao, S., and D. Ma. (2010). Preparation of CoFe2O4 nano crystallites by solvothermal process and its catalytic activity on the thermal decomposition of ammonium perchlorate. *J. Nanomater., 2010,* 1–5.

Zhao, X., W. Wang, Y. Zhang, S. Wu, F. Li, and J.P. Liu. (2014). Synthesis and characterization of gadolinium doped cobalt ferrite nanoparticles with enhanced adsorption capability for Congo Red. *Chem. Eng. J., 250,* 164– 74.

Zhao, Y., Z. Qiu, and J. Huang. (2008). Preparation and analysis of Fe3O4 magnetic nanoparticles used as targeted-drug carriers. *Chin. J. Chem. Eng., 16,* 451–55.

Zhi, J., Y. Wang, Y. Lu, J. Ma, and G. Luo. (2006). In situ preparation of magnetic chitosan/Fe3O4 composite nanoparticles in tiny pools of water-in-oil microemulsion. *React. Funct. Polym., 66,* 1552–58.

Zhou, Z., F. Jiang, T.-C. Lee, and T. Yue. (2013). Two-step preparation of nano-scaled magnetic chitosan particles using Triton X-100 reversed-phase water-in-oil microemulsion system. *J. Alloys Compd., 581,* 843–48.

Zi, Z., Y. Sun, X. Zhu, Z. Yang, J. Dai, and W. Song. (2009). Synthesis and magnetic properties of CoFe2O4 ferrite nanoparticles. *J. Magn. Magn. Mater., 321,* 1251–55.

Zijlstra, H. (1982). Permanent magnets: Theory. *Handbook of Ferromagnetic Materials,* North Holland, (Vol. 3, pp. 37–105).

14

Implications of Fungal Synthesis of Nanoparticles and Its Various Applications

Monika Gupta, Rajesh Singh Tomar, Vinay Dwivedi, and Raghvendra Kumar Mishra

CONTENTS

14.1 Introduction

In all of the processes developed to date, the production of metal nanoparticles is highly popular, innocuous, inexpensive, and eco-friendly. A biological method of plant extract-based procurement provides an advantage of not leaving dangerous residue which pollutes the environment, post-synthesis [1]. Chemical methods are more prevalent as far as metal nanoparticle synthesis is concerned, but their use is limited primarily due to the aforesaid concerns. Therefore, biogenic synthesis is one of the best alternatives that is natural and does not leave pollutants and is very safe for human health and the atmosphere [2–7]. Microorganisms play a vital role in the production of nanoparticles as they reduce metal ions and exhibit extracellular or intracellular synthesis achieved by biomineralization, biosorption, rain, and bio-accumulation [8–17]. Fungi employ enzymes and protein-reducing agents; they can always be used to synthesize metal nanoparticles from salts, thereby laying the foundation for the green synthesis of nanoparticles using microorganisms. However, the pathogenic nature of some fungi deters their utilization which could prove to be biologically noxious, if ignored. Fungal biomass increases at much faster rates as compared to bacterial biomass, under the same conditions [18]. Since fungal mycelia provide a large surface area for the interaction, nanoparticle synthesis is far more beneficial by the fungus vis-à-vis the bacterial synthesis (Figure 14.1), [19]. Reports suggest that the production of protein in the fungus is greater than in bacteria [26]. Metal salts thus chemically turn into metal nanoparticles in the cycle of nanoparticles synthesis. Different metal nanoparticles shapes and sizes originated from fungi and are also listed with their applications. Synthesis of microbial classifies within the extracellular and intracellular synthesis. On the other hand, the extracellular synthesis manner includes immobilization of metallic ions on the surface of the cells and thereby lowers the ion in the presence of enzymes [20]. Fungus is used to secrete extracellular proteins which have been used to take away metal ions in the form of nanoparticles. In current times, big scale packages of steel nanoparticles are seen in diverse fields along with agriculture, biomedical, cosmetics [21, 22]. Many of the metallic nanoparticles synthesized by way of fungi are antimicrobial and still have medicinal utility [23–25]. As soon as the metal nanoparticles are used in mixture with metals along with gold and silver, their antibacterial hobby increases [26].

Hence, this review also emphasizes the biogenic synthesis of metal nanoparticles and its mechanisms in order to control the morphology and size of the particles. They provide various avenues of synthesis of green nanoparticles employed in various sectors including nanomedicine.

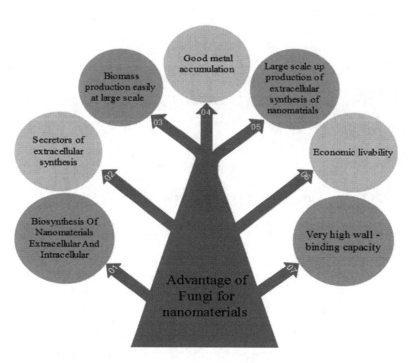

FIGURE 14.1 Advantage of fungi used as bio factories for nanomaterials.

14.2 Factors Influencing the Biosynthesis of Nanomaterials

Amalgamation-directing components – for example, pH, temperatures, grouping of metal salt, and biomass – assume a key job in the biosynthesis of nanoparticles. The shape and size of nanoparticles are gathered as synthetic and physical variables.

14.3 Role of pH

The pH esteem assumes a significant job in the response medium during the arrangement of nanoparticles. As per reports and different research papers which were distributed on this investigation, it shows a variety of very strong nanoparticles during union on changing the pH of the response medium as appeared. Small-sized particles were acquired at higher ph, whereas enormous sizes were secured at lower pH. For example, Silver(Ag) nanoparticles of 10 nm at pH 11, got from Sclerotinia sclerotiorum, was a strong circular, though 15 nm nanoparticle was acquired utilizing Cladosporium sphaerospermum at pH 7. The investigation likewise pointed out that at pH 11, utilitarian gatherings in the parasitic biomass, which are basic for molecule nucleation, are accessible in higher sums. Meanwhile, at pH 7, useful gatherings were accessible in smaller sums which brings about molecule combination, subsequently framing bigger silver nanoparticles.

14.4 Impact of Concentration

The biomolecule focus found in a contagious biomass gradually impacts the development of metal nanoparticles. The size of nanoparticles began diminishing quickly as there is an expansion in the centralization of silver particles (Ag+) from 2mM to 100mM in the response blend, which is because of the nearness of useful gatherings in the contagious biomass present during nanoparticle amalgamation. At the point when metal salt fixation was expanded further, huge size nanoparticles were framed because of the agglomeration process, which is because of the nearness of high grouping of silver particles in the response blend. For instance, utilization of organism Penicillium chrysogenum, when the metal salt (AgNO3 silver nitrate) was presented in the response blend and convergence of metal salt was expanded to 100mM, the nanoparticle size delivered was of size 70 nm. Thus utilizing Trichoderma viride parasites [31] and diminishing the measure of metal salt (2mM), brought about a lot smaller size of nanoparticles (16 nm).

14.5 Effect of Temperature

At a response temperature of 30°C, utilizing Hypocrea lixii biomass [29], Ni nanoparticles were blended with a decreased size of around 3.8 nm. Yet, when the temperature was expanded to 80°C and utilizing Sclerotinia sclerotiorum biomass, the molecule size was found to increment to 50 nm. Squeezes and Gericke demonstrated that the arrangement rates for nanoparticles increments with an increment in temperature. Similarly, plate molded and pole-molded nanoparticles were framed at higher temperatures, while lower temperatures yielded round formed nanoparticles.

14.6 Portrayal Techniques of Nanoparticles

The portrayal of nanomaterials is especially important to comprehend their properties, especially in surveying surface zone, porosity, solvency, pore size, accumulation, molecule size dispersion, zeta limit, adsorption limit, and crystalline nature. Direction, move and scattering of nanoparticles and nanotubes in nanocomposite materials can be comprehended from the nanoparticle portrayal [100]. A few strategies can be utilized to assess Brownian movement and molecule size examination, including UV-noticeable spectroscopy, AFM, TEM, SEM, DLS, XRD, FTIR [101–103]. Subsequently, during the portrayal of nanoparticles, it is very essential to assess the charge on their surface. Different portrayal strategies and methods for nanoparticles are laid out.

14.7 UV-Visible Range

The most essential portrayal strategy is UV-visible spectrometry. Different sorts of metal nanoparticles are recognized and broken down utilizing this procedure (Figure 14.2).

At the point when silver nanoparticles were created by means of Phoma glomerata organisms, it rendered a particular absorption band at 440 nm, yet biosynthesized and balanced out gold (Au) nanoparticles, by utilizing Fusarium semitectum parasites, and gave a surface plasmon reverberation ingestion band at 545 nm. Absorption groups were gotten roughly at 260 nm [33], in a combination of platinum nanoparticles, by utilizing Fusarium oxysporum. Other metal nanoparticles viz., zinc oxide, silica nanoparticles, titanium dioxide, and iron oxide were likewise blended by parasitic operators [33–37].

14.7.1 X-Ray Diffraction (XRD)

XRD is utilized to assess the crystallization of integrated nanoparticles. This procedure decides the oxidation conditions of these nanoparticles as a component of time [38]. It is intended to distinguish different crystalline structures and look at the crucial structure of common and counterfeit nanoparticles [39, 68, 104]. Nonetheless, XRD has numerous vantage highlights, yet there are a few burdens too. Right off the bat, single precious stone

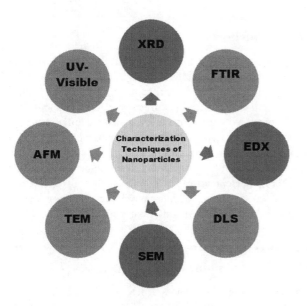

FIGURE 14.2 Common characterization techniques of nanoparticles (Adapted and modified)

development and structure investigation is troublesome [40]. The second drawback is that when contrasted with electron diffractions [41], the force of XRD is lower. It depends on Bragg's law which principally utilizes wide-edge flexible dispersing marvels of X-beams [42, 43].

14.8 Transmission Electron Microscopy (TEM)

TEM is another significant and important system to describe nanoparticles. This method is utilized to evaluate molecule size estimations and its conveyance and morphology [44]. These are resolved with the assistance of the proportion of separation between the target focal point and the picture plane, just as an example [45]. TEM has two significant points of interest. To start with, it has the ability of the extra investigative estimations and second it gives better spatial goals [46]. In any case, the fundamental downside of TEM is that it requires slender example planning and high vacuum to accomplish higher goals pictures.

14.9 Scanning Electron Microscopy (SEM)

SEM is the surface imaging method. It has the ability to determine the size appropriation of various particles, surface morphology, and integrated nanoparticles [47], while vitality dispersive X-beam (EDX) spectroscopy is an element of the produced X-beam vitality component, and this component breaks down subjectively and quantitatively [105, 106]. The mix of EDX and SEM can be utilized to watch and break down the powder morphology of metal nanoparticles and furthermore their concoction arrangement. SEM has a favorable position in that it can decide inward structure and holds important data about the virtue of particles and their total [48].

14.10 FTIR Spectroscopy

This is a non-intrusive procedure. Utilizing FTIR spectroscopy, identification of small retentive changes to the request for 10–3 nm is conceivable, which helps in performing differential spectroscopy. Small ingestion of practically dynamic buildups can recognize the band, when contrasted with the enormous foundation assimilation of an entire protein [49, 50]. This procedure is utilized to get biomolecules during the combining of nanoparticles. FTIR has additionally been done to contemplate different parts of nanomaterials, for example, biomolecules' affirmation [51, 52]. FTIR spectroscopy has an assortment of favorable circumstances viz. less example heat-up, solid sign, enormous sign to commotion proportion, and quick information assortment [53, 54]. As of late, FTIR spectroscopy is known as ATR (weakened complete reflection) FTIR spectroscopy. ATR-FTIR spectroscopy is utilized to decide the substance properties on the outside of polymer, and test accreditation is far more advantageous than traditional FTIR [55, 56]. Notwithstanding that, FTIR is likewise economical and non-intrusive and furthermore assists with distinguishing the job of biomolecules in the decrease of metal particles [57, 58].

14.11 Atomic Force Microscopy (AFM)

This procedure can be utilized to describe the nanoparticles progressively, with bolstered lipid bilayers, which is not achievable by electron microscopy methods [59, 60]. Furthermore, it can quantify in nanometer (nm) scale in fluid suspension. AFM can likewise be utilized to contemplate the morphology, size, and retention of nanoparticles and additionally decide their agglomeration. Three distinct methods of checking are currently utilized: discontinuous example contact mode, contact mode, and non-contact mode [61–64]. However, this procedure has a significant downside which is the overestimation of the sidelong components of the examples, due to the size of the cantilever. Accordingly, it requires a great deal of accuracy to stay away from incorrect estimations [65–68].

14.12 Dynamic Light Scattering (DLS)

Different portrayal strategies are utilized to break down nanoparticles [69, 70]. Dynamic light dispersing strategy is particularly used for steadfast circulation of size of particles in 2–1000 nm run and to decide the size of particles in fluids [71–74]. DLS chips away at the rule collaboration of light with particles. The size acquired from Dynamic Light Scattering (DLS) is typically bigger than transmission electron microscopy (TEM) which is fundamental because of the impact of Brownian movement [75, 76]. This one-of-a-kind capacity of DLS to investigate enormous amounts of nanoparticles simultaneously is especially advantageous, yet a few examples of explicit restrictions with it have emerged [77].

14.13 Utilization of Nanoparticles

A blend of parasitic nanoparticles from metal salts is attractive overall, basically because of its favorable qualities and certain different points of interest over a bacterial combination. For a huge scope of nanoparticle creations, parasites are viewed as highly valuable over different living beings. Parasites additionally deliver significant decreasing operators, for example, anthraquinones and naphthoquinones [78–81]. Thus, a particular protein can follow up on a particular metal. Reports have demonstrated that to diminish a metal particle, catalyst is important, yet additionally, an electron transport is similarly critical [24]. It is necessary to know the kind or defame conduct of nanomaterials towards living things [1, 15, 21, 82–84]. There exist various uses of nanomaterials viz. as a nanocatalyst, in agro-substance enterprises, beauty care products, materials, nourishment bundling, fuel-added substances, ointments, quality conveyance operators, drugs, nanomedicine, optics, attractive reverberation imaging, and so forth [26, 85–89]. A few papers have been distributed on intracellular and extracellular silver nanoparticle combinations utilizing organisms [90, 91]. The detailed extracellular blend of silver (Ag) nanoparticles, utilizing F. oxysporum, can be presented in material-assembling ventures which can keep a contamination from S. aureus [92]. Namasivayam and Avimanyu announced that silver (Ag) nanoparticles, utilizing Lecanicillium lecanii, which were covered on the faded cotton texture with the utilization of acrylic

folio, got safe towards E. coli and S. aureus contaminations [93]. Bhainsa et al. revealed that extracellular amalgamation of silver nanoparticles, which are in the size scope of 5–25 nm and are integrated from Aspergillus treats, was fit for decreasing silver particle (Ag+) into silver nanoparticles inside 10 minutes. Along these lines, it is hard to unmistakably anticipate the reactant action of the delivered nanoparticles if their sizes vary in each clump [97]. Nanowire-based paper is utilized to tidy up oil and natural poisons present in water and soil silt. These nanowires were acquired from permeable gold microwires through warmth treatment forms [26]. Pinakin C. Dhandhukia et al. have revealed that extracellular amalgamation of gold nanoparticles, which are in the size scope of 22 nm and integrated from F. oxysporum f. sp. Cubense, and which can be fused in the cotton industry, displayed a sort of antibacterial movement against Pseudomonas sp. [94].

Studies relating to bio-blended gold nanoparticles uncovered that presentation of gold nanoparticles yielded HBL-100 and HBL-100 apoptosis (human bosom disease cells) [95]. Mishra et al. depicted gold nanoparticle union by means of supernatant, live cell filtrate, and biomass of the organism Penicillium brevicompactum [96]. Taking another instance of a study, Jeyaraj et al. assessed the effect of silver nanoparticles on malignancy cell lines [97]. It was accounted for that the bio-blend of silver nanoparticles utilizing Trichoderma viride, and after laser excitation, a few photoluminescence estimations got transmitted which were in the scope of 320–520 nm. This made such a sort of silver nanoparticles appropriate for up-and-coming use in imaging and naming. Sarkar et al. likewise directed a comparative examination for the above discussion [98]. These models were not examined for the utilization of biosynthesis in a total and careful way. These applications were seen as dependent on trial ramifications of nanomaterials by utilizing contagious biomass. Therefore, it is fundamental that the biosynthesis of nanomaterials done by utilizing parasites, as of now have a functionalized surface while protein, polysaccharides, and natural ligands were not discovered when these nanoparticles were combined by physical and compound techniques [99].

14.14 Conclusion

The green blend of metal nanoparticles utilizing parasites has ended up being ecofriendly, nontoxic, full of potential and can be utilized in a few fields, for example, nano-prescriptions. As of late, unique organic procedures assumed a key job in lessening, topping, and settling operators, which are utilized during the union of nanomaterials. The metal nanoparticles blended with organically utilizing parasites have set better stage for their execution in different fields including material designing, attractive reverberation, elements of nutrition, ecological detecting, malignant growth treatment, sedate conveyance, imaging, catalysis, antibacterial factor, and quality treatment.

14.15 Future Point of View

Green amalgamation of nanoparticles has been an examination cynosure in the last couple of years because of its nontoxic traits. Contagious and bacterial biosynthesis of nanoparticles offers comparative procedure execution. Be that as it may, because of the huge measure of redox proteins, metal nanoparticle amalgamation utilizing parasites fits the bill for a scale-up. At long last, the chance of animating the resistant reaction of a parasitic starting point to the topping operators is more uncertain when contrasted with bacterial proteins. An ongoing report on the extracellular union of nanoparticles utilizing parasites shows that the enzymatic hardware required for their profile combination is indistinguishable from the utilization of metal detoxification. Intracellular metal nanoparticle biosynthesis is accounted for when there is satisfactory accessibility to high centralization of metal particles and furthermore in the condition when the integrity of the layer is undermined in order to allow the dispersion of metal particles inside the cell. In any case, there are no considerable confirmations accessible yet; parasites utilize a biosynthetic pathway in their digestion for nanoparticle amalgamation. Metal bioreduction happens in two primary stages: [1] M+3 particles are diminished to M+, and [2] M+ particle is decreased to M0ion. A single-step decrease of M+3 particles to M0ion was often observed, which might be because of the short life expectancy of an M+ particle shaped at an encompassing temperature. Test conditions are required to be controlled so as to approve the metal bioreduction framework. For metallic nanoparticle biosynthesis, the proteins and other auxiliary metabolites can go about as topping and balancing out operators. As of now, for bioreduction and topping specialists, not very many biomolecules are incorporated. More research is required to unequivocally distinguish different species associated with this procedure. For metal particle bio-blend of a huge scope, the most attractive competitors are the growths which wrap a lot of metabolites and outside compounds which brings about scattered metal nanoparticles in the extracellular space. Different up-and-comers include the filamentous parasites Aspergillus and Trichoderma sp. These produce an assortment of metabolites and extracellular catalysts which have wide modern use [140], and these are very notable. Subsequently, they end up being phenomenal model life forms that can be utilized for top-to-bottom hereditary and biochemical examination of the bio-decrease process.

Metal nanoparticles' combination of enough high dependability and of a characterized molecule size appropriation and shape will require the enhancement of a few parameters which incorporate biomass/M+3 proportion, pH, and working states of the bio-reactor. Be that as it may, if these parameters are properly enhanced, the after said issue can be effortlessly tended to. The very moderate energy of the development of metal nanoparticles can be tended to utilizing cell separating instead of feasible biomass. On the off chance that there should arise an occurrence of the use of reasonable biomass, high surface bioreactors would be required to accelerate the procedure of biosynthesis.

REFERENCES

1. Durán, N., P.D. Marcato, M. Durán, A. Yadav, A. Gade, and M. Rai. (2011). Mechanistic aspects in the biogenic synthesis of extracellular nanomaterialsby peptide, bacteria, fungi, and plants. *Applied Microbiology and Biotechnology*, *90*, 1609–1624.
2. Raveendran, P., J. Fu, and S.L. Wallen. (2003). Completely "green" synthesis and stabilization of metal nanoparticles. *Journal of the American Chemical Society*, *125*, 13940–13941.
3. Narayanan, K.B., and N. Sakthivel. (2010). Biological synthesis of nanomaterials by microbes. *Advances in Colloid and Interface Science*, *156*, 1–13.
4. Narayanan, S., B.N. Sathy, U. Mony, M. Koyakutty, S.V. Nair, and D. Menon. (2012). Biocompatible magnetite/gold nanohybrid contrast agents via green chemistry for MRI and CT bioimaging. *ACS Applied Materials & Interfaces*, *4*, 251–260.
5. Kumar, B., K. Smita, L. Cumbal, and A. Debut. (2017). Sacha inchi (Plukenetia volubilis L.) shell biomass for synthesis of silver nanocatalyst. *Journal of Saudi Chemical Society*, *27*, 293–298.
6. Haverkamp, R.G., and A.T. Marshall. (2009). The mechanism of metal nanoparticle formation in plants: Limits on accumulation. *Journal of Nanoparticle Research*, *11*, 1453–1463.
7. Kumar, B., K. Smita, L. Cumbal, and A. Debut. (2014). Biogenic synthesis of iron oxidenanoparticles for 2-arylbenzimidazole fabrication. *Journal of Saudi Chemical Society*, *18*, 364–369.
8. Gericke, M., and A. Pinches. (2006). Microbial production of gold nanoparticles. *Gold Bulletin*, *39*, 22–28.
9. Li, X., H. Xu, Z.S. Chen, and G. Chen. (2011). Biosynthesis of nanoparticles by microorganisms and their applications. *Journal of Nanomaterials*, *16*, 270974.
10. Prabhu, S., and E.K. Poulose. (2012). Silver nanoparticles: Mechanism of antimicrobial action, synthesis, medical applications, and toxicity effects. *International Nano Letters*, *2*, 32.
11. Husen, A., and K.S. Siddiqi. (2014). Plants and microbes assisted selenium nanoparticles: Characterization and application. *Journal of Nanobiotechnology*, *12*, 28.
12. Iravani, S. (2014). Bacteria in nanoparticle synthesis: Current status and future prospects. *International Scholarly Research Notices*, *18*, 359316.
13. Singh, R., U.U. Shedbalkar, S.A. Wadhwani, and B.A. Chopade. (2015). Bacteriagenic silver nanoparticles: Synthesis, mechanism, and applications. *Applied Microbiology and Biotechnology*, *99*, 4579–4593.
14. Sastry, M., A. Ahmad, M.I. Khan, and R. Kumar. (2003). Biosynthesis of nanomaterials using fungi and actinomycete. *Current Science*, *85*, 162–170.
15. Durán, N., P.D. Marcato, O.L. Alves, G. De Souza, and E. Esposito. (2005). Mechanistic aspects of biosynthesis of silver nanoparticles by several Fusarium oxysporum strains. *Journal of Nanobiotechnology*, *3*, 1–8.
16. Maliszewska, I., A. Juraszek, and K. Bielska. (2014). Green synthesis and characterization of silver nanoparticles using ascomycota fungi Penicillium nalgiovense AJ12. *Journal of Cluster Science*, *25*, 989–1004.
17. Vágó, A., G. Szakacs, G.Sáfrán, R. Horvath, B.Pécz, and I. Lagzi. (2016). One-step green synthesis of gold nanoparticles by mesophilic filamentous fungi. *Chemical Physics Letters*, *645*, 1–4.
18. Kitching, M., M. Ramani, and E. Marsili. (2015). Fungal biosynthesis of gold nanoparticles: Mechanism and scale up. *Microbial Biotechnology*, *8*, 904–917.
19. Taherzadeh, M.J., M. Fox, H. Hjorth, and L. Edebo. (2003). Production of mycelium biomass and ethanol from paper pulp sulfite liquor by Rhizopus oryzae. *Bioresource Technology*, *88*, 167–177.
20. Zhang, X., S. Yan, R.D. Tyagi, and R.Y. Surampalli. (2011). Synthesis of nanoparticles by microorganisms and their

application in enhancing microbiological reaction rates. *Chemosphere*, 82(4) 489–494.

21. Nair, R., S.H. Varghese, B.G. Nair, T. Maekawa, Y. Yoshida, and D.S. Kumar. (2010). Nanoparticulate material delivery to plants. *Plant Science, 179,* 154–163.

22. Mondal, A., R. Basu, S. Das, and P. Nandy. (2011). Beneficial role of carbon nanotubes on mustard plant growth: An agricultural prospect. *Journal of Nanoparticle Research, 13,* 4519–4528.

23. Alam, M.N., N. Roy, D. Mandal, and N.A. Begum. (2013). Green chemistry for nanochemistry: Exploring medicinal plants for the biogenic synthesis of metal NPs with fine-tuned properties. *RSC Advances, 3,* 11935–11956.

24. Husen, A., and K.S. Siddiqi. (2014). Phytosynthesis of nanoparticles: Concept, controversy and application. *Nanoscale Research Letters, 9,* 229.

25. Kim, J.S., E. Kuk, K.N. Yu, J.H. Kim, S.J. Park, H. J. Lee, S.H. Kim, Y.K. Park, Y.H. Park, C.Y. Hwang, Y.K. Kim, Y.S. Lee, D.H. Jeong, and M.H. Cho. (2007). Antimicrobial effects of silver nanoparticles. *Nanomedicine: Nanotechnology Biology and Medicine, 3,* 95–101.

26. Sperling, R.A., G.P. Rivera, F. Zhang, M. Zanella, and W.J. Parak. (2008). Biological applications of gold nanoparticles. *Chemical Society Reviews, 37,* 1896–1908.

27. Ahmed, A., S. Senapati, M.I. Khan, R. Kumar, and M. Sastry. (2003). Extracellular biosynthesis of monodisperse gold nanoparticles by a novel extremophilic actinomycete, Thermomonospora sp. *Langmuir, 19,* 3550–3553.

28. Gericke, M., and A. Pinches. (2006). Biological synthesis of metal nanoparticles. *Hydrometallurgy, 83,* 132–140.

29. Saxena, J., P.K. Sharma, M.M. Sharma, and A. Singh. (2016). Process optimization for green synthesis of silver nanoparticles by Sclerotinia sclerotiorum MTCC 8785 and evaluation of its antibacterial properties. *SpringerPlus, 5,* 861.

30. Salvadori, M.R., R.A. Ando, C.A.O. Nascimento, and B. Corrêa. (2015). Extra and intracellular synthesis of nickel oxide nanoparticles mediated by dead fungal biomass. *PLoS One, 10,* e0129799.

31. Patil, H.B.V., K.S. Nithin, S. Sachhidananda, Siddaramaiah, K.T. Chandrashekara, and B.Y. Sathish Kumar. (2016). Mycofabrication of bioactive silver nanoparticle: Photo catalysed synthesis and characterization to attest its augmented bio-efficacy. *Arabian Journal of Chemistry, 12(8),* 4596–4611..

32. Othman, A.M., A. Elsayed, A.M. Elshafei, and M.M. Hassan. (2017). Application of response surface methodology to optimize the extracellular fungal mediated nanosilver green synthesis. *Journal of Genetic Engineering and Biotechnology, 15,* 497–504.

33. Sawle, B.D., B. Salimath, R. Deshpande, M.D. Bedre, B.K. Prabhakar, and A. Venkataraman. (2008). Biosynthesis and stabilization of Au and Au–Ag alloy nanoparticles by fungus, Fusarium semitectum. *Science and Technology of Advanced Materials, 9(3),* 035012.

34. Syed, A., and A. Ahmad. (2012). Extracellular biosynthesis of platinum nanoparticles using the fungus Fusarium oxysporum. *Colloids and Surfaces B: Biointerfaces, 97,* 27–31.

35. Srivastava N., M. Srivastava, P.K. Mishra, and P.W. Ramteke. (2016). Application of ZnO Nanoparticles for improving the thermal and pH stability of crude cellulase obtained from aspergillus fumigates. *Frontiers in Microbiology, 7,* 514.

36. Mohamed Y.M., A.M. Azzam, B.H. Amin, and N.A. Safwat. (2015). Mycosynthesis of iron nanoparticles by Alternaria alternate and its antibacterial activity. *African Journal of Biotechnology, 14,* 1234–1241.

37. Subramanyam, S.G., and K. Siva. (2016). Bio-synthesis, characterization and application of titanium oxide nanoparticles by fusarium oxysporum. *International Journal of Life Sciences Research, 4,* 69–75.

38. Khan, S.A., I. Uddin, S. Moeez, and A. Ahmad. (2014). Fungus-mediated preferential bioleaching of waste material such as fly-ash as a means of producing extracellular, protein capped, fluorescent and water soluble silica nanoparticles. *PLoS ONE, 9(9),* e107597.

39. Strasser, P., S. Koh, T. Anniyev, J. Greeley, K. More, C. Yu, Z. Liu, S. Kaya, D. Nordlund, H. Ogasawara, and M.F. Toney. (2010). Lattice-strain control of the activity in dealloyed core–shell fuel cell catalysts. *Nature Chemistry, 2,* 454–460.

40. Das, R., E. Ali, and S.B. Hamid. (2014). Current applications of X-ray powder diffraction: A review. *Reviews on Advanced Materials Science, 38,* 95–109.

41. Cao, G. (2011). *Nanostructures and Nanomaterials: Synthesis, Properties, and Applications.* Hackensack, NJ: World Scientific.

42. Chapman, H.N., P. Fromme, A. Barty, T.A. White, R.A. Kirian, A. Aquila, M.S. Hunter, J. Schulz, D.P. DePonte, and U. Weierstall. (2011). Femtosecond X-ray protein nanocrystallography. *Nature, 470,* 73–77.

43. Cantor, C.R. (1980). *Techniques for the Study of Biological Structure and Function* (Schimmel, P.R., Ed.). San Francisco, CA: W.H. Freeman.

44. Joshi, M., and A. Bhattacharyya. (2008). Characterization techniques for nanotechnology applications in textiles. *Indian Journal of Fiber & Textile Research, 33,* 304–317.

45. Williams, D.B., and C.B. Carter. (2009). *The Transmission Electron Microscope.* New York: Springer.

46. Lin, P.C., S. Lin, P.C. Wang, R. Sridhar. (2014). Techniques for physicochemical characterization of nanomaterials. *Biotechnology Advances, 32,* 711–726.

47. Hall, J.B., M.A. Dobrovolskaia, A.K. Patri, and McNeil, S.E. (2007). Characterization of nanoparticles for therapeutics. *Nanomedicine: Nanotechnology Biology and Medicine, 2,* 789–803.

48. Ranter, B.D., A.S. Hoffman, F.J. Schoen, J.E. Lemons. (2004). *Biomaterials Science: An Introduction to Materials in Medicine.* San Diego, CA: Elsevier.

49. Zscherp, C., and Barth, A. (2001). Reaction-induced infrared difference spectroscopy for the study of protein reaction mechanisms. *Biochemistry, 40,* 1875–1883.

50. Shang, L., Y., Wang, J. Jiang, and S. Dong. (2007). pH-dependent protein conformational changes in albumin: Gold nanoparticle bioconjugates: A spectroscopic study. *Langmuir, 23,* 2714–2721.

51. Perevedentseva, E.V., F.Y. Su, T.H. Su, Y.C. Lin, Cheng, C.L., A.V. Karmenyan, A.V. Priezzhev, and A.E. Lugovtsov. (2010). Laser-optical investigation of the effect of diamond nanoparticles on the structure and functional properties of proteins. *Quantum Electronics, 40,* 1089–1093.

52. Baudot, C., C.M. Tan, and J.C. Kong. (2010). FTIR spectroscopy as a tool for nano-material characterization. *Infrared Physics & Technology, 53,* 434–438.

53. Barth, A., and C. Zscherp. (2002). What vibrations tell us about proteins. *Quarterly Reviews of Biophysics, 35*, 369–430.

54. Goormaghtigh, E., V. Raussens, and J.M. Ruysschaert. (1999). Attenuated total reflection infrared spectroscopy of proteins and lipids in biological membranes. *Biochimica et Biophysica Acta, 1422*, 105–185.

55. Harrick, N.J., and K.H. Beckmann. (1974). Internal reflection spectroscopy. In Kane, P., and Larrabee, G. (Eds.), *Characterization of Solid Surfaces* (pp. 215–245). New York: Springer.

56. Hind, A.R., S.K. Bhargava, and A. McKinnon. (2001). At the solid/liquid interface: FTIR/ATR: The tool of choice. *Advances in Colloid and Interface Science, 93*, 91–114.

57. Johal, M.S. (2011). *Understanding Nanomaterials.* Boca Raton, FL: CRC Press.

58. Kazarian, S.G., and K.L.Chan. (2006). Applications of ATR-FTIR spectroscopic imaging to biomedical samples. *Biochimica et Biophysica Acta, Biomembranes, 1758*, 858–867.

59. Hinterdorfer, P., M.F. Garcia-Parajo, and Y.F Dufrene. (2012). Single-molecule imaging of cell surfaces using near-field nanoscopy. *Accounts of Chemical Research, 45*, 327–336.

60. Koh, A.L., W. Hu, R.J. Wilson, S.X. Wang, and R. Sinclair. (2008). TEM analyses of synthetic anti-ferromagnetic (SAF) nanoparticles fabricated using different release layers. *Ultramicroscopy, 108*, 1490–1494.

61. Mavrocordatos, D., W. Pronk, and M. Boiler. (2004). Analysis of environmental particles by atomic force microscopy, scanning and transmission electron microscopy. *Water Science and Technology, 50*, 9–18.

62. Picas, L., P.E. Milhiet, and B. Hernandez. (2012). Atomic force microscopy: A versatile tool to probe the physical and chemical properties of supported membranes at the nanoscale. *Chemistry and Physics of Lipids, 165*, 845–860.

63. Song, J., H. Kim, Y. Jang, and J. Jang. (2013). Enhanced antibacterial activity of silver/polyrhodanine-composite-decorated silica nanoparticles. *ACS Applied Materials & Interfaces, 5*, 11563–11568.

64. Parot, P., Y.F. Dufrene, P. Hinterdorfer, G. Le, C. Rimellec, D. Navajas, J.L. Pellequer, and S. Scheuring. (2007). Past, present and future of atomic force microscopy in life sciences and medicine. *Journal of Molecular Recognition, 20*, 418–431.

65. Yang, L., and D.J. Watts. (2005). Particle surface characteristics may play an important role in phytotoxicity of alumina nanoparticles. *Toxicology Letters, 158*, 122–132.

66. Tiede, K., A.B. Boxall, S.P. Tear, J. Lewis, H. David, and M. Hassellov. (2008). Detection and characterization of engineered nanoparticles in food and the environment. *Food Additives & Contaminants Part A: Chemistry Analysis Control Exposure & Risk Assessment, 25*, 795–821.

67. Gmoshinski, I.V., S.A. Khotimchenko, V.O. Popov, B.B. Dzantiev, A.V. Zherdev, V.F. Demin, and Y.P. Buzulukov. (2013). Nanomaterials and nanotechnologies: Methods of analysis and control. *Russian Chemical Reviews, 82*, 48–76.

68. Bhushan, B., and O. Marti. (2004). *Scanning Probe Microscopy: Principle of Operation, Instrumentation, and Probes.* In *Springer Handbook of Nanotechnology* (pp. 325–369).

69. Stephan, T.S., E.M. Scott, K.P. Anil, and A.D. Marina. (2006). Preclinical characterization of engineered nanoparticles intended for cancer therapeutics. In *Nanotechnology for Cancer Therapy* (pp. 105–137). Boca Raton, FL: CRC Press.

70. Jans, H., X. Liu, L. Austin, G. Maes, and Q. Huo. (2009). Dynamic light scattering as a powerful tool for gold nanoparticle bioconjugation and biomolecular binding studies. *Analytical Chemistry, 81*, 9425–9432.

71. Khlebtsov, B.N., and N.G. Khlebtsov. (2011). On the measurement of gold nanoparticle sizes by the dynamic light scattering method. *Colloid Journal, 73*, 118–127.

72. Zanetti-Ramos, B.G., M.B. Fritzen-Garcia, C.S. de Oliveira, A.A. Pasa, V. Soldi, R. Borsali, and T. Creczynski-Pasa. (2009). Dynamic light scattering and atomic force microscopy techniques for size determination of polyurethane nanoparticles. *Materials Science and Engineering C, 29*: 638–640.

73. Fissan, H., S. Ristig, H. Kaminski, C. Asbach, and M. Epple. (2014). Comparison of different characterization methods for nanoparticle dispersions before and after aerosolization. *Analytical Methods, 6*, 7324–7334.

74. Berne, B.J., R. Pecora. (2000). *Dynamic Light Scattering: With Applications to Chemistry, Biology, and Physics.* New York: Courier Corporation.

75. Koppel, D.E. (1972). Analysis of macromolecular polydispersity in intensity correlation spectroscopy: The method of cumulants. *Journal of Chemical Physics, 57*, 4814–4820.

76. Dieckmann, Y., H. Cölfen, H. Hofmann, and A. Petri-Fink. (2009). Particle size distribution measurements of manganese-doped ZnS nanoparticles. *Analytical Chemistry, 81*, 3889–3895.

77. Lange, H. (1995). Comparative test of methods to determine particle size and particle size distribution in the submicron range. *Particle & Particle System Characterization, 12*, 148–157.

78. Medentsev, A.G., and V.K. Alimenko. (1998). Naphthoquinone metabolites of the fungi. *Photochemistry, 47*, 935–959.

79. Baker, R.A., and J.H. Tatum. (1998). Novel anthraquinones from stationary cultures of Fusarium oxysporum. *Journal of Fermentation and Bioengineering, 85*, 359–361.

80. Durán, N., M.F.S. Teixeira, R De Conti, and E. Esposito. (2002). Ecological-friendly pigments from fungi. *Critical Reviews in Food Science and Nutrition, 42*, 53–66.

81. Bell, A.A., M.H. Wheeler, J. Liu, R.D. Stipanovic, L.S. Puckhaber, and H. Orta. (2003). United States Department of Agriculture: Agricultural Research Service studies on polyketide toxins of Fusarium oxysporum f sp Vasinfectum: potential targets for disease control. *Pest Management Science, 59*, 736–747.

82. Song, J.Y., and B.S. Kim. (2009). Rapid biological synthesis of silver nanoparticles using plant leaf extracts. *Bioprocess and Biosystems Engineering, 32*, 79–84.

83. Rico, C.M., S. Majumdar, M. Duarte-Gardea, J.R. Peralta-Videa, and J.L. Gardea-Torresdey. (2011). Interaction of nanoparticles with edible plants and their possible implications in the food chain. *Journal of Agricultural and Food Chemistry, 59*, 3485–3498.

84. Husen, A., and K.S. Siddiqi. (2014). Carbon and fullerene nanomaterials in plant system. *Journal of Nanobiotechnology, 12*, 16.

85. Salata, O.V. (2004). Applications of nanoparticles in biology and medicine. Journal of Nanobiotechnology, 2, 3.

86. Yadav, A., K. Kon, G. Kratosova, N. Duran, A.P. Ingle and M. Rai. (2015). Fungi as an efficient mycosystem for the synthesis of metal nanoparticles: Progress and key aspects of research. *Biotechnology Letters, 37*, 2099–2120.

87. Khan, M.M., S. Kalathil, T.H. Han, J. Lee, and M.H. Cho. (2013). Positively charged gold nanoparticles synthesized by electrochemically active biofilm: A biogenic approach. *Journal of Nanoscience and Nanotechnology, 13,* 6079–6085.

88. Gajbhiye, M.B., J.G. Kesharwani, A.P. Ingle, A.K. Gade, and M.K. Rai. (2009). Fungus mediated synthesis of silver nanoparticles and its activity against pathogenic fungi in combination of fluconazole. *Nanomedicine, 5,* 382–386.

89. Jelveh, S., and D.B. Chithrani. (2011). Gold nanostructures as a platform for combinational therapy in future cancer therapeutics. *Cancers, 3,* 1081–1110.

90. Bhainsa, K.C., S.F. D'Souza. (2006). Extracellular biosynthesis of silver nanoparticles using the fungus Aspergillus fumigatus. *Colloids and Surfaces B: Biointerfaces, 47,* 160–164.

91. Ahmad, A., P. Mukherjee, S. Senapat, D. Mandal, M.I. Khan, R. Kumar, and M. Sastry. (2003). Extracellular biosynthesis of silver nanoparticles using the fungus Fusarium oxysporum. *Colloids and Surfaces B: Biointerfaces, 28,* 313–318.

92. Durán, N., P.D. Marcato, I.H. De S. Gabriel, O.L. Alves, and Esposito E. (2007). Antibacterial effect of silver nanoparticles produced by fungal process on textile fabrics and their effluent treatment. *Journal of Biomedical Nanotechnology, 3,* 203–208.

93. Namasivayam, S.K.R., and Avimanyu. (2011). Silver nanoparticle synthesis from Lecanicillium lecanii and evalutionary treatment on cotton fabrics by measuring their improved antibacterial activity with antibiotics against Staphylococcus aureus (ATCC 29213) and E. coli (ATCC 25922) strains. *International Journal of Pharmacy and Pharmaceutical Sciences, 3,* 190–195.

94. Janki, N.T., P. Dalwadi, and P.C. Dhandhukia. (2012). Biosynthesis of gold nanoparticles using Fusarium oxysporum f. sp. cubense JT1, a plant pathogenic fungus.

95. Amarnath, K., N.L. Mathew, J. Nellore, C.R.V. Siddarth and J. Kumar. (2011). Facile synthesis of biocompatible gold nanoparticles from Vites vinefera and its cellular internalization against HBL-100 cells. *Cancer Nanotechnology, 2,* 121–132.

96. Mishra, A., S. Tripathy, R. Wahab, S.H. Jeong, I. Hwang, Y.B. Yang, Y.S. Kim, H.S. Shin. and S.I. Yun. (2011). Microbial synthesis of gold nanoparticles using the fungus Penicillium brevicompactum and their cytotoxic effects against mouse mayo blast cancer C2C12 cells. *Applied Microbiology and Biotechnology, 92,* 617–630.

97. Jeyaraj, M., G.S. Kumar, G. Sivanandhan, D.M. Ali, M. Rajesh, R. Arun, G. Kapildev, M. Manickavasagam, N. Thajuddin, and K. Premkumar. (2013). Biogenic silver nanoparticles for cancer treatment: An experimental report. *Colloids and Surfaces B: Biointerfaces, 106,* 86–92.

98. Sarkar, R., P. Kumbhakar, and A.K. Mitra. (2010). Green synthesis of silver nanoparticles and its optical properties. *Digest Journal of Nanomaterials and Biostructures, 5,* 491–496.

99. Gurunathan, S. (2009). Biosynthesis, purification and characterization of silver nanoparticles using Escherichia coli. *Colloids and Surfaces B: Biointerfaces, 74,* 328–335.

100. A.G. Ingale, and A.N. Chaudhari. (2013). Biogenic synthesis of nanoparticles and potential applications: An eco-friendly approach. *Journal of Nanomedicine and Nanotechnology, 4.*

101. Khomutov, G.B., and S.P. Gubin. (2002). Interfacial synthesis of noble metal nanoparticles. *Materials Science and Engineering, 22,* 141–146.

102. Choi, Y., N.H. Ho, and C.H. Tung. (2007). Sensing phosphatase activity by using gold nanoparticles. *Angewandte Chemie, 46,* 707–709.

103. Gupta, V., A.R. Gupta, and V. Kant. (2013). Synthesis, characterization and biomedical application of nanoparticles. *Science International, 5,* 167–174.

104. Chauhan, R.P.S., C. Gupta, and D. Prakash. (2012). Methodological advancements in green nanotechnology and their applications in biological synthesis of herbal nanoparticles. *International Journal of Bioassays, 7,* 6–10.

105. Prasad, K.S., D. Pathak, and A. Patel. (2011). Biogenic synthesis of silver nanoparticles using Nicotiana tobaccum leaf extract and study of their antibacterial effect. *African Journal of Biotechnology, 41,* 8122–8130.

106. Ali D.M., M. Sasikala, M. Gunasekaran, and N. Thajuddin. (2011). Biosynthesis and characterization of silver nanoparticles using marine cyanobacterium, Oscillatoria willei ntdm. *Digest Journal of Nanomaterials and Biostructures, 6,* 385–390.

107. Meyer, B., B. Wu, and Ram, A.F.J. (2011). Aspergillus as a multi-purpose cell factory: Current status and perspectives. *Biotechnology Letters, 33,* 469–476.

15

Genotoxicity of Functionalized Nanoparticles

Varsha Dogra, Gurpreet Kaur, Rajeev Kumar, and Sandeep Kumar

CONTENTS

15.1 Introduction

With advances of science and technology, the nanotechnology field is creating a revolution in commercial industries because of nanotechnology's unique properties, giving benefits to a range of fields such as aerospace engineering, nanoelectronics, in ecological remediation, and therapeutic healthcare. Nanomaterials are substances with a size of less than 100nm and they possess different morphologies such as spherical particles, tubes, rods,

fibers, or wires; they also have more complicated structures such as nanopeapods and nano-onions (Imasaka et al. 2006; Warner et al. 2008). With investigation in the field of nanotechnology, the exclusive properties of nanosubstances have been identified, such as improved catalytic, magnetic, electrical, optical, and mechanical properties (Ferrari 2005; Qin et al. 1999; Vasir et al. 2005; Webster et al. 1999, 2000). Nanomaterials' physicochemical properties, together with a small size and large surface area, render them exceedingly reactive, and therefore, may also be responsible for adverse toxic impacts on flora and fauna (Oberdorster et al. 2005b). As a result, the environmental and health safety of nanomaterials have attracted increasing concern. The foremost report released by the 'Royal Society and Royal Academy of Engineering' in 2004, highlighted the need to explore the genotoxicity of engineered nanosubstances in order to combat the environmental and health risks. Ever since, many reports have been published indicating the involvement of nanomaterials in prompting the cytotoxicity, genotoxicity, oxidative stress, and responses towards inflammation but such information is still limited and needs more in-depth analysis (Nel et al. 2006; Xia et al., 2006; Sayes et al. 2007; Stone et al. 2006).

Currently, nanomaterials are already being used by several industries for commercial purposes in various fields like catalysts, fillers, opacifiers, semiconductors, water filteration, microelectronics, cosmetics, etc., and causing direct or indirect exposure to humankind (Nel et al. 2006). The cosmetics industry utilizes nanosubstances in anti-aging creams, sunscreen creams, and lipstick (Brown et al. 2006; Ballestri et al. 2001; Hart et al. 2006). Nanomaterials are also employed in medical sources such as orthopaedic implant wear debris and dental prosthesis, which is leading to a high risk of exposure to humans, without a complete investigation of the genotoxicity (Brown et al. 2006; Ballestri et al. 2001; Hart et al. 2006). Nanomaterials are reported to have importance in biomedicine fields, such as in therapeutic treatments for patients, drug-delivery agents, diagnostic aids, biosensors, or imaging contrast agents (Ferrari 2005; Sahoo et al. 2007; Vasir et al., 2005). Because of the extensive usage of nanomaterials in commercial goods and also in the medical field, different issues have been raised by the toxicity of nanosubstances (Figure 15.1); there

should also be investigation before administering nanosubstances in different applications.

In addition, there is a need to inspect the link between the genotoxicity of nanosubstances and their physicochemical properties to ensure the possible reason for toxicity and to set the standards of clear genotoxic trends. This information may be useful for industries to mold the design and synthesis method of new nanosubstances for their biocompatibility.

Nanotoxicology is a field of toxicology that addresses the toxicity study of nanomaterials and to investigate the impact of nanosubstances on health and the environment (Donaldson et al. 2004). Nanotoxicology includes physicochemical characterization, exposure paths, biodistribution, cytotoxicity, genotoxicity, and controlling aspects. Furthermore, nanotoxicology comprises of consistent, robust, and reliable test procedures to examine the risk assessment on flora and fauna (Donaldson et al. 2004; Lewinski et al. 2008).

Nanomaterials cause a release of reactive oxygen species (ROS), which are extremely sensitive radicals that react with the nucleic acids, proteins, lipids, and carbohydrates, leading to apoptosis or necrosis of cells. Nanomaterials destabilize the cell balance between the tendency to synthesis and detoxify the reactive oxygen species (ROS); nanomaterials may enhance the synthesis of pro-oxidants (Curtis et al. 2006; Kabanov 2006). Physicochemical traits such as size, shape, superficial charge, and aggregation also result in the formation of ROS (Shvedova et al. 2005a, b). Reactive oxygen species (ROS) involves free radicals such as superoxide, hydroxyl radical ($\cdot$OH), peroxide, and singlet oxygen. ROS generates in cells as a by-product of the regular aerobic metabolism but the generation of ROS is enhanced in the presence of any stress factors (Luo et al. 2002).

Cells have the ability to reduce the ROS with an antioxidant defense system, either by interrupting the ROS formation or by turning back the oxidative damage. Antioxidants synthesized by the cells are separated into a primary and secondary defense mechanism. The primary defense mechanism includes enzymes such as glutathione peroxidase, superoxide dismutase, catalase, and thioredoxin reductase; secondary defense mechanisms include the reduced glutathione (Stahl et al. 1998).

Superoxide radicals which are extremely reactive are converted into less reactive peroxide (H_2O_2) with the help of superoxide dismutase (SOD), which is further degraded by the action of glutathione peroxide or catalase (Fridovich 1995). Catalase enzymes help in the conversion of H_2O_2 into water and oxygen (Mates and Sanchez-Jimenez 1999).

15.2 Nanomaterials' Physicochemical Characterization

Nanogenotoxicity is not well understood, but it is evident from many reports that it is associated with the physical and chemical characters of nanomaterials that make them cytotoxic and genotoxic. More exploration is needed to ensure such generalizations are correct, like which factor has the most influence on the genome and causes more genotoxicity. This section takes into account a few examples pertaining to correlation between physicochemical parameters and toxicity.

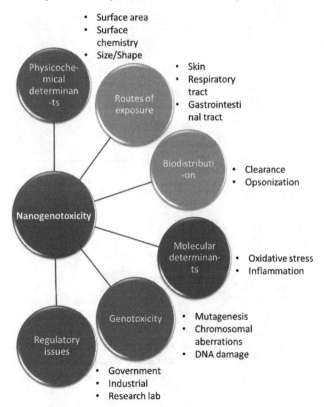

- Surface area
- Surface chemistry
- Size/Shape

Physicoche-mical determinan-ts

Routes of exposure
- Skin
- Respiratory tract
- Gastrointesti nal tract

Biodistributi-on
- Clearance
- Opsonization

Nanogenotoxicity

Molecular determinan-ts
- Oxidative stress
- Inflammation

Genotoxicity
- Mutagenesis
- Chromosomal aberrations
- DNA damage

Regulatory issues
- Government
- Industrial
- Research lab

FIGURE 15.1 Different issues caused by nanotoxicity (Arora et al. 2012).

15.2.1 Size, Shape, and Surface Area

The extremely small size of nanomaterials with a size array of 1–100nm provides them with unique properties and behavior, as compared to larger materials composed of the same constituents because with a decrease in size, a particle's number per unit mass increases. Small-size nanosubstances have the potential to cause a health hazard because they are more probable to traverse the cells and pass in the body and therefore, size governs the kinetics of nanomaterials like incorporation, biodistribution, metabolism, and excretion (Baek et al. 2012).

There are various mechanisms by which nanomaterials enter the cells, possibly the most well-known is diffusion through a plasma membrane. Diffusion occurs either directly crossing the cell wall or by passing the 10–30nm-wide membrane channels and also by endocytosis. Clathrin or caveolae-mediated endocytosis also promotes the entry of nanomaterials by forming 120nm or up to 80nm pits respectively that control the materials' entry according to size (Patel et al. 2007). After gaining entry into the cells, nanosubstances interact with the biomolecules and the cell organelles, thereby resulting in destabilization of cellular functions.

Furthermore, the shape of nanosubstances can highly influence the uptake of nanomaterials. Spherical nanomaterials have a higher rate of uptake than nanorods, although these cylindrical materials are influenced by their dimensions; for instance, the uptake of high-aspect-ratio particles is more rapid than the low-aspect ratio particles (Chithrani et al. 2007; Gratton et al. 2008). Pal et al. 2007 studied the shape-dependent interaction of silver nanoparticles with *E. coli*. It was proven by Chithrani et al. (2006) that sphere-shaped gold nanoparticle incorporation in the cell is better than nanorods in HeLa cells.

Nanomaterials' shape also influences the surface area, like spherical shape nanoparticles have a smaller surface area than a hexagonal shape of the same size. Large surface area nanomaterials are reported to have a high catalytic activity because of unstable high energy bonds, thus to stabilize themselves they react with other molecules.

15.2.2 Purity

The purity of nanomaterials is an important concern, as contaminants may be a reason for genotoxicology rather than the authentic nanomaterials. Post-synthesis of nanomaterials removes contaminants and catalysts but still, approximately 15% of residual metal remains by mass (Singh et al. 2009). Therefore researchers are attempting to enhance the purification methods to minimize the effects of contaminants on toxicity.

15.2.3 Conglomeration

Many nanomaterials are hydrophobic in nature and thus they have a tendency to aggregate under some physical conditions. Generally, fibrous nanomaterials represent complex behavior affecting agglomeration as they change their dimensions and surface area. The physiology and rigidity of fibrous nanomaterials is determined by the method of fabrication, and they are drawn to each other by van der Waals forces (Maynard et al. 2004; Shvedova et al. 2005b). Currently many modified methodologies are adopted to fabricate the hydrophilic nanomaterials by altering the surfactants or chemicals, consequently the nanogenotoxicity will also change. The aggregation of nanomaterials leads to an increase in the size range of the nanosubstance that further results in the reduction of a nanomaterial's cellular uptake.

15.2.4 Surface Chemistry and Charge

It is important to interpret the surface chemistry of nanosubstances to investigate the nature and behavior of nanomaterials under diverse experimental conditions. The administration of nanomaterials to form aggregation is done by surface chemistry, charge, ionic strength, or pH of their environment (Jiang et al. 2009). This leads to gaining information about the dispersion of nanomaterials or aggregation kinetics. The cellular uptake of nanomaterials is also governed by the surface charge.

Phospholipids present in the plasma membrane of the cell have a negative charge and therefore facilitate endocytosis of cationic nanomaterials (positively charged) at a higher rate than anionic nanomaterials. On the contrary, cationic nanomaterials are more cytotoxic and genotoxic which leads to cell death; although the reason is not clear, one possibility could be the high intake of cationic nanomaterials into the cell or interaction with the negatively charged DNA (Nan et al. 2008). Nanomaterials also interact with the biological tissue (Nel et al. 2006) and enter the cells because of their physicochemical properties (Chithrani et al. 2006; Sonavane et al. 2008).

15.3 Exposure Routes of Nanomaterials

The human body interacts with the nanosubstance through many routes such as the dermal layer of the body, respiratory tract, and gastrointestinal tract.

15.3.1 Dermal Layer

The dermal layer is the outer covering of the human body which plays an important role in providing defense against antigens. The dermal layer comprises three main layers i.e. epidermis, dermis, and a subcutaneous layer which acts as a chief defense organ. The stratum corneum is the prime defensive layer of the epidermis that provides protection from the particles of a micron size and other antigens. Many face creams, lotions, cosmetics, and several drugs are comprised of nanomaterials, and the deliberate use of these commercial products exposes the skin to nanoscale particles (Curtis et al. 2006; Hagens et al. 2007; Oberdorster et al. 2005b). According to an investigation by Gontier et al. (2008), titanium dioxide (TiO$_2$) nanoparticles (size array 20–100nm) penetrate the skin of humans, human grafted skin, and porcine. It was observed that titanium dioxide nanoparticles entered a few layers of the corneocyte layer of the stratum corneum of the epidermis. In contradiction to this study, there are many investigations that proved more in-depth infiltration of nanoparticles. Lademann et al. (1999) confirmed that nanomaterials can enter many layers of skin such as the stratum corneum, epidermis, and dermis. Oberdorster et al. (2005b) verified many nanoparticles and their tendency to penetrate the dermis, and nanoparticles translocate to the lymphatic system and lymph. According to one clinical study, nanosilver-covered dressings used for the treatment of burns led to an increased blood silver level and argyrosis (a condition caused by excessive silver deposition in the skin and leading to a purple or purple-gray color to the dermal layer) (Trop et al. 2006).

15.3.2 Respiratory Tract

Nanomaterials enter the respiratory tract by inhalation, because of their extremely small size they passed down the lung's alveoli which further leads to affecting the respiratory system (Donaldson et al. 2006; Lam et al. 2006; Nel et al. 2006; Oberdorster et al. 2005a). Particles of a size greater than 2.5 µm normally get trapped and are removed by the mucociliary system. Nanoparticles are absorbed in the lung epithelium layer and enter body fluids such as blood and lymph, leading them to reach various organs such as lymph nodes, spleen, heart, and bone marrow (Hagens et al. 2007; Oberdorster et al. 2005a). Nanoscale particles also get accumulated in the alveoli of the lungs (Curtis et al. 2006; Hagens et al. 2007). Some nanomaterials such as silver nanoparticles are cytotoxic to lung epithelial cells and macrophage cells (Soto et al. 2007).

15.3.3 Gastrointestinal Tract

Nanomaterials enter the intestinal tract by food, water, drugs, and cosmetics, or via the respiratory tract after getting passed down from the mucociliary system (Hagens et al. 2007; Oberdorster et al. 2005b). According to the work of Chen et al. (2006a), nanocopper was confirmed to be toxic and can cause damage to the kidney, spleen, and liver. Fluorescently labeled polystyrene nanoparticles were confirmed to penetrate the peyer's patches, which are small masses of lymphatic tissue (Smith et al. 1995). According to Chung et al. (2010), silver nanoparticles translocate from the intestinal tract after getting ingested and cause a condition called argyrosis.

15.3.4 Uptake and Biodistribution of Nanomaterials

Some nanomaterials enter the gastrointestinal tract and are removed in the faecal matter and urine (Curtis et al. 2006; Oberdorster et al. 2005b). On the other hand, during metabolism some nanoparticles get deposited in the liver (Oberdorster et al. 2005b). According to a few investigations, nanoparticles enter the human body via inhalation and get distributed in organs such as the lungs, intestines, heart, liver, kidney, and spleen (Oberdorster et al. 2002; BeruBe et al. 2007; Hagens et al. 2007; Medina et al. 2007) after entering by way of phagocytosis through macrophages in the alveoli of lungs (Oberdorster et al. 2005b; Curtis et al. 2006; Garnett and Kallinteri 2006). De Jong et al. (2008) verified the biodistribution of gold nanoparticles according to the size; according to the study, small-size nanoparticles of range 1–10nm confirmed the extensive distribution such as in blood, the liver, heart, kidney, spleen, thymus, lung, brain, and testis, while particles of size 50, 100 and 250nm were found only in blood, the spleen, and liver.

15.4 Interaction of Nanomaterials with Cell

Nanomaterials interact with a cell due to the difference in their charges and this leads to cytotoxicity or genotoxicity (Forest et al. 2015). There are many impacts of nanomaterials on the cell, plasma membrane, cellular organelles, and on their genome. There are four common routes of cellular uptake of nanomaterials, as evident from the study by Silverstein and Steinman (1977), i.e. phagocytosis, macropinocytosis, clathrin-mediated endocytosis, and non-clathrin-mediated endocytosis. Cell-nanomaterials' interaction impacts on intercellular transport or transcytosis, signaling of cells, cellular electron transfer cascades, membrane perturbation, synthesis of reactive oxygen species (ROS), chemokines and cytokines, gene regulation, and leads to cell apoptosis or necrosis (Jones and Grainger 2009).

15.4.1 Phagocytosis

This is the process of engulfment of a solid particle, typically by macrophages, neutrophils, dendritic cells, monocytes, and mast cells which is instigated by particle opsonization and leads to the stimulation of F-actin-driven pseudopods that ingest the solid particle and forms phagosome. This process can be inhibited by the action of cytochalasin D (Geiser et al. 2005) which blocks actin polymerization (Bronson 1998) that stops the action of pseudopodia. Neutrophils, macrophages, and monocytes are specialized phagocytes, other cells like epithelial cells, fibroblast, etc. can also uptake the solid particles by the process called pinocytosis (Mellman 1996). This process takes places at the site of clathrin-coated pits that are found in most animal cells and includes the intake of particles having a size less than 200nm, along with fluids surrounding the particles.

15.4.2 Macropinocytosis

This process is also known as cell drinking, and involves the internalization of large gulps of nutrients, solute molecules, and antigens. This is a fluid phase endocytosis that leads to the formation of vacuoles with a size range of 1–5 µm and this process is generally seen in fibroblasts, dendritic cells, and macrophages (Greish 2007).

15.4.3 Caveolae-Mediated Endocytosis

This type of endocytosis occurs after the inhibition of clathrin-dependent internalization pathways. Caveolae help in the transportation of materials across the endothelial cells and the process is called transcytosis (Mellman 1996).

15.4.4 Clathrin-Mediated Endocytosis

This process is also known as receptor-mediated endocytosis. In this type of endocytosis, cells absorb the hormones, metabolites, etc. by forming an inward infolding of plasma membrane vesicles enclosing the protein, i.e. clathrin. Clathrin-coated pits are approximately of size 100nm in diameter (Marsh and McMohan 1999). Genotoxicity is of two types, primary and secondary, primary genotoxicity is associated with the direct exposure of toxic material, whereas secondary genotoxicity is linked with the exposure of toxic material with the cells or tissues and cell organelles, leading to oxidative stress and inflammation (Vallyathan and Shi 1997; Knaapen et al. 2004; MacNee and Donaldson 2003).

15.4.5 Interactional Mechanism of Nanomaterials with DNA

There is a number of ways by which a nanosubstance enters the body, e.g. by respiratory, dermal, or oral routes, and subsequently to the body cells by interacting with it. After entering the cell, nanomaterials may cross the nuclear membrane and interact with the genome; there are several direct and indirect mechanisms that cause DNA damage (Geiser et al. 2005; Liu et al. 2007).

15.4.5.1 Direct Mechanism

Many reports have revealed that direct interaction of nanomaterials with the DNA or DNA associated proteins leads to its impairment by changing the conformation or by degrading it. According to a few reports (Geiser et al. 2005; Liu et al. 2007), it has been testified that titanium dioxide and silica nanoparticles can pass in the nucleus and lead to the aggregation of intranuclear proteins and cause inhibition of DNA replication, transcription, translation, and cell division (Chen and Von 2005).

15.4.5.2 Indirect Mechanism

In this mechanism, nanosubstances do not directly interact with the DNA molecules but with the other cell proteins which are engaged in the cell cycle. Furthermore, nanomaterial interaction with cellular protein leads to the induction of cellular responses that can cause genotoxicity, oxidative stress, disturbed signaling, and inflammation (Figure 15.2) (Singh et al. 2009). In order to explain a few factors, oxidative stress is the indirect key method related to the redox imbalance in the cells caused by nanomaterials leading to genotoxicity. This further results in the reduction of antioxidants and increase in the number of reactive oxygen species (ROS). ROS are chemically unstable and reactive molecules containing oxygen radicals that interfere with the cellular functioning such as transcription, translation, etc. by interacting with the DNA, proteins, and lipids. ROS induce DNA degradation by causing a cleavage of hydrogen bonds in the DNA, by fragmentations of the DNA strands, modifying base pairs (e.g. formation of 8-hydroxydeoxyguanosine adducts); all these effects tend to cause carcinogenesis (Toyokuni 1998).

Certain nanomaterials release transition metal ions (such as copper, iron, nickel, cadmium, cobalt, chromium, titanium, and zinc), which have the tendency to produce hydroxyl radicals ($^{\cdot}OH$) by converting oxygen metabolic products such as H_2O_2 and superoxide anion. Hydroxyl radicals are a key species which cause DNA damage. According to a few reports, free iron ions produces hydroxyl radicals, which can induce a modification in the purines (adenine and guanine) and pyrimidines (thymine and cytosine) (Zastawny et al. 1995) and can cause cross-links of thymine-tyrosine (i.e. DNA-histone protein) (Valko et al. 2006). Composition, size, and a large surface area of nanomaterials can also magnify the synthesis of ROS. According to reports, smaller size nanoparticles induce higher oxidative stress (Brown et al. 2001; Knaapen et al. 2004). There are many investigations indicating the release of ROS by nanomaterials (Gurr et al. 2005; Papageorgiou et al. 2007; Karlsson et al. 2008). DNA damage caused by oxidative stress can be determined by the micronucleus assay and comet assay (Karlsson et al. 2008; Papageorgiou et al. 2007; Gurr et al. 2005). Oxidative stress triggers the specific pathway of cell signaling involving mitogen-activated protein kinase and NF-kB (Bonvallot et al. 2001), which facilitate the release of pro-inflammatory cytokines (Abe et al. 2000). A signaling cascade activates the process of inflammation which is a defensive reaction of inflammatory cells such as neutrophils, which leads to the release of ROS (Donaldson and Tran 2004; Waldman et al. 2007).

Inflammation is a biological response to injured body tissue mediated by inflammatory cells that secretes several factors which are soluble, such as cytokines (tumor necrosis factor and interleukins protein families), reactive nitrogen species, ROS,

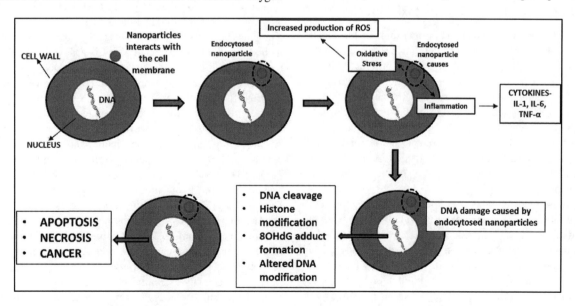

FIGURE 15.2 Indirect mechanism of nanogenotoxicity.

and migration inhibition factors. Although these factors are defense reactions toward the injury or infection of body tissue, they also cause damage to the genetic material by causing fragmentation of chromosomes, mutations, DNA cleavage; in addition, they cause problems in proofreading or repair of DNA and induce unusual methylation patterns, thus causing change in gene profiles (Jaiswal et al. 2000; Valinluck et al. 2007). As a result, chronic inflammation is also linked with carcinogenesis (Blanco et al. 2007; Federico et al. 2007; O'Byrne and Dalgleish 2001; Ohshima et al. 2005). Many investigations demonstrated that the small size and wide surface area of nanomaterials elevate the inflammation responses and induce DNA damage by oxidative stress through excessive synthesis of ROS (Jaiswal et al. 2000; Valinluck et al. 2007).

15.5 Methods to Assess Nanogenotoxicity

There are many methods to validate the nanogenotoxicity; it can be confirmed by *in vitro* or by *in vivo* investigation. The *in vitro* investigation (by using cell lines) has many advantages, such as: 1) they reveal the effects of nanomaterials on the specific cell lines; 2) exclude the secondary effects of nanomaterials such as inflammation; 3) detect the chief mechanism related to toxicity; 4) are cost-effective 5) effective and rapid method (Huang et al. 2010).

Dye-based assays determine the toxicity but the results are not reliable. According to the study of Monteiro-Riviere et al. (2009), dye analysis such as methylthiazolyldiphenyl-tetrazolium bromide (MTT) and neutral red (NR) can verify the cell viability; however, in a few cases it can give invalid results. Dhawan and Sharma (2010) discussed the problem of using *in vivo* and *in vitro* assay and as per their investigation, the chief challenge for *in vivo* assay is optimization of dispersion, dosimetry, and interaction evaluation, whereas for *in vitro* assay, the unique physicochemical characters of the nanomaterials is the hindrance.

15.5.1 *In vitro* Methods for Nanotoxicology

In vitro methods have become very important for the identification of nanotoxicity. There are many *in vitro* toxicity assessment methods like: i) examination of cell viability and division; ii) analyzing DNA damage, ROS generation, apoptosis, and necrosis; iii) using microscopes such as transmission electron microscopy (TEM), scanning electron microscopy-energy-dispersive X-ray spectroscopy (SEM-EDS), atomic force microscopy (AFM), Video-Enhanced Differential Interference Contrast microscopy (VEDIC microscopy), magnetic resonance imaging (MRI) and fluorescence spectroscopy; iv) hemolysis; v) analysis by gene expression.

Cell culture investigations are important to have a knowledge about the mechanism of nanotoxicity inside the body. Cell culture studies are easy to reproduce, are less expensive, and are uncomplicated to manage as compared to do a study on animals. It is important to check whether the toxicity is caused by nanomaterials, or by the residual contamination. Cell cultures are sensitive to environmental change such as a change in the nutrients, temperature, pH, and contamination. Therefore, it is

essential to ensure the environment or agent causing the cytotoxicity and manifold experiments are also essential (Lewinski et al. 2008).

15.5.2 Hemolysis

This is a test to check the biocompatibility of nanomaterials. In this method, the interaction of nanomaterials with the erythrocytes/red blood cells of human is examined. The impact of nanomaterials' physicochemical characteristic on the red blood cells is determined by enumerating the hemoglobin release. Yu et al. (2011) investigated hemolysis by nanomaterials; it was reported that mesoporous SiO_2 and amine-modified SiO_2 led to decreased hemolysis than bare SiO_2.

15.5.3 Nanogenotoxicity Assay

Toxicity of nanosubstance is carried out *in vitro* on cell lines. There is much research which confirmed the toxicity of different nanomaterials such as by metal, metal oxide, metal hydroxide, semiconductor, polymeric, and carbon nanoparticles. Before commercializing the use of nanomaterials, it is important to check or prove the non-toxicity. There are many genotoxic assays that confirm the toxic effects of nanomaterials.

15.5.3.1 *Micronucleus (MN) Assay*

This is a test to screen the genotoxicity of nanoparticles; several types of nanosubstances are reported to cause an increase in the micronucleus frequency. To confirm the cyto/genotoxicity of nanomaterials, the proper protocol should be followed to get the validate results and by maximizing the sensitivity. Exposure time, cellular dose, serum levels, cytochalasin-B treatment, and type of cytotoxic assay should be considered to gain a better understanding of MN frequency.

15.5.3.2 *Comet Assay*

This method is also called single-cell gel electrophoresis (SCGE) assay. It is a broadly used technique to assess the genotoxicity of nanomaterials, chemicals, and drugs. It is also used in environmental biomonitoring. Comet assay is a flexible, sensitive, and rapid approach for the identification of any DNA change in the form of a DNA single- or double-strand break. Karlsson (2010) reported the use of comet assay for the validation of nanomaterial cytotoxicity. This method is quite sensitive at lower level toxicity caused by any toxic agents, which gives reliable results. According to Karlsson (2010), many nanomaterials are highly reactive and cause DNA cleavage, change in nitrogenous base pairs of DNA, DNA damage by oxidative stress, or by the generation of ROS. This technique can also be used for the investigation of base pair damage by oxidative stress, cross-linking of DNA–protein or DNA–DNA, abasic sites, and alkali-labile sites (Pavanello and Clonfero, 2000; Collins et al. 2008). In this technique, a cell is compressed in a thin deposit of an agarose gel on a microscopic slide and are lysed. Electric charge is passed through the gel, DNA remains intact in place, but if DNA is fragmented into small pieces they are able to migrate towards the anode and take comet shape. DNA is analyzed using ethidium bromide or propidium iodide stain. The total DNA impairment is directly proportional to length and fluorescence intensity of the comet tail (Collins et al. 2008; Tice et al. 2000).

15.5.3.3 Ames Test

This is an *in vitro* test for the detection of nanogenotoxicity and is also called as bacterial reversion mutation test. It was first explained in 1972 (Ames et al. 1972) and used for the detection of mutation caused by any substance. This test uses salmonella typhimurium bacteria strains, which are histidine-dependent, these mutant bacteria generally grow on the medium supplemented with histidine, as they are unable to produce histidine amino acid by itself. In this test, mutant bacteria are cultured on the agar plate containing nanomaterials and lacking histidine, only those bacteria will grow which have gone under some change in the nucleic acid or reverse mutations, resulting in the regaining of the functionality of genes which produces histidine. Total bacterial colony count is equal to the reverse mutation caused by the nanomaterials. This method can give vague results also, as some nanomaterials are bactericidal and some cannot penetrate the bacterial cell wall.

15.5.3.4 Chromosomal Aberration Test

This test analyzes the structural changes of the chromosome and also their count, any change in the chromosomes or any kind of deletion/insertion is detected by this test. This examination can be done both in *in vitro* and also *in vivo* conditions. For laboratory study, the culturing of mammalian cells is done and it is treated with the test substance. Furthermore, it is exposed to a cell cycle-arresting chemical that seizes the metaphase stage. Culture cells are collected on the glass slide and stained with Giemsa to analyze the chromosome's banding patterns. Examination of cells is done by optical microscope to check the chromosomal aberrations.

For *in vivo* study, rodents are treated with the test substance and further with metaphase arresting chemical. Bone marrow cells are used for the metaphase chromosome preparations, they are collected on the slide, stained, and analyzed: the same as explained for in the *in vitro* study. This method requires a skilled cytogeneticist for all the careful work.

15.5.4 ROS Effector Assays

To detect the ROS generation in the cell, immunocytochemistry is the direct methodology. This technique determines the DNA lesion at a specific region such as 8-hydroxydeoxyguanosine lesion caused by OH$^-$ radical (Schins et al. 2002). Enhancement in the ROS synthesis causes membrane peroxidation of the cell and their organelles, leading to damaging effects such as mitochondrial injury, which is also a common indicator of high ROS level in the cell (Choi et al. 2007).

15.5.5 Detection of ROS

ROS can be detected by fluorescein-compound based test or by electron paramagnetic resonance (EPR). ROS, when coming into contact with fluorescein-compounds such as

2',7'- difluorescein-diacetate (Wilson et al. 2002; Lin et al. 2006) and dichlorodihydrofluorescein diacetate (Hussain et al. 2005), it results in oxidization and gives fluorescence (Wilson et al. 2002; Long et al. 2007). EPR is a method of detecting materials with unpaired electrons and it is a helpful technique for the detection of radicals such as reactive oxygen and nitrogen species (Fubini and Hubbard 2003).

15.6 Eco-Genotoxicology

Nanomaterials cause damage to genetic materials that is not just related to humans but also to other species. Any change in the human genetic material caused by nanomaterials leads to the induction of cancer, mutation, or genetic aberrations, but not much stress has been given to the study of the impact of nanosubstances on environmental species. The environment is exposed more to nanomaterials as compared to humans, because of the contamination of biotic and abiotic components of the ecosystem. There are only a few datasets available for the study for nanomaterials' impact on flora and fauna (Baun et al. 2008). Therefore, more stress should be given to the investigation of such impact on environmental species.

15.7 Future Research Needs

The commercialization and synthesis of nanomaterials are increasing day by day and nanomaterials' physicochemical characteristics and their unique properties are becoming more complex; therefore, to ensure the nanomaterial safety, it is required to set standard protocols to keep a check on cytotoxicity, genotoxicity, and ecotoxicity. Biomagnification of nanomaterials also needs strict considerations and it has to be checked before commercialization. There is a necessity to check the fate of nanomaterials inside the body; however, their effects on flora, fauna, and abiotic components of the ecosystem are equally important.

15.8 Conclusion

The commercialization of nanotechnology is growing at an exponential rate before proper analysis of their genotoxic effects on flora and fauna. Nanotoxicology is an emerging branch that provides a safe and sound way to use the nanomaterials without causing any harm to biotic and abiotic components of the ecosystem. A depth knowledge, understanding, and proper testing help in the detection of toxic impacts of nanosubstances on the population health and on other environment species. With an increase in the fabrication of anthropogenic nanosubstances, there is a growing necessity for the standard protocol of testing before their commercialization. Besides standard testing protocols, improved and fast screening methodologies are also required that will inhibit the risk factors by recognizing toxicity causing specific characteristic of nanomaterials and also by controlling the exposure. Screening methodologies also lead to better predictive toxicology that will help in designing procedure and parameters to eradicate the toxicity.

Acknowledgments

G. K. is highly grateful to DST for an Inspire Faculty award (IFA-12-CH-41) and PURSE grant II. R.K. is gratified to DST, SERB/F/8171/2015-16, as well as UGC (F. No. 194-2/2016 IC) for providing financial support. S.K. is thankful to DST grant (SERB/ET-0038/ 2013 dated 16-08-2013). V.D. is grateful to UGC for SRF.

REFERENCES

Abe S, Takizawa H, Sugawara I, Kudoh S. (2000). Diesel exhaust (DE)-induced cytokine expression in human bronchial epithelial cells: A study with a new cell exposure system to freshly generated DE in vitro. *Am J Respir Cell Mol Biol*, 22, 296–303.

Ames BN, Gurney EG, Miller JA, Bartsch H. (1972). Carcinogens as frameshift mutagens: Metabolites and derivatives of 2-acetylaminofluorene and other aromatic amine carcinogens. *Proc Natl Acad Sci USA*, 69, 3128–3132.

Arora S, Rajwade JM, Paknikar KM. (2012). Nanotoxicology and in vitro studies: The need of the hour. *Toxicol Appl Pharmacol*, 258(2), 151–165.

Baek M, Chung HE, Yu J, Lee JA, Kim TH, Oh JM, Lee WJ, Paek SM, Lee JK, Jeong J, Choy JH. (2012). Pharmacokinetics, tissue distribution, and excretion of zinc oxide nanoparticles. *Int J Nanomedicine*, 7, 3081.

Ballestri M, Baraldi A, Gatti AM, Furci L, Bagni A, Loria P, Rapanà RM, Carulli N, Albertazzi A. (2001). Liver and kidney foreign bodies granulomatosis in a patient with malocclusion, bruxism, and worn dental prostheses. *Gastroenterology*, 121(5), 1234–1238.

Baun A, Hartmann NB, Grieger K, Kusk KO. (2008). Ecotoxicity of engineered nanoparticles to aquatic invertebrates: A brief review and recommendations for future toxicity testing. *Ecotoxicology*, 17, 387–395.

BeruBe K, Balharry D, Sexton K, Koshy L, Jones T. (2007). Combustion derived nanoparticles: Mechanisms of pulmonary toxicity. *Clin Exp Pharmacol Physiol*, 34, 1044–1050.

Blanco D, Vicent S, Fraga MF, Fernandez-Garcia I, Freire J, Lujambio A, Esteller M, Ortiz-de-Solorzano C, Pio R, Lecanda F, Montuenga LM. (2007). Molecular analysis of a multistep lung cancer model induced by chronic inflammation reveals epigenetic regulation of p16, activation of the DNA damage response pathway. *Neoplasia*, 9(10), 840-IN12.

Bonvallot V, Baeza-Squiban A, Baulig A, Brulant S, Boland S, Muzeau F, Barouki R, Marano F. (2001). Organic compounds from diesel exhaust particles elicit a proinflammatory response in human airway epithelial cells and induce cytochrome p450 1A1 expression. *Am J Respir Cell Mol Biol*, 25(4), 515–521.

Bronson R. (1998). Is the oocyte a non-professional phagocyte? *Hum Reprod Update*, 4(6), 763–775.

Brown C, Fisher J, Ingham E. (2006). Biological effects of clinically relevant wear particles from metal-on-metal hip prostheses. *Proc Inst Mech Eng*, 220, 355–369.

Brown DM, Wilson MR, MacNee W, Stone V, Donaldson K. (2001). Size-dependent proinflammatory effects of ultrafine polystyrene particles: A role for surface area and oxidative stress in the enhanced activity of ultrafines. *Toxicol Appl Pharmacol*, 175, 191–199.

Chen M, von MA. (2005). Formation of nucleoplasmic protein aggregates impairs nuclear function in response to SiO2 nanoparticles. *Exp Cell Res*, 305, 51–62.

Chen Z, Meng H, Xing G, Chen C, Zhao Y, Jia G, Wang T, Yuan H, Ye C, Zhao F, Chai Z. (2006a). Acute toxicological effects of copper nanoparticles in vivo. *Toxicol Lett*, 163(2), 109–120.

Chithrani BD, Chan WC. (2007). Elucidating the mechanism of cellular uptake and removal of protein-coated gold nanoparticles of different sizes and shapes. *Nano Lett*, 7, 1542–1550.

Chithrani BD, Ghazani AA, Chan WC. (2006). Determining the size and shape dependence of gold nanoparticle uptake into mammalian cells. *Nano Lett*, 6(4), 662–668.

Choi AO, Cho SJ, Desbarats J, Lovrić J, Maysinger D. (2007). Quantum dot-induced cell death involves Fas upregulation and lipid peroxidation in human neuroblastoma cells. *J Nanobiotechnol*, 5(1), 1.

Chung IS, Lee MY, Shin DH, Jung HR. (2010). Three systemic argyria cases after ingestion of colloidal silver solution. *Int J Dermatol*, 49, 1175–1177.

Collins AR, Oscoz AA, Brunborg G, Gaivao I, Giovannelli L, Kruszewski M, Smith CC, Štětina R. (2008). The comet assay: Topical issues. *Mutagenesis*, 23(3), 143–151.

Curtis J, Greenberg M, Kester J, Phillips S, Krieger G. (2006). Nanotechnology and nanotoxicology. *Toxicol Rev*, 25(4), 245–260.

De Jong WH, Hagens WI, Krystek P, Burger MC, Sips AJ, Geertsma RE. (2008). Particle size-dependent organ distribution of gold nanoparticles after intravenous administration. *Biomaterials*, 29(12), 1912–1919.

Dhawan A, Sharma V. (2010). Toxicity assessment of nanomaterials: Methods and challenges. *Anal Bioanal Chem*, 398(2), 589–605.

Donaldson K, Aitken R, Tran L, Stone V, Duffin R, Forrest G, Alexander A. (2006). Carbon nanotubes: A review of their properties in relation to pulmonary toxicology and workplace safety. *Toxicol Sci*, 92(1), 5–22.

Donaldson K, Stone V, Tran CL, Kreyling W, Borm PJA. (2004). Nanotoxicology. *Occup Environ Med*, 61, 727–728.

Donaldson K, Tran CL. (2004). An introduction to the short-term toxicology of respirable industrial fibres. *Mutat Res*, 553, 5–9.

Federico A, Morgillo F, Tuccillo C, Ciardiello F, Loguercio C. (2007). Chronic inflammation and oxidative stress in human carcinogenesis. *Int J Cancer*, 121, 2381–2386.

Ferrari M. (2005). Cancer nanotechnology: Opportunities and challenges. *Nat Rev Cancer*, 5, 161–171.

Forest V, Cottier M, Pourchez J. (2015). Electrostatic interactions favor the binding of positive nanoparticles on cells: A reductive theory. *Nano Today*, 10(6), 677–680.

Fridovich I. (1995). Superoxide radical and superoxide dismutases. *Annu Rev Biochem*, 64, 97–112.

Fubini B, Hubbard A. (2003). Reactive oxygen species (ROS) and reactive nitrogen species (RNS) generation by silica in inflammation and fibrosis. *Free Radic Biol Med*, 34(12), 1507–1516.

Garnett MC, Kallinteri P. (2006). Nanomedicines and nanotoxicology: Some physiological principles. *Occup Med*, 56(5), 307–311.

Geiser M, Rothen-Rutishauser B, Kapp N, Schürch S, Kreyling W, Schulz H, Semmler M, Im Hof V, Heyder J, Gehr P. (2005). Ultrafine particles cross cellular membranes by nonphagocytic mechanisms in lungs and in cultured cells. *Environ Health Perspect*, 113(11), 1555–1560.

Gontier E, Ynsa MD, Biro T, Hunyadi J, Kiss B, Gáspár K, Pinheiro T, Silva JN, Filipe P, Stachura J, Dabros W. (2008) Is there penetration of titania nanoparticles in sunscreens through skin? A comparative electron and ion microscopy study. *Nanotoxicology, 2*(4), 218–231.

Gratton SE, Ropp PA, Pohlhaus PD, Luft JC, Madden VJ, Napier ME, DeSimone JM. (2008). The effect of particle design on cellular internalization pathways. *Proc Natl Acad Sci USA 105*, 11613–11618.

Greish K. (2007). Enhanced permeability and retention of macromolecular drugs in solid tumors: A royal gate for targeted anticancer nanomedicines. *J Drug Target, 15*(7–8), 457–464.

Gurr JR, Wang AS, Chen CH, Jan KY. (2005). Ultrafine titanium dioxide particles in the absence of photoactivation can induce oxidative damage to human bronchial epithelial cells. *Toxicology, 213,* 66–73.

Hagens WI, Oomen AG, de Jong WH, Cassee FR, Sips AJ. (2007). What do we (need to) know about the kinetic properties of nanoparticles in the body? *Regul Toxicol Pharm, 49,* 217–219.

Hart AJ, Hester T, Sinclair K, Powell JJ, Goodship AE, Pele L, Fersht NL, Skinner J. (2006). The association between metal ions from hip resurfacing and reduced T-cell counts. J Bone Joint Surg, *88,* 449–454.

Huang YW, Wu CH, Aronstam RS. (2010). Toxicity of transition metal oxide nanoparticles: recent insights from in vitro studies. *Materials, 3*(10), 4842–4859.

Hussain SM, Hess KL, Gearhart JM, Geiss KT, Schlager JJ. (2005). In vitro toxicity of nanoparticles in BRL 3A rat liver cells. *Toxicol In Vitro, 19*(7), 975–983.

Imasaka K, Kanatake Y, Ohshiro Y, Suehiro J, Harashima H. (2006). Production of carbon nanoonions and nanotubes using an intermittent discharge in water. *Thin Solid Films, 506,* 250–254.

Jaiswal M, LaRusso NF, Burgart LJ, Gores GJ. (2000). Inflammatory cytokines induce DNA damage and inhibit DNA repair in cholangiocarcinoma cells by a nitric oxide-dependent mechanism. *Cancer Res, 60,* 184–190.

Jiang J, Oberdorster G, Biswas P. (2009). Characterisation of size, surface charge and agglomeration state of nanoparticle dispersions for toxicological studies. *J Nanopart Res, 11,* 77–89.

Jones CF, Grainger DW. (2009). In vitro assessments of nanomaterial toxicity. *Adv Drug Delivery Rev, 61*(6), 438–456.

Kabanov AV. (2006). Polymer genomics: An insight into pharmacology and toxicology of nanomedicines. *Adv Drug Delivery Rev, 58,* 1597–1621.

Karlsson HL. (2010). The comet assay in nanotoxicology research. *Anal Bioanal Chem, 398,* 651–666.

Karlsson HL, Cronholm P, Gustafsson J, Moller L. (2008). Copper oxide nanoparticles are highly toxic: A comparison between metal oxide nanoparticles and carbon nanotubes. *Chem Res Toxicol, 21,* 1726–1732.

Knaapen AM, Borm PJ, Albrecht C, Schins RP. (2004). Inhaled particles and lung cancer. Part A: Mechanisms. *Int J Cancer, 109,* 799–809.

Lademann J, Weigmann HJ, Rickmeyer C, Barthelmes H, Schaefer H, Mueller G, Sterry W. (1999). Penetration of titanium dioxide microparticles in a sunscreen formulation into the horny layer and the follicular orifice. *Skin Pharmacol Appl Skin Physiol, 12,* 247–256.

Lam CW, James JT, McCluskey R, Arepalli S, Hunter RL. (2006). A review of carbon nanotube toxicity and assessment of potential occupational and environmental health risks. *Crit Rev Toxicol, 36,* 189–217.

Lewinski N, Colvin V, Drezek R. (2008). Cytotoxicity of nanoparticles. *Small, 4*(1), 26–49.

Lin W, Huang YW, Zhou XD, Ma Y. (2006). In vitro toxicity of silica nanoparticles in human lung cancer cells. *Toxicol Appl Pharmacol, 217*(3), 252–259.

Liu L, Takenaka T, Zinchenko AA, Chen N, Inagaki S, Asada H, Kishida T, Mazda O, Murata S, Yoshikawa K. (2007). Cationic silica nanoparticles are efficiently transferred into mammalian cells. In International Symposium on Micro-NanoMechatronics and Human Science (Vol. 1–2, pp. 281–285).

Long TC, Tajuba J, Sama P, Saleh N, Swartz C, Parker J, Hester S, Lowry GV, Veronesi B. (2007). Nanosize titanium dioxide stimulates reactive oxygen species in brain microglia and damages neurons in vitro. *Environ Health Perspect, 115*(11), 1631–1637.

Luo J, Borgens R, Shi R. (2002). Polyethylene glycol immediately repairs neuronal membranes and inhibits free radical production after acute spinal cord injury. *J Neurochem, 83,* 471–480.

MacNee W, Donaldson K. (2003). Mechanism of lung injury caused by PM10 and ultrafine particles with special reference to COPD. *Eur Respir J, 21,* 47s–51s.

Marsh M, McMahon HT. (1999). The structural era of endocytosis. *Science, 285*(5425), 215–220.

Mates JM, Sanchez-Jimenez F. (1999). Antioxidant enzymes and their implications in pathophysiologic processes. *Front Biosci, 4,* 339–345.

Maynard AD, Baron PA, Foley M, Shvedova AA, Kisin ER, Castranova V. (2004). Exposure to carbon nanotube material: Aerosol release during the handling of unrefined single-walled carbon nanotube material. *J Toxicol Environ Health A, 67,* 87–107.

Medina C, Santos-Martinez MJ, Radomski A, Corrigan OI, Radomski MW. (2007). Nanoparticles: Pharmacological and toxicological significance. *Br J Pharmacol, 150,* 552–558.

Mellman I. (1996). Endocytosis and molecular sorting. *Annu Rev Cell Dev Biol, 12*(1), 575–625.

Monteiro-Riviere NA, Inman AO, Zhang LW. (2009). Limitations and relative utility of screening assays to assess engineered nanoparticle toxicity in a human cell line. *Toxicol Appl Pharmacol, 234,* 222–235.

Nan A, Bai X, Son SJ, Lee SB, Ghandehari H. (2008). Cellular uptake and cytotoxicity of silica nanotubes. *Nano Lett, 8,* 2150–2154.

Nel A, Xia T, Mädler L, Li N. (2006). Toxic potential of materials at the nanolevel. *Science 311*(5761), 622–627.

Oberdorster G, Maynard A, Donaldson K, Castranova V, Fitzpatrick J, Ausman K, Carter J, Karn B, Kreyling W, Lai D, Olin S. (2005a). Principles for characterizing the potential human health effects from exposure to nanomaterials: elements of a screening strategy. *Part Fibre Toxicol, 2,* 8–43.

Oberdorster G, Oberdorster E, Oberdorster J. (2005b). Nanotoxicology: An emerging discipline evolving from studies of ultrafine particles. *Environ Health Perspect, 113,* 823–839.

Oberdörster G, Sharp Z, Atudorei V, Elder A, Gelein R, Lunts A, Kreyling W, Cox C. (2002). Extrapulmonary translocation of ultrafine carbon particles following whole-body inhalation exposure of rats. *J Toxicol Environ Health, 65*, 1531–1543.

O'Byrne KJ, Dalgleish AG. (2001). Chronic immune activation and inflammation as the cause of malignancy. *Br J Cancer, 85*, 473–483.

Ohshima H, Tazawa H, Sylla BS, Sawa T. (2005). Prevention of human cancer by modulation of chronic inflammatory processes. *Mutat Res, 591*, 110–122.

Pal S, Tak YK, Song JM. (2007). Does the antibacterial activity of silver nanoparticles depend on the shape of the nanoparticle? A study of the gram-negative bacterium *Escherichia coli*. *Appl Environ Microbiol, 73*(6), 1712–1720.

Papageorgiou I, Brown C, Schins R, Singh S, Newson R, Davis S, Fisher J, Ingham E, Case CP. (2007). The effect of nano-and micron-sized particles of cobalt–chromium alloy on human fibroblasts in vitro. *Biomaterials, 28*(19), 2946–2958.

Patel LN, Zaro JL, Shen WC. (2007). Cell penetrating peptides: Intracellular pathways and pharmaceutical perspectives. *Pharm Res, 24*, 1977–1992.

Pavanello S, Clonfero E. (2000). Biological indicators of genotoxic risk and metabolic polymorphisms. *Mutat Res, 463*, 285–308.

Qin XY, Kim JG, Lee JS. (1999). Synthesis and magnetic properties of nanostructured γ-Ni-Fe alloys. *Nanostruct Mater, 11*, 259–270.

Royal Society and Royal Academy of Engineering Report. (2004). Nanoscience and nanotechnologies: Opportunities and uncertainties. Available from: http://www.nanotec.org.uk/finalReport.htm.

Sahoo SK, Parveen S, Panda JJ. (2007). The present and future of nanotechnology in human health care. *Nanomedicine, 3*, 20–31.

Sayes CM, Reed KL, Warheit DB. (2007). Assessing toxicity of fine and nanoparticles: Comparing in vitro measurements to in vivo pulmonary toxicity profiles. *Toxicol Sci, 97*, 163–80.

Schins RP, Duffin R, Höhr D, Knaapen AM, Shi T, Weishaupt C, Stone V, Donaldson K, Borm PJ. (2002). Surface modification of quartz inhibits toxicity, particle uptake, and oxidative DNA damage in human lung epithelial cells. *Chem Res Toxicol, 15*(9), 1166–1173.

Shvedova AA, Kisin ER, Mercer R, Murray AR, Johnson VJ, Potapovich AI, Tyurina YY, Gorelik O, Arepalli S, Schwegler-Berry D, Hubbs AF. (2005b). Unusual inflammatory and fibrogenic pulmonary responses to single-walled carbon nanotubes in mice. *Am J Physiol Lung Cell Mol Physiol, 289*, 698–708.

Silverstein SC, Steinman RM, Cohn ZA. (1977). Endocytosis. *Annu Rev Biochem, 46*(1), 669–722.

Singh N, Manshian B, Jenkins GJ, Griffiths SM, Williams PM, Maffeis TG, Wright CJ, Doak SH. (2009). NanoGenotoxicology: The DNA damaging potential of engineered nanomaterials. *Biomaterials, 30*(23–24), 3891–3914.

Smith MW, Thomas NW, Jenkins PG, Miller NG, Cremaschi D, Porta C. (1995). Selective transport of microparticles across Peyer's patch follicle-associated M cells from mice and rats. *Exp Physiol, 80*, 735–743.

Sonavane G, Tomoda K, Makino K. (2008). Biodistribution of colloidal gold nanoparticles after intravenous administration: Effect of particle size. *Colloids Surf B, 66*, 274–280.

Soto K, Garza KM, Murr LE. (2007). Cytotoxic effects of aggregated nanomaterials. *Acta Biomater, 3*, 351–358.

Stahl W, Junghans A, de Boer B, Driomina ES, Briviba K, Sies H. (1998). Carotenoid mixtures protect multilamellar liposomes against oxidative damage: Synergistic effects of lycopene and lutein. *FEBS Lett, 427*, 305–308.

Stone V, Kinloch I, Clift M, Fernandes T, Ford A, Christofi N, Griffiths A, Donaldson K. (2006). Nanoparticle toxicology and ecotoxicology: The role of oxidative stress. In Zhao Y, Nalwa H (Eds.), *Nanotoxicology: Interactions of Nanomaterials with Biological Systems*. Valencia, CA: American Scientific.

Tice RR, Agurell E, Anderson D, Burlinson B, Hartmann A, Kobayashi H, Miyamae Y, Rojas E, Ryu JC, Sasaki YF. (2000). Single cell gel/comet assay: Guidelines for in vitro and in vivo genetic toxicology testing. *Environ Mol Mutagen, 35*, 206–221.

Toyokuni S. (1998). Oxidative stress and cancer: The role of redox regulation. *Biotherapy, 11*, 147–154.

Trop M, Novak M, Rodl S, Hellbom B, Kroell W, Goessler W. (2006). Silver-coated dressing acticoat caused raised liver enzymes and argyria-like symptoms in burn patient. *J Trauma, 60*, 648–652.

Valinluck V, Sowers LC. (2007). Inflammation-mediated cytosine damage: A mechanistic link between inflammation and the epigenetic alterations in human cancers. *Cancer Res 67*, 5583–5586.

Valko M, Rhodes CJ, Moncol J, Izakovic M, Mazur M. (2006). Free radicals, metals and antioxidants in oxidative stress-induced cancer. *Chem Biol Interact, 160*, 1–40.

Vallyathan V, Shi X. (1997). The role of oxygen free radicals in occupational and environmental lung diseases. *Environ Health Perspect, 105*, 165–177.

Vasir JK, Reddy MK, Labhasetwar VD. (2005). Nanosystems in drug targeting: Opportunities and challenges. *Curr Nanosci, 1*, 47–64.

Waldman WJ, Kristovich R, Knight DA, Dutta PK. (2007). Inflammatory properties of iron-containing carbon nanoparticles. *Chem Res Toxicol 20*, 1149–1154.

Warner JH, Ito Y, Zaka M, Ge L, Akachi T, Okimoto H, Porfyrakis K, Watt AA, Shinohara H, Briggs GA. (2008). Rotating fullerene chains in carbon nanopeapods. *Nano Lett, 8*(8), 2328–2335.

Webster TJ, Ergun C, Doremus RH, Siegel RW, Bizios R. (2000). Enhanced functions of osteoblasts on nanophase ceramics. *Biomaterials, 21*(17), 1803–1810.

Webster TJ, Siegel RW, Bizios R. (1999). Osteoblast adhesion on nanophase ceramics. *Biomaterials, 20*(13), 1221–1227.

Wilson MR, Lightbody JH, Donaldson K, Sales J, Stone V. (2002). Interactions between ultrafine particles and transition metals in vivo and in vitro. *Toxicol Appl Pharmacol, 184*(3), 172–179.

Xia T, Kovochich M, Brant J, Hotze M, Sempf J, Oberley T, Sioutas C, Yeh JI, Wiesner MR, Nel AE. (2006). Comparison of the abilities of ambient and manufactured nanoparticles to induce cellular toxicity according to an oxidative stress paradigm. *Nano Lett, 6*(8), 1794–1807.

Yu T, Malugin A, Ghandehari H. (2011). Impact of silica nanoparticle design on cellular toxicity and hemolytic activity. *ACS Nano, 5*(7), 5717–5728.

Zastawny TH, Altman SA, Randers-Eichhorn L, Madurawe R, Lumpkin JA, Dizdaroglu M, Rao G. (1995). DNA base modifications and membrane damage in cultured mammalian cells treated with iron ions. *Free Radic Biol Med, 18*, 1013–1022.

Index